90 0915628 0

D1759904

Vitamin A and Carotenoids
Chemistry, Analysis, Function and Effects

Food and Nutritional Components in Focus

Series Editors:
Professor Victor R Preedy, *School of Medicine, King's College London, UK*

Titles in the Series:
1: Vitamin A and Carotenoids: Chemistry, Analysis, Function and Effects

How to obtain future titles on publication:
A standing order plan is available for this series. A standing order will bring delivery of each new volume immediately on publication.

For further information please contact:
Book Sales Department, Royal Society of Chemistry, Thomas Graham House, Science Park, Milton Road, Cambridge, CB4 0WF, UK
Telephone: +44 (0)1223 420066, Fax: +44 (0)1223 420247
Email: booksales@rsc.org
Visit our website at http://www.rsc.org/Shop/Books/

Vitamin A and Carotenoids

Chemistry, Analysis, Function and Effects

Edited by

Victor R Preedy
School of Medicine, King's College London, UK
Email: victor.preedy@kcl.ac.uk

RSC Publishing

Food and Nutritional Components in Focus No. 1

ISBN: 978-1-84973-368-7
ISSN: 2045-1695

A catalogue record for this book is available from the British Library

Published by The Royal Society of Chemistry,
Thomas Graham House, Science Park, Milton Road,
Cambridge CB4 0WF, UK

Registered Charity Number 207890

For further information see our web site at www.rsc.org

Printed in the United Kingdom by Henry Ling Limited, at the Dorset Press, Dorchester, DT1 1HD

Preface

In the past three decades there have been major advances in our understanding of the chemistry and function of nutritional components. This has been enhanced by rapid developments in analytical techniques and instrumentation. Chemists, food scientists and nutritionists are, however, separated by divergent skills, and professional disciplines. Hitherto this transdisciplinary divide has been difficult to bridge.

The series **Food and Nutritional Components in Focus** aims to cover in a single volume the chemistry, analysis, function and effects of components in the diet or its food matrix. Its aim is to embrace scientific disciplines so that information becomes more meaningful and applicable to health in general.

The series **Food and Nutritional Components in Focus** imparts holistic information and covers the latest knowledge with a structured format.

Vitamin A and Carotenoids has four major sections, namely:

Vitamin A and Carotenoids in Context
Chemistry and Biochemistry
Analysis
Function and Effects

The first section covers vitamin A terminology, vitamin A in context of other vitamins, β-carotene and supplementation. The Chemistry and Biochemistry section covers basic features, nomenclature, metabolites, structural analysis, complexes with DNA and RNA, encapsulation, thermal degradation and bioavailability. The section on Analysis includes blood, nutritional status, plants and fruit, milk, dairy products and many specific foods. There are various assay techniques such as HPLC, capillary and thin-layer chromato-

Food and Nutritional Components in Focus No. 1
Vitamin A and Carotenoids: Chemistry, Analysis, Function and Effects
Edited by Victor R Preedy
© The Royal Society of Chemistry 2012
Published by the Royal Society of Chemistry, www.rsc.org

graphy, NMR, mass spectrometry and other methods. Finally, the section on Function and Effects includes the distribution of Vitamin A and their metabolites in human tissue, deficiencies, receptors, developmental growth, signalling, 9-*cis*-retinoic acid, cancer risk, the immune system, brain and lung.

Each Chapter transcends the intellectual divide with a novel cohort of features namely by containing:

- Summary Points
- Key Facts (areas of focus explained for the lay person)
- Definitions of Words and Terms

It is designed for chemists, biochemists, food scientist and nutritionists, as well as healthcare workers and research scientists. Contributions are from leading national and international experts, including contributions from world-renowned institutions.

Professor Victor R. Preedy
King's College London

Contents

Vitamin A and Carotenoids in Context

Food and Nutritional Components in Focus No. 1
Vitamin A and Carotenoids: Chemistry, Analysis, Function and Effects
Edited by Victor R Preedy
© The Royal Society of Chemistry 2012
Published by the Royal Society of Chemistry, www.rsc.org

Chemistry and Biochemistry

            Alessandra Gentili

            5.1   Introduction   73
            5.2   Physicochemical Properties   76
                   5.2.1   Appearance and Solubility   76
                   5.2.2   Chemical Stability   76
            5.3   Spectral Properties   77
            5.4   Mass Spectrometry   82
            Summary Points   82
            Key Facts   84
            List of Abbreviations   86
            References   86

Chapter 6   Nomenclature of Vitamin A and Related Metabolites   90
            Niketa A. Patel

            6.1   Introduction   90
            6.2   Nomenclature   91
            Summary Points   93
            Key Facts   93
            Definition of Words and Terms   93
            List of Abbreviations   94
            Acknowledgements   94
            References   94

Analysis

            and RNA   97**
                   H. A. Tajmir-Riahi and P. Bourassa

            7.1   Introduction   97
            7.2   Analytical Methods   98
                   7.2.1   FTIR Spectroscopy   98
                   7.2.2   CD Spectroscopy   98
                   7.2.3   Fluorescence Spectroscopy   99
                   7.2.4   Molecular Modelling   100
            7.3   Structural Characterization   100
                   7.3.1   FTIR Spectra of Retinoid–DNA and Retinoid–
                            RNA Complexes   100

                        Bioavailability**   **142**
                Torsten Bohn

                10.1   Introduction   142
                10.2   Occurrence of Provitamin A Carotenoids in the Diet   145
                10.3   Dietary Intake of Provitamin A Carotenoids   147
                10.4   Detection of Provitamin A Carotenoids in Food Items
                         and Body Tissues   147
                10.5   Aspects of Bioavailability of Provitamin A
                         Carotenoids   149
                         10.5.1   Overview of Provitamin A Carotenoid
                                      Absorption   149
                Summary Points   156
                Key Facts   156
                Definition of Words and Terms   157
                List of Abbreviations   158
                References   158

Chapter 11   Vitamin A – Serum Vitamin A Analysis   **162**
                Ronda F. Greaves

                11.1   Introduction   162
                         11.1.1   Preamble   162
                         11.1.2   Definitions, Nomenclature and Terminology 163
                         11.1.3   Role of Vitamin A in the Body   164
                         11.1.4   Pathophysiology   164
                11.2   Measurement   166
                         11.2.1   Overview of Method   166
                         11.2.2   Pre-analytical Considerations   167
                         11.2.3   Sample Preparation for Analysis   167
                         11.2.4   Chromatographic Analysis   168
                11.3   Standardisation   169
                         11.3.1   Reference Measurement System   169
                         11.3.2   Primary Calibrators   170
                         11.3.3   Secondary Calibrators   170
                11.4   Interpretation of Results   170
                         11.4.1   Reference Intervals   170
                         11.4.2   Biological Variation   172
                         11.4.3   Additional Analytes   172
                11.5   Method Validation   172
                         11.5.1   Analytical Range   172
                         11.5.2   Imprecision   173
                         11.5.3   Recovery   173
                         11.5.4   Interference   173

Function and Effects

Vitamin A and Carotenoids in Context

CHAPTER 1

Retinol, Retinoic Acid, Carotenes and Carotenoids: Vitamin A Structure and Terminology

GERALD WOOLLARD

Department of Chemical Pathology, Lab Plus, Auckland City Hospital, Auckland, New Zealand
E-mail: geraldw@adhb.govt.nz

1.1 Introductory Remarks

The fact that the terminology vitamin A is used colloquially in everyday conversations and in commercial products within the cosmetic industry tends to belie the fascinating nature of this compound and to understate the importance of retinol (and the carotenoids) in the biological world. There can hardly be a more intriguing set of compounds which are intrinsically related to so many fundamental biological processes. Any discussions concerning the structure of vitamin A are never complete without due regard to the carotenoids themselves.

To discuss the chemical and biochemical behavior of vitamin A and the carotenoids takes the reader on a journey from fundamental photosynthetic processes in plants and into the realm of human nutrition and pathology. Vitamin A is born out of these plant-derived products and transposed into a set of animal compounds which have their own specific carrier proteins [retinol-binding protein (RBP)] and nuclear receptors [retinoic acid receptor (RAR)]. This notion in itself is remarkable and models the idea of the

Food and Nutritional Components in Focus No. 1
Vitamin A and Carotenoids: Chemistry, Analysis, Function and Effects
Edited by Victor R Preedy
© The Royal Society of Chemistry 2012
Published by the Royal Society of Chemistry, www.rsc.org

interdependence of the natural world. The evolutionary process leading to retinoid/carotenoid biological complicity can be bewildering to consider in that parallel chemistries can be used by unrelated species for unrelated purposes.

1.2   Structure and Function of Carotenoids

It is instructive to consider briefly what carotenoids are and how their intended function dictates their structure. It gives an appreciation of their general chemical structure and what characteristics are essential. An example of a very familiar carotenoid, β-carotene is shown in Figure 1.1. A cursory glance at its basic structure shows the obvious feature of a long conjugated central chain with two rings (identical in β-carotene) at each end.

1.2.1   Central Carotenoid Chain

β-Carotene represents a convenient prototype carotenoid to assist with the appreciation of carotenoid structure. It is ubiquitous in the natural plant and animal world, physiologically and nutritionally important in itself and as a precursor carotenoid for the production of other compounds. The general properties of carotenoids can be discussed by consideration of β-carotene.

1. The conjugated polyene structure is paramount to carotenoid function because the electrons in the double bonds are delocalized and have a lower ground energy state. This allows visible light to be absorbed.
2. Carotenoids act as chromophores with high extinction coefficients. They confer colour to fruit or flowers to attract birds and insects (for seed propagation) or by the birds themselves to enhance dichromatic behavior between the genders. Animals may modify this basic structure to extend the chromophore to make other carotenoids such as astaxanthin, the intense red pigment evident in salmon.
3. The ability to absorb light is at the very heart of the photosynthetic apparatus. The process is enormously complicated and will be discussed in detail later in this book. Basically, β-carotene itself or the other two important photosynthetic carotenoids lutein and zeaxanthin play multiple roles:

    (a) Capture incoming photons and passing on this energy for use in photosynthesis (carotenoids contribute 20–30% of the absorbed light energy).

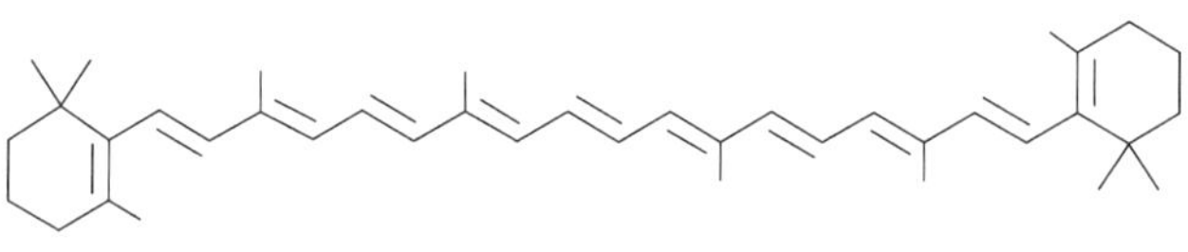

Figure 1.1   Structure of a typical carotene as illustrated by β-carotene.

(b) Broaden the absorption spectrum of the photosystem because carotenoids have a wider spectrum than chlorophyll and the hydroxylated carotenoids (lutein and zeaxanthin) specifically have a bathochromic red shift in their absorbance characteristics.

(c) Absorb excess light energy (the intensity of sunlight obviously varies greatly) and remove it by dissipation as heat (*i.e.* by increased vibrational energy of the carotenoid chain).

(d) Quench the high energy of other excited molecules such as singlet-state oxygen and triplet-state chlorophyll which protects the photosynthetic molecules from damage.

(e) Confer chemical protection by capture of excited singlet oxygen with chemical attachment to the carotenoid (the polyene structure is able to dissipate the free radical). This is a sacrificial action in which the carotenoid is chemically altered.

(f) Carotenoids are involved with membrane stabilization and may also conduct electrons (molecular wires) between other molecules such as cytochromes and chlorophylls.

4. In order that carotenoids can perform these photosynthetic tasks, they must maintain an unsubstituted polyene chain and all-*trans* geometry. Bending the central polyene chain has a profound effect on the geometry of β-carotene and also changes the ground state energy of the delocalized electrons. Hence, although there are potentially a large number of *cis*-isomers, they are less common in nature.

5. The central chain in most common carotenoids is unaltered. Any alteration causes a change of function or chain cleavage.

1.2.2 The End Ring Systems

The unsubstituted end ring systems have a single double bond. In β-carotene it extends the conjugation of the chain at both identical ends. A complete set of compounds have the double bond shifted along one position (see biochemical pathway in Figure 1.3) to produce a set of geometric isomers based on α-carotene. The double bond can also occur on the methyl group. These ring systems are occasionally referred to as ionones. The three variations are α-ionone, β-ionone and γ-ionone depending on the position of the double bond. The names are derived from the volatile fragrant compounds produced by cleavage of the carotene central chain near the end ring (see Figure 1.2). They contribute to the aroma of rose petals. The principle reason for pointing out the ionone structures is to emphasis that in order for a carotenoid to be a precursor to vitamin A it must have at least one β-ionone end ring.

In contrast to the central chain system, substitution of the end rings is common. Most reactions of carotenoids are oxidations, with reductions being fairly rare. The end chains are usually altered by the additions of a hydroxy,

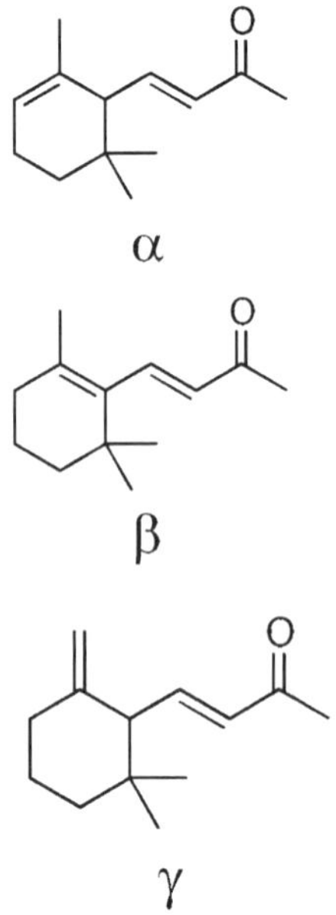

Figure 1.2   Structure of the unsubstituted ionone rings.

keto or epoxy group. This can occur at either end and these transformations can be different forming non-symmetrical products.

1.3   Biosynthesis

Perhaps the best way to understand the structure and the relationship between the various carotenoids is to let nature be the teacher. The biosynthetic assembly of the carotenoids is from an intermediary metabolic precursor isoprene (2-methyl-1,3-butadiene). Isoprene is a fundamental plant building block of many plant-derived compounds which includes sterols, tocopherols (vitamin E), phylloquinone (vitamin K) and countless other terpenoids with characteristic odours and flavours. Isoprene is a branched five carbon molecule and therefore carotenoids can be expected to have multiples of five carbon atoms. Although C30 carotenoids exist, by far the largest group is the C40 carotenoids, *i.e.* they are synthesized from eight isoprenoid precursor molecules.

1.3.1   Biosynthetic Pathway

The basic biosynthetic pathway is shown in Figure 1.3. This occurs only in photosynthetic plants and microorganisms. Animals are incapable of *de novo* synthesis of carotenoids but are capable of modification of dietary carotenoids to a range of compounds of importance to animal physiology.

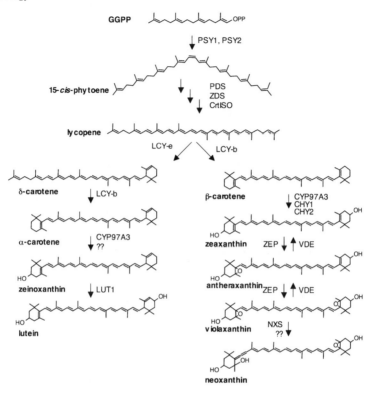

Figure 1.3 Biosynthetic pathway (taken from open access BioMed Central with acknowledgement Diretto 2006).

The first committed step of carotenoid biosynthesis is the C20 compound geranylgeranylphosphate in which the isoprenes are linked in a head to tail arrangement. The subsequent transitions are catalyzed by diiron proteins or cytochrome p450 enzymes.

1.3.2 Key Observations for Carotenoid Biosynthesis

The details are unimportant for this discussion, except to make some observations concerning carotenoid structure. Here are several specific key features of this pathway:

1. The coupling of the two C20 precursors takes place in a 'tail to tail' arrangement so that the central position reverses the chains to produce the colourless compound phytoene. Hence the central two methyl groups are in 1:6 positions, whereas all the others methyl groups have 1:5 relation-ships. This is common to all carotenoids as well as squalene (precursor to cholesterol) but not to all higher terpenoids, *e.g.* phytol.
2. The synthetic pathway divides at lycopene into an alpha and a beta pathway. These two series are geometric isomers with the solitary shift of

a single unsaturated position on the ring system. This is not an insignificant phenomenon because each series leads to a distinct set of downstream products of separate importance which are not necessarily interconvertible by isomerization.

3. The trivial names (see below for discussions on trivial names) for isomers of carotene can at times be confusing. Consider α-carotene, β-carotene, δ-carotene and ζ-carotene: it may be construed from their common names that they are intimately related in structure, biosynthetic origin and/or function but this is not really the case.

4. The biosynthetic pathway initially produces unsubstituted hydrocarbon carotenes, principally α-carotene and β-carotene. These are both very important compounds in themselves but this pathway proceeds to metabolize these carotenes to hydroxylated, keto or epoxy carotenoids. Arbitrarily, carotenoids can be classified in two classes:

    (a) Carotenes which are strictly hydrocarbons.
    (b) Xanthophylls which are carotenes that have one or more oxy-substitutions anywhere on their structure (usually end rings). The xanthophylls are more polar than the carotenes.

5. Most reactions involving carotenoids are oxidations and are irreversible. However, the 'xanthophyll cycle' on the lower right of Figure 1.3 is reversible. It is oxidative recycling of zeaxanthin after conversion into two epoxy carotenoids antheraxanthin and violaxanthin. This is an important plant process.

Both plants (including microorganisms) and animals (including marine species) can convert the basic carotenoids shown in Figure 1.3 into a bewildering array of more complex carotenoids. Over 700 are currently known and they serve a range of essential metabolic functions in a range of environmental habitats. Together with the porphyrins, carotenoids are one of the great colour chemistries in the biological world.

1.4 Trivial Names of Carotenoids

1.4.1 Origins of Trivial Names

Not surprisingly carotenoids were named by the original discoverer which reflects the plant from which they were originally extracted. The most obvious of these is carotene from carrots, zeaxanthin from wheat (the botanical name is *Zea mays*) and lutein the principle yellow pigment in the macula lutea in the retina. The terminology xanthophyll refers to the more yellow colour of the polar pigments from autumn leaves (from the Greek words for yellow leaf). Lutein and zeaxanthin are both xanthophylls. These non-systematic names are still in common use but are inexact because they do not reveal the stereochemistry and often cannot differentiate the multitude of isomeric forms of substituted carotenoids.

1.4.2   Major Nutritional Carotenoids

It is worthwhile retaining the trivial names for the most important nutritional carotenoids. Indeed there is nothing to be gained by deviating from them because their simplicity outweighs any confusion that is created in assigning structure. In plasma of higher animal (including humans) the profile of carotenoids reflects their respective diets. Quantitatively, the most common carotenoids observed in plasma are the major plant carotenoids β-carotene, α-carotene, lycopene, β-cryptoxanthin, β-canthaxanthin, lutein and zeaxanthin. There may be up to 40 measurable carotenoids including a range of *cis*-isomers but many are minor dietary components. Carotenoids in humans have been studied extensively for their relationship to various diseases.

1.4.3   Provitamin A Carotenoids

The trivial names of the five most common pro-vitamin A carotenoids are β-carotene, α-carotene, β-cryptoxanthin, β-canthaxanthin and β-echinenone. These trivial names are retained in nutritional literature because of their widespread acceptance. β-Cryptoxanthin, β-canthaxanthin and β-echinenone are pro-vitamin A xanthophylls. This is because these three xanthophylls have hydroxyl (and keto) groups attached at one end ring only. The most common xanthophylls, lutein and zeaxanthin, are substituted at both ends and are not pro-retinol precursors. The α-carotenoids can be pro-vitamin A because they can still possess a β-ionone end ring.

1.4.4   Ambiguities in Trivial Names

Trivial nomenclature can be ambiguous or slightly deceiving. For instance, with β-carotene there is a β-ionone ring at both ends. However, α-carotene does not have two α-ionone rings, it has one α- and one β-ring. Moreover, in the semi-systematic naming system α-carotene is not named α- because it is called the end ring and has a ε- descriptor not α- (see Table 1.1). It is advisable to be aware of this. Even more unusual is that lutein and zeaxanthin are geometric isomers with exactly the same relationship as α-carotene and β-carotene but they have completely disparate names.

1.5   IUPAC Definitions

International Union of Pure and Applied Chemistry (IUPAC) systematic naming of all organic compounds is very rigorous and is based solely on exact assignment of structure and not biological rationale or relationships (IUPAC 1978).

1.5.1   Systematic Names

IUPAC systematic naming generates systematic names. The same rules are applicable to carotenoids and retinoids. These are higher terpenoids, consequently their names are long and clumsy and the relationship between these compounds can be obscured. A selection of the rules most applicable to carotenoids is listed in the Key Facts Section. Rules for heterocyclic compounds are omitted for brevity because they are not required in common carotenoids.

The application of IUPAC systematic names to carotenoids can be aptly illustrated for two important photosynthetic xanthophylls, lutein and zeaxanthin. Their structures are shown in Figure 1.4. They are geometric isomers differing only by the position of a single double bond. Systematic rules require both zeaxanthin and lutein to be formally named as derivatives of a cyclohexene. In zeaxanthin the molecule is symmetrical and hence numbering can start at either end. The assignment of locants on the cyclohexene follows the IUPAC suffix priority rules, *i.e.* OH > ene > alkyl, and the count then gives the hydroxyl locant position 1 the double bond the lowest number, *i.e.* locant position 3. The attached alkyl group is named as a derivative of a fully saturated 18 carbon alkane chains with four attached methyl groups, nine conjugated double bonds and another attached unsaturated cyclic ring structures (hydroxylated cyclohexenyl) at the other end. The first carbon in the central chain is counted as locant position number 1 and the last being counted as locant position number 18. The systematic name for zeaxanthin (without due regard to stereochemistry) then becomes 4-[18-(4-hydroxy-2,6,6-trimethyl-1-cyclohexenyl)-3,7,12,16-tetramethyl-octadeca-1,3,5,7,9,11,13,15,17-nonaenyl]-3,5,5-trimethyl-cyclohex-3-en-1-ol.

Note that the cyclohexene at the distal end is numbered differently because it is formally regarded as a substituent on position 18 of the nonaenyl chain. Hence the priority order is changed and the prefixed OH group must have locant position 4, whereas the numbering direction of the other substituents will give the ene group the lowest locant, *i.e.* position 1.

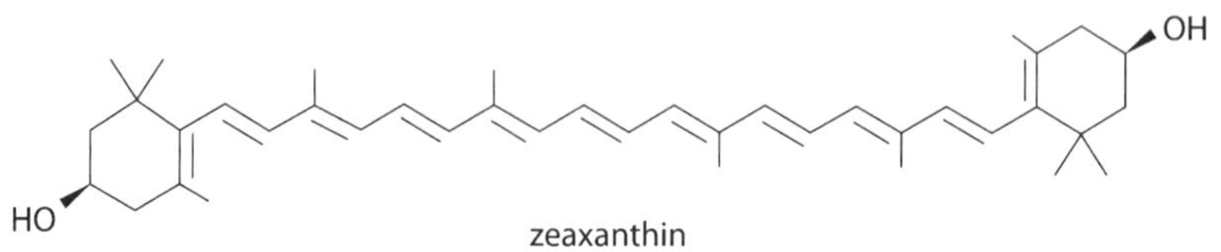

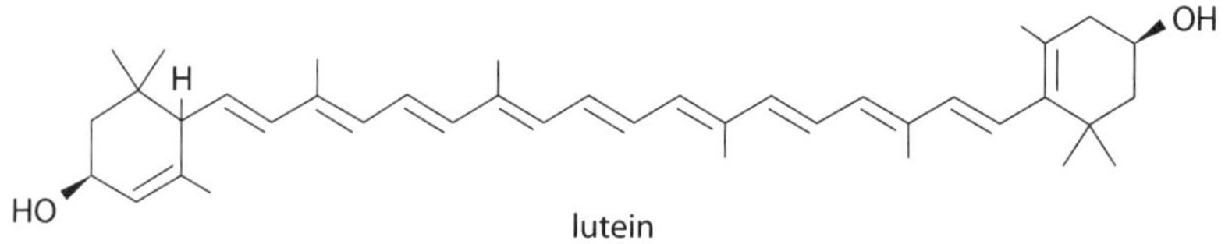

Figure 1.4   Structures of zeaxanthin and lutein.

Conversely, lutein is asymmetrical and IUPAC naming must start at the end with the end ring double bond in the lowest locant. Hence it is considered as a derivative of cyclohex-2-ene not cyclohex-3-ene namely 4-[18-(4-Hydroxy-2,6,6-trimethyl-1-cyclohexenyl)-3,7,12,16-tetramethyloctadeca-1,3,5,7,9,11,13,15,17-nonaenyl]-3,5,5-trimethyl-cyclohex-2-en-1-ol. Careful scrutiny of zeaxanthin and lutein will show that there is an asymmetric carbon only on lutein. Asymmetry rules are considered separately in section 1.6 because they apply to both naming systems. Both hydroxyl groups on these xanthophylls are also on asymmetric centres.

1.5.2 Semi-systematic Names

IUPAC Commission on the Nomenclature of Organic Chemistry (CNOC) and IUPAC–International Union of Biochemistry (IUB) Commission on Biochemical Nomenclature (CBN) discussed a set of tentative rules for the nomenclature of carotenoids multiple times from 1967 until 1973. The rules were approved in 1974 (IUPAC 1974). These IUPAC carotenoid rules are a very elegant method for naming carotenoids. They are systematic in that the structure can be unequivocally assigned follow a set of strict rules. The resultant semi-systematic names are not as unwieldy as the systematic names. Moreover, the IUPAC carotenoids rules are easier to apply and retain a sense of stoichiometric relationship between carotenoids and carotenoid-like compounds which might not otherwise be recognized as carotenoids derivatives, such as apocarotenoids.

1.5.2.1 Chemical Alterations to Carotenoids

All carotenoids-like compounds can be considered as derivatives of a long acyclic hydrocarbon chain of conjugated double bonds $C_{40}H_{56}$. This base compound (see Figure 1.5) is chemically identical to lycopene (although not referred to as such in the IUPAC document). This base structure can be chemically altered by:

(a) hydrogenation
(b) dehydrogenation
(c) cyclization
(d) oxidation
(e) or any combination of these to make a huge variety of carotenoids.

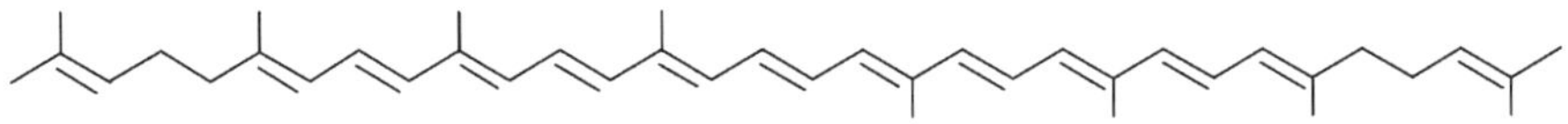

Figure 1.5 The stereo-parental base carotenoid structure as defined by IUPAC 1974. The methyl groups are shown for completeness.

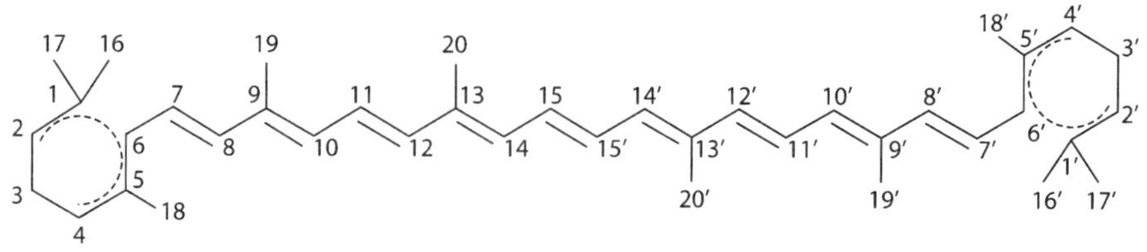

Figure 1.6   The semi-systematic numbering of carotenoids using acyclic γ-carotene as an example where the dotted lines denote an unspecified double bond locant (from IUPAC 1974).

1.5.2.2   Stem Name and Numbering

All carotenoid specific names are based on the stem 'carotene' in Figure 1.5. The two most elementary rules are in the numbering and the rules governing the end groups. The numbering of this stem carotenoid is shown in Figure 1.6 where the circular dotted line represents unassigned double bonds. It can be seen the numbering begins at both ends of the chain with primed and unprimed numbers, 1–15 and 1'–15', symmetrically to the central position in the chain. The attached methyl groups are then numbered in a similar fashion from 16–20 and 16'–20' from the outside to the centre. This semi-systematic numbering does not follow the underlying tetraterpenoid structure.

1.5.2.3   End Ring Systems

Carotenoids can have none, one or (mostly) two ring systems at the end of the central chain. The central chain is usually more invariant than the end ring systems. This is certainly the situation in the major dietary carotenoids and the major photosynthetic pigments. Geometrical isomerism and substitutions occurs mainly on the end rings.

IUPAC defines the naming of these end groups by adding a Greek letter as a prefix to the stem name to indicate whether it is acyclic or cyclic and also indicating the position of the single double bond within the ring system. There is also naming for a cyclopentyl and fully saturated aryl end group. The Greek letters follow their natural order in the Greek alphabet.

To see this systematic naming in action, consider the following structure in Figure 1.7. The carotenoid is asymmetric and the Greek letter prefixes follow

Table 1.1   Prefixes denoting the nature of the end groups of a carotenoid.

Type	Prefix
Acyclic	(psi) ψ
Cyclohexene	(beta, epsilon) β,ε
Methylenecyclohexane	(gamma) γ
Cyclopentane	(kappa) κ
Aryl	(phi, chi) Φ,χ

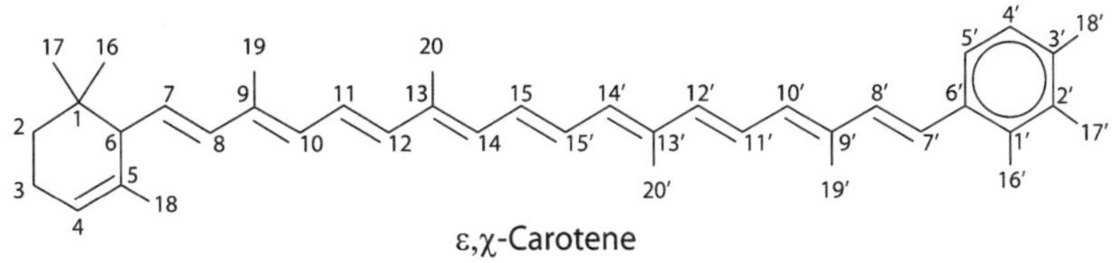

ε,χ-Carotene

Figure 1.7   The semi-systematic naming of an aromatic carotenoid showing the use of
Greek letter prefixes are per Table 1.1. Note the order follows the Greek
alphabet and prioritizes the primed numbering.

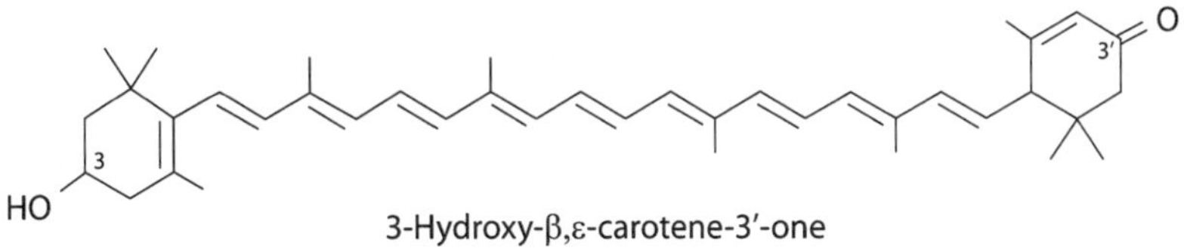

3-Hydroxy-β,ε-carotene-3'-one

Figure 1.8   The semi-systematic naming of a xanthophyll showing the use of locants
for a hydroxy and keto attachment. Note an alcohol takes priority over
keto and therefore occupies unprimed locant 3.

the alphabet and the structure is shown with the unprimed numbering to the
left (by convention).

This IUPAC ring naming can be at variance with some trivial naming of
carotenoids. For instance, there is no α position on end ring (see Table 1.1)
and α-carotene is systematically named β,ε-carotene. The same applies to δ-
carotene (a biosynthetic intermediate between lycopene and α-carotene) which
is systematically named ε,ψ-carotene because one end is still acyclic.

The other useful rule governs the xanthophylls containing hydroxyl, keto or
epoxy carotenoids. They are applied according to the normal conventions of
substituents as prefixes or suffixes. A good example is shown in Figure 1.8
whereby the keto group on 3' position is a suffix and the hydroxyl group is a
prefix on position 3.

1.5.2.4   Cleaved Central Chain Systems and Apocarotenoids

The central chain can be cleaved at a number of positions. The semi-synthetic
IUPAC rules are still applicable to this situation and the resulting compound
are still named as derivatives of carotenoids called apocarotenoids. The
important of central chain cleavage is that it is can result in the formation of
retinol (vitamin A) if:

(a) at least one end ring system is β-ionone (see Figure 1.2) and
(b) the chain cleavage is at the central position between the 15 and 15'
carbons.

For further rules the reader is referred to the IUPAC Nomenclature of Carotenoids. The subject of apocarotenoids will be discussed with reference to retinol.

1.6 Stereochemistry of Carotenoids

IUPAC carotenoid rules dictate the systematic naming of carotenoid structures but they do not account for stereoisomerism and geometric isomerism.

1.6.1 Stereoisomers

Many carotenoids have asymmetric carbon atoms, although the common β-carotene does not. Consider the important xanthophylls zeaxanthin and lutein in Figure 1.4. Zeaxanthin has two asymmetric centres with attached hydroxyl groups at the 3 and 3′ locants, whereas lutein has additional asymmetric carbon at position 6. The normal Cahn–Ingold–Prelog rules apply to the assignment of stereochemistry. These rules are well known and need not be discussed here. Figure 1.4 shows the spatial arrangement of the asymmetric centres *via* wedge shaped and dotted bonds. All three have R configurations. Hence the semi-systematic name of lutein becomes 3R,3′R,6′R-β,ε-carotene-3,3′-diol. Note that the β-ionone ring takes priority because it has the lower alphabetic rank and hence attracts an unprimed number. This means the asymmetric carbon is locant position 6′ and not 6. Also note that the unprimed prefix 3 is written before the primed 3′ prefix. The 6′ must follow the 3 prefixes in numerical order. An astute observer will note that the lutein is deliberately presented the wrong way in Figure 1.4 because unprimed numbers should be displayed on the left.

1.6.2 Geometric Isomers

Any double bond can have a *cis*- or *trans*-conformation. The all-*trans* conformation is energetically favoured based on the greater delocalization of the π-electrons of conjugate double bonds and the lower steric strain by the methyl groups. Formation of a *cis*-conformation causes a gross change in the geometry of the central chain and a di-*cis*-conformation would distort further.

Assignment of conformation across double bonds follows the same priority rules as for asymmetric centres. If the two priority substituents are on the same side of the double bond, i.e. *cis*-, then the assignment is given the Z descriptor. Conversely, if the two priority substituents are on opposite side of the double bond, *i.e. trans*-, then the descriptor is E. In carotenoids the *trans*-conformation is consistent with an E designation but only in the absence of any substitution. Hence, the common all-*trans* central chain is synonymous with all-E, provided there are no substituents on the chain. This may not be the case in more complicated carotenoids. Lutein (in Figure 1.4) is then fully described as all-E 3R,3′R,6′R-β,ε-carotene-3,3′-diol.

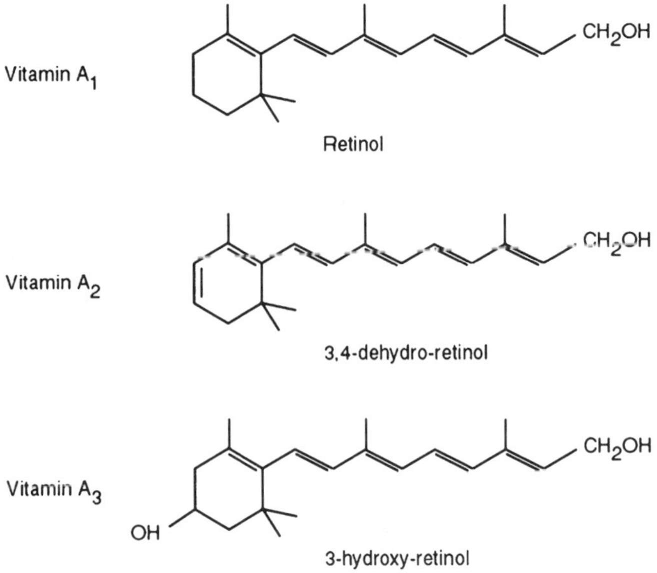

Figure 1.9   Structures of three natural forms of vitamin A.

1.7   Structure of Retinol

1.7.1   Prototypical Vitamin A Compounds

Any compound that has the activity of vitamin A can be referred to as such. By far the most common vitamin A is vitamin A_1 which is chemically retinol. There are two other forms vitamin A_2 and vitamin A_3 (refer to structures in Figure 1.9). These are found in the freshwater fish livers and various mollusks respectively and may find their way into human nutrition *via* diet. Vitamin A activity is defined by the ability of the compound to support vision. Both have much lower vitamin A activity in higher animals and can be mostly discounted. The referral of retinol as vitamin A_1 and especially as vitamin A_1 alcohol is usually redundant in most nutritional discussions.

1.7.2   Chemical Nature of Retinol

Vitamin A_1 has the all-*trans* structure equivalent to the stable all-*trans* geometry of most carotenoids. Chemically retinol is a diterpenoid with an end ring and a polyunsaturated chain which is half the length of a carotenoid central chain. It is clearly produced by oxidative cleavage of a carotenoid at the middle position 15,15′ of the central carotenoid chain and has the β-ionone ring structure, although the prefix β- is never used. Retinol has an alcohol at the end of the polyunsaturated chain as implied by its name and its name is derived from the retina in the eye where it was first chemically and biochemically investigated. Retinol is transported by a specific carrier protein, retinol binding protein (RBP), in animal circulation.

1.7.3   Retinol Systematic Name

The full IUPAC systematic name of retinol is (2E,4E,6E,8E)-3,7-dimethyl-9-(2,6,6-trimethyl-1-cyclohexen-1-yl)-2,4,6,8-nonatetraen-1-ol. The all-*trans* chain has an all-E stereochemistry and is formally regarded as a dimethyl derivative of fully unsaturated nonane whereby the substituted cyclohexene is at the opposite end to the hydroxyl group in position 9 (the hydroxyl group takes precedence in its locant placement). The features to note are the priorities given to the locants of the cyclohexene ring itself giving the -ene priority and allocating locant 1 to the nonyl side chain. Retinol does not have any asymmetric centres and hence stereoisomers do not need defining.

1.7.4   Retinol Semi-systematic Name

Because retinoids are derived from carotenoids by oxidative cleavage they can be considered as an apocarotene. Retinol semi-systematic name using the carotenoid nomenclature is 15-apo-β-caroten-15-ol. There is a set of IUPAC rules available for use with retinoids (IUPAC 1983). As with the carotenoids, they define a base structure which can be regarded as a stereo-parent compound which requires no further descriptive terminology. All alterations to this structure such as change of the functional group, hydrogenation, substitutions, removal of methyl groups, ring fission *etc.* will require a prefix (suffixes are not used). The examples given already of the three known vitamin A compounds (see Figure 1.9) show this naming system in operation. Figure 1.10 illustrates the numbering of the retinol molecule which closely follows the carotenoid system. The locant of any derivative uses this system.

1.7.5   Retinol Trivial Names

Retinol had a large number of trivial names applied by various researchers during its discovery and throughout its early investigations. None of these are used any more. Retinol is now the WHO-approved non-proprietary name of vitamin A. The use of the terminology vitamin A is very widespread and is used interchangeably with retinol. It gained acceptance historically and its major scientific merit is to indicate its biological and nutritional importance. Its use should be confined to a general descriptor of a molecule exhibiting the biological qualitative activity of retinol. This term can also used in medical and nutritional context such as vitamin A activity, vitamin A deficiency, vitamin A antagonism *etc.*

1.7.6   Retinol *cis–trans* Isomerism

All-*trans*-retinol is the most stable and most prevalent form in foods and tissues for the same reasons governing the more stable π-bonds which favour all-*trans* carotenoids. However, small amounts of other geometric isomers such as 9-*cis*-retinol and 13-*cis*-retinol co-exist with all-*trans*.

1.7.7 Retinol from Symmetrical Cleavage of Carotenoids

For a long time the enzyme responsible for cleavage of pro-vitamin A carotenoids remained elusive. Symmetrical cleavage produces retinal directly whereas asymmetric cleavage does not. The gene for symmetrical cleavage is now known as *Bcmo1* and the gene product is ββ-carotene-15,15′-mono-oxygenase (BCMO1). It has homology to the enzyme retinal pigment epithelium specific protein 65kDa (RPE65) that isomerizes all-*trans*-retinol into 11-*cis*-retinol in the visual cycle. BCMO1 is a soluble protein with a catalytic ferrous atom and is differentially expressed in many tissues. Asymmetric cleavage by ββ-carotene-9,10-mono-oxygenase (BCMO2) produces two dissimilar apocarotenes β-apo-10′-carotenal and β-ionone, Olsen (2004).

1.8 Retinoids: Derivatives of Retinol

Natural retinoids are derivatives of retinol which bear the same structural motifs but differ by the conversion of the hydroxyl group (see Figure 1.10). The retinal hydroxyl group behaves in exactly the same way as all aliphatic alcohols. It can be converted into an ester (more stable less toxic storage form) by lecithin:retinol acyltransferase (LRAT) (Harrison 2005) and it can be oxidized to an aldehyde or a carboxylic acid. All these forms exist in animals. Retinol itself should best be considered as the parent molecule for other active retinoids. Its two major metabolic products retinal and retinoic acid are not referred to as actual

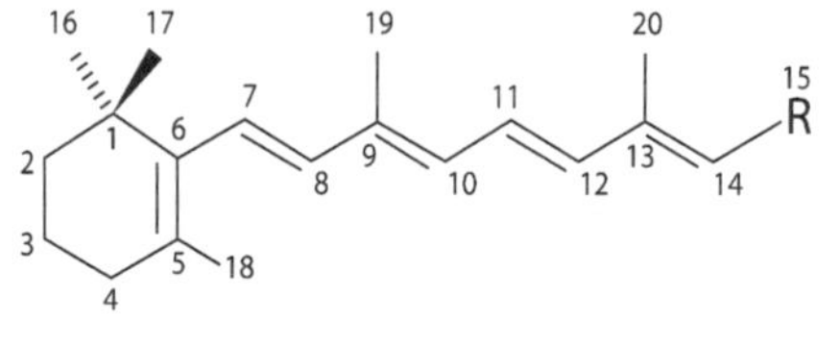

Figure 1.10   The numbering system is the same for all retinoids [R = functional group, *e.g.* CH_2OH (retinol), CHO (retinal), CO_2H (retinoic acid) and $CO_2C_2H_5$ (retinyl acetate)].

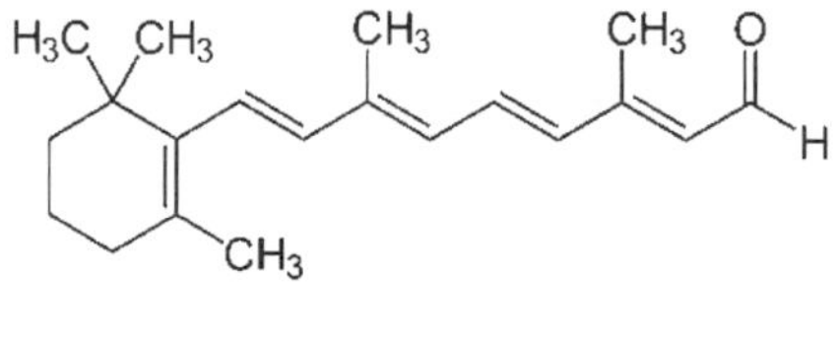

Figure 1.11   Structure of retinal.

vitamins, reflecting the notion that the vitamin terminology refers to its dietary form. Retinol parallels several other vitamins in this regard.

1.8.1   Retinal

Retinal (retinyl aldehyde) has exactly the same structure as retinol except the terminal alcohol is oxidized to an aldehyde (Figure 1.11). Its systematic name is as expected: (2E,4E,6E,8E)-3,7-dimethyl-9-(2,6,6-trimethylcyclohexen-1-yl)nona-2,4,6,8-tetraen-1-al.

When carotenoids are oxidized by 15,15′-mono-oxygenases they irreversibly produce retinal. Retinal can be reduced to retinol for transport and storage. Retinol can be oxidized back into retinal when required. The enzyme required for these reversible reactions is retinol dehydrogenase (retinol:NAD$^+$ oxidoreductase) requiring NAD$^+$ or NADP$^+$ as the donor or acceptor (eqns 1 and 2). The reversible inter-conversion of retinol and retinal is exemplified in the visual cycle (Fulton 2001).

$$\text{retinol} + \text{NAD}^+ \rightleftharpoons \text{retinal} + \text{NADH} + \text{H}^+ \tag{1}$$

$$\text{retinal} + \text{NADPH} + \text{H}^+ \rightleftharpoons \text{retinol} + \text{NADP}^+ \tag{2}$$

1.8.2   Retinoic Acid

All-*trans*-retinoic acid has the systematic name (2E,4E,6E,8E)-3,7-dimethyl-9-(2,6,6-trimethylcyclohexen-1-yl)nona-2,4,6,8-tetraenoic acid and is the most biologically important retinoid (Figure 1.12). It has a nuclear receptor (RAR).

Retinoic acid is produced *in vivo* from retinal by irreversible oxidation (eqns 3 and 4). Two enzymes can use different electron donors. Retinaldehyde dehydrogenases (RALDHs) can use NAD$^+$ as a donor.

$$\text{retinal} + \text{NAD}^+ + \text{H}_2\text{O} \rightarrow \text{retinoic acid} + \text{NADH} + \text{H}^+ \tag{3}$$

Alternatively, retinal oxidase (retinal:oxygen oxidoreductase) can use molecular oxygen as a donor.

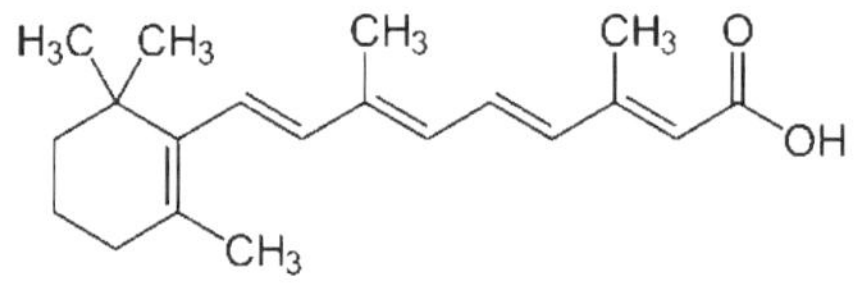

Figure 1.12   Structure of retinoic acid.

$$retinal + O_2 + H_2O \rightleftharpoons retinoic\ acid + H_2O_2 \qquad (4)$$

Retinoic acid is a remarkably important compound and should be considered as a hormone. It embodies all the fundamental activities of the retinoid biochemistries. Retinoic acid or its *cis*-isomers are ligands for the DNA-binding RAR subtypes (RARα, RARβ and RARγ) and the retinoid X receptors (RXRα, RXRβ and RXRγ). RARs regulate transcription of specific genes which control the programmed differentiation of epithelial cells in the digestive tract, skin, bone, respiratory, nervous and immune system. RXR heterodimerizes with other nuclear receptors, such as peroxisome proliferation activation receptor (PPAR), as well as RAR itself. The complexity of retinoic acid and its nuclear receptors should be appreciated. RAR is causal in acute promyelocytic leukaemia (APML) by chromosomal translocation of the *RARα* gene with the *PML* gene. Levels of intracellular retinoic acid are regulated by cytochrome P450 (CYP26) which oxidizes (inactivates) it at the 4 and 18 positions (White 2000).

1.8.3   Retinoid Geometric Isomers

Geometric isomers of retinoic acid are 9-*cis* retinoic acid (aliretinoin) and 13-*cis* retinoic acid (isotretinoin) (Figure 1.13). They are also ligands for RAR and are marketed as chemotherapeutic medications.

The dietary requirements of vitamin A (and its precursor carotenoids) are all based on their ability to produce retinoic acid. The deficiency states (and toxicity states) of vitamin A are all explainable by the actions of retinoic acid.

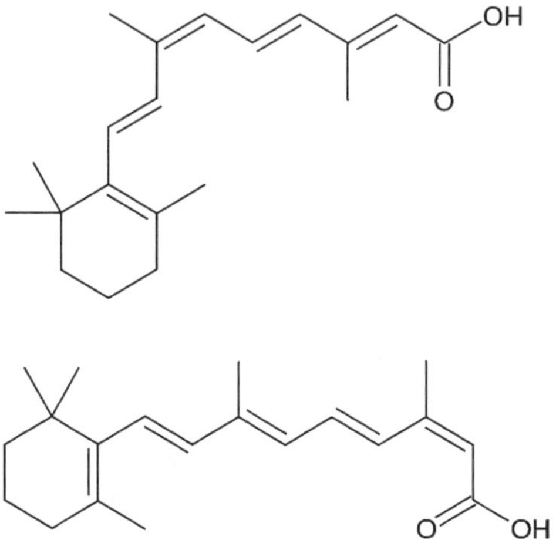

Figure 1.13   9-*cis*-Retinoic acid and 13-*cis*-retinoic acid are two biologically and pharmaceutically important geometric isomers of retinoic acid.

The terminology vitamin A remains familiar to us all in casual discussions about diet but few would appreciate the complexity and fundamental relevance of these essential biological compounds. For those of us geographically removed from the nutritionally impoverished countries we probably barely even recognize the pervasive characteristics of vitamin A deficiency and the disastrous consequences to affected populations. The intricacies of the metabolism of retinoids continues to impress (D'Ambrosio 2011) and their involvement in the regulation of greater than 500 genes ensures retinoid their vital role in the pathophysiology of animals.

Key Facts

IUPAC nomenclature rules relevant to the semi-systematic naming of carotenoids (IUPAC 1974):

1. Class of compound
      Octa-isoprenoids oxygenated (xanthophylls) or not (carotenes)
2. Stem name
      Carotene in which the double-bond positions are assumed (see Section 1.5.2.2) with acyclic and/or cyclic end groups and numbered according to Figure 1.6
3. End group designation
      Addition of two Greek letters as prefixes
      Denoting acyclic (psi – ψ)
      Aryl (phi or chi – ϕ or χ)
      Cyclopentane (kappa – κ)
      Cyclohexene (with the point of unsaturation as beta, alpha or gamma – β, ε or γ)
      There is no α designation
      Cited in alphabetical order
      Separated by comma
      Second prefix with hyphen
4. Two end groups dissimilar
      One end is numbered with prime
      Unprimed numbers given to end with first listed Greek prefix
      Unprimed locants are cited before primed
      Formulae drawn with unprimed to the left
5. Demethylation
      Indicated with nor prefix preceding locant of eliminated methyl group
      Nor locant is given lowest possible number
6. Fission (excepting C atoms at 1 and 6 positions)
      Denoted by seco with the two carbons locants named
7. Hydrogenation change
      Use prefix hydro- or dehydro- with the two carbon locants named
      Must be multiple of two, *e.g.* didehydrotetrahydro

>   Non-detachable and immediately preceding Greek letter prefix with hyphen

8. Oxygenation

>   Denoted by suffixes or prefixes as per usual nomenclature
>   Group priority sequence carboxylic acid > ester > aldehyde > ketone > alcohol
>   Principle group is suffix, lesser order as prefix
>   Ether named as alkoxy substituent with locant
>   Bridging ether use prefix epoxy preceded by locants of C atoms (they may not be adjacent)
>   Epoxide across double bond must be epoxydihydro
>   Formal addition of water (or methanol) across double bond is hydroxydihydro (or methoxydihydro).

List of Abbreviations

BCMO1	β-carotene-15,15'-monooxygenase
E	Entgegen, isomeric descriptor for priority groups on opposite sides of a double bond (usually equivalent to *trans-*)
IUPAC	International Union of Pure and Applied Chemistry
RBP	retinol-binding protein
RAR	retinoic acid receptor
RXR	retinoid X receptor
Z	Zusammen, isomeric descriptor for priority substituents same side of a double bond (usually equivalent to *cis-*)

References

D'Ambrosio, D. N., Clugston, R. D., and Blaner, W. S., 2011. Vitamin A metabolism: an update. Nutrients. 3: 63–103.

Diretto, G., Tavazza, R., Welsch, R., Pizzichini, D., Mourgues, F., Papacchioli, V., Beyer, P., and Giuliano, G., 2006. Metabolic engineering of potato tuber carotenoids through tuber-specific silencing of lycopene epsilon cyclase. BMC Plant Biology. 6: 13.

Fulton, J. T., 2001. Processes in Biological Vision (online). Vision Concepts, Corona Del Mar, CA. USA., published 2000-08-01, revised 2000-08-01, cited 2000-08-01. Available at: http://neuronresearch.net/vision/

Harrison E. H., 2005. Mechanisms of digestion and absorption of dietary vitamin A. Annual Review of Nutrition 25: 87–103.

International Union of Biochemistry (IUB), 1978. Biochemical nomenclature and related documents. The Biochemical Society, London.

International Union of Pure and Applied Chemistry and International Union of Biochemistry Nomenclature of Carotenoids (IUPAC_IUB), 1974. Butterworths, London.

International Union of Pure and Applied Chemistry and International Union of Biochemistry Nomenclature of Retinoids, 1983. Pure Applied Chemistry. 55: 721–726. Moss, G.P., 1983. Archives Biochemistry and Biophysics. 224: 728–731 http://www.chem.qmul.ac.uk/iupac/misc/ret.html

Olsen J. A., 1989. Provitamin A function of carotenoids: the conversion of β-carotene into vitamin A. J. Nutr. 119: 105–108.

White J. A., *et al.*, 2000. Identification of the human cytochrome P450, P450RAI-2, which is predominantly expressed in the adult cerebellum and is responsible for all-trans-retinoic acid metabolism. Proceedings of the National Academy of Sciences. 97: 6403–6408.

CHAPTER 2

Vitamin A in the Context of Other Vitamins and Minerals

JENNIFER H. LIN*[a] AND KUANG-YU LIU[b]

[a] Division of Preventive Medicine, Department of Medicine, Brigham and Women's Hospital, Harvard Medical School, 900 Commonwealth Ave. East, Boston, MA 02215, USA; [b] Department of Anesthesiology, Brigham and Women's Hospital, Harvard Medical School, 75 Francis Street, Boston, MA. 02115, USA
*E-mail: jhlin@rics.bwh.harvard.edu

2.1 Source of Vitamin A and Other Vitamins and Minerals

Micronutrients including both vitamins and trace minerals play an essential role in a variety of physiological processes including hormonal and antioxidant properties, regulation of tissue growth and differentiation, regulation of DNA methylation, promotion of embryonic development, and nutrient metabolism (Figure 2.1) (Mora *et al* 2008). Most vitamins and minerals are derived in minute amounts from natural foods of animal and/or plant sources, although supplements also appear to be commonly used in many countries. It is estimated that one-half to two-thirds of the adult population in the developed countries such as the USA are likely taking multivitamin and mineral supplements that contain at least 10 vitamins or minerals with a wide range of doses.

Vitamin A, a fat-soluble nutrient, has an important role in embryonic growth, vision, and resistance to infection. Dietary vitamin A is obtained in foods mainly from two sources: *preformed vitamin A* as retinyl esters in foods

Food and Nutritional Components in Focus No. 1
Vitamin A and Carotenoids: Chemistry, Analysis, Function and Effects
Edited by Victor R Preedy
© The Royal Society of Chemistry 2012
Published by the Royal Society of Chemistry, www.rsc.org

Fat-soluble vitamins:

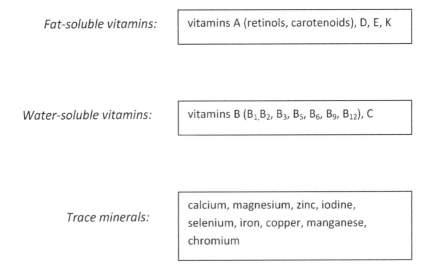

vitamins A (retinols, carotenoids), D, E, K

Water-soluble vitamins:

vitamins B (B_1, B_2, B_3, B_5, B_6, B_9, B_{12}), C

Trace minerals:

calcium, magnesium, zinc, iodine, selenium, iron, copper, manganese, chromium

Figure 2.1   Key micronutrients for human health.

of animal origins such as eggs, dairy products, and liver; *provitamin A carotenoids* found in plant-derived foods including β-carotene, α-carotene, lycopene, lutein, and β-crypotxanthin. Preformed vitamin A is efficiently absorbed and utilized by humans at absorption rates of 70–90%. In contrast, provitamin A carotenoids are absorbed much less efficiently, at rates of 20–50%, depending on an individual's vitamin A status and other factors. Up to 75% of dietary vitamin A in several developed countries is preformed vitamin A, largely derived from multivitamins, fish liver oil, and fortified foods such as milk, butter, margarine, breakfast cereals, and some snack foods (Penniston and Tanumihardjo 2006). In developing countries, however, 70–90% of vitamin A is obtained from provitamin A carotenoids in plant foods.

2.2   Intakes of Vitamin A as well as Other Vitamins and Minerals Around the World

2.2.1   In the Malnourished Countries

It has been estimated by the United Nations System Standing Committee on Nutrition that over 2 billion people around the world remain at risk for deficiencies of several essential vitamins and minerals with serious health consequences. Iron, iodine, and vitamin A deficiencies are the most common forms of micronutrient malnutrition, particularly in women and children in the developing world. Despite the ongoing efforts to control the deficiencies in these countries, they continue to be the major public health problems.

Iron deficiency, the most widespread nutritional disorder is more likely to occur in populations that rely on plant-based diets which have low iron bioavailability. Severe iron deficiency which may result in anemia, increases maternal mortality and ill-health, and affects normal development of the cognitive function in children (Silverberg *et al* 2001). Iodine deficiency disorders, more likely to occur in populations who do not consume seafood and fortified iodine, are easily recognized by two well-known clinical signs, endemic goiter and cretinism. It has been known that iodine deficiency during pregnancy and early postnatal life causes mental retardation, stunted growth, and other developmental abnormalities which are largely irreversible. The control of iodine deficiency has been successful since the past decade with the widespread adoption and coverage of iodinized salt worldwide, whereas it has been slow to prevent and control iron deficiency due likely to the less obvious clinical symptoms resulting from iron deficiency and the lack of suitable supplements with less side effects (Cook *et al* 1994).

Severe vitamin A deficiency entails clinical signs of xerophthalmia and night blindness. In the early 1960s, the World Health Organization (WHO), which undertook the first global survey of xerophthalmia, observed a significant prevalence of the disorder in several South and East Asian countries. It was later estimated in the 1980s that 13 million children worldwide suffered from xerophthalmia and another 40–80 million children were at risk of subclinical deficiency (Sommer *et al* 1981). Since 1995, the risk indicator for vitamin A deficiency in the population has been redefined by the serum vitamin A levels of <0.7 μmol L. With this new definition, it was found that the burden of vitamin A deficiency has far surpassed the previous estimates, with 3 million preschool children having clinical signs of vitamin A deficiency annually and another 140–250 million preschool children being at risk for the disorder (Underwood 2004).

Ever since the revelation by the WHO of the widespread problem of vitamin A deficiency, several intervention trials were subsequently conducted, including the trials performed in Jordan and India in the late 1960s, which demonstrated the feasibility of periodic vitamin A supplementation for the prevention of vitamin A deficiency (Reddy 2002). During the early decades, primary attention was focused on the outcomes of xerophthalmia and blindness. It was also later noted that there was an increased mortality associated with xerophthalmia as children with xerophthalmia died at higher rates than their non-xerophthalmic peers. However, the link between vitamin A and mortality remains ignored until 1986 when the first large-scale randomized trial of high-dose vitamin A supplementation every 6 months for the prevention of child mortality in Indonesia reported a 34% reduction in mortality among children aged 1 to 5 years (Sommer *et al* 1986). Several trials have also been conducted in other countries with similar results. Although vitamin A may prevent mortality through its impacts on adaptive immune responses, vitamin A deficiency is also common in the poor and malnourished countries where infectious diseases such as measles, diarrhea, and respiratory disease are prevalent. Thus, the combination of poor diet and infection leads to a vicious cycle that particularly affects young children and pregnant or lactating mothers.

A recent meta-analysis review of 17 trials reported a 24% reduction in risk for all-cause mortality [relative risk (RR) = 0.76; 95% confidence interval (CI), 0.69–0.83] among 194 795 children aged 6 months to 5 years (Imdad *et al* 2010). Specifically, there was a 28% reduction in diarrheal mortality (RR = 0.72; 95% CI, 0.57–0.91) and a 50% reduction in measles morbidity (RR = 0.50; 95% CI, 0.37–0.67). Although the review reported no effects of vitamin A supplementation in the mortality of measles, respiratory disease, and meningitis, another recent meta-analysis review of 4 randomized trials suggested that higher dose of vitamin A supplementation is necessary for reducing mortality; a dose of 200 000 IU of vitamin A in children aged ≥1 year and a dose of 100 000 of vitamin A in infants were associated with >60% reduction in measles mortality (RR = 0.38; 95% CI, 0.18–0.81) (Sudfeld *et al* 2010). These observations require confirmation in large randomized trials.

2.2.2   In the USA and Europe

Deficiency in micronutrients is less of a public health concern in most developed countries such as the USA and most European countries, although inadequate intakes of certain nutrients have been reported in some subgroups of the population. The USA National Health and Nutrition Examination Survey (NHANES), the first national survey of the estimation of total nutrient intake using food and beverage intake data from 24 h dietary recalls and detailed information on dietary supplement, has reported inadequate intakes of several vitamins and minerals in the elderly. According to the third NHANES data in 1988–1994 and the 2010 recommended daily allowance (RDA) guide, approximately 20–60% elderly people aged ≥60 years were more likely to have insufficient intakes of calcium and zinc as compared with other nutrients such as iron (Ervin and Kennedy-Stephenson 2002). In a separate report using the same data, inadequate zinc intake was also observed in ∼80% of younger children aged 1–3 years and ∼60% of adolescent females aged 11–18 years (Briefel *et al* 2000). In addition, there was an averaged prevalence of 7% of vitamin C deficiency in the USA, particularly among smokers and low-income people (Schleicher *et al* 2009). Moreover, it has been reported that vitamin D insufficiency affects 36% of the USA population whose serum vitamin D (25-hydroxy vitamin D levels) levels were <20 ng mL (Ginde *et al* 2009), the levels of which are required to prevent adverse outcomes such as fractures and falls. Folate insufficiency affects 15%–19% of the USA population, whereas an approximately 5% of men and women aged ≥50 years may exceed the tolerable upper intake levels (Bailey *et al* 2010). Finally, the estimated prevalence of low vitamin B_{12} levels was approximately 4% in the USA population (Evatt *et al* 2010).

According to the European Prospective Investigation into Cancer and Nutrition (EPIC) study of 27 centers from 10 countries, intakes of water-soluble vitamins B and C do not appear to vary by countries, except that higher vitamin C intake was shown in the southern centers compared with the northern ones (Olsen *et al* 2009). In contrast, fat-soluble nutrients including vitamin A, β-carotene, vitamin D, and vitamin E differ to some extent in European countries; consumption of β-carotene and vitamin E was higher from the southern to central European countries, whereas vitamin D intake was higher in the Northern Europe (Jenab *et al* 2009). Despite the difference in intake behaviors which is likely a consequence of difference and heterogeneity in dietary patterns across Europe, there is some consensus among European countries regarding inadequate intakes of nutrients. For instance, it has been suggested that in most European countries, inadequate intakes of folate and vitamin D appear to affect many age groups (Fabian and Elmadfa 2008; Tabacchi *et al* 2009). In two recent reviews of general population, 21–76% of British women of South Asian origin and 15–30% of the general German population are deficient in vitamin D with serum vitamin D levels of <10 ng mL (Ashwell *et al* 2010; Zittermann 2010). An additional report has suggested that children, adolescents, and young women in many European countries were at a greater risk of suboptimal intakes for folate, iron, zinc, and calcium (Fletcher *et al* 2004). In addition, according to the European Health and Nutrition Report in 2004 evaluating 18 vitamins and minerals, elderly people aged ≥55 years were also more likely to have inadequate intakes of calcium and iodine (Fabian and Elmadfa 2008).

There is little evidence that vitamin A deficiency is of public health significance in most developed countries (Tabacchi *et al* 2009). In a cross-sectional study of serum vitamin A distribution in NHANES III, there was a suboptimal status among some children of 4 to 8 years of age and some minority groups, but few (approximately 1%) had values that require clinical attention (*i.e.*, serum vitamin A of <0.7 μmol L) (Ballew *et al* 2001). It is also not common to find vitamin A deficiency in elderly individuals in the developed countries, as vitamin A is easily obtained from food and dietary supplements. However, excessive amounts of intake of preformed vitamin A equivalents may cause vitamin A toxicity. In individuals who have asymptomatic hepatic dysfunction, the risk for vitamin A toxicity increases (Penniston and Tanumihardjo 2006). Several cohort studies have also suggested that the chronic high intake of preformed vitamin A (≥10 000 IU day) is associated with an increased risk for fractures or osteoporosis as compared with those with lower vitamin A intake (<4200 IU day) (Feskanich *et al* 2002). In contrast, the adverse effects on bone and liver function have not been shown with higher intake of β-carotene, one of the provitamin A carotenoids (Hathcock 1997).

2.3 Interaction of Vitamin A with Other Vitamins and Minerals

2.3.1 Vitamin A and Zinc

The absorption and metabolism of vitamin A depend, in part, on the status of other nutrients. Experimental studies in animals have shown that zinc mediates intracellular and intercellular transport of retinal by the retinol-binding protein (RBP) and serves as a cofactor for the synthesis of enzymes responsible for vitamin A absorption (Christian and West 1998). However, given that zinc-deficient rats also gained less weight than the zinc-sufficient rats, it is hard to isolate the specific effects of zinc deficiency from general malnutrition on the changes in RBP concentrations and the enzymatic activity. In humans, findings from both observational studies and intervention trials have been inconsistent about the relationship between vitamin A and zinc status. Some but not all studies have reported a positive association between the concentrations of the two nutrients in malnourished children. In a small trial of 24 USA preterm babies, those who were supplemented with zinc had elevated levels of RBP and vitamin A as compared with those in the placebo group. However, other trials among pregnant teenagers or younger children in Asia or South America did not show the beneficial effects of zinc supplementation in increasing vitamin A status. While the results in observational studies and intervention trials are inconclusive, it remains plausible that the interaction between vitamin A and zinc may be relevant to certain populations. For instance, as reported in some observational studies, the positive relationship between the two nutrients was seen in severely malnourished populations who are deficient in both nutrients. In addition, positive serum levels of vitamin A and zinc were seen in patients with hepatic dysfunction such as alcoholic cirrhosis who are deficient in zinc or vitamin A, or both (Abdu-Gusau *et al* 1989). The potential impact of the interaction between vitamin A and zinc on health outcomes remains unknown.

2.3.2   Vitamin A and Iron

It has been suggested that vitamin A metabolism is affected by iron status. Rats fed a diet low in iron had significantly reduced serum vitamin A but increased hepatic vitamin A levels as compared with the rats fed a normal diet. Interestingly, rats fed adequate vitamin A and low iron also had lower circulating and higher hepatic vitamin A levels as compared with animals with adequate iron but low vitamin A intakes (Strube *et al* 2002). These observations suggest that iron deficiency results in reduction in vitamin A mobilization from the liver to the circulation. In humans, intervention trials conducted in developing countries have found that iron supplementation elevated serum retinol and/or RBP levels in pregnant women and children. One among these trials also reported that more women supplemented with iron and riboflavin than those in the placebo had an improvement in dark adaption (Graham *et al* 2007). In addition, women with particularly poor iron status at baseline who received iron supplementation benefited most with a significant improvement in pupillary threshold. Data from these studies suggest potential synergistic effects between vitamin A and iron, and large randomized trials will help confirm their combined effects on health outcomes.

2.3.3 Vitamin A and Vitamin D

Recent studies of bone health have further suggested that vitamins A and D interact at the molecular level as vitamin A may antagonize the positive influence of vitamin D on bone mineralization and calcium uptake (Crandall 2004). The molecular mechanism underlying the actions of vitamins A and D has been linked to the retinoid X receptor (RXR). The signaling of both vitamin A, which binds retinoic acid receptors (RARs), and vitamin D, through binding to the vitamin D receptor (VDR), requires RXR to function as a partner in heterodimer formation (Mora *et al* 2008). As such, the request of the RXR protein results in the competition of the two vitamin signaling pathways. It is likely that in the presence of high vitamin A, vitamin D action may be compromised. Rats administered high dose of vitamin A resulted in significant reduction in bone mineralization and serum calcium levels, regardless of whether these rats were also administered high dose of vitamin D. In humans, large cohort studies conducted in Scandinavia and USA have reported a positive association between total vitamin A intake and risk for hip fracture or osteoporosis. In a small intervention trial evaluating the acute effects of vitamins A and D on calcium homeostasis among middle-aged men, participants receiving supplementation of vitamin A alone or combined with vitamin D had lower serum calcium levels or reduced responses of calcium to vitamin D relative to those in the placebo group (Johansson and Melhus 2001). Other observational studies have also shown an inverse association of serum levels of vitamin A (retinol) with osteocalcin, a marker for bone formation, or with risk for osteoporosis. The inverse association between vitamin A levels and bone mineral density was also found to be more pronounced in those with inadequate vitamin D levels (<20 ng mL). Nevertheless, the magnitude of association between levels of vitamins A and D was low (correlation coefficient = 0.15) (Mata-Granados *et al* 2010), suggesting there may be other factors linking both vitamins A and D to bone health.

2.4 Chronic Disease Prevention with Supplementation of Vitamin A Alone or in Combination with Other Vitamins and Minerals

Although vitamin deficiency disorders are rarely a concern in developed countries, evidence suggests that certain groups of individuals are more likely to be at risk for suboptimal vitamin status. Acquiring sufficient amounts of micronutrients are believed to be necessary for normal function of a variety of biological processes. Several vitamins and minerals including vitamins A, C, E, and selenium contain antioxidant properties which scavenge free radicals and prevent oxidative damage to cells. Vitamins A, C, D, and E, iron, zinc, and selenium also modulate immune responses through the regulation of many processes such as T-cell activity and inflammation (Figure 2.2). Calcium and vitamin D promote bone mineralization which provides strength and hardness

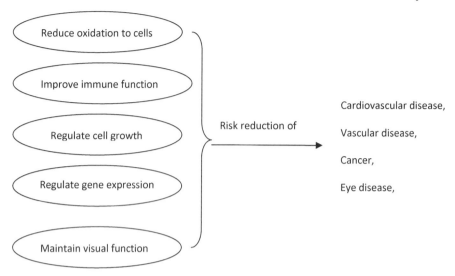

Figure 2.2   Possible mechanisms by which vitamin A prevents diseases.

Table 2.1   Dietary RDAs for vitamins and minerals in adults.

Vitamin (unit day^{-1})	*Adequate value*	*Tolerable upper intake levels*
Vitamin A (IU day^{-1})	2333–3000	10 000
Vitamin C (mg day^{-1})	75–90	2000
Vitamin D (IU day^{-1})	600–800	4000
Vitamin E (IU day^{-1})	22	1500
Vitamin B$_6$ (mg day^{-1})	1.3–1.7	100
Vitamin B$_{12}$ (μg day^{-1})	2.4	Not determinable
Folate (μg day^{-1})	400	1000
Selenium (μg day^{-1})	55	400
Zinc (mg day^{-1})	8–11	40
Iron (mg day^{-1})	8–18	45
Iodine (μg day^{-1})	150	1100

to bones and teeth. The B vitamins such as folate, vitamins B$_6$, and B$_{12}$ help maintain DNA integrity and metabolize homocysteine, a protein which is involved in the atherosclerotic process (Fairfield and Fletcher 2002).

Since the late 1970s, numerous intervention trials have been conducted for the study of the efficacy of supplementation with vitamins and minerals, alone or in combination, for the primary or secondary prevention of various chronic diseases including cancer, cardiovascular disease, osteoporosis, and age-related macular disease. The vitamins and minerals that are central in the preventive care of adults are the provitamin A carotenoids, vitamin A, folate, vitamins B$_6$, B$_{12}$, C, D, and E, and selenium. The dosage of these vitamins and minerals tend to be high in most intervention trials, which may be several times higher than the recommended values; for instance, vitamin C (*e.g.*, ≥500 mg daily) and vitamin E (≥300 IU daily) (Table 2.1). As of 2011, the majority of most intervention trials of vitamin and mineral supplementation are complete. The duration of the treatment and follow-up in these trials varies from as short as one day to 10 or longer years. The mean age of participants in the trials are >40 years. The findings from most of these trials do not support a benefit of vitamin A, carotenoids, and other vitamins and minerals in the disease prevention (Bjelakovic *et al* 2007).

2.4.1 Cancer Prevention

Owing to its potential effects on the epithelium and on immunity, retinol has been investigated as a chemopreventive agent for many cancers. However, much to the surprise of the research community, the Beta-Carotene and Retinol Efficacy Trial (CARET) evaluating the efficacy of 25 000 IU of vitamin A supplementation daily combined with 30 mg of β-carotene daily among smokers reported an increased risk for lung cancer incidence in the treatment group after 5 years of follow-up as compared with the placebo group (RR = 1.28; 95% CI, 1.04–1.57). In the same trial, a 17% increase in risk for total mortality was seen (RR = 1.17; 95% CI, 1.03–1.33) in the treatment group. Similar results were also shown in the Alpha-Tocopherol, Beta-Carotene Cancer Prevention (ATBC) Study, which reported a 16% excess in incident lung cancer among smokers in the β-carotene supplement group (20 mg daily) as compared with those in the placebo group for an average of 6 years of follow-up (RR = 1.16; 95% CI, 1.02–1.33).

The elevation of lung cancer incidence or death as a result of β-carotene and/or vitamin A supplementation, however, does not seem to appear in the general population tested in large randomized trials including the Physician Health Study (PHS), Women's Health Study (WHS), and the Women's Antioxidant Cardiovascular Study (WACS). A recent meta-analysis pooling 12 randomized trials of β-carotene supplementation has concluded a slight increase in cancer incidence (RR = 1.06; 95% CI, 1.00–1.12), with the increased risk mostly seen among smokers (RR = 1.10; 95% CI, 1.03–1.18) (Bardia *et al* 2008). In addition, supplementation of β-carotene or vitamin A combined with other vitamins and minerals (*i.e.*, vitamins C and E, and selenium) have shown no beneficial effects on total cancer incidence (RR = 0.99; 95% CI, 0.94–1.04), including lung cancer incidence (RR = 1.05; 95% CI, 0.94–1.17) and mortality (RR = 0.99; 95% CI, 0.92–1.15).

Despite the disappointing findings of vitamin and mineral supplementation from most prevention trials, two additional trials, the Linxian General Population Trial and the Supplémentation en Vitamines et Minéraux Antioxydants (SUVIMAX) have presented beneficial effects of supplementation of β-carotene (6 mg or 15 mg daily) combined with other supplements in reducing cancer development, with an approximately 20–30% risk reduction in stomach cancer or total cancer incidence. It is noteworthy that participants in these two trials are either of poor nutrition or of much lower baseline status of the supplemented nutrients, suggesting that micronutrient supplementation including β-carotene may be more relevant to those who are susceptible to deficiencies in these nutrients.

2.4.2 Prevention of vascular events

The potential protection of antioxidants including vitamin A against oxidative damage to lipids and proteins also leads to many intervention trials for the prevention of atherosclerotic vascular disease. However, reports regarding the effects of vitamin A and/or β-carotene supplementation against cardiovascular events have been disappointing. Of the two trials among smokers, the CARET study reported an increase of 26% of death from cardiovascular disease (RR = 1.26; 95% CI, 0.99–1.61) and the ATBC study reported more deaths from fatal coronary heart disease in the β-carotene arm than in the placebo group (RR = 1.75; 95% CI, 1.16–2.64). Similarly, another small randomized trial of High-density Lipoprotein (HDL) Atherosclerosis Treatment Study (HATS) in the secondary prevention of patients with established coronary artery disease reported a 22% decreased HDL levels in the treatment group after 12 months of supplementation with β-carotene (25 mg day) and other antioxidants. However, several other large trials of supplementation with β-carotene, alone or combined with vitamins C, E, and selenium, in the general populations for >4 years of follow-up including the heart protection study, the PHS, the WHS, the Linxian General Population Trial, and the SUVIMAX trial have reported neither beneficial nor harmful effects on the prevention of cardiovascular events and/or death. The WACS trial also reported no effects of β-carotene on the secondary prevention among women at high risk for cardiovascular disease.

2.4.3 Prevention of Other Diseases

Several intervention trials have also evaluated the influence of vitamin A or β-carotene supplementation on other chronic diseases among the elderly and children. Some trials have reported no benefits of vitamin A and/or β-carotene combined with other vitamins in preventing infectious diseases (*e.g.*, acute respiratory tract infection and influenza) in the elderly. A meta-analysis review pooling three randomized trials also did not support the use of vitamin A and/or β-carotene in preventing eye-related diseases including atrophic age-related macular degeneration. Finally, it has been suggested that lower vitamin A intake among children may be associated with increased risk for asthma and wheeze, which, however, was not supported by a recent randomized trial showing no effects of the long-term supplementation of vitamin A on subsequent asthma risk among >6400 Nepal children.

In a recent meta-analysis review of 47 clinical trials with >180 000 participants, antioxidant supplementation was associated with an increased risk for total mortality (RR = 1.16; 95% CI, 1.05–1.29) (Bjelakovic *et al* 2007).

Specifically, β-carotene and vitamin A were associated with 7% and 16%, respectively, increased risk for mortality (RRs = 1.07–1.16; 95% CIs, 1.02–1.24), raising the concern of the adverse effects of excess intakes of antioxidants including vitamin A and β-carotene.

2.5 Conclusion

Deficiency in vitamin A continues to affect several parts of the world where dairy products and supplements are not widely available and there is a need to increase the intake to avoid the well-known clinical symptoms. In contrast, deficiency in vitamin A is less of a concern in the developed countries; there are also no clear clinical signs among individuals who suffer from inadequate intake of vitamin A or provitamin A. Accumulating evidence have suggested that excess intakes of vitamin A may pose health risks in specific populations including smokers. In addition, the elderly with excess vitamin A levels may suffer from the adverse effects of vitamin A toxicity including increasing the risk for bone loss. Finally, supplementation of vitamin A has no effects in preventing chronic diseases.

Summary Points

- Numerous individuals, especially children and women, in the developing countries remain at a greater risk for vitamin A deficiency which causes adverse outcomes of xerophthalmia and blindness, and perhaps, increases mortality.
- Vitamin A deficiency is less of a public health concern in the well-nourished countries, and there is no clear clinical significance with suboptimal vitamin A intake. However, excess intake of vitamin A may be harmful in the elderly due to the adverse effects of vitamin A toxicity on bone loss.
- The absorption and metabolism of vitamin A rely, to some extent, on several nutrients including zinc, iron, and vitamin D. However, the interaction of vitamin A with these nutrients in relation to health outcomes remains unclear.
- Supplementation with vitamin A or β-carotene, alone or combined with other vitamins and minerals, does not appear to have beneficial effects in adults on the prevention of chronic diseases including cancers, vascular disease, infectious disease, and eye-related disease.
- Vitamin A supplementation is most likely beneficial for individuals who are deficient in the nutrient.

Key Facts of Micronutrients and their Importance

- Vitamins and trace minerals are micronutrients that play an essential role in a variety of physiological processes including hormonal and antioxidant properties, regulation of tissue growth and differentiation, regulation of DNA methylation, promotion of embryonic development, and nutrient metabolism.

- Most vitamins and minerals are derived in minute amounts from natural foods of animal and/or plant sources, although supplements of these nutrients are commonly used in many parts of the world.
- In the malnourished countries, vitamin A, iron, and iodine deficiencies are the most common forms of micronutrient deficiencies. However, deficiencies in micronutrients are not a health concern in well-nourished countries.
- Although deficiencies in vitamins and minerals may cause clinical symptoms, excess intake of these micronutrients including vitamins A and E do not appear to benefit health outcomes.

Definition of Words and Terms

Antioxidants: Including vitamins A, C, E, carotenoids, selenenium, and among others, which are present mostly in plant-derived foods. Antioxidants counterbalance the production of reactive oxygen species, prevent oxidative damage to cells, and modify cell growth regulatory pathways.

B vitamins: Water-soluble vitamins including B_1 (thiamin), B_2 (riboflavin), B_3 (niacin), B_5 (pantothenic acid), B_6 (pyridoxine), B_7 (biotin), B_9 (folic acid), and B_{12} (cobalamin) that are essential for DNA synthesis, immune and metabolic function, and skin health. These vitamins can be found in cereals, rice, nuts, dairy products, meat, and plant foods.

Incidence: The number of new cases of the disease within a specified time period.

Intervention trial of supplements: A prospective biomedical research study of human subjects that is designed to determine the efficacy, effectiveness, and safety of supplements for the prevention of disease outcomes of interest.

Iron: A mineral which plays a vital role in oxygen transport, oxidative metabolism, cellular growth and function. Dietary sources of iron include red meat, poultry, fish, beans, peas, tofu, fortified foods, and leaf vegetables.

Preformed vitamin A: Retinol esters that are present in foods of animal origins such as eggs, dairy products, and liver. They are efficiently absorbed and used by humans at absorption rates of 70–90%.

Prevalence: The total number of existing cases of the disease outcome in the population at a given time.

Provitamin A: Carotenoids in plant-derived foods including β-carotene, α-carotene, lycopene, lutein, and β-crypotxanthin. They are absorbed less efficiently, at rates of 20–50%, by humans.

Vitamin C: A water-soluble vitamin which acts as an antioxidant by protecting cells from oxidative stress and as a cofactor for several enzymatic reactions which are important for wound healing. Foods of plant origin such as fruits and vegetables are rich sources of vitamin C.

Vitamin D: A fat-soluble vitamin which is essential for bone health and immune system. Vitamin D can be obtained from dietary sources including dairy products, fatty fish, and supplements, or synthesized in the skin from 7-dehydroxy cholesterol.

Vitamin E: A fat-soluble vitamin which has specific biological activities including the regulation of gene expression, signaling, cell proliferation, and reproduction. It also has antioxidant properties that stops the production of reactive oxygen species formed when fat undergoes oxidation. α-Tocopherol, the most biologically active form of vitamin E, can be obtained from sunflower, safflower oils, nuts, and plant foods.

Zinc: A mineral which plays a critical role in the metabolism of RNA and DNA, signal transduction, and gene expression. Red meats, especially beef, lamb, and liver are rich in zinc. Zinc is also found in beans, nuts, and pumpkin and sunflower seeds.

List of Abbreviations

ATBC	Alpha-Tocopherol, Beta-Carotene Cancer Prevention
CARET	Beta Carotene and Retinol Efficacy Trial
CI	confidence interval
HDL	high-density lipoprotein
NHANES	National Health and Nutrition Examination Survey
PHS	Physicians' Health Study
RBP	retinol binding protein
RDA	recommended daily allowance
RXR	retinoid X receptor
RR	relative risk
SUVIMAX	Supplémentation en Vitamines et Minéraux Antioxydants
WACS	Women's Antioxidant and Cardiovascular Study
WHO	World Health Organization
WHS	Women's Health Study

References

Abdu-Gusau K., Elegbede J. A. and Akanya H. O., 1989. Serum zinc, retinol and retinol-binding protein levels in cirrhotics with hypogonadism. European Journal of Clinical Nutrition. 43: 53–57.

Ashwell M., Stone E. M., Stolte H., Cashman K. D., Macdonald H., Lanham-New S., Hiom S., Webb A. and Fraser D., 2010. UK Food Standards Agency Workshop Report: an investigation of the relative contributions of diet and sunlight to vitamin D status. British Journal of Nutrition. 104: 603–611.

Bailey R. L., Dodd K. W., Gahche J. J., Dwyer J. T., McDowell M. A., Yetley E. A., Sempos C. A., Burt V. L., Radimer K. L. and Picciano M. F., 2010. Total folate and folic acid intake from foods and dietary supplements in the United States: 2003–2006. The American Journal of Clinical Nutrition. 91: 231–237.

Ballew C., Bowman B. A., Sowell A. L. and Gillespie C., 2001. Serum retinol distributions in residents of the United States: third National Health and Nutrition Examination Survey, 1988–1994. The American Journal of Clinical Nutrition. 73: 586–593.

Bardia A., Tleyjeh I. M., Cerhan J. R., Sood A. K., Limburg P. J., Erwin P. J. and Montori V. M., 2008. Efficacy of antioxidant supplementation in reducing primary cancer incidence and mortality: systematic review and meta-analysis. Mayo Clinic Proceedings. 83: 23–34.

Bjelakovic G., Nikolova D., Gluud L. L., Simonetti R. G. and Gluud C., 2007. Mortality in randomized trials of antioxidant supplements for primary and secondary prevention: systematic review and meta-analysis. The Journal of American Medical Association. 297: 842–857.

Briefel R. R., Bialostosky K., Kennedy-Stephenson J., McDowell M. A., Ervin R. B. and Wright J. D., 2000. Zinc intake of the U.S. population: findings from the third National Health and Nutrition Examination Survey, 1988–1994. Journal of Nutrition. 130: 1367S–1373S.

Christian P. and West Jr, K. P.,, 1998. Interactions between zinc and vitamin A: an update. The American Journal of Clinical Nutrition. 68: 435S–441S.

Cook J. D., Skikne B. S. and Baynes R. D., 1994. Iron deficiency: the global perspective. Advances in Experimental Medicine and Biology. 356: 219–228.

Crandall C., 2004. Vitamin A intake and osteoporosis: a clinical review. Journal of Women's Health. 13: 939–953.

Ervin R. B. and Kennedy-Stephenson J., 2002. Mineral intakes of elderly adult supplement and non-supplement users in the third national health and nutrition examination survey. Journal of Nutrition. 132: 3422–3427.

Evatt M. L., Terry P. D., Ziegler T. R. and Oakley G. P., 2010. Association between vitamin B12-containing supplement consumption and prevalence of biochemically defined B12 deficiency in adults in NHANES III (third National Health and Nutrition Examination Survey). Public Health Nutrition. 13: 25–31.

Fabian E. and Elmadfa I., 2008. Nutritional situation of the elderly in the European Union: data of the European Nutrition and Health Report (2004). Annals of Nutrition and Metabolism. 52 (Supplement) 1: 57–61.

Fairfield K. M. and Fletcher R. H., 2002. Vitamins for chronic disease prevention in adults: scientific review. The Journal of the American Medical Association. 287: 3116–3126.

Feskanich D., Singh V., Willett W. C. and Colditz G. A., 2002. Vitamin A intake and hip fractures among postmenopausal women. The Journal of the American Medical Association. 287: 47–54.

Fletcher R. J., Bell I. P. and Lambert J. P., 2004. Public health aspects of food fortification: a question of balance. Proceedings of the Nutrition Society. 63: 605–614.

Ginde A. A., Liu M. C. and Camargo Jr, C. A., 2009. Demographic differences and trends of vitamin D insufficiency in the US population, 1988–2004. Archives of Internal Medicine. 169: 626–632.

Graham J. M., Haskell M. J., Pandey P., Shrestha R. K., Brown K. H. and Allen L. H., 2007. Supplementation with iron and riboflavin enhances dark adaptation response to vitamin A-fortified rice in iron-deficient, pregnant, nightblind Nepali women. The American Journal of Clinical Nutrition. 85: 1375–1384.

Hathcock J. N., 1997. Vitamins and minerals: efficacy and safety. The American Journal of Clinical Nutrition. 66: 427–437.

Imdad A., Herzer K., Mayo-Wilson E., Yakoob M. Y. and Bhutta Z. A., 2010. Vitamin A supplementation for preventing morbidity and mortality in children from 6 months to 5 years of age. Cochrane Database System Review. CD008524.

Jenab M., Salvini S., van Gils C. H., Brustad M., Shakya-Shrestha S., Buijsse B., Verhagen H., Touvier M., Biessy C., Wallstrom P., Bouckaert K., Lund E., Waaseth M., Roswall N., Joensen A. M., Linseisen J., Boeing H., Vasilopoulou E., Dilis V., Sieri S., Sacerdote C., Ferrari P., Manjer J., Nilsson S., Welch A. A., Travis R., Boutron-Ruault M. C., Niravong M., Bueno-de-Mesquita H. B., van der Schouw Y. T., Tormo M. J., Barricarte A., Riboli E., Bingham S. and Slimani N., 2009. Dietary intakes of retinol, beta-carotene, vitamin D and vitamin E in the European Prospective Investigation into Cancer and Nutrition cohort. European Journal of Clinical Nutrition. 63 (Supplement) 4: S150–S178.

Johansson S. and Melhus H., 2001. Vitamin A antagonizes calcium response to vitamin D in man. Journal of Bone and Mineral Research. 16: 1899–1905.

Mata-Granados J. M., Cuenca-Acevedo R., Luque de Castro M. D., Sosa M. and Quesada-Gomez J. M., 2010. Vitamin D deficiency and high serum levels of vitamin A increase the risk of osteoporosis evaluated by Quantitative Ultrasound Measurements (QUS) in postmenopausal Spanish women. Clinical Biochemistry. 43: 1064–1068.

Mora J. R., Iwata M. and von Andrian U. H., 2008. Vitamin effects on the immune system: vitamins A and D take centre stage. Nature Reviews in Immunology. 8: 685–698.

Olsen A., Halkjaer J., van Gils C. H., Buijsse B., Verhagen H., Jenab M., Boutron-Ruault M. C., Ericson U., Ocke M. C., Peeters P. H., Touvier M., Niravong M., Waaseth M., Skeie G., Khaw K.T., Travis R., Ferrari P., Sanchez M. J., Agudo A., Overvad K., Linseisen J., Weikert C., Sacerdote C., Evangelista A., Zylis D., Tsiotas K., Manjer J., van Guelpen B., Riboli E., Slimani N. and Bingham S., 2009. Dietary intake of the water-soluble vitamins B1, B2, B6, B12 and C in 10 countries in the European Prospective Investigation into Cancer and Nutrition. European Journal of Clinical Nutrition. 63 (Supplement) 4: S122–S149.

Penniston K. L. and Tanumihardjo S. A., 2006. The acute and chronic toxic effects of vitamin A. The American Journal of Clinical Nutrition. 83: 191–201.

Reddy V., 2002. History of the International Vitamin A Consultative Group 1975–2000. Journal of Nutrition. 132: 2852S–2856S.

Schleicher R. L., Carroll M. D., Ford E. S. and Lacher D. A., 2009. Serum vitamin C and the prevalence of vitamin C deficiency in the United States: 2003–2004 National Health and Nutrition Examination Survey (NHANES). The American Journal of Clinical Nutrition. 90: 1252–1263.

Silverberg D. S., Iaina A., Wexler D. and Blum M., 2001. The pathological consequences of anaemia. Clinical and Laboratory Haematology. 23: 1–6.

Sommer A., Tarwotjo I., Hussaini G., Susanto D. and Soegiharto T., 1981. Incidence, prevalence, and scale of blinding malnutrition. Lancet. 1: 1407–1408.

Sommer A., Tarwotjo I., Djunaedi E., West Jr, K. P., Loeden A. A., Tilden R. and Mele L., 1986. Impact of vitamin A supplementation on childhood mortality. A randomised controlled community trial. Lancet. 1: 1169–1173.

Strube Y. N., Beard J. L. and Ross A. C., 2002. Iron deficiency and marginal vitamin A deficiency affect growth, hematological indices and the regulation of iron metabolism genes in rats. Journal of Nutrition. 132: 3607–3615.

Sudfeld C. R., Navar A. M. and Halsey N. A., 2010. Effectiveness of measles vaccination and vitamin A treatment. International Journal of Epidemiology. 39 (Supplement) 1: i48–i55.

Tabacchi G., Wijnhoven T. M., Branca F., Roman-Vinas B., Ribas-Barba L., Ngo J., Garcia-Alvarez A. and Serra-Majem L., 2009. How is the adequacy of micronutrient intake assessed across Europe? A systematic literature review. British Journal of Nutrition. 101 (Supplement) 2: S29–S36.

Underwood B. A., 2004. Vitamin A deficiency disorders: international efforts to control a preventable "pox". Journal of Nutrition. 134: 231S–236S.

Zittermann A., 2010. The estimated benefits of vitamin D for Germany. Molecular Nutrition and Food Research. 54: 1164–1171.

CHAPTER 3

The Importance of β-Carotene in the Context of Vitamin A

HANS K. BIESALSKI* AND DONATUS NOHR

Department of Biological Chemistry and Nutrition, University Hohenheim, Garbenstr. 30, D-70593 Stuttgart, Germany
*E-mail: biesal@uni-hohenheim.de

3.1 Introduction

Vitamin A is an important ingredient of human diet and thus of worldwide interest. Vitamin A deficiency (VAD) can result in various diseases, mostly starting with night-blindness and xerophthalmia. It can also impair immune functions leading to increased numbers and severity of gastrointestinal and/or respiratory infections. It has also been found that vitamin A levels are higher in HIV-positive children as compared with healthy children. However, VAD is mainly a problem in developing countries, particularly in children. One reason might be, that vitamin A has to be obtained from the diet and in developing countries provitamin A from carotenoids of fruits and vegetables provides approximately 70% of daily vitamin A intake, whereas in Western countries (USA and Europe) it represents less than 30% because here much more sources of preformed vitamin A (*e.g.* from liver, eggs, fish and dairy products) are available and part of the daily food (*cf.* Tang and Russell 2009).

In developing countries, extensive programs had been started to supplement people at risk with synthetic vitamin A, which is an efficient and safe strategy. However, logistic problems and high costs very often interrupted the programs. Alternative programs trying to prevent VAD by (local) food-based

Food and Nutritional Components in Focus No. 1
Vitamin A and Carotenoids: Chemistry, Analysis, Function and Effects
Edited by Victor R Preedy
© The Royal Society of Chemistry 2012
Published by the Royal Society of Chemistry, www.rsc.org

interventions with foods rich in provitamin A have been questioned with regards to their efficacy.

In Western countries, interest in provitamin A and carotenoids grew intensively after first results showing a correlation between a reduced risks for a number of diseases such as cancer, coronary heart disease (CHD) and cataract and the intake of foods rich in carotenoids (Johnson and Krinsky 2009; Rock 2009; Schalch *et al* 2009).

In contrast to all other carotenoids, β-carotene has a β-ionone structure as the terminal ring system on each end of the polyene chain (*cf.* Chapter 1). Its central oxidative cleavage in intestinal cells results in the formation of two molecules of vitamin A. The oxidative cleavage of β-carotene, the major carotenoid of human diets, is achieved by the β-carotene 15,15′-oxygenase (BCO1), which cleaves β-carotene into two molecules of all-*trans*-retinal (retinal-aldehyde) followed by a conversion into retinol and esterification with fatty acids to form retinyl esters (Figure 3.1). (Fierce *et al* 2008).

Numerous studies have shown that a substantial amount of the absorbed carotenoids are not cleaved by the intestinal BCO1 enzyme. Some portions of the carotenoids (up to 60% of dietary intake) that escape BCO1 enzyme activity are incorporated in chylomicrons together with lipids, and circulate in association with very low-density lipoprotein (VLDL), low-density lipoprotein (LDL) and high-density lipoprotein (HDL), and hence can be taken up by the

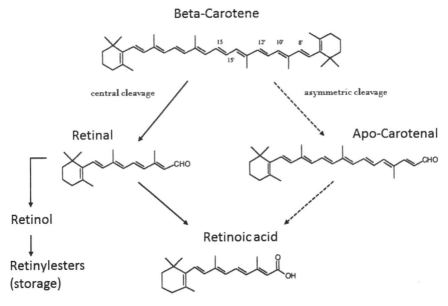

Figure 3.1   Schematic drawing of β-carotene cleavage The Figure shows the original molecule of β-carotene and its possible cleavages: central cleavage resulting in two molecules of vitamin A (retinol) or eccentric cleavage resulting in one of molecule vitamin A and one of molecule apo-carotenal. Figure is modified from Chichili 2005.

respective receptors for these lipoproteins. The recent demonstration that the β-carotene-cleaving enzyme BCO1 is also present in tissues other than intestine opens new questions about the tissue-specific function of this enzyme, as well as on the physiological potential of vitamin A production in tissues other than intestine, including the retinal pigment epithelial cells or other specific targets for β-carotene (reviewed in Biesalski *et al* 2007). The fact that β-carotene can be metabolized into vitamin A in many tissues raises the question of the efficacy and control of vitamin A formation following dietary intake.

3.2 β-Carotene as Antioxidant and/or Pro-oxidant

Before the major role of β-carotene as provitamin A will be discussed in more details, a short overview about its other functions will be given.

3.2.1 Antioxidant/Pro-oxidant Activities

Due to its structure and the observation of experimental data, β-carotene is regarded as an antioxidant (Burton and Ingold 1984; Chichili *et al* 2006; Yeum *et al* 2009), however, pro-oxidant activities have also been found *in vitro* (Palozza *et al* 2003; Yeum *et al* 2009). It still is very difficult to precisely show those effects, as very different types of studies with in part contradictory effects have been performed and even very often used tests like the determination of thiobarbituric acid reactive substances (TBARS), such as 4-hydroxynonenal (4-HNE) or malondialdehyde (MDA), have to be considered very carefully, because β-carotene interferes with this type of assay (Yeum *et al* 2009). In the meantime, many experts no longer accept β-carotene as an important antioxidant (personal communications).

3.2.2 Radical Scavenging Properties

Although we do not yet understand completely the interaction of β-carotene with radicals, it has been postulated to be an important chain-breaking antioxidant which scavenges lipid oxide and lipid peroxide radicals (Krinsky and Johnson 2005). However, pro-oxidant actions in the context of lipid peroxidation have also been described, *i.e.* an exposure of β-carotene to oxidizing conditions in high oxygen levels led to increased markers of lipid peroxidation (*e.g.* MDA) (El-Agamey *et al* 2004).

3.2.3 Singlet Oxygen Quenching

The core system of conjugated carbon–carbon bonds makes β-carotene an efficient singlet oxygen quencher – at least in plants. In humans, studies with patients suffering from erythropoietic protoporphyria indicate a quenching effect of β-carotene (Mathews-Roth 1993), and in healthy humans there are some indications that β-carotene offers some protection against UV-light-induced effects in the skin, where it can be enriched after supplementation and can ameliorate sunburn (Köpcke and Krutmann 2008).

3.3 Necessity of Vitamin A

Vitamin A is essential for normal growth and development, immune system, vision and some other functions in the human body. Vitamin A, as such, is only present in animal products (*e.g.* liver, eggs and milk products); thus, in countries where the intake of animal products is low, dietary vitamin A supply is mostly met by carotenoids.

Even in developed countries carotenoids traditionally contribute to a key part of the vitamin A supply. In some physiological situations (pregnancy and lactation) vitamin A has a particularly important role in the healthy development of the child, and an increase in vitamin A intake has been recommended. Low vitamin A intake by the mother increases the risk for inadequate lung maturation of the child (for details, see Chapter 31). However, at the same time it is currently recommended that women who are planning to become pregnant or who are pregnant should not consume cooked animal liver or other organ meats which are rich sources of vitamin A. The tolerable upper intake levels (UL) for preformed vitamin A (retinol and retinyl esters) has been set at 3000 μg of retinol equivalents (RE) day^{-1} (IOM 2001). However, there is no clear evidence that natural vitamin A, in contrast to synthetic vitamin A (retinoic acid), will indeed exert teratogenic effects. Even in cases of an intake in high doses of natural vitamin A (a 10-fold recommendation), the formation of retinoic acid within the target cells is tightly controlled and does not lead to critical levels. Similarly, the cleavage of β-carotene is regulated through feedback mechanisms, therefore only the required amount is metabolized to retinol. All in all, it can be recommended that a proportion of vitamin A requirements should be in the form of β-carotene.

The recent demonstration that BCO1 is also present in tissues other that the intestine led to numerous investigations of the molecular structure and function of this enzyme in several species, including the human (Biesalski *et al* 2007; Chichili *et al* 2005). Retinal pigment epithelial cells were shown to contain BCO1 and to be able to cleave β-carotene into retinal *in vitro*, offering a new pathway for vitamin A production in another tissue than the intestine, possibly explaining the more mild vitamin A deficiency symptoms of two human siblings lacking the retinol-binding protein for the transport of hepatic vitamin A to the target tissues (reviewed in Biesalski *et al* 2007).

β-Carotene contributes to balance inadequate retinol supply in significant parts of the population. Based on the calculated RE and using a conversion factor of 1:6, the equivalent intake of β-carotene would be as follows:

Men

Vegans: total RE 1.2 mg, or 7.2 mg of β-carotene
Vegetarians: total RE 0.7 mg, or 4.2 mg of β-carotene
Omnivores: total RE 0.5 mg, or 3.0 mg of β-carotene

Women

Vegans: total RE 1.5 mg, or 9.0 mg of β-carotene
Vegetarians: total RE 0.9 mg, or 5.4 mg of β-carotene
Omnivores: total RE 0.8 mg, or 4.8 mg of β-carotene

The data show that high intakes of provitamin A are possible under very special conditions, but the data also show that the intake of preformed vitamin A is sometimes critically low and does not reach the recommendations. According to national surveys in UK, US and Europe the intake of β-carotene in the general population is between 1–2 mg day^{-1}.

Mean daily intake data of total vitamin A and carotenoids in Germany is available from the National Consumption Study (1985–1988) on food and nutrient intake in Germany. Based on these data, 15–20% in the age group 10 to 25 years old does not reach recommendations for total vitamin A intake. However, in this survey recommended vitamin A intake levels will be met by less than 25% (depending on age-group) by preformed vitamin A only, showing the importance of provitamin A in assuring adequate vitamin A intake levels. Daily vitamin A intakes are notably insufficient in children and adolescents in particular when recommended intakes of β-carotene are not being reached. The second German National Consumption Survey (Nationale Verzehrsstudie II, NVS II) showed different results from the first survey. Based on the calculation of RE, the German National Nutrition Survey showed that approximately 33% comes from preformed retinol (meat, meat products, dairy products and fat) and 48% from β-carotene sources such as vegetables, vegetable soups and non-alcoholic beverages, and 19% from mixed carotenoids. These sources contribute to the total of retinol equivalents of 1.7 mg day^{-1} (median). The β-carotene conversion factor used was 1:6 which results in a mean uptake of 4.5 mg of β-carotene per day. This, however, is in strong contrast to other studies from Germany and Europe, showing a mean intake of 1.5–1.8 mg day^{-1} β-carotene (Elmadfa *et al* 2005). If the data are taken as realistic, approximately 15% of the female participants and 10.3% of male participants do not reach the recommendations for total vitamin A. Compared to other countries, β-carotene intakes in Germany are not higher than, for example, in Austria (0.8–2.1 mg of β-carotene day^{-1} comes from vegetables and 0.3 mg from fruits) (Elmadfa and Freisling 2004), Ireland (0.5 mg day^{-1}) or even Spain (0.3 mg day^{-1}).

Estimation of food intakes in Spain using a 24 h recall using photographic food models to estimate portion sizes (4728 males, 5480 females, aged 25–60 years) showed the prevalence for inadequate intake [<⅔ recommended daily intake (RDI) of vitamin A based on RE was 60.5% in men and 48.5% in women] (Aranceta *et al* 2001). The major reason for these findings seems the unexpectedly low intake of preformed vitamin A (means, 293 μg in men and

276 μg in women). Mean β-carotene intake was 1.7 mg in men and 2 mg in women.

A nutrition survey in UK based on a 7 day weighed intake dietary record with 1724 responders showed an intake below the lower reference nutrient intake (LRNI) in 16% of men and 30% of women aged 19–34 years (Swan 2004). The LRNI is the amount of a nutrient that is enough for only the small number of people who have low requirements (2.5%). The majority need more. It is of great importance to realize that 30% women at child-bearing age belong to the group that does not receive an adequate supply with vitamin A! In the situation of pregnancy this might create problems.

3.3.1   How Much β-Carotene is Needed to Ensure Sufficient Vitamin A Supply?

Provitamin A is a major source for vitamin A. Based on the intake data as discussed above, it is suggested that, if the intake of provitamin A falls below 2 mg day^{-1} (equivalent to 0.3 mg RE), the gap between the inadequate vitamin A intake and the recommended dietary intake of 1 mg day^{-1} cannot be bridged. This is of special importance for pregnant women due to two reasons: daily intake should be higher (1.1 mg day^{-1}) due to a higher demand and the majority of women of child-bearing age avoid liver as the major source for preformed vitamin A.

The recommended daily intake of vitamin A for children is between 0.6 mg and 1.1 mg, for adults, between 0.8 mg and 1 mg for pregnant women, and 1.5 mg and for breastfeeding women. According to the above-mentioned surveys, β-carotene intake of almost half of the German population is less than 1 mg day^{-1}, while 64.1% have intakes less than the recommended intake of 2 mg day^{-1}. The β-carotene intake of 75% of the population is less than 3 mg day, even when nutritional supplements and fortified foods are included.

3.3.2   For Embryonic Development and Pregnant Women

Vitamin A plays an important role during pregnancy and the breast-feeding period (Biesalski and Nohr 2003; Strobel *et al* 2011). The intake should be around a third higher during pregnancy and during breast feeding intake and should be 0.7 mg day^{-1} higher than for non-pregnant, non-breast-feeding women. A much more detailed overview on this topic is given in Chapter 31.

3.4   β-Carotene as Provitamin A

3.4.1   Natural Sources of β-Carotene

As no animals, including humans, are able to synthesize carotenoids, these have to be taken up with the diet. Some carotenoids appear in animal-derived food – dependent how animals were fed – such as lutein or zeaxanthin in eggs,

Table 3.1   Concentration of β-carotene in selected nutrients. The Table shows some selected food sources for β-carotene and its concentration in 100 g of each.

Food name	Content of β-carotene (range 100 g^{-1})
Apricot	0.5–>2 mg
Broccoli	>2 mg
Brussel sprouts	0.5–2 mg
Carrot	>2 mg
Green leafy vegetables	0.1–>2 mg
Kale	>2 mg
Mango	0.5–>2 mg
Peach	0.5–2 mg
Pepper (various colors)	0.5–2 mg
Red palm oil	>2 mg
Spinach	>2 mg
Sweet potato	>2 mg
Tomato	0.1–0.5 mg
Tomato "high beta" (GMO)	>2 mg
West Indian Cherry	0.5–2 mg

some little β-carotene in dairy products and astaxanthin in pink-fleshed fish (salmon and trout), almost exclusively plant-derived food such as fruits and vegetables are good sources for β-carotene (Britton and Khachik 2009). A brief overview in alphabetical order of those sources with really high concentrations of β-carotene (0.5–>2 mg 100 g^{-1}) is presented below (modified from Britton and Khachik 2009).

Although there are some sources really rich in β-carotene, that does not automatically mean that this amount of carotenoid reaches its final destination in the body, as storage, cooking, processing and also bioavailability influence this amount.

3.4.2   Bioavailability of β-Carotene

As lipophilic substances, carotenoids are taken up by the small intestine together with other lipids. They reappear in the lipoprotein fractions of the plasma, in erythrocytes and leucocytes (for detailed references, see Strobel *et al* 2007). The largest portion of the lipoprotein fraction is found in LDL ($\sim 80\%$), then HDL ($\sim 15\%$) and VLDL ($\sim 10\%$). Serum levels after carotenoid intake differ inter-individually depending on the respective bioavailability; β-carotene levels normally range from 0.2 to 0.5 μM. Non-absorbed β-carotene is exclusively found in the faeces, whereas other fragments such as apo-carotenals have also been found in the urine.

SLAMENGHI is a mnemonic grouping the factors that determine bioavailability and bioconversion of carotenoids (*S*pecies of carotenoid, molecular *L*inkage, *A*mount of carotenoids consumed in a meal, *M*atrix in which the carotenoid is incorporated, *E*ffectors of absorption and bioconversion, *N*utrient status of the host, *G*enetic factors, *H*ost-related factors and mathematical *I*nteractions; West and Castenmüler 1998). The bioavailability of carotenoids depends on the form in which they are found in the food matrix, *e.g.* crystalline, esterified or emulsified in fat, the content of fat in the meal leading to absorption from plant food ranges from 30 to 60%, however, still being subject to large inter-individual variations. Several studies could show that the intake of β-carotene supplements led to greater increases in plasma levels than comparable intakes from plant food.

In addition to fat, other components also influence absorption, metabolism and bioconversion when taken up with carotenoids. Proteins can stabilize fat emulsion in the small intestine, along with lecithin they support micelle formation. The pH of the stomach is involved and also genetic factors (*e.g.* polymorphisms) may play a role as they do in many other stages of metabolism. Taking up a given daily dose in three single doses improves absorption, as does the presence of bile acids.

Absorption rates are best when carotenoids are taken up with some oil. Calculations ended up in a so-called "conversion factor" of "6" for calculating retinol equivalents from carotenoids (FAO/WHO 1967). However, in the meantime new factors were calculated: "12" for β-carotene from fruits and "26" for those from vegetables, much less than the "6" before. A list of various conversion factors and the respective references can be found in Strobel *et al* (2007).

In 2001, in the USA, a new conversion factor was published by the Food and Nutrition Board together with new recommendations for vitamin A intakes. The "RE" was replaced by the "retinol activity equivalent" (RAE). As a consequence, the vitamin A activity of carotenoids as provitamin A from foods is now half of its former value; the RAE of β-carotene was defined as 12 μg of carotenoids (1 μg of retinol). Another consequence is that using the conversion factor "12" instead of "6" would lead to a calculated 15% reduction of the total vitamin A intake in Germany, meaning that also the number of people who reach just 50% of the recommended uptake would increase markedly.

3.4.3   The Role of β-Carotene as Provitamin A for Vegetarians

As vegetarians do not take up preformed vitamin A (except maybe from supplements), their almost exclusive source for vitamin A is β-carotene. Due to its limited bioavailability from plant-derived foods and due to individual differences, vegetarians should be aware of this problem and should take up the relevant amounts from their meals, using as many ways as possible to enhance bioavailability, *e.g.* by always adding some oil to their meals or mixing their diet in a way that enhances bioavailability. In times of enhanced needs, such as pregnancy or breast-feeding, amounts should be enhanced and supplements should perhaps be used.

3.4.4 Supply of the Population with Preformed Vitamin A

Interestingly, the levels of minimal requirement of vitamin A, established by various regulatory authorities are somewhat different. This has to be taken into account; since nutritional surveys, in fact, are using the level established in the country in which they were performed. Consequently, even if 100% of the required/recommended levels are achieved, different populations might have different intakes. Since the nutritional habits in various populations are quite different, it is to be expected that the total intake of vitamin A might be met mostly by the intake of preformed vitamin A, or in other words by a high intake of animal products. On the other hand, other populations or subgroups not having such a high intake of animal products might fill the gap between their actual requirement for vitamin A and their intake of preformed vitamin A by consuming β-carotene. Consequently, the required β-carotene intake strongly depends on the amount of vitamin A consumed by other food sources. Strict vegans, not consuming any animal products, therefore need to meet their vitamin A requirement exclusively from β-carotene. In people for whom a high percentage of their vitamin A intake is realized from β-carotene consumption, the question of the conversion factor (β-carotene to retinol) is of huge importance.

Nutrition surveys, as performed in the US, Germany, Austria, UK, Ireland, Spain, and other countries, demonstrate that the general intake of total vitamin A is in the range of the actual recommendation (see Grune *et al* 2010). The NVS II shows that the median of the total vitamin A intake is between 140–170% of the DACH reference value [DACH reference values are given by German (D), Austrian (A) and Swiss (CH) societies for nutrition] (DACH 2008). Also the Austrian nutrition report gives for all adult population groups a median intake level above 100% The same is true for Ireland, as demonstrated in 'The North/South Ireland Food Consumption Survey' and also in the UK in the 'British National Diet and Nutrition Survey'–the latter with the exception of 19–24 year old males and 19–34 year old females. Interestingly, the eVe study in Spain reports a median intake of RE considerably below the recommended Spanish DRI, as well as for males and females 25–60 year old adults. As most of the data in these nutrition surveys are given in median intake levels, it should be taken into consideration that the total retinol equivalent intake varies considerably. Although very high total vitamin A intakes are reported in Germany, some 10–15% of the population (14–80 year olds) has an intake below the recommended DACH value The same was also reported in Ireland. In the US a median daily intake of 744–811 µg RAE was reported for adult males and 530–716 µg for females. However, it is worth notifying that the DRI for the US population is based on a conversion factor of 12, whereas in Germany and Austria a conversion factor of 6 is used. Other sources report an adequate intake of total vitamin A in the US as measured by retinol equivalents. Thus, the intake of vitamin A in some countries, as in Germany and Austria, might be overestimated due to the use of a low conversion factor, and a significantly higher percentage of the population might not meet their biological requirements for vitamin A. In the Austrian nutrition report it is clearly evident that the median vitamin A intake for the entire female adult population (19–64 years) and for young adult males (18–25 years) will be below the DACH value if a conversion factor of 12 is applied. Interestingly, in a study of 154 Vienna inhabitants a subdivision by nutritional preferences revealed that vegetarians and vegans take in very high β-carotene levels. This enables them to meet the recommendations for vitamin A intake.

Studies in Germany and Austria also investigated the vitamin A and provitamin A intake in children or adolescents. In Germany the situation for adolescents is the same as found in adults, where the mean intake is approximately at the level of the recommended intake, revealing that a considerable percentage of this population group is not meeting the recommendations. Interestingly, the percentage of the population not reaching the DACH reference level is highest in 14–18 year olds. This is true also for Austria where the median intake of the 7–15 year old population (boys and girls) does not reach the reference intake, even using a conversion factor of 6 for β-carotene. The same report did not find a difference between ethnic backgrounds (Austrian *vs.* non-Austrian parents) for the children, but there was a clear difference in the total vitamin A intake due to educational background (depending on school type) in adolescent boys and girls.

The elderly population is, in general, meeting the total vitamin A reference values very well in Germany and Austria. As the intake of preformed vitamin A is, in general, high, the conversion factor of β-carotene seems not to be so important for this group. Also Austrian inhabitants of retirement or nursing homes seem to be well nourished regarding total vitamin A.

3.4.5 Basic Need for β-Carotene to Ensure a Sufficient Intake to Meet the Vitamin A Requirement?

The National Diet and Nutrition Survey (NDNS) data clearly indicates that men aged 19–24 years and women aged 19–34 years were the only age groups where mean intake of vitamin A from food sources was below the reference nutrient intake (RNI). Although a large portion of the UK population seems to consume enough vitamin A, the NDNS data are based on the underlying assumptions that: (a) 6 µg of provitamin A is equivalent to 1 µg retinol and (b) provitamin A conversion into vitamin A has no inter-individual variation. These assumptions were recently rectified, as the bioefficacy of provitamin A sources is lower than the assumed conversion efficiency of 1:6 and has recently been altered to 1:12 for β-carotene and 1:24 for both α-carotene and β-cryptoxanthin. More importantly, the vitamin A activity of provitamin A sources such as β-carotene, even when measured under controlled conditions, is highly variable and surprisingly low (Lin *et al* 2000; Wang *et al* 2004). Approximately 27–45% of volunteers in double-tracer studies have been classified as poor converters (Hickenbottom *et al* 2002; Lin *et al* 2000; Wang *et al* 2004). These individuals have a capacity to form only 9% vitamin A from β-carotene compared with those who are classified as normal converters (Wang *et al* 2004). Genetic variability in β-carotene metabolism may provide an explanation for the molecular basis of the poor converter phenotype within the population, as two non-synonymous single nucleotide polymorphisms (SNPs) (R267S and A379V) in the human β-carotene 15,15′-oxygenase (BCO1) gene with variant allele frequencies of 42 and 24% have recently been identified (Leung *et al* 2009). Responsiveness to a pharmacological dose of β-carotene in a human intervention study revealed that these SNPs were associated with significant alterations in β-carotene metabolism in female volunteers with reduced β-carotene conversion efficiency of 32 and 69% for the A379V and R267S/A379V variant carriers, respectively (Leung *et al* 2009).

Even if the current recommendation of 4 mg of β-carotene could be adhered to, female volunteers with either the A379V and R267S/A379V variant would only obtain 72 to 51% of the recommended intake, respectively, given the conversion efficiency of 1:12 for β-carotene and assuming an intake of 201 µg of pre-formed retinol.

If we assume no inter-individual differences in the ability to convert β-carotene into retinol and no changes in current pre-formed retinol intakes, the amount of β-carotene needed to cover between 95 to 97.5% of the US or UK population with recommended RE would be 7 mg, respectively, based on a conversion efficiency of 1:12 (West *et al* 2002). If, however, we correct the conversion efficiency factor for reduced conversion efficiencies of 69% for the R267S/A379V variant carriers, the recommended intake of β-carotene would need to be 22 mg to cover between 95 to 97.5% of the US or UK population with recommended RE. Due to the possible health risks associated with high intakes of β-carotene in smokers (Albanes *et al* 1995), it might be advisable to screen these individuals and to advice personalised nutrition in these circumstances. On the other hand, if preformed retinol intakes could be increased to a concentration of 430 µg at the 5th percentile, a daily intake of 7 mg of β-carotene, which is considered generally safe, would be enough for those individuals who have the double SNP to enable a coverage of 95 to 97.5% of the US or UK population with recommended RE.

Supplements and fortified foods contribute significantly to achieving the RNI and DRI for total vitamin A intake, as, for example, 25.9% of US adults take vitamin and mineral supplements and fortified foods themselves contributed 26% of the DRI for vitamin A (Berner *et al* 2001).

In summary, current recommended amounts of consumed β-carotene should be increased to 7 mg to ensure that at least 95% of the population consumes the recommended intakes of total vitamin A. Individuals with reduced conversion efficiencies due to a genetic variability in β-carotene metabolism might need to increase their daily β-carotene intakes up to 22 mg. However, much more research is needed to define new recommended intake levels if the large inter-individual variation in β-carotene conversion efficiency is taken into account.

Summary Points

- β-Carotene plays important roles in the human body as an antioxidant in the skin and some other tissues.
- Its main function is that of provitamin A.
- In its role as provitamin A, its efficacy (bioavailability and bioconversion) depends on its source in the food, the matrix of the food, the nutrient status of the host, genetic factors and host-related factors, but note that this list may not be complete.
- Vitamin A and its functional form as retinoic acid is very important for the proliferation and differentiation of mucosal and other quickly proliferating tissues,
- It is important not only in ontogenesis but also through the whole of human life.
- As vitamin A can only be taken up from animal-derived food, β-carotene is important as provitamin A to fill the gap of carnivores. It is the almost only provitamin A and thus source for vitamin A in vegetarians and vegans.
- Its importance in human nutrition thus cannot be underestimated.

Key Facts

Key Facts for β-Carotene

The basic facts for the plant-derived pigment β-carotene and its main functions in humans are:

- One out of more than 600 carotenoids, a red-to-orange pigment from plants.
- Can act as an antioxidant, thus reducing oxidative stress, which would be harmful for cells and tissues.
- Is a provitamin A as it can be cleaved by a special enzyme into two molecules of vitamin A (retinol/retinal).

Key Facts for Provitamin A

The basic facts for provitamin A and examples for it are listed as follows:

- Precursor of vitamin A.
- Central or excentral cleavage of the molecule results in 1 (excentric) or 2 (centric) molecules of vitamin A.
- Members: several carotenoids (α-, β- and γ-carotene and β-cryptoxanthin).

Key Facts for Vitamin A

The main basic facts regarding vitamin A, its sources, main functions and risks and symptoms of deficiency are:

- Can be taken up only from animal-derived food.
- Can be converted from β-carotene.
- Is important for vision.
- Plays a major role in tissues with high rates of proliferation and differentiation (*e.g.* gut mucosa or lung respiratory epithelium (see Chapter 31).
- Regulates gene expression *via* specific retinioic acid receptors (RAR and RXR).
- Is deficient in millions of children in developing countries; deficiency leads first to night-blindness, later on several tissue-defects and total blindness can appear.
- Should be in an optimal range during pregnancy.

Definitions of Words and Terms

Bioavailability: is that fraction of a substance which is available for the normal physiological functions or for storage.

Bioconversion: (of β-carotene) refers to the amount of β-carotene that is converted into retinol.

Retinol activity equivalent (RAE): replaced retinol equivalent (RE) after it was found out that the conversion factor for β-carotene from food was much lower than thought previously. One RAE (1 μg of retinol) is equivalent to 12 μg of (all-*E*)-β-carotene or 24 μg of other carotenoids.

Retinol equivalent (RE): is defined as equivalent to 1 μg of retinol, 6 μg of β-carotene or 12 μg of other provitamin A carotenoids.

List of Abbreviations

BCO1	β-carotene 15,15′-oxygenase-1
DACH	reference values for nutrition given by German (D), Austrian (A) and Swiss (CH) societies for nutrition
DRI	daily recommended intake
HDL	high-density lipoprotein
LDL	low-density lipoprotein
LRNI	lower reference nutrient intake
MDA	malondialdehyde
NDNS	National Diet and Nutrition Survey
RAE	retinol activity equivalent
RE	retinol equivalents
RNI	reference nutrient intake
SNP	single nucleotide polymorphism
UL	tolerable upper intake level
VAD	vitamin A deficiency
VLDL	very low-density lipoprotein

References

Albanes, D., Heinonen, O. P., Huttunen, J. K., Taylor, P. R., Virtamo, J., Edwards, B. K., Haapakoski, J., Rautalahti, M., Hartman, A. M. and Palmgren, J. 1995. Effects of alpha-tocopherol and beta-carotene supplements on cancer incidence in the Alpha-Tocopherol Beta-Carotene Cancer Prevention Study. American Journal of Clinical Nutrition. 62: 1427S–1430S.

Aranceta, J., Serra-Majem, L., Pérez-Rodrigo, C., Llopis, J., Mataix, J., Ribas, L., Tojo, R., and Tur, J. A. 2001. Vitamins in Spanish food patterns: the eVe Study. Public Health Nutrition. 4: 1317–1323.

Berner, L. A., Clydesdale, F. M. and Douglass, J. S. 2001. Fortification contributed greatly to vitamin and mineral intakes in the United States, 1989–1991. Journal of Nutrition. 131: 2177–2183.

Biesalski, H.K. and Nohr, D. 2003. Importance of vitamin A for lung function and development. Molecular Aspects of Medicine 24: 431–440.

Biesalski, H. K., Chichili, G. R., Frank, J., von Lintig, J. and Nohr, D. 2007. Conversion of β-carotene to retinal pigment. Vitamins and Hormones. 75: 117–130.

Britton, G. and Khachik, F. 2009. Carotenoid in Food. In: Britton, G., Liaaen-Jensen, S. and Pfander, H. (eds.) Carotenoids. Nutrition and Health (Vol. 5). Birkhauser, Boston, Basel, Berlin. pp 13–147.

Burton, G. W. and Ingold, K. U. 1984. Beta-Carotene: an unusual type of lipid antioxidant. Science. 224: 569–573.

Castenmiller, J. J. and West, C. E. 1998. Bioavailability and bioconversion of carotenoids. Annual Review of Nutrition. 18: 19–38.

Chichili, G. R., Nohr, D., Schäffer, M., von Lintig J. and Biesalski H. K. 2005. β-Carotene conversion into vitamin A in human retinal pigment epithelial cells. Investigative Ophthalmology and Visual Science. 46: 3562–3569.

Chichili, G. R., Nohr, D., Frank, J., Flaccus, A., Fraser, P. D., Enfissi, E. M. A. and Biesalski H. K. 2006. Protective effect of tomato with elevated beta-carotene levels on oxidative stress in ARPE-19 cells. British Journal of Nutrition 96: 643–649.

DACH Deutsche Gesellschaft für Ernährung, Österreichische Gesellschaft für Ernährung, Schweizerische Gesellschaft für Ernährungsforschung, Schweizerische Vereinigung für Ernährung 2008. Referenzwerte für die Nährstoffzufuhr. 1. Auflage, 3. korrigierter Nachdruck, Umschau/Braus.

El-Agamey, A., Lowe, G. M., McGarvey, D. J., Mortensen, A., Phillip, D. M., Truscott, T. G. and Young, A. J. 2004. Carotenoid radical chemistry and antioxidant/pro-oxidant properties. Archives of Biochemistry and Biophysics 430: 37–48.

Elmadfa, I. and Freisling, H. 2004. Austrian Nutrition Report 2003. Institute of Nutritional Sciences (IfEW), University of Vienna.

Elmadfa, I. and Weichselbaum, E., (eds.) 2005. European Nutrition and Health Report 2004. Forum of Nutrition. Karger, Basel, Switzerland. pp 224.

FAO/WHO. 1967. Requirements of vitamin A, thiamine, riboflavin and niacin. Vol. 8, FAO Food and Nutrition Series.

Fierce, Y., de Morais Vieira M., Piantedosi, R., Wyss, A., Blaner, W.S. and Paik, J. 2008. *In vitro* and *in vivo* characterization of retinoid synthesis from beta-carotene. Archives of Biochemistry and Biophysics. 472: 126–138.

Grune, T., Lietz, G., Palou, A., Ross, A.C., Stahl, W., Tang, G., Thurnham, D., Yin, S. and Biesalski, H. K. 2010. β-Carotene is an important vitamin A source for humans. Journal of Nutrition. 140: 2268S–2285S.

Hickenbottom, S. J., Follett, J. R., Lin, Y., Dueker, S. R., Burri, B. J., Neidlinger, T. R. and Clifford, A. J. 2002. Variability in conversion of beta-carotene to vitamin A in men as measured by using a double-tracer study design. American Journal of Clinical Nutrition. 75: 900–907.

IOM; Institute of Medicine. 2001. Dietary reference intakes for vitamin A, vitamin K, arsenic, boron, chromium, copper, iodine, iron, manganese, molybdenum, nickel, silicon, vanadium, and zinc. National Academy Press, Washington D.C.

Johnson, E. J. and Krinsky, N. I. 2009. Carotenoids and coronary heart disease. In: Britton, G., Liaaen, S and Pfander, H. (ed.) Carotenoids, Vol. 5, Nutrition and Health. Birkhäuser Verlag, Basel, Switzerland, pp. 287–300.

Köpcke, W. and Krutmann, J. 2008. Protection from sunburn with β-carotene: a meta-analysis. Photochemistry and Photobiology 84: 284–288.

Krinsky, N. I. and Johnson, E. J. 2005. Carotenoid actions and their relation to health and disease. Molecular of Aspects Medicine. 26: 459–516.

Leung, W.C., Hessel, S., Meplan, C., Flint, J., Oberhauser, V., Tourniaire, F., Hesketh, J. E., von Lintig, J. and Lietz, G. 2009. Two common single nucleotide polymorphisms in the gene encoding beta-carotene 15,15′-monoxygenase alter beta-carotene metabolism in female volunteers. FASEB Journal. 23: 1041–1053.

Lin, Y., Dueker, S. R., Burri, B. J., Neidlinger, T. R. and Clifford, A. J. 2000. Variability of the conversion of beta-carotene to vitamin A in women measured by using a double-tracer study design. American Journal of Clinical Nutrition 71: 1545–1554.

Mathews-Roth, M. M. 1993. Carotenoids in erythropoietic protoporphyria and other photosensitivity diseases. Annals of the New York Academy Science. 691: 127–138.

Palozza, P., Serini, S., Torsello, A., DiNicuolo, F., Piccioni, E., Ubaldi, V., Pioli, C., Wolf, F. I. and Calviello, G. 2003. Beta-Carotene regulates NF-kappaB DNA binding activity by a redox mechanism in humanleukemia and colon adenocarvcinoma cells. Journal of Nutrition 133: 381–388.

Rock, C. L. 2009. Carotenoids and cancer. In: Britton, G., Liaaen, S and Pfander, H. (ed.) Carotenoids, Vol. 5, Nutrition and Health. Birkhäuser Verlag, Basel, Switzerland, pp. 269–286.

Schalch, W., Landrum, J. T. and Bone, R. A. 2009. The Eye. In: Britton, G., Liaaen, S and Pfander, H. (eds.) Carotenoids, Vol. 5, Nutrition and health. Birkhäuser Verlag, Basel, CH, pp. 301–334.

Strobel, M., Tinz, J. and Biesalski, H. K. 2007. The importance of β-carotene as a source of vitamin A with special regard to pregnant and breastfeeding women. European Journal of Nutrition 46: I/1–I/20.

Swan, G. 2004. Findings from the latest National Diet and Nutrition Survey. Proceedings of the Nutrition Society. 63: 505–512.

Tang, G. and Russell, R. M. 2009. Carotenoids as provitamin A. In: Britton, G., Liaaen, S and Pfander, H. (ed.) Carotenoids, Vol. 5, Nutrition and Health. Birkhäuser Verlag, Basel, Switzerland, pp. 149–172.

Wang, Z., Yin, S., Zhao, X., Russell, R.M. and Tang, G. 2004. Beta-Carotene-vitamin A equivalence in Chinese adults assessed by an isotope dilution technique. British Journal of Nutrition. 91: 121–131.

West, C. E. and Castenmiller, J. J. 1998. Quantification of the "SLAMENGHI" factors for carotenoid bioavailability and bioconversion. International Journal of Vitamin and Nutrition Research 68: 371–377.

West, C. E., Eilander, A. and van Lieshout, M. 2002. Consequences of revised estimates of carotenoid bioefficacy for dietary control of vitamin A deficiency in developing countries. Journal of Nutrition 132: 2920S–2926S.

Yeum K-J., Aldini, G., Russell, R. M. and Krinski, N. I. 2009. Antioxidant/pro-oxidant actions of carotenoids. In: Britton, G., Liaaen, S and Pfander, H. (eds.) Carotenoids, Vol. 5, Nutrition and Health. Birkhäuser Verlag, Basel, Switzerland, pp. 235–268.

CHAPTER 4
Vitamin A in the Context of Supplementation

FRANK T. WIERINGA*[a], MARJOLEINE A. DIJKHUIZEN[b]
AND JACQUES BERGER[a]

[a] Institute de Recherche pour le Développement (IRD), UMR 204 "Prévention des malnutritions et des pathologies associées", IRD-UM2-UM1, Montpellier, BP 64501, 34394 Montpellier cedex 5, France; [b] Department of Human Nutrition, Faculty of Life Sciences, University of Copenhagen, Rolighedsvej 30, DK-1958 Frederiksberg C, Denmark
*E-mail: franck.wieringa@ird.fr

4.1 The Rational for Vitamin A Supplementation

As early as 1500 BC, clinical symptoms of vitamin A deficiency such as night blindness were described by the Egyptians (Wolf 1978). Furthermore, the importance of vitamin A for resistance to infectious diseases has been known for almost hundred years. In 1913, McCollum and Davis published their work on essential fat-soluble factors in the diet (McCollum and Davis, 1913), and the Danish pediatrician Bloch described the effects of vitamin A and Vitamin C deficiency on immunity as early as 1924 (Bloch, 1924). Indeed, vitamin A was named the 'anti-infective vitamin'. However, with the disappearance in Europe and the United States of xerophthalmia, a serious eye condition due to vitamin A deficiency leading to blindness, the interest in the role of vitamin A in health and disease waned, and only returned in the early 1980's when studies in Indonesia showed that half-yearly high-dose vitamin A supplementation of children dramatically reduced child mortality (Sommer *et al* 1986). This

Food and Nutritional Components in Focus No. 1
Vitamin A and Carotenoids: Chemistry, Analysis, Function and Effects
Edited by Victor R Preedy
© The Royal Society of Chemistry 2012
Published by the Royal Society of Chemistry, www.rsc.org

brought vitamin A deficiency back on to the agenda of international public health, and high-dose vitamin A supplementation was embraced as a 'golden bullet' against child mortality. Overall, undernutrition causes over 3.5 million child deaths a year, and it is estimated that vitamin A deficiency alone is responsible for almost 0.6 million of these child deaths per year (Black *et al* 2008). With close to 200 million children having vitamin A deficiency (plasma retinol concentrations < 0.70 µmol L) [World Health Organization (WHO) 2009], improving vitamin A status is likely to have an impact on child mortality substantially. However, the exact mechanisms behind the reduction in mortality after high-dose vitamin A supplementation are still not fully understood, and the wide variation in effectiveness of vitamin A supplementation in different settings and different target groups causes great confusion among policy makers. Therefore this Chapter will take a closer look at vitamin A supplementation and its context to better understand the effects and the evidence we have so far to recommend supplementation or not.

4.2 Vitamin A Supplementation for Children between 6 and 59 Months of Age

In the early 1990's, Beaton and colleagues reviewed the then available evidence for the impact of vitamin A supplementation on child morbidity and mortality. From eight large trials, the overall reduction in mortality in children receiving vitamin A between 6 months and 5 years of age was estimated to be $\sim 23\%$ (Beaton *et al* 1992). The authors found an especially strong effect on mortality due to diarrhea, but no effect on respiratory infection-related mortality. These findings have been the driving force behind the push by many international organizations such as UNICEF and WHO to implement universal half-yearly high-dose vitamin A supplementation for children under the age of 5 years in all developing countries. Indeed, in 2004, almost 200 million children worldwide received at least 1 vitamin A capsule. This strong focus on supplementation by international organizations is understandable from a policy point of view: The provision of relatively low-cost vitamin A capsules to as many children as possible is an intervention with clearly definable objects and targets. Interestingly, however, Beaton and colleagues noticed already in 1993 that the effect on mortality was not dependent on the high dose of vitamin A or on the route of administration. Indeed, two out of the eight trials in their meta-analyses provided low dosing, either vitamin A in a weekly physiological dose (Rahmathullah *et al* 1990) or in the form of vitamin A-fortified monosodium glutamate (Muhilal *et al* 1988), a popular food seasoning agent in South-East Asia. Both trials showed significant reductions in child mortality, whereas two large trials in India and Sudan providing half-yearly high-dose vitamin A supplementation to children did not find any effect on mortality. Almost 20 years later, these findings are re-confirmed by an WHO-sponsored meta-analysis of the effects of vitamin A supplementation on childhood morbidity and mortality, now comprising over 40 different trials with a total of >200 000 children included (Mayo-Wilson *et al* 2011). This latest meta-analysis estimates a 24% reduction in overall mortality after vitamin A supplementation, a figure very similar to the 23% reported by Beaton and colleagues (1992). However, despite this clear effect of half-yearly vitamin A supplements on child mortality, the evidence is not unequivocal and less straightforward than it seems at first sight. Results remain unpredictable, effects not well understood and maybe more questions are raised than answered for now. It follows that not everybody is in favour of high-dose vitamin A supplementation (Latham 2010). One reason is that the focus on high-dose vitamin A capsules has led to a lack of research into other approaches to improve vitamin A status, such as through food fortification and dietary diversification. Furthermore, the evidence of vitamin A supplementation in different age groups is not unequivocal, and even confusing when examined more thoroughly, with studies reporting widely varying results and even evidence for potentially negative effects. In the early 1990's, high-dose vitamin A supplements were intended to be a short-term solution, however, 20 years later there is still no end in sight to the worldwide vitamin A capsule distribution programs, giving rise to concerns about sustainability. Also, the latest meta-analysis of vitamin A supplementation in children between 6 and 59 months of age does not include the results of the largest trial ever with vitamin A supplementation, the DEVTA (Deworming and Enhanced Vitamin A) trial. Results of this trial have until now not been published, with the exception of a congress abstract (Awasthi *et al* 2007). In the DEVTA trial, over 1 million children received deworming and high-dose vitamin A supplementation half-yearly or placebo. No significant effect on mortality was found. Including the DEVTA trial in the meta-analysis would result in a halving of the mortality effect; that is, the overall effect on mortality would only be a 12% reduction, but the total effect would still be a significant reduction in child mortality. However, this trial stresses the need for a better understanding of the effects of vitamin A on morbidity and mortality.

Intriguingly, whereas the effect of vitamin A supplementation on mortality is clearly significant, there appears to be no effect on morbidity. A 2003 meta-analysis of eight studies, including over 45 000 children, found no effect of vitamin A supplementation in children <7 years of age on diarrheal disease incidence and even reported a significant increase in respiratory infections [relative risk (RR) 1.08; 95% confidence interval (CI) 1.05–1.11] (Grotto *et al* 2003). In contrast, a meta-analysis in 2010 that included 13 studies on diarrhea-related morbidity found a beneficial effect of vitamin A supplementation on diarrheal incidence (RR 0.85, 95% CI 0.82–0.87) (Mayo-Wilson *et al* 2011). Inclusion or exclusion of certain studies appears to underlie these differences, as, for example, in the 2010 meta-analysis results from two large studies in Haiti and Ghana on diarrheal disease incidence were not included in the analysis.

So it must be concluded that, although there is clear evidence for a large impact of vitamin A supplementation on mortality, this evidence does not specifically point to high-dose supplementation, but also includes low-dose supplementation and food-based approaches such as fortification. Also, the evidence is far from homogeneous and the reasons behind the incongruity are not well understood. This incongruity is often not sufficiently appreciated or conveyed, in order to promote a single highly effective approach to solve the whole problem at once, a so-called 'golden bullet'. Unfortunately, the complex, interacting and dynamic nature of the nutrition–health inter-relationship make such approaches unrealistic and inadequate.

4.3   Vitamin A Supplementation for Whom? And does it Work?

Buoyed by the evidence of the large impact of vitamin A supplementation on child mortality, researchers and policymakers started looking for other target groups in which vitamin A supplementation could be beneficial: young infants under the age of 6 months, neonates, pregnant and lactating women, and women just after delivery. Also specific target groups such as HIV-infected adults, children with respiratory infections or HIV-infected pregnant women were studied. However, results have been overall disappointing, and this expanding field of potential beneficiaries has not experienced the expected health benefits of vitamin A supplementation, as will be discussed below.

4.3.1   Vitamin A Supplementation of Newborns

The effect of vitamin A supplementation in newborns on morbidity and mortality is not clear. Three studies in India, Indonesia and Bangladesh showed reductions in neonatal mortality of 22% (Rahmathullah *et al* 2003), 64% (Humphrey *et al* 1996) and 15% (Klemm *et al* 2008) respectively, but two trials in Africa showed no effect (Benn *et al* 2008; Malaba *et al* 2005). Moreover, the beneficial effect in the trial in India was restricted to low-birth-weight infants, whereas the effect in Indonesia was only observed in infants with a birth weight >2500 g. Therefore, the WHO is currently supporting new trials in Africa (Ghana and Tanzania) and Asia (Pakistan and India), hoping to obtain more clear results to guide policy. For the moment, the WHO does not recommend vitamin A supplementation for infants under the age of 6 months. Several hypotheses have been put forward to explain the differences among the studies, such as differences in initial vitamin A status, differences in health care facilities and different vaccination coverage, but none has been widely accepted.

4.3.2   Vitamin A Supplementation for Women Directly after Delivery

Improving vitamin A status of lactating mothers by supplementing them with vitamin A would be an obvious strategy to improve vitamin A status of infants, with the potential to significantly reduce infant mortality. Moreover, maternal supplementation is more acceptable in certain parts of the world than supplementation of young infants. Indeed, one study in Bangladesh and one study in Indonesia showed not only improved maternal vitamin A status, but also a higher content of vitamin A in breast milk, and improved vitamin A status of infants at 6 months of age (Rice *et al* 1999; 2000). In view of the expected health benefits of improved neonatal vitamin A status, the WHO therefore recommended that lactating women should receive a high dose [200 000 international units (IU)] of vitamin A within the first 6 weeks post-partum. However, more recent trials, in which women received the WHO recommended 200 000 IU of vitamin A directly post-partum, reported no significant improvements in vitamin A status of either the mother or infant, not even after doubling the dose to 400 000 IU (Ayah *et al* 2007; Malaba *et al* 2005). Therefore, recently the WHO revised their guidelines, and the WHO is no longer recommending high-dose vitamin A supplementation after delivery. More research is needed to understand why a high dose of vitamin A directly post-partum is not effective in improving vitamin A status of either mother or infant. Raising the dose even higher is not recommended, as the 400 000 IU dose, or 100 times the recommended dietary intake (RDI), already failed to show a difference with 200 000 IU (Tchum *et al* 2006). Instead, a better understanding of vitamin A metabolism and utilization, and the implications of high dosing is needed to develop a more effective strategy in an evidence-based manner.

4.3.3   Vitamin A Supplementation of Pregnant Women

Supplementing pregnant women with high-dose vitamin A is undesirable, as vitamin A can be teratogenic, and intake levels above 10 000 IU per day or 25 000 IU per week should be avoided (WHO 1998). However, a large trial in Nepal in the 1990's suggested that vitamin A or β-carotene (a precursor to Vitamin A) supplementation during pregnancy could reduce maternal mortality by as much as 40% (West *et al* 1999). This generated strong interest, as each year more than 350 000 women die due to pregnancy-related complications, which is an equivalent to almost 1000 women per day. A subsequent trial in Ghana, recruiting and supplementing over 200 000 women weekly with 25 000 IU vitamin A or placebo, and following almost 80 000 pregnancies, failed, however, to find a beneficial effect of vitamin A supplementation on either maternal, neonatal or infant mortality up to 1 year of age (Kirkwood *et al* 2010). And another large trial in Bangladesh, following >60 000 pregnant women also showed no effect on maternal mortality, although results have not been published in peer-reviewed journals until now (JiVitA-1 2011). As a result of these disappointing findings, and in view of the potential harmful effects of vitamin A in pregnancy, the WHO recommends *not* to include vitamin A supplementation in the routine natal care programs for pregnant women, unless vitamin A deficiency is a severe health problem.

4.3.4   Vitamin A Supplementation and HIV Infection

In the 1990's, it was hypothesized that vitamin A could reduce the prevalence of mother-to-child transmission (MTCT) of HIV, as vitamin A is essential for maintaining gut integrity. Surprisingly, a trial in Tanzania found the opposite: vitamin A supplementation of pregnant women increased MTCT by 30% (Fawzi *et al* 2002). Indeed, adding vitamin A to a multiple micronutrient supplement containing B-vitamins, vitamins C and E, and folic acid seems to reduce the beneficial effects of the supplement (Fawzi *et al* 2004; Villamor *et al* 2005) and even increase overall mortality (Fawzi and Msamanga 2004). In a recent meta-analysis, five randomized controlled trials were included comprising in total over 7 500 HIV-infected pregnant women. The authors found no overall effect of vitamin A supplementation during pregnancy on the risk of MTCT, either positive or negative. As such, the WHO does not recommend vitamin A for the prevention of MTCT.

4.3.5   Vitamin A Supplementation and Measles Infection

One infectious disease which is clearly linked to vitamin A is measles. Already in 1932, Ellison described a protective role of vitamin A in the treatment of measles, with 11 children dying of measles in a cohort of 300 children receiving vitamin A as compared to 26 children dying of measles in the 300 children receiving placebo (Ellison 1932). In 2008, >150 000 children still died of measles, although this is already much less than 10 years earlier, with measles mortality falling by 80% between 2000 and 2008. This may partly be due to the better availability of vaccines against measles with 85% of children having had a measles vaccine by the age of 1 year. However, in 2000 the vaccination coverage for measles was already approximately 72% of children worldwide, and the modest further increase in vaccination coverage cannot completely explain the large drop in measles mortality and mechanisms such as herd immunity, infection load and changes in virulence have been proposed. Measles infection reduces circulating retinol concentrations as all infections do due to the acute-phase response (Thurnham *et al* 2003; Wieringa *et al* 2002). However, in addition, it appears that measles infection increases the risk for functional vitamin A deficiency in the months thereafter, as has been shown by an increased incidence of xerophthalmia after measles infection (Hussey and Klein 1992). In the 1980's and 1990's, studies confirmed the reduction in mortality in children with measles receiving vitamin A, and, since 1998, the WHO recommends high-dose vitamin A supplementation for children with severe measles. The current recommendation is two times 200 000 IU of vitamin A for children over 1 year of age. Surprisingly, a recent Cochrane review on the effect of vitamin A in measles infection concluded that, although there was an 30% reduction in mortality in measles infection after vitamin A supplementation, the combined results of the eight trials with overall >2500 children were not statistically significant. As the three studies which provided two doses of 200 000 IU of vitamin A did show a significant 60% reduction in mortality, some have argued that the one high dose of vitamin A is not enough. However, these three studies comprise only 429 children in total, and it is not clear from a physiological perspective why one high dose of vitamin A should not be enough. Far more important, but not emphasized, is the fact that vitamin A supplementation in measles infection was only effective in children <2 years of age (RR 0.12; 95% CI 0.07–0.66), whereas there was no benefit after 2 years of age (RR 0.98). Clearly, despite the fact that it is almost 80 years since Ellison described the effects of vitamin A supplementation in measles infection, we are still far from understanding the underlying mechanism, and seemingly straightforward WHO recommendations are rooted in uncertainty (Table 4.1).

4.4 Time for Reflection

As will be clear from above, high-dose vitamin A supplementation is not the 'golden bullet' that many people had hoped for. Although high-dose vitamin A supplementation definitely has a place as a short-term solution to reduce mortality in children between 6 months and 5 years of age, other solutions to improve vitamin A status of risk groups need to be developed as sustainable long-term interventions. Food fortification, for example, could be a very cost-effective long-term strategy to reduce the prevalence of vitamin A deficiency (Darnton-Hill and Nalubola 2002). It should be remembered that the disappearance of vitamin A deficiency in Denmark in the 1930's as a public health problem coincided with the introduction of mandatory vitamin A (and D) fortification of margarine, the 'poor man's butter' (Dary and Mora 2002). Moreover, food fortification has the potential to address the deficiency of several micronutrients simultaneously, whereas the current supplementation programs focus on only vitamin A. Concurrent deficiency of several micronutrients is very common (Dijkhuizen *et al* 2001). Hence, addressing only one micronutrient is less advantageous and less cost effective, as the effectiveness of the intervention is reduced by the other deficiencies (Dijkhuizen *et al* 2004). On a larger scale, more integrated interventions to reduce morbidity and overall health are essential. Indeed, Beaton and colleagues already argued in the early 1990's that vitamin A is *not* a panacea to control young child morbidity but that one should focus upon the environment in which morbidity occurs (Beaton *et al* 1992). A fuller appreciation of the complex, interactive and dynamic inter-relationship between nutrition and health is needed. This will also help to put the evidence into context and lead to a better interpretation and a fuller understanding of the effects of vitamin A and the place of supplementation in improving health. The present focus on high-dose vitamin A supplementation programs has drawn away efforts and money from other much-needed, more sustainable interventions. It has also led to a sense of false security in that the solution, the 'golden bullet', is available, whereas in reality in-depth research into the mechanisms by which vitamin A affects mortality is urgently needed as the present evidence is unequivocal, confusing and incompletely understood.

Indeed, as already noted in the discussion of measles above, it is strange and highly unsatisfactory that more than 20 years after the first large trials on vitamin A supplementation and mortality, we still do not understand the exact mechanisms by which vitamin A reduces mortality, seemingly without affecting morbidity.

4.5   Vitamin A and the Immune System

Clearly, effects of vitamin A supplementation on disease severity, with vitamin A reducing the severity of infectious diseases, are implicated in the observed reduction in child mortality after vitamin A supplementation. Vitamin A is a highly potent immune function modulator (Stephensen 2001), and the modulating effects depend on the dose given, the age and gender of the subject, and perhaps other factors such as whether the vitamin A is given together with vaccination, and whether the subject is vitamin A deficient or not. Therefore, providing a high dose of vitamin A every half year will certainly affect immune function differently from providing a low dose of vitamin A on a daily or weekly basis. A further complication with vitamin A supplementation for infants is that the immune system of an infant is immature at birth, and develops rapidly during the first year of life. Immuno-modulation during this time can have profound and long-lasting, perhaps even life-long, effects. The mechanisms by which vitamin A affects the immune system have not been fully elucidated, but are discussed in more detail in a later Chapter. Most research seems to indicate that vitamin A affects the so-called Th1 (T-helper cell 1; cellular)/Th2 (T-helper cell 2; humoral) immune balance, with vitamin A deficiency causing a dominance of Th1 responses. For example, vitamin A-replete mice react to helminth infection with a strong Th2 response. In contrast, vitamin A-deficient mice responded with a Th1 response (Cantorna *et al* 1995). It is clear that this bias in immune response, with a reduction in the humoral immune response, will be unfavourable in infections requiring a strong humoral response such as for extracellular pathogens. Other studies suggest a reduced immune response of both the Th1 and Th2 pathways in vitamin A deficiency. We also reported diminished immune responses in vitamin A-deficient infants, although it was the production of the Th1 cytokine interferon-γ, in particular, that was reduced after stimulation (Wieringa *et al* 2004). In contrast, there was evidence of increased Th1 activity *in vivo* in vitamin A-deficient infants. Interestingly, in young infants, the Th1 response is almost absent, and only develops during the first months of life, whereas the Th2 response is present from birth. Therefore, age is a very important determinant of immune responses in infancy, and the development of an appropriate Th1/Th2 balance is deemed crucial for health and survival (Levy 2007). This immune balance develops over the first year of life, and probably includes programming of set-points that determine long-term immune function. For example, prolonged Th2 (humoral) immune responses during the first years of life have been associated with asthma and allergy (Levy 2007).

Indeed, in mice fed a vitamin A-deficient diet, pulmonary hyper-reactivity was absent, which was in contrast to mice fed a high vitamin A diet (Schuster *et al* 2008). In this context, it is important to realize that the development of the immune system of a newborn starts *in utero*, and the fetal environment already affects immune development. Recently, we reported that supplementation of pregnant women affected the immune response of the newborn at 6 months of age (Wieringa *et al* 2010). Infants born from mothers receiving β-carotene had a >30% reduction in the production of interferon-γ after stimulation of whole blood, but β-carotene supplementation had no effect on morbidity during the first 6 months of life. These findings are in agreement with the animal studies on vitamin A deficiency and supplementation, and suggest that, in the neonate, the Th2 dominance during the first months of life is extended due to the supplementation of β-carotene during pregnancy, which might render the newborn more susceptible to, for example, asthma later in life (Wieringa *et al* 2010). Few data exist on the effect of high-dose vitamin A given directly to neonates on immune function. High-dose vitamin A given together with the BCG vaccination appears not to diminish the immune response to the BCG vaccine, and, indeed, even results in higher interferon-γproduction (Th1) (Diness *et al* 2007). In contrast, retinoic acid supplementation of neonatal mice resulted in a bias towards Th2 responses (Ma *et al* 2005).

Therefore, currently it is unclear whether vitamin A supplementation given to pregnant and lactating women or to infants and children modulates the immune response of the recipient, what kind of changes occur and what the consequences of the immune modulation could be. That this is not only of theoretical importance was shown by Long *et al* (2006), who investigated pathogen-specific responses of vitamin A supplementation. One of their findings was that vitamin A supplementation of children reduced the duration of *Escherichia coli*-associated diarrhea, but increased the duration of diarrhea due to *Giardia lamblia* (Long *et al* 2006). In addition, the group of Peter Aaby has reported strong differences between boys and girls in the response to vitamin A supplementation in combination with routine childhood vaccinations. For example, boys benefited from receiving vitamin A with a DTP vaccine with a significant reduction in mortality, whereas in girls mortality increased after the same intervention (Benn *et al* 2007; 2009). These patterns were also seen for vitamin A supplementation in combination with BCG vaccination (Benn *et al* 2008). Clearly, in-depth research into this very important topic is long overdue.

To conclude, improving vitamin A status of children between 6 and 59 months of age and at risk for vitamin A deficiency, clearly benefits the child, with a strong reduction in child mortality. Whether this improvement is done through supplementation, fortification or dietary changes appears not to be relevant. One likely link between vitamin A supplementation and reduction in mortality is through improved immune function, with the child being more resistant for infection. However, the evidence for reduced morbidity in children after vitamin A supplementation is weak. The effects of vitamin A may be more basic, indirect and subtle, as vitamin A is a strong immuno-modulator, and it is also likely that effects of vitamin A supplementation on immune function are modified by underlying vaccination status, gender, age and baseline vitamin A status. Finally, whether these modifying effects on immune response brought about by vitamin A supplementation prove to be beneficial for the individual or not might depend on the type of pathogen encountered. Therefore the end effect on public health outcomes such as morbidity is difficult to interpret with our current incomplete understanding of vitamin A–immune system interactions.

Summary Points

- Vitamin A supplementation is given to children to reduce vitamin A deficiency. This strategy has been shown to reduce childhood mortality by $\sim 23\%$. High-dose vitamin A supplementation is the most publicized strategy, but the protective effect is also attributable to low-dose and fortification interventions.
- Much research has focused on other target groups such as pregnant and lactating women and neonates. However, until now, no consistent evidence of benefit has emerged for other age groups, except children with measles under the age of 2 years.
- Therefore, the WHO recommends vitamin A supplementation only in children between 6 and 59 months of age, and in children with severe measles infection.
- Despite the clear reduction in mortality in children, effects on morbidity are much less clear.
- Vitamin A is an immuno-modulating agent, and supplementation affects the immune responses. However, whereas this might be beneficial for the response against some pathogens, it could be disadvantageous in infections with other pathogens. This might underlie the observations that vitamin A supplementation tends to reduce diarrheal disease but increase respiratory infections.
- More research into this field is highly warranted, as this may well be a key to understanding the public health effects of vitamin A, allowing evidence-based development of interventions and accurate prediction of effects.

Key Facts

Key Facts of Vitamin A Deficiency

- Vitamin A deficiency is still prevalent in many developing countries, with $\sim 600\ 000$ children dying each year due to vitamin A deficiency

- Good dietary sources of vitamin A include eggs and liver. Although plant sources contain carotenoids, some of which can be converted into vitamin A in the human body, low bioavailability of carotenoids make it difficult to fulfill vitamin A requirements on a monotonous low-cost rice-based diet.
- Severe vitamin A deficiency leads to a blindness (a condition known as xerophthalmia), and is nowadays not often seen.
- Less severe or marginal vitamin A deficiency is much more common and affects the immune system, resulting in reduced resistance to infection. This is reflected in public health outcomes of increased mortality and, less clearly, morbidity.

Key Facts of Vitamin A Supplementation

- To reduce the prevalence of vitamin A deficiency different strategies exist, including supplementation, fortification of food items and dietary diversification.
- Vitamin A supplementation can be given in daily low doses, in weekly medium doses or in half-yearly high doses.
- Vitamin A supplementation has been shown to reduce the mortality in children between 6 and 59 months of age by almost one-quarter. However, this effect is attributable to high dose as well as low dose and even fortification interventions.

Table 4.1 Current recommendations by the WHO for Vitamin A supplementation. (Current World Health Organization recommendations on vitamin A supplementation and other micronutrients can be found at http://www.who.int/elena/titles/en/).

Target group	WHO recommendation
Children between 6 and 59 month	High dose of vitamin A every 4 – 6 months recommended
Infants between 1–5 months of age	Not recommended. Mothers should be encouraged to breastfeed exclusively.
Neonates (<1 month of age)	Currently not recommended. Recommendations will be reviewed in 2013.
Children with measles	Two high doses of vitamin A, with at least 24 h between the two doses is recommended.
Pregnant women	Not recommended as routine antenatal care. Recommended for prevention of night blindness
Post-partum women directly after delivery	Not recommended.
HIV-positive pregnant women	Not recommended for the prevention of MTCT.
Children with lower respiratory infection/pneumonia	No guidelines. More research is needed.

- The WHO therefore recommends half-yearly high-dose vitamin A supplementation for all children in developing countries. However, this strategy might not be sustainable in the long-term.
- Benefits of vitamin A supplementation in other age groups (neonates) or other target groups (pregnant women and lactating women) have been less clear and routine supplementation therefore cannot be recommended.
- Little is known how vitamin A supplementation affects the immune system and more research is needed in this important area.

List of Abbreviations

CI	confidence interval
IU	international units
MTCT	mother-to-child transmission
RR	relative risk
Th1	T-helper cell 1
Th2	T-helper cell 2
WHO	World Health Organization

References

Awasthi, S., Peto, R., Read, S. and Bundy, D., 2007. Six-monthly vitamin A from 1 to 6 years of age. DEVTA: cluster-randomised trial in 1 million children in North India. Micronutrient Forum, Istanbul, Turkey: ILSI.

Ayah, R.A., Mwaniki, D.L., Magnussen, P., Tedstone, A.E., Marshall, T., Alusala, D., Luoba, A., Kaestel, P., Michaelsen, K.F. and Friis, H, 2007. The effects of maternal and infant vitamin A supplementation on vitamin A status: a randomised trial in Kenya. British Journal of Nutrition 98: 422–430.

Beaton, G.H., Martorell, R., L'Abbe, K.A., Edmonston, B., McCabe, G.P., Ross, A.C. and Harvey, B., 1992. Effectiveness of vitamin A supplementation in the control of young child morbidity and mortality in developing countries. Final report to CIDA (University of Toronto, Canada).

Benn, C.S., Fisker, A.B., Jorgensen, M.J. and Aaby, P., 2007. Why worry: vitamin A with DTP vaccine? Vaccine 25: 777–779.

Benn, C.S., Diness, B.R., Roth, A., Nante, E., Fisker, A.B., Lisse, I.M., Yazdanbakhsh, M., Whittle, H., Rodrigues, A. and Aaby, P., 2008. Effect of 50,000 IU vitamin A given with BCG vaccine on mortality in infants in Guinea-Bissau: randomised placebo controlled trial. British Medical Journal 336: 1416–1420.

Benn, C.S., Aaby, P., Nielsen, J., Binka, F.N. and Ross, D.A., 2009. Does vitamin A supplementation interact with routine vaccinations? An analysis of the Ghana Vitamin A Supplementation Trial. American Journal of Clinical Nutrition 90: 629–639.

Black, R.E., Allen, L.H., Bhutta, Z.A., Caulfield, L.E., de Onis, M., Ezzati, M., Mathers, C. and Rivera, J., 2008. Maternal and child undernutrition: global and regional exposures and health consequences. Lancet 371: 243–260.

Bloch, C.E., 1924. Further clinical investigations into the diseases arising in consequence of a deficiency in fat-solubable A factor. Am. J. Dis. Child. 28: 659–667.

Cantorna, M.T., Nashold, F.E. and Hayes, C.E., 1995. Vitamin A deficiency results in a priming environment conducive for Th1 cell development. European Journal of Immunology 25, 1673–1679.

Darnton-Hill, I. and Nalubola, R., 2002. Fortification strategies to meet micronutrient needs: successes and failures. The Proceedings of the Nutrition Society 61: 231–241.

Dary, O. and Mora, J.O., 2002. Food fortification to reduce vitamin A deficiency: International Vitamin A Consultative Group recommendations. Journal of Nutrition 132: 2927S–2933S.

Dijkhuizen, M.A., Wieringa, F.T., West, C.E., Muherdiyantiningsih and Muhilal, 2001. Concurrent micronutrient deficiencies in lactating mothers and their infants in Indonesia. American Journal of Clinical Nutrition 73: 786–791.

Dijkhuizen, M.A., Wieringa, F.T., West, C.E. and Muhilal, 2004. Zinc plus β-carotene supplementation of pregnant women is superior to β-carotene supplementation alone in improving vitamin A status in both mothers and infants. American Journal of Clinical Nutrition 80: 1299–1307.

Diness, B.R., Fisker, A.B., Roth, A., Yazdanbakhsh, M., Sartono, E., Whittle, H., Nante, J.E., Lisse, I.M., Ravn, H., Rodrigues, A., *et al.*, 2007. Effect of high-dose vitamin A supplementation on the immune response to Bacille Calmette-Guerin vaccines. American Journal of Clinical Nutrition 86: 1152–1159.

Ellison, J.B., 1932. Intensive vitamin therapy in measles. British Medical Journal 2, 708–711.

Fawzi, W. and Msamanga, G., 2004. Micronutrients and adverse pregnancy outcomes in the context of HIV infection. Nutrition Reviews 62: 269–275.

Fawzi, W.W., Msamanga, G.I., Hunter, D., Renjifo, B., Antelman, G., Bang, H., Manji, K., Kapiga, S., Mwakagile, D., Essex, M., *et al.*, 2002. Randomized trial of vitamin supplements in relation to transmission of HIV-1 through breastfeeding and early child mortality. AIDS 16: 1935–1944.

Fawzi, W.W., Msamanga, G.I., Spiegelman, D., Wei, R., Kapiga, S., Villamor, E., Mwakagile, D., Mugusi, F., Hertzmark, E., Essex, M., *et al.*, 2004. A randomized trial of multivitamin supplements and HIV disease progression and mortality. New England Journal of Medicine 351: 23–32.

Grotto, I., Mimouni, M., Gdalevich, M. and Mimouni, D., 2003. Vitamin A supplementation and childhood morbidity from diarrhea and respiratory infections: a meta-analysis. Journal of Pediatrics 142: 297–304.

Humphrey, J.H., Agoestina, T., Wu, L., Usman, A., Nurachim, M., Subardja, D., Hidayat, S., Tielsch, J., West, Jr, K.P. and Sommer, A., 1996. Impact of neonatal vitamin A supplementation on infant morbidity and mortality. Journal of Pediatrics 128, 489–496.

Hussey, G.D. and Klein, M., 1992. Measles-induced vitamin A deficiency. Annals of the New York Academy of Sciences 669: 188–194.

JiVitA-1, 2011. Maternal vitamin A or β-carotene supplementation trial to reduce maternal and infant mortality (Johns Hopkins University, Baltimore, MD, USA). Available at: http://www.jhsph.edu/chn/research/jivita/jivita1.html. Last accessed: 30 April 2012.

Kirkwood, B.R., Hurt, L., Amenga-Etego, S., Tawiah, C., Zandoh, C., Danso, S., Hurt, C., Edmond, K., Hill, Z., Ten Asbroek, G., *et al.*, 2010. Effect of vitamin A supplementation in women of reproductive age on maternal survival in Ghana (ObaapaVitA): a cluster-randomised, placebo-controlled trial. Lancet 375: 1640–1649.

Klemm, R.D., Labrique, A.B., Christian, P., Rashid, M., Shamim, A.A., Katz, J., Sommer, A. and West, Jr, K.P., 2008. Newborn vitamin A supplementation reduced infant mortality in rural Bangladesh. Pediatrics 122, e242–e250.

Latham, M., 2010. The great vitamin A fiasco. World Nutrition 1: 12–45.

Levy, O., 2007. Innate immunity of the newborn: basic mechanisms and clinical correlates. Nature Reviews in Immunology 7: 379–390.

Long, K.Z., Santos, J.I., Rosado, J.L., Lopez-Saucedo, C., Thompson-Bonilla, R., Abonce, M., DuPont, H.L., Hertzmark, E. and Estrada-Garcia, T., 2006. Impact of vitamin A on selected gastrointestinal pathogen infections and associated diarrheal episodes among children in Mexico City, Mexico. Journal of Infectious Diseases 194: 1217–1225.

Ma, Y., Chen, Q. and Ross, A.C., 2005. Retinoic acid and polyriboinosinic:polyribocytidylic acid stimulate robust anti-tetanus antibody production while differentially regulating type 1/type 2 cytokines and lymphocyte populations. Journal of Immunology 174: 7961–7969.

Malaba, L.C., Iliff, P.J., Nathoo, K.J., Marinda, E., Moulton, L.H., Zijenah, L.S., Zvandasara, P., Ward, B.J. and Humphrey, J.H., 2005. Effect of postpartum maternal or neonatal vitamin A supplementation on infant mortality among infants born to HIV-negative mothers in Zimbabwe. American Journal of Clinical Nutrition 81: 454–460.

Mayo-Wilson, E., Imdad, A., Herzer, K., Yakoob, M.Y. and Bhutta, Z.A., 2011. Vitamin A supplements for preventing mortality, illness, and blindness in children aged under 5: systematic review and meta-analysis. British Medical Journal 343: d5094.

McCollum, E.V. and Davis, M., 1913. The necessity of certain lipids in the diet during growth. J. Biol. Chem. 15: 167–175.

Muhilal, Permeisih, D., Idjradinata, Y.R., Muherdiyantiningsih and Karyadi, D., 1988. Vitamin A-fortified monosodium glutamate and health, growth, and survival of children: a controlled field trial. American Journal of Clinical Nutrition 48: 1271–1276.

Rahmathullah, L., Underwood, B.A., Thulasiraj, R.D., Milton, R.C., Ramaswamy, K., Rahmathullah, R. and Babu, G., 1990. Reduced mortality among children in southern India receiving a small weekly dose of vitamin A. New England Journal of Medicine 323: 929–935.

Rahmathullah, L., Tielsch, J.M., Thulasiraj, R.D., Katz, J., Coles, C., Devi, S., John, R., Prakash, K., Sadanand, A.V., Edwin, N., *et al.*, 2003. Impact of supplementing newborn infants with vitamin A on early infant mortality: community based randomised trial in southern India. British Medical Journal, 327: 254.

Rice, A.L., Stoltzfus, R.J., de Francisco, A., Chakraborty, J., Kjolhede, C.L. and Wahed, M.A., 1999. Maternal vitamin A or β-carotene supplementation in lactating Bangladeshi women benefits mothers and infants but does not prevent subclinical deficiency. Journal of Nutrition 129: 356–365.

Rice, A.L., Stoltzfus, R.J., de Francisco, A. and Kjolhede, C.L., 2000. Evaluation of serum retinol, the modified-relative-dose-response ratio, and breast-milk vitamin A as indicators of response to postpartum maternal vitamin A supplementation. American Journal of Clinical Nutrition 71: 799–806.

Schuster, G.U., Kenyon, N.J. and Stephensen, C.B., 2008. Vitamin A deficiency decreases and high dietary vitamin A increases disease severity in the mouse model of asthma. Journal of Immunology 180: 1834–1842.

Sommer, A., Tarwotjo, I., Djunaedi, E., West, Jr, K.P., Loeden, A.A., Tilden, R. and Mele, L., 1986. Impact of vitamin A supplementation on childhood mortality. A randomised controlled community trial. Lancet 327: 1169–1173.

Stephensen, C.B., 2001. Vitamin A, infection and immune function. Annual Review of Nutrition 21: 167–192.

Tchum, S.K., Tanumihardjo, S.A., Newton, S., de Benoist, B., Owusu-Agyei, S., Arthur, F.K. and Tetteh, A., 2006. Evaluation of vitamin A supplementation regimens in Ghanaian postpartum mothers with the use of the modified-relative-dose-response test. American Journal of Clinical Nutrition 84: 1344–1349.

Thurnham, D.I., McCabe, G.P., Northrop-Clewes, C.A. and Nestel, P., 2003. Effects of subclinical infection on plasma retinol concentrations and assessment of prevalence of vitamin A deficiency: meta-analysis. Lancet 362: 2052–2058.

Villamor, E., Saathoff, E., Bosch, R.J., Hertzmark, E., Baylin, A., Manji, K., Msamanga, G., Hunter, D.J. and Fawzi, W.W., 2005. Vitamin supplementation of HIV-infected women improves postnatal child growth. American Journal of Clinical Nutrition 81: 880–888.

West, K.P.J., Katz, J., Khatry, S.K., LeClerq, S.C., Pradhan, E.K., Shrestha, S.R., Connor, P.B., Dali, S.M., Christian, P., Pokhrel, R.P., *et al.*, 1999. Double blind, cluster randomised trial of low dose supplementation with vitamin A or β-carotene on mortality related to pregnancy in Nepal. NNIPS-2 Study Group. British Medical Journal 318: 570–575.

Wieringa, F.T., Dijkhuizen, M.A., West, C.E., Northrop-Clewes, C.A. and Muhilal, 2002. Estimation of the effect of the acute phase response on indicators of micronutrient status in Indonesian infants. Journal of Nutrition 132: 3061–3066.

Wieringa, F.T., Dijkhuizen, M.A., West, C.E., van der Ven-Jongekrijg, J. and van der Meer, J.W., 2004. Reduced production of immunoregulatory cytokines in vitamin A- and zinc-deficient Indonesian infants. European Journal of Clinical Nutrition 58: 1498–1504.

Wieringa, F.T., Dijkhuizen, M.A., Muhilal and Van der Meer, J.W., 2010. Maternal micronutrient supplementation with zinc and β-carotene affects morbidity and immune function of infants during the first 6 months of life. European Journal of Clinical Nutrition 64: 1072–1079.

Wolf, G., 1978. A historical note on the mode of administration of vitamin A for the cure of night blindness. American Journal of Clinical Nutrition 31: 290–292.

World Health Organization, 1998. Safe vitamin A dosage during pregnancy and lactation. Recommendations and report from a consultation. Micronutrient Series. World Health Organization, Geneva.

World Health Organization, 2009. Global prevalence of vitamin A deficiency in populations at risk 1995–2005. WHO global database on vitamin A deficiency. http://www.who.int/vmnis/vitamina/prevalence/report/en/. Last accessed: 30 April 2012.

Chemistry and Biochemistry

CHAPTER 5
The Chemistry of Vitamin A

ALESSANDRA GENTILI

Department of Chemistry, Faculty of Mathematical, Physical and Natural Sciences, University of Rome "La Sapienza", Piazzale Aldo Moro no. 5, P.O. Box 34, Posta 62, 00185 Roma, Italy
E-mail: alessandra.gentili@uniroma1.it

5.1 Introduction

Vitamin A is an essential micronutrient involved in the regulation of various physiological functions such as vision, gene expression, maintenance of the immune system (it is also known as anti-infective vitamin), embryonic development, and red blood cell production (Bennasir *et al* 2010).

Two are the groups of compounds with vitamin A activity: retinoids (preformed vitamin A) and provitamin A carotenoids (vitamin A precursors) (Bai *et al* 2011). Natural retinoids are retinol, retinal, retinoic acid, and retinyl esters of saturated and unsaturated fatty acids; they are derived from animal sources where mainly occur in the form of retinyl esters (Ball 2006). Plants synthesize hundreds of carotenoids, but only approximately 50 of them can act as vitamin A precursors (Olson 1994). A stringent requisite is that the carotenoid contains at least one molecule of retinol, *i.e.* an unsubstituted β-ionone ring and a polyene chain containing 11 carbons (Bai *et al* 2011; Ball 2006). β-Carotene is the most important precursor because it is very widespread and incorporates two molecules of retinol; other provitamins A (for example, α-carotene, β-cryptoxanthin, γ-carotene) have only one retinyl group and approximately half of the β-carotene biological activity (Bai *et al* 2011; Ball 2006). The synthetic analogues of retinoids (retinyl palmitate, retinyl acetate) are used for supplementation purpose, whereas those of carotenoids (β-carotene, apocar-

Food and Nutritional Components in Focus No. 1
Vitamin A and Carotenoids: Chemistry, Analysis, Function and Effects
Edited by Victor R Preedy
Published by the Royal Society of Chemistry, www.rsc.org

otenal, ethyl ester of apocarotenic acid, and the non-provitamin carotenoid canthaxanthin) mainly as food color additives (Ball 2006; Paust 1991).

A large number of geometrical isomers are theoretically possible for retinoids and carotenoids, but many mono-*cis* and poli-*cis* isomers exhibit steric hindrance and are rarely encountered in nature (*e.g.*, 7-*cis*- and 11-*cis*-β-carotene) (Ball 2006; Britton 1995). All-*trans* forms are predominant in natural foodstuffs and are often found along with *cis* isomers which, characterised by a small steric hindrance, are thermodynamically stable (*e.g.* 9-*cis*-, 13-*cis*-, 9,13-di-*cis*-, and, only for carotenoids, 15-*cis*-) (Ball 2006; Chandler and Schwartz 1987; Godoy and Rodriguez-Amaya 1994; Lessin *et al* 1997).

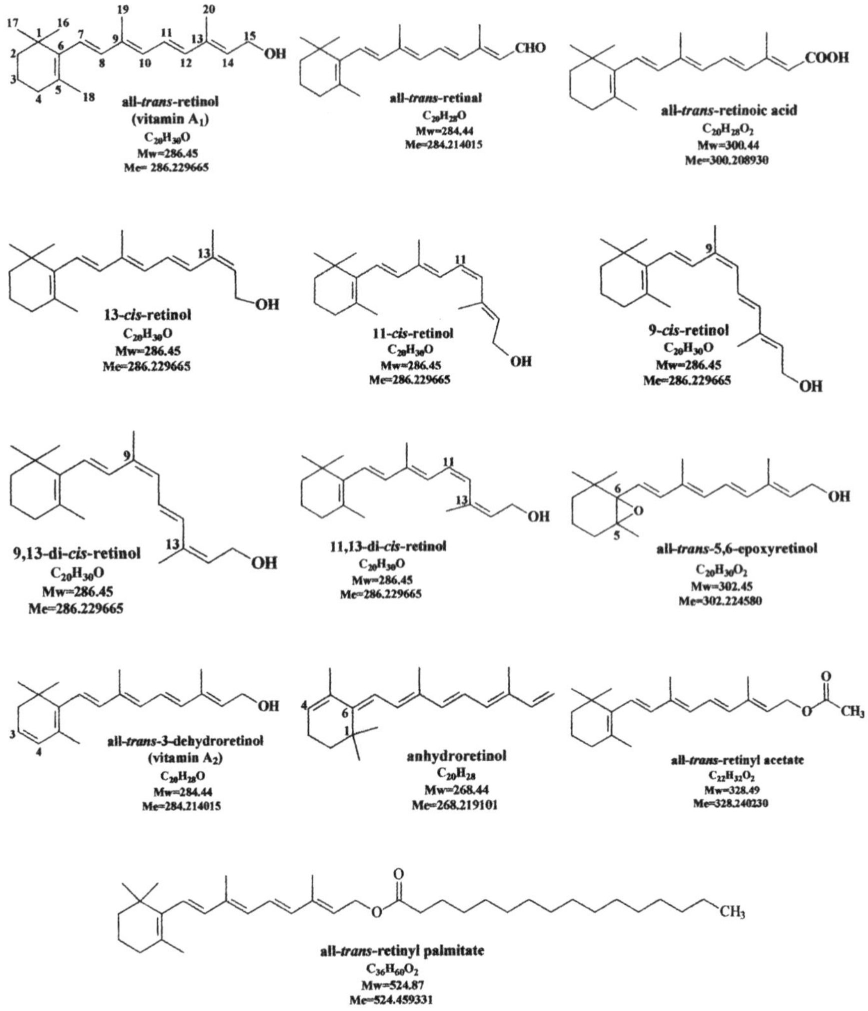

Figure 5.1   Structures, formulas, molecular masses (M_w), and exact masses (M_e) of some important retinoids.

Structures, formulas, molecular masses (M_w), and exact masses (M_e) of selected retinoids and provitamin A carotenoids are shown in Figure 5.1 and Figure 5.2, respectively.

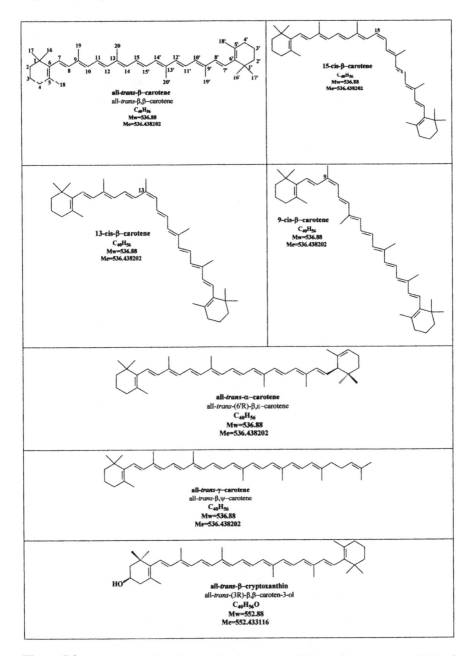

Figure 5.2 Structures, formulas, molecular masses (M_w), and exact masses (M_e) of the main provitamin A carotenoids.

5.2   Physicochemical Properties

5.2.1   Appearance and Solubility

Retinoids appear as yellow-to-light orange crystalline or amorphous solids (Barua and Furr 1998; Gundersen and Blomhoff 2001). Retinol and retinyl esters are low-melting compounds (T_m at approximately 30°C for retinyl palmitate and approximately 60°C for the others) that may change into oil when the room temperature becomes warm (Combs 2007; Friedrich 1988). Retinoids are insoluble in water and soluble in organic solvents, especially diethyl ether, petroleum ether, chloroform, and acetone (Barua and Furr 1998; Gundersen and Blomhoff 2001). Retinoic acid solubility depends on the pH of the solvent: at pH values greater than its pK_a ($pK_a \approx 6$–8) retinoic acid is highly soluble in water (Gundersen and Blomhoff 2001).

Provitamin A carotenoids are pigments forming reddish-brown crystals and having melting points between 139 and 212°C (Combs 2007; Friedrich 1988). They are insoluble in water, very sparingly soluble in alcohol, slightly soluble in acetone and ether, and rather soluble in chloroform and benzene (Ball 2006).

5.2.2   Chemical stability

Light, heat, oxygen, and Lewis acids degrade both retinoids and carotenoids due to the extensive system of conjugated double bonds. Moreover, their decomposition in solution is faster than in the solid state.

UVA radiation (315−400 nm) is more destructive than UVB radiation (280−315 nm) (Failloux *et al* 2004). The photodegradation of retinol is associated with a strong yellow-green fluorescence and it is amplified by oxygen (Tolleson *et al* 2005). The main oxidation products are 5,6-epoxyretinol, 13,14-epoxyretinol, photocyclized products, and anhydrovitamin A (Failloux *et al* 2004). Retinyl esters are more stable towards oxidation than retinol, for this reason both retinyl palmitate and retinyl acetate have been commonly added in cosmetic and food products (Ball 2006). Oxidation of carotenoids involves epoxidation, cleavage to apocarotenals and apocarote-nones, and subsequent breakdown in low-molecular-weight compounds, some of which are volatile (Britton 1995; Rodriguez-Amaya and Kimura 2004). Oxidative interruption of the conjugated double-bond system is accompanied by a loss of color ("bleaching" of the carotenoid) and biological activity (Krinsky 1994).

Retinoids and carotenoids can undergo *cis−trans* isomerisation, especially at positions 9, 11, and 13 of their polyene chain (Ball 2006; Britton 1995). Acids, heat treatment, and exposure to light can promote this process (Britton 1995; Krinsky 1994; Rodriguez-Amaya and Kimura 2004). The traces of hydrochloric acid occurring in the chlorinated solvents as well as the amounts of organic acids released during the fruit squeezing are sufficient to induce stereoisomerisation (Ball 2006; Rodriguez-Amaya and Kimura 2004).

However, this conversion takes place to a larger extent during thermal processing: for instance, it has been verified that the isomerisation degree of all-*trans*-retinol to 13-*cis*-retinol increases with the severity of milk heat treatment (up to approximately 16% for UHT milk and 34% for sterilised milk) (Panfili *et al* 1998).

Strong acids cause quick dehydration of retinoids into anhydroretinoids (hydrocarbons with a *retro*-double-bond structure); this reaction, producing an intense blue color, is at the basis of the Carr–Price assay for vitamin A (Blatz and Estrada 1972). In alkaline environment, retinoids and carotenoids are stable enough to use hot saponification (Ball 2006) for their isolation from real matrices: retinyl esters are hydrolysed to free retinol for the determination of the total vitamin A, but a prolonged contact should be avoided. Unlike carotenes, xanthophylls are degraded by alkalis at high temperatures; overnight cold saponification could be a possible solution for their extraction from food matrices. In order to preserve retinoids and carotenoids as much as possible, it is always advisable to use an antioxidant (butylated hydroxytoluene, pyrogallol, or ascorbic acid) during their analysis (Ball 2006; Barua and Furr 1998).

5.3 Spectral Properties

The conjugated polyene system of retinoids and carotenoids is responsible for $\pi \rightarrow \pi^*$ transitions (K-bands), characterised by high molar-extinction coefficients (ε range: 30 000–100 000). All these compounds absorb strongly in the UV-Vis region of the spectrum: approximately 325–380 nm for retinoids (see Table 5.1) and 450 nm for carotenoids (see Table 5.2).

Both the maximum absorption wavelength (λ_{max}) and the molar-extinction coefficient (ε_{max}) of retinoids and carotenoids increase with the number of double bonds. For all-*trans* polyenes containing more than four conjugated double bonds, λ_{max} and ε_{max} can be predicted using the Fieser rule (Fieser 1950):

$$\lambda_{max} = 114 + 5\,M + n\,(48 - 1.7n) - 16.5\,R_{endo} - 10\,R_{exo} \qquad (1)$$

$$\varepsilon_{max} = \left(1.74 \times 10^4\right)n \qquad (2)$$

where n is the number of double conjugated bonds; M is the number of alkyl or alkyl-like substituents on the conjugated system; R_{endo} is the number of rings with endocyclic double bonds in the conjugated system; and R_{exo} is the number of rings with exocyclic double bonds.

The energy of the electronic transition is highly dependent on the solvent polarizability: the bathochromic shift becomes more significant as the refractive index of solvent increases. Spectra of polyenes are essentially the same in pentane, hexane, petroleum ether, diethyl ether, methanol, ethanol,

Table 5.1   Maximum wavelength (λ_{max}), molar extinction coefficient (ε), and percent extinction coefficient ($E_{1cm}^{1\%}$) of some important retinoid.

Retinoid	Solvent	λ_{max}	ε	$E_{1cm}^{1\%}$	Reference
all-*trans* retinol (Vitamin A₁)	2-propanol	325	52300	1830	(Boldingh *et al* 1951)
	hexane	325	51770	1810	(Hubbard *et al* 1971)
13-*cis* retinol	ethanol	328	48305	1689	(Robeson *et al* 1955b)
11-*cis* retinol	ethanol	319	34890	1220	(Hubbard *et al* 1971)
	hexane	318	34320	1200	(Hubbard *et al* 1971)
9-*cis* retinol	ethanol	323	42300	1477	(Robeson *et al* 1955)
11,13-di-*cis* retinol	ethanol	311	29240	1024	(Barua and Furr 1998)
9,13-di-*cis* retinol	ethanol	324	39500	1379	(Robeson *et al* 1955b)
Anhydroretinol	ethanol	371	97820	3650	(Gundersen and Blomhoff 2001)
all-*trans* retinal	ethanol	383	42880	1510	(Boldingh *et al* 1951)
	hexane	368	48000	1690	(Boldingh *et al* 1951)
13-*cis* retinal	ethanol	375	35500	1250	(Boldingh *et al* 1951)
	hexane	363	38770	1365	(Boldingh *et al* 1951)
11-*cis* retinal	ethanol	380	24935	878	(Boldingh *et al* 1951)
	hexane	365	26360	928	(Boldingh *et al* 1951)
9-*cis* retinal	ethanol	373	36100	1270	(Robeson *et al* 1955a)
11,13-*cis* retinal	ethanol	373	19880	700	(Barua and Furr 1998)
9,13-*cis* retinal	ethanol	368	32380	1140	(Barua and Furr 1998)
all-*trans* retinoic acid	ethanol	350	45300	1510	(Robeson *et al* 1955b)
13-*cis* retinoic acid	ethanol	354	39750	1325	(Robeson *et al* 1955b)
9-*cis* retinoic acid	ethanol	345	36900	1230	(Robeson *et al* 1955b)
11,13-di-*cis* retinoic acid	ethanol	346	25890	863	(Gundersen and Blomhoff 2001)
9,13-di-*cis* retinoic acid	ethanol	346	34450	1150	(Gundersen and Blomhoff 2001)
5,6-epoxyretinol	ethanol	310	73140	2422	(Gundersen and Blomhoff 2001)
5,6-epoxyretinal	ethanol	365	45330	1511	(Gundersen and Blomhoff 2001)
5,6-epoxyretinoic acid	ethanol	338	45280	1442	(Gundersen and Blomhoff 2001)
all-*trans* retinyl acetate	ethanol	325	51180	1560	(Barua and Furr 1998)
all-*trans* retinyl palmitate	ethanol	325	49260	940	(Gundersen and Blomhoff 2001)
all-*trans* 3,4-didehydroretinol (Vitamin A₂)	ethanol	350	41320	1455	(Gundersen and Blomhoff 2001)
13-*cis* 3,4-didehydroretinol	ethanol	352	39080	1376	(Koefler and Rubin 1960)
9-*cis* 3,4-didehydroretinol	ethanol	348	32460	1143	(Koefler and Rubin 1960)
9,13-di-*cis* 3,4-didehydroretinol	ethanol	350	29950	1030	(Koefler and Rubin 1960)

Table 5.1 (*Continued*)

Retinoid	Solvent	λ_{max}	ε	$E_{1cm}^{1\%}$	Reference
all-*trans* 3,4-didehydroretinal	ethanol	401	41450	1470	(Gundersen and Blomhoff 2001)
all-*trans* 3,4-didehydroretinoic acid	ethanol	370	41570	1395	(Gundersen and Blomhoff 2001)

Table 5.2 UV-Vis data for the identification of the main provitamin A carotenoids and their *cis* isomers.

Provitamin A carotenoid	λ_{cis} (nm)	λ_I, $\lambda_{II}/\lambda_{max}$, λ_{III} (nm)	%(III/II)	Q-ratio	Reference
all-*trans* β-carotene	–	431, 455, 482	26	–	(Gentili and Caretti 2011)[a]
9-*cis* β-carotene	347	428, 449, 478	23	5.7	(Gentili and Caretti 2011)[a]
13-*cis* β-carotene	345	428, 448, 472	5	3	(Gentili and Caretti 2011)[a]
15-*cis* β-carotene	337	420, 449, 472	10	1.6	(De Rosso and Mercadante 2007)[b]
all-*trans* β-carotene 5,6-epoxide	–	422, 445, 472	–	–	(Britton *et al* 2004)[c]
all-*trans* β-cryptoxanthin	–	425, 449, 476	25	–	(Britton *et al* 2004)[d]
all-*trans* β-cryptoxanthin 5,6-epoxide	–	418, 444, 473	–	–	(Britton *et al* 2004)[e]
all-*trans* α-carotene	–	422, 445, 473	55	–	(Britton *et al* 2004)[d]
9-*cis* α-carotene	330	418, 441, 467	–	9.8	(Emenhiser *et al* 1996)[f]
13-*cis* α-carotene	332	416, 438, 465	–	2.1	(Emenhiser *et al* 1996)[f]
all-*trans* α-cryptoxanthin	–	421, 445, 475	60	–	(Britton *et al* 2004)[c]
all-*trans* β-zeacarotene	–	404, 428, 452	–	–	(Britton *et al* 2004)[c]
all-*trans* γ-carotene	–	437, 462, 494	40	–	(Britton *et al* 2004)[c]
all-*trans* γ-carotene 1′,2′-epoxide	–	431, 456, 488	–	–	(Britton *et al* 2004)[c]
all-*trans* β-apo-8′-carotenal	–	456	–	–	(Britton *et al* 2004)[e]

[a]Linear gradient of methanol and isopropanol: hexane (50:50, v/v); [b]linear gradient of methanol and methyl-terbutyl ether; [c]hexane; [d]hexane or petroleum; [e]ethanol; [f]mobile phase of methyl-terbutyl ether: methanol (11:89, v/v)

and acetonitrile, but are proportionally displaced when the solvent is acetone, chloroform, dichloromethane, benzene, pyridine, and carbon disulphide (Britton *et al* 2004; Fieser 1950). The calculated λ_{max} shows a good agreement with the observed λ_{max} when the UV-Vis spectra are recorded in aliphatic hydrocarbons and in solvents with comparable refractive indexes. The calculated λ_{max} of all-*trans*-retinol, with five conjugated double bonds, is 325 nm ($n = 5$, $M = 6$, $R_{endo} = 1$), whereas that of all-*trans*-β-carotene, with 11 conjugated double bonds, is 453 nm ($n = 11$, $M = 10$, $R_{endo} = 2$). These values are very close to those listed in Tables 5.1 and 5.2. Spectra of retinol and retinyl esters are indistinguishable as the polyene chromophore is not disturbed. Also the presence of hydroxyl groups on the molecule does not influence the λ_{max} value; thus, the absorption spectrum of all-*trans*-β-cryptoxanthin is the same as all-*trans*-β-carotene. Nevertheless, all-*trans*-retinoic acid, possessing five double bonds conjugated with C=O of the

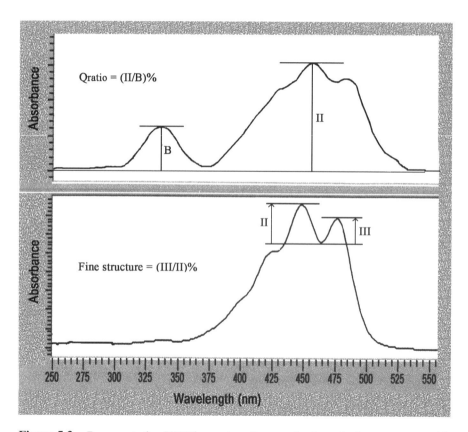

Figure 5.3   Representative UV-Vis spectra of a generic *cis*- and all-*trans*-carotenoid. The intensity of *cis*-peak (peak B) and the spectral fine structure of a carotenoid can be numerically expressed as indicated in Figure. The calculated values are used for the tentative identification of geometric isomers (see also Table 5.2).

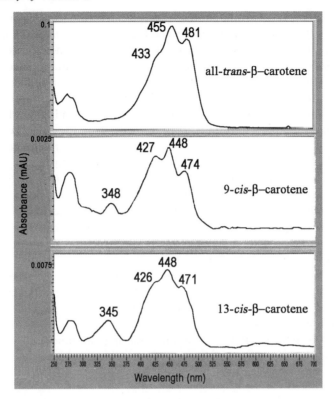

Figure 5.4 Absorption spectra of some geometric isomers of β-carotene in hexane. The Figure shows the characteristic UV-Vis spectra of all-*trans*-, 9-*cis*-, and 13-*cis*-β-carotene, provitamin A carotenoids occurring in bovine milk (A paper on the vitamin and carotenoid characterisation of milk samples from different animal species will be submitted for publication in June). Their tentative identification is mainly based on the Q-ratio (relative intensity of *cis*-peak), hypsochromic shift, and reduction in the spectral fine structure.

carboxyl group, absorbs at approximately 350 nm in ethanol (see Table 5.1). Moreover, its absorption spectrum is affected by the pH value of solution: above pH 7.0, a hypsochromic shift displaces λ_{max} at 337 nm (Barua and Furr 1998). For polyenes with an aldehyde or acid functional group, the λ_{max} values can be calculated using a modified Fieser rule (Fieser 1950):

$$\lambda_{max} = 114 + 5\ M + n\ (55.5 - 2.1n) - 16.5\ R_{endo} \qquad (3)$$

The λ_{max} value is instead lowered by an epoxide substituent (see Tables 5.1 and 5.2). In this case, the empirical equation used for its calculation is (Fieser 1950):

$$\lambda_{max} = 118 + 5\ M + n\ (48 - 1.7n) - 23\ R_{endo} \qquad (4)$$

Most carotenoids exhibit a typical three-banded absorption spectrum: its shape (fine structure) and absorption maxima (λ_I, λ_{II} or λ_{max}, λ_{III}) allow the identification of the chromophore. The fine structure can be numerically expressed as the percentage ratio between the heights of the peaks II and III, *i.e.* (III/II)%, measured from minimum (Figure 5.3). Geometrical isomers can be identified comparing their spectra with that of the related all-*trans* carotenoid (Britton *et al* 2004). The differences consist of a reduction in fine structure, a hypochromic effect, a hypsochromic shift of λ_{max} (approximately 2–6 nm for mono-*cis* isomers, 10 nm for di-*cis* isomers, and 50 nm for poli-*cis* isomers), and the appearance of a *cis* peak (peak B) in the near-UV spectrum between 330–350 nm. The relative intensity of the *cis* peak is represented by the Q ratio (Corts *et al* 2004), *i.e.* the percentage ratio between the heights of the peaks II and B, measured from the baseline of spectrum (Figure 5.3). These parameters are all relevant to establish the possible *cis* double-bond position in the chromophore; every one of them is a function of the molecular shape, becoming more appreciable as the double bond is closer to the centre of the polyene chain (see Figure 5.4 and compare the structures of all-*trans*-, 9-*cis*-, and 13-*cis*-β-carotene depicted in Figure 5.2).

Under UV irradiation, retinol, 3,4-didehydroretinol, and retinyl esters show a strong native fluorescence (Sobotka *et al* 1943; 1944); in particular, vitamin A_1 fluoresces brilliant yellow and vitamin A_2 brownish orange. The wavelength of excitation maxima is in the range 324–328 nm, whereas emission occurs in the range 470–490 nm (Ball 2006; Gundersen and Blomhoff 2001). Retinal, retinoic acid, most synthetic retinoids, and provitamin A carotenoids do not fluoresce to any significant extent (Ball 2006).

5.4   Mass Spectrometry

The technique of choice for the ionisation of retinoids and carotenoids is the atmospheric pressure chemical ionisation (APCI) source in positive ion mode.

The most abundant ion species observed for retinol is the dehydrated pseudomolecular ion $[MH-H_2O]^+$ at *m/z* 269. Dehydrated retinol is also base peak for retinyl esters which, in the APCI source, undergo a cleavage of the ester linkage; other detected ions with less intensity are $[M+H]^+$, $[M]^{+\cdot}$, $[M-H]^+$ (Figure 5.5). Retinal and retinoic acid generate a series of peaks related to $[M+H]^+$, $[M]^{+\cdot}$, $[MH-H_2O]^+$, and $[MH-H_2O-CH_3]^+$ (data unpublished).

Carotenoids give a series of ions ($[M+H]^+$, $[M]^{+\cdot}$, $[M-H]^+$), but the protonated pseudomolecular ion is often base peak on the APCI spectrum; moreover, the provitamin A xanthophylls loss their hydroxyl group to give an intense $[MH-H_2O]^+$ ion (Gentili and Caretti 2011).

Summary Points

- This Chapter focuses on is the physicochemical properties of the vitamin A-active compounds.
- Vitamin A-active compounds are fat-soluble and low-melting compounds.

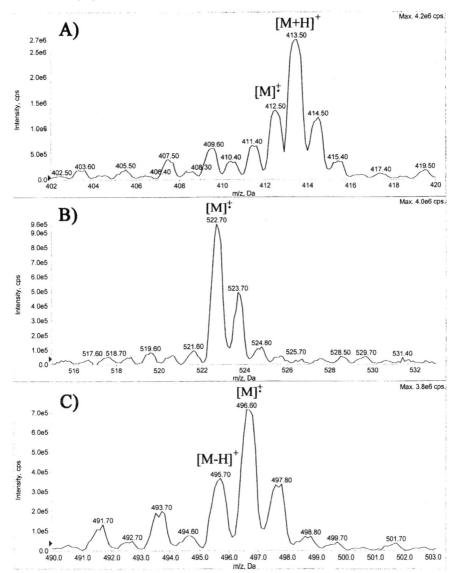

Figure 5.5 Mass spectra of some retinyl esters, acquired by the atmospheric pressure chemical ionisation (APCI) source in positive ion mode. Besides dehydrated retinol, retinyl esters generate different ion species characterized by lower intensity. (A–C) The mass spectra of retinyl caprylate, retinyl palmitoleate and retinyl myristate along with the generated ion species.

- Retinoids are diterpenes (C_{20} compounds) absorbing in the UV region.
- Provitamin A carotenoids are tetraterpenes (C_{40} compounds) having melting points higher than retinoids and showing a characteristic three-peak UV-Vis spectrum.

- Retinoids and carotenoids are thermolabile, photosensitive and easily attacked by oxidants because of their electron-rich polyene chain.
- The identification of a provitamin A *cis* isomer is based on the appearance of a *cis* peak in the near-UV spectrum (330–350 nm): its intensity gets greater as the *cis* double bond is nearer to the centre of the chromophore.
- Retinoids and provitamin A carotenoids are all detectable in positive atmospheric pressure chemical ionisation (APCI) with excellent intensity.

Key Facts of Retinoids and Carotenoids

- The discovery of vitamin A dates back to 1914 when Elmer V. McCollum and M. Davis found a fat-soluble growth factor in butter and egg yolk.
- In 1916, McCollum introduced the term "fat-soluble A" to distinguish it from other nutritional factors named "water-soluble B".
- In 1917, its involvement in eye diseases in children was already known.
- In 1920, vitamin A activity was observed in β-carotene
- In 1937, J.R. Edisbury, R.A. Morton, and G.W. Simpkins isolated vitamin A_2 from liver of freshwater fishes.
- In 1985, vitamin A_2 was identified in human skin.
- Preformed vitamin A is represented by a group of compounds known as retinoids, which are found in foods from animals (cod liver oil, eggs, butter, and milk).
- Precursors of vitamin A are the provitamin A carotenoids, occurring in yellow, orange, and green vegetables (sweet potato, carrot, spinach, and broccoli).
- Ingested retinyl esters and provitamin A carotenoids are converted into retinol in the small intestine, stored in the liver and from here into the visual pigment 11-*cis*-retinal and the humoral effector all-*trans*-retinoic acid.
- An inadequate vitamin A nutritional status is the leading cause of xerophthalmia, reduced immunity, increased risk of respiratory infections, and hematological disorders.
- In animals, vitamin A deficiency has been associated with a higher incidence of cancer and increased susceptibility to chemical carcinogens.

Definitions of Words and Terms

Atmospheric Pressure Chemical Ionisation (APCI): APCI is a technique used in mass spectrometry to ionise polar compounds *via* acid–base reactions in gas-phase. Liquid effluent containing the sample is vaporized by a pneumatic heated nebulizer and ionised by electron impact in the corona discharge region in order to generate plasma (gas consisting of ions from nebulizer gas and solvent). Plasma ions react with molecules of the interested compound *via* a mechanism of proton transfer (positive ions) or proton abstraction (negative ions). Substances with basic or acidic characteristics produce protonated or deprotonated molecular ions, respectively.

Bathochromic shift: Bathochromic shift is also known as red shift. It is the shift of absorption to a longer wavelength due to a substitution or solvent effect.

Carotenoids: Carotenoids include tetraterpenoids (with a C_{40} skeleton), polyterpenoids (with C_{45} and C_{50} skeleton), and terpenoids in which the C_{40} skeleton has been shortened by the removal of carbon atoms from the ends (apocarotenoids) or from the middle (norcarotenoids) of the molecule. Hydrocarbon carotenoids, known as carotenes, can be acyclic (*e.g.* lycopene), monocyclic (*e.g.* γ-carotene), or dicyclic (*e.g.* α-carotene, β-carotene). Oxygenated derivates are named xanthophylls; common oxygen functions are hydroxy (*e.g.* α-cryptoxanthin, β-cryptoxanthin), keto (*e.g.* canthaxanthin), epoxy (*e.g.* violaxanthin), aldheyde (*e.g.* β-apo-8'-carotenal), and carboxy groups (*e.g.* crocetin).

Carr-Price assay: This is a test for the quantitative determination of vitamin A based on its reaction with antimony trichloride in chloroform. The first step is dehydration of retinol to anhydrovitamin A, followed by the addition of the Lewis acid to the terminal ($\lambda_{max} = 619$ nm) and endocyclic ($\lambda_{max} = 586$ nm) double bond. The two brilliant blue complexes can be measured by colorimetry.

Chromophore: A functional group able to have characteristic electronic transitions. It is a covalently unsaturated group such as $C=C$, $C=O$, and NO_2.

Hyperchromic effect: Hyperchromic effect is an increase in absorption intensity.

Hypochromic effect: Hypochromic effect is a decrease in absorption intensity.

Hypsochromic shift: Hypsochromic shift is also known as blue shift. It is the shift of absorption to a shorter wavelength due to a substitution or solvent effect.

Molar extinction coefficient: The molar extinction coefficient (ε) is also known as molar absorptivity or molar absorption coefficient. It is a measurement of how intensely a chromophore absorbs at a given wavelength: ε values larger than 10^4 correspond to high-intensity absorptions, whereas ε values less than 10^3 are related to low-intensity absorptions. The Lambert–Beer law, $A = \varepsilon bc$, establishes a relationship between the absorbance (A), the concentration of solute (c), and path length through the sample (b). If c is expressed in *mol* L^{-1} and b in *cm*, units of $\varepsilon = \dfrac{A}{bc}$ will be *mol* $^{-1}$ *L cm* $^{-1}$. When the molecular weight of an absorbing species is unknown, the intensity of absorption may be expressed as $E_{1cm}^{1\%} = A_{1cm}^{1\%} = \dfrac{A}{cb}$, *i.e.* the absorbance (at a specific wavelength) for a 1% solution ($c = 1$ g/100 mL) measured in a 1 cm cuvette. $E_{1cm}^{1\%}$ is called the percentage extinction coefficient; sometimes is represented as $A_{1cm}^{1\%}$, specific absorbance coefficient. The relationship between the molar extinction coefficient and percent extinction coefficient is $\varepsilon = E_{1cm}^{1\%} \dfrac{\text{Molecular weight}}{10}$.

Retinoids: Retinoids are isoprenoids (diterpenes) with an 11-carbon polyene chain attached to a trimethylated cyclohexenyl ring (β-ionone ring) at the carbon-6 position in common. They differ in the functional group at the end of the side chain: retinol (alcohol group), retinal (aldehyde group), retinoic acid (carboxylic acid group), and retinyl esters (ester group with a fatty acid).

Terpenes: Terpenes are organic compounds containing isoprene (2-methylbutane) as repetitive unit. For this reason, they are also known as isoprenoids. In nature, the 30 000 known terpenes occur predominantly as hydrocarbons, but alcohols, aldehydes, and ketones were also found (terpenoids). Terpene family includes homologues differing in the number of subunits $(C_5)_n$: hemi- (C_5), mono- (C_{10}), sesqui- (C_{15}), di- (C_{20}), sester- (C_{25}), tri- (C_{30}), tetra- (C_{40}), and polyterpenes $(C_5)_n$ with $n > 8$. The isopropyl part of 2-methylbutane is defined as the head, and the ethyl residue as the tail. In mono-, sesqui-, di- and sesterterpenes, the isoprene units are linked to each other in a head-to-tail fashion; tri- and tetraterpenes contain two C_{15} or C_{20} units connected tail-to-tail in the centre.

List of Abbreviations

APCI	atmospheric pressure chemical ionisation
BHT	butylated hydroxytoluene
ε	molar extinction coefficient
$E_{1cm}^{1\%}$	percentage extinction coefficient
M	symbol used in the Fieser rule to represent the number of alkyl or alkyl-like substituents on a conjugated polyene
M_e	exact mass
M_w	molecular mass
R_{endo}	symbol used in the Fieser rule to represent number of rings with endocyclic double bonds in the conjugated system
R_{exo}	symbol used in the Fieser rule to represent number of rings with exocyclic bonds
T_m	melting point
UHT	ultra-high temperature
UVA	ultraviolet A
UVB	ultraviolet B
UV-Vis	ultraviolet-visible

References

Bai, C., Twyman, M., Farré, G., Sanahuja, G., Christou, P., Capell, T., and Zhu, C., 2011. A golden era-pro-vitamin A enhancement in diverse crops. In Vitro Cellular and Developmental Biology – Plant. 47: 205–221. Available at: http://www.writescience.com/RMT%20PDFs/Bai_2011_IVCDBP.pdf. Accessed 26 December 2011.

Ball G. F. M., 2006. Vitamin A: Retinoids and the Provitamin A Carotenoids. In: Vitamins in Foods. Analysis, Bioavailability, and Stability.CRC Press, Taylor & Francis Group, Boca Raton, FL, 785 pp.

Barua, A.B. and Furr, H.C., 1998. Properties of Retinoids. Structure, Handling, and Preparation. Molecular Biotechnology. 10: 167–182.

Bennasir, H., Sridhar, S., and Abdel-Razek, T.T., 2010. Vitamin a ... From physiology to disease prevention. International Journal of Pharmaceutical Sciences Review and Research. 1: 68–73. Available at: www.globalresearchonline.net. Accessed 26 December 2011.

Blatz, P.E., and Estrada, A., 1972. Molecular behaviour of the Carr-Price reaction. Analytical Chemistry. 44: 570–573.

Boldingh, J., Cama, H.R., Collins, F.D., Morton, R.A., Gridgeman, N.T., Isler, O., Kofler, M., Taylor, R.J., Welland, A.S., and Bradbury, T., 1951. Pure all-*trans* vitamin A acetate and the assessment of vitamin A potency by spectrophotometry. Nature. 168: 598–600.

Britton, G., 1995. Structure and properties of carotenoids in relation to function. The FASEB Journal. 9: 1551–1558. Available at: http://www.fasebj.org/content/9/15/1551.full.pdf+html. Accessed 26 December 2011.

Britton, G., Liaaen-Jensen, S., and Pfander, H., 2004. Carotenoids Handbook. Birkähuser Verlag, Basel Switzerland, 660 pp.

Chandler, L.A., and Schwartz, S.J., 1987. HPLC Separation of cis-trans carotene isomers in fresh and processed fruits and vegetables. Food Chemistry. 52: 517–836.

Combs Jr, G.F., 2007. Chemical and physiological properties of vitamins. In: The vitamins: fundamental aspects in nutrition and health. Elsevier Academic Press, Burlington, MA, 608 pp.

Corts, C., Esteve, M.J., Frgola, A., and Torregrosa, F., 2004. Identification and quantification of carotenoids including geometrical isomers in fruit and vegetable juices by liquid chromatography with ultraviolet−diode array detection. Journal of Agricultural and Food Chemistry. 52: 2203–2212.

De Rosso, V.V., and Mercadante, A.Z., 2007. Identification and quantification of carotenoids, by HPLC-PDA-MS/MS, from Amazonian fruits. Journal of Agricultural and Food Chemistry. 55: 5062–5072.

Emenhiser, C, Englert, G., Lane C. Sander, L.C., Ludwig, B., and Schwartz, S.J., 1996. Isolation and structural elucidation of the predominant geometrical isomers of α-carotene. Journal of Chromatography A. 719: 333–343

Failloux, N., Bonnet, I., Perrier, E., and Baron, M.-H., 2004. Effects of light, oxygen and concentration on vitamin A_1. Journal of Raman Spectroscopy. 35: 140–147.

Fieser, L.F., 1950. Absorption spectra of carotenoids; structure of vitamin A. The Journal of Organic Chemistry. 15: 930–943.

Friedrich, W., 1988. Vitamin A and its provitamins. In: Vitamins. De Gruyter, Berlin, Germany, 1063 pp.

Furr, H.C., 2004. Analysis of retinoids and carotenoids: problems resolved and unsolved. The Journal of Nutrition. 134: 281S–285S.

Gentili, A. and Caretti, F., 2011. Evaluation of a method based on liquid chromatography-diode array detector-tandem mass spectrometry for a rapid and comprehensive characterization of the fat-soluble vitamin and carotenoid profile of selected plant foods. Journal of Chromatography A. 1218: 684–97.

Godoy, H.T., and Rodriguez-Amaya, D.B., 1994. Occurrence of cis-isomers of provitamin A in Brazilian fruits. Journal of Agricultural and Food Chemistry. 42: 1306–1313.

Gundersen, T.E., and Blomhoff, R., 2001. Qualitative and quantitative liquid chromatographic determination of natural retinoids in biological samples. Journal of Chromatography A. 935: 13–43.

Hubbard, R., Brown, P.K., and Bownds, D., 1971. Methodology of vitamin A and visual pigments. Methods in Enzymology. 18C: 615–653.

Koefler, M. and Rubin, S.H., 1960. Physicochemical assay of vitamin A and related compounds. Vitamins & Hormones. 18: 315–339.

Krinsky, N.I., 1994. The biological properties of carotenoids. Pure & Applied Chemistry. 66: 1003–1010

Lessin, W.J., Catigani, G.L., and Schwartz, S.J., 1997. Quantification of *cis-trans* isomers of provitamin A carotenoids in fresh and processed fruits and vegetables. Journal of Agricultural and Food Chemistry. 45: 3728–3732.

Olson, J.A., 1994. Absorption, transport, and metabolism of carotenoids in humans. Pure & Applied Chemistry. 66: 1011–1016. Available at: https://iupac.org/publications/pac/1994/pdf/6605x1011.pdf. Accessed 26 December 2011.

Panfili, G., Manzi, P., and Pizzoferrato, L., 1998. Influence of thermal and other manufacturing stresses on retinol isomerization in milk and dairy products. Journal of Dairy Research. 65: 253–260.

Paust, J., 1991. Recent progress in commercial retinoids and carotenoids. Pure and Applied Chemistry. 63: 45–58. Available at: http://www.iupac.org/publications/pac/pdf/1991/pdf/6301x0045.pdf. Accessed 26 December 2011.

Rodriguez-Amaya, D.B., and Kimura, M., 2004. Harvestplus handbook for carotenoid analysis. HarvestPlus Technical Monograph 2. Washington, DC and Cali, 57 pp. Available at: http://www.ifpri.org/sites/default/files/publications/hptech02.pdf. Accessed 27 December 2011.

Robeson, C.D., Blum, W.P., Dieterle, J.M., Cawley, J.D., and Baxter, J.G., 1955a. Chemistry of vitamin A. XXV. Geometrical isomers of vitamin A aldehyde and an isomer of its α-ionone analog. Journal of American Chemical Society. 77: 4120–4125.

Robeson, C.D., Cawley, J.D., Weisler, L., Stern, M.H., Edinger, C.C., and Checkak, A.J., 1955b. Chemistry of vitamin A. XXIV. The synthesis of geometric isomers of vitamin A via methyl-methylglutaconate1. Journal of American Chemical Society. 77: 4111–4119.

Sobotka, H., Kann, S., and Loewenstein, E., 1943. The fluorescence of vitamin A. Journal of the American Chemical Society. 65: 1959–1961.

Sobotka, H., Kann, S., Winternitz, W., and Brand, E., 1944. The fluorescence of vitamin A. II. Ultraviolet absorption of irradiated vitamin A. Journal of the American Chemical Society. 66: 1162–1164.

Tolleson, W.H., Cherng, S.-H., Xia, Q., Boudreau, M., Yin, J.J., Wamer, W.G., Howard, P.C., Yu, H., and Fu, P.P., 2005. Photodecomposition and phototoxicity of natural retinoids. International Journal of Environmental Research and Public Health. 2: 147–155.

CHAPTER 6

Nomenclature of Vitamin A and Related Metabolites

NIKETA A. PATEL

Research Scientist, James A. Haley Veterans Hospital And Assistant
Professor, Department of Molecular Medicine, College of Medicine,
University of South Florida, 13000 Bruce B. Downs Blvd., Research Service,
VAR 151, Tampa, FL 33612, USA
E-mail: npatel@health.usf.edu

6.1 Introduction

Vitamins are organic substances which may be fat-soluble or water-soluble and
are required in microquantities for the normal body growth, development and
function. Vitamins are not synthesized in the human body but occur naturally
in specific foods. In 1912, Casimir Funk (Funk 1912) discovered a substance
which when ingested prevented the disease beriberi. He named this substance
"vitamine" derived from the latin word *vita* for life and the chemical *amine* as
he presumed the substance was derived from an amine. The "e" was later
dropped from the name as these substances did not contain nitrogen and hence
were not amines.

In 1913, McCollum set up his laboratory and gave rats controlled diets to
explore the nutritional aspects which led to the discovery and isolation of a fat-
soluble substance A. By 1920, several substances had been identified which
were essential for growth and occurred naturally. The substances of this series
of factors were called vitamins and were given letter names such as A, B, *etc.*
by Drummond. Vitamin A and its active metabolites have crucial function in
vision, embryonic development, cell proliferation and differentiation, cell

Food and Nutritional Components in Focus No. 1
Vitamin A and Carotenoids: Chemistry, Analysis, Function and Effects
Edited by Victor R Preedy
© The Royal Society of Chemistry 2012
Published by the Royal Society of Chemistry, www.rsc.org

survival and apoptosis, as well as gene expression (Russell 2000). It regulates the transcription of its target genes as well as post-transcriptional alternative splicing of mRNA (Apostolatos *et al* 2010).

6.2 Nomenclature

The nomenclature policy stated below is based on the recommendations by the International Union of Pure and Applied Chemistry (IUPAC)-International Union of Biochemistry and Molecular Biology (IUB) Joint Commission on Biochemical Nomenclature (JCBN), Nomenclature of Retinoids (IUPAC-IUB 1983; 1982; 1984) and the skeletal formulae and rules are reproduced here with permission from IUPAC, the copyright holder.

The dietary forms of vitamin A are derived from either retinyl ester or caratenoids which serve as precursors for retinoids. The term provitamin A caratonoid is the generic descriptor of caratonoids exhibiting the biological activity of β-carotene and are named based on the rules of nomenclature of caratenoids (IUPAC-IUB 1971).

Vitamin A is a term used as a generic descriptor for retinoids that qualitatively exhibit the biological activity of retinol (IUPAC-IUB 1972). Chemically, vitamin A belongs to a class of compounds called retinoids. Retinoids consist of four isoprenoid units joined in a head-to-tail manner. Retinoids are derived from a monocyclic parent compound containing a six-carbon ring (B-ionone), five carbon=carbon double bonds and an isoform-specific side chain which contains a functional group at the terminus of the acyclic portion.

Vitamin A or retinol is the precursor of its two active metabolites: retinal (or retinaldehyde) and retinoic acid. The skeletal formula and numbering system is shown in Figure 6.1 in which C and H atom representation is omitted. It is further understood that positions 16, 17, 18, 19 and 20 are $-CH_3$.

The nomenclature of retinoids and its metabolites is based on the group present on position 15 (R) as shown in Table 6.1.

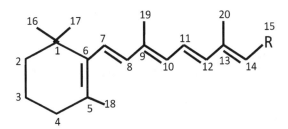

Figure 6.1 Representation of skeletal formula and numbering of all-*trans* retinoids. The designation of groups at "R" at position 15 are shown in Table 6.1. The *cis* double bonds are also indicated. Adapted from IUPAC-IUB guidelines (4-6) with permission.

Table 6.1 Nomenclature of vitamin A and its metabolites is based on the group (R) on position 15 of the parent compound.

	R=
Retinol (vitamin A; axerophtol)	-CH$_2$OH
Retinal (retinaldehyde)	-CHO
Retinoic acid (tretinoin)	-CO$_2$H

The retinoids shown above in Figure 6.1 have the polyene chain in the *trans* configuration. Other biologically active isoforms include the 9-*cis*, 11-*cis* and 13-*cis* configurations. An example of the 13-*cis* configuration is shown in Figure 6.2.

Nomenclature of the compounds which have modifications to the stereoparent are named with the prefix as "hydro" or dehydro" along with the position number of the modified carbon atom on the parent compound. An example is shown in Figure 6.3. If both modifications occur together, then "dehydro" is cited before "hydro".

Nomenclature of compounds with functional substitution of position 15 "R" group of the parent compound is notated with the group name. For example, when R is –CH$_2$, it is named retinyl, or when R is =CH, it is named retinylidene. Retinoids with other modifications or which have substituted

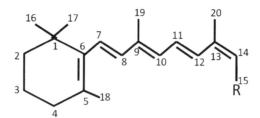

Figure 6.2 Representation of skeletal formula and numbering of 13-*cis* retinoids. The "R" group name is as shown in Table 6.1. Adapted from IUPAC-IUB guidelines (4-6) with permission.

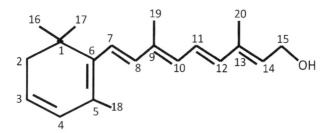

Figure 6.3 Nomenclature and structure of 3,4-didehydroretinol. Adapted from IUPAC-IUB guidelines (4-6) with permission.

derivatives use appropriate prefixes in accordance with the rules of organic chemistry. The prefix "seco" denotes the fission of the ring and is indicated by the position of the locant in the original numbering of the retinoid. The prefix "retro" denotes a shift in one position and is indicated by the position of the locant in the original numbering of the retinoid. The prefix "nor" denotes elimination of a CH_3, CH_2, CH or C group from the original retinoid and is denoted by the locant of the removed carbon atom.

Summary Points

- Vitamin A is a term used as a generic descriptor for retinoids that qualitatively exhibit the biological activity of retinol.
- Vitamin A or retinol is the precursor of its two active metabolites: retinal (or retinaldehyde) and retinoic acid.
- The group (R) on position 15 in the parent retinoid class of compounds specifies the name. Hence, retinol has $-CH_2OH$ on this position; retinal (or retinaldehyde) has $-CHO$ on this position; and retinoic acid has $-COOH$ on this position.
- Retinoids may have functional substitutions on position 15.
- Retinoids can differ in hydrogenation levels.
- Modified and substituted derivatives of retinoids are named by use of prefixes.

Key Facts of IUPAC

- The IUPAC was founded in 1919 and is an international scientific body addressing universal chemical nomenclature and terminology.
- The IUPAC nomenclature is the world's most accepted and used nomenclature rules for chemical substances.
- The IUPAC nomenclature rules generate a systemic chemical name for each compound such that it is unique to that compound.
- 2011 was the "Year of Chemistry" and IUPAC organized several programs to involve scientists worldwide.
- The IUPAC partnered with the IUB to form the JCBN. This committee gathers expert advice from leading scientists around the world and recommends common policies for biochemical nomenclature.

Definitions of Words and Terms

cis: Atoms or groups are on the same side of the double bond.
Nomenclature: Name; designation; set or system of names or terms.
trans: Atoms or groups are on the opposite side of the double bond.

Vitamin: Group of organic substances essential in minute quantities for normal metabolism.

List of Abbreviations

IUB   International Union of Biochemistry and Molecular Biology
IUPAC   International Union of Pure and Applied Chemistry
JCBN   Joint Commission on Biochemical Nomenclature

Acknowledgements

Dr Renner, Executive Director of the IUPAC, kindly gave the permission to cite the nomenclature rules and skeletal formulas for this Chapter.

References

Apostolatos, H., Apostolatos, A., Vickers, T., Watson, J. E., Song, S., Vale, F., Cooper, D. R., Sanchez-Ramos, J. and Patel, N. A., 2010. Vitamin A metabolite, *all-trans* retinoic acid, mediates alternative splicing of protein kinase C δVIII isoform via the splicing factor SC35. Journal of Biological Chemistry. 285: 25987–25995.

Funk, C., 1912. The effect of a diet of polished rice on the nitrogen and phosphorus of the brain. Journal of Physiology. 44: 50–53.

IUPAC-IUB. 1971. IUPAC Commission on the Nomenclature of Organic Chemistry and IUPAC-IUB Commission on Biochemical Nomenclature: tentative rules for the nomenclature of carotenoids. Biochemistry. 10: 4827–4837.

IUPAC-IUB. 1972. Nomenclature policy: generic descriptors and trivial names for vitamins and related compounds. Journal of Nutrition. 102: 157–163.

IUPAC-IUB. 1982. Nomenclature of retinoids. Recommendations 1981. European Journal of Biochemistry. 129: 1–5.

IUPAC-IUB. 1983. Nomenclature of retinoids: recommendations 1981. IUPAC-IUB Joint Commission on Biochemical Nomenclature (JCBN). Archives of Biochemistry and Biophysics. 224: 728–731.

IUPAC-IUB. 1984. Nomenclature policy: generic descriptors and trivial names for vitamin A and related compounds. Journal of Nutrition. 114: 643–644.

Russell, R. M. 2000. The vitamin A spectrum: from deficiency to toxicity. American Journal of Clinical Nutrition. 71: 878–884.

Analysis

CHAPTER 7

Structural Analysis of Vitamin A Complexes with DNA and RNA

HEIDAR-ALI TAJMIR-RIAHI* AND PHILIPPE BOURASSA

Department of Chemistry-Biology, University of Québec at Trois-Rivières, C. P. 500, Trois-Rivières, Québec G9A 5H7, Canada
*E-mail: tajmirri@uqtr.ca

7.1 Introduction

DNA adducts have been widely used to identify the health hazards and to evaluate the dose–response relationship in human exposed to carcinogens and mutagenic compounds (Palli *et al* 2003). Dietary constituents of fresh fruits and vegetables might play a relevant role in DNA adduct formation by inhibiting enzymatic activities (Block 1992; Wargoviche, 1997). Antioxidant micronutrients have been shown to prevent DNA damage by polycyclic aromatic hydrocarbon and other carcinogens and to alter the expression of metabolic enzymes (Block 1992; Murata and Kawanishi 2000). Many studies have addressed the protective role of the antioxidant vitamins A, B and C against cancer and cardiovascular diseases (Slaga 1995). It has been suggested that the antioxidant activity of retinol, retinoic acid (Scheme 1) and β-carotene includes scavenging free radicals and preventing DNA damage (Omenn *et al* 1996; Willet 1994; Zheng *et al* 2004). The effect of vitamin A on cleavage of plasmid DNA has been recently reported (Szekely and Gates 2006). β-Carotene radical was found to intercalate DNA duplex (Kleinjans *et al* 2004). Vitamin A components, retinol and retinoic acid, are fat-soluble micronu-

Food and Nutritional Components in Focus No. 1
Vitamin A and Carotenoids: Chemistry, Analysis, Function and Effects
Edited by Victor R Preedy
© The Royal Society of Chemistry 2012
Published by the Royal Society of Chemistry, www.rsc.org

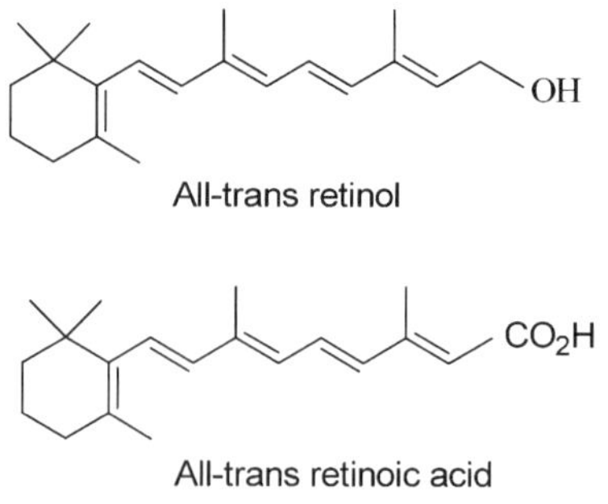

All-trans retinol

All-trans retinoic acid

Scheme 1

trients and critical for many biological processes, including vision, reproduction, growth and regulation of cell proliferation and differentiation (Wang *et al* 1997). As the components of vitamin A play a role in DNA adduct formation, the interactions of DNA and RNA with retinol and retinoic acid are of a major biological importance.

In this Chapter, we report the structural analysis of calf thymus–DNA and – tRNA complexes with retinol and retinoic acid by Fourier-transform infrared (FTIR), circular dichroism (CD) and fluorescence spectroscopic methods, as well as molecular modelling. Structural information regarding the retinoid-binding mode, binding constant and the effects of retinoid on DNA and RNA stability and secondary structure are provided here.

7.2 Analytical Methods

7.2.1 FTIR Spectroscopy

Infrared spectroscopy is widely used to determine the binding sites of a ligand to nucleic acids and their components. Drug-binding modes to different sites of DNA and RNA, such as major and minor grooves, external binding and intercalation, are determined using infrared spectroscopy and its derivative methods (Ahmed Ouameur and Tajmir-Riahi 2004; Ahmed Ouameur *et al* 2010). Infrared spectroscopy is used to locate the binding mode of vitamin A components (retinol and retinoic acid) to DNA and tRNA.

7.2.2 CD Spectroscopy

CD spectroscopy is often used to characterize the nature of DNA and RNA conformational changes upon ligand complexation. Conformational changes such as B to A to Z to C for DNA and RNA are detected in the presence of different ligands (N'soukpoé-Kossi *et al* 2008; 2009). CD spectroscopy is applied to analyse the structural changes such as DNA and RNA conformational transitions and biopolymer aggregation and particle formation upon vitamin A interaction.

7.2.3 Fluorescence Spectroscopy

Fluorescence quenching is considered as a powerful technique for measuring the binding affinity between ligands and nucleic acids. Fluorescence quenching is the decrease in the quantum yield of fluorescence from a fluorophore, induced by a variety of molecular interactions with quencher molecule(s) (Lakowicz 1999).

Assuming that there are (n) substantive binding sites for quencher (Q) on DNA or RNA (B_0), the quenching reaction can be shown as follows:

$$nQ + B \Leftrightarrow Q_n B \tag{1}$$

The binding constant (K_A), can be calculated as:

$$K_A = [Q_n B]/[Q]^n[B] \tag{2}$$

where $[Q]$ and $[B]$ are the quencher and retinoid concentration, respectively, $[Q_n B]$ is the concentration of non-fluorescent fluorophore–quencher complex and $[B_0]$ gives total retinoid concentration:

$$[Q_n B] = [B_0] - [B] \tag{3}$$

$$K_A = ([B_0] - [B])/[Q]^n[B] \tag{4}$$

The fluorescence intensity is proportional to the retinoid concentration as described:

$$[B]/[B_0] \propto F/F_0 \tag{5}$$

Results from fluorescence measurements can be used to estimate the binding constant of the retinoid–polynucleotide complex from eqn (4):

$$\log\,[(F_0 - F)/F] = \log\,K_A + n\log\,[Q] \tag{6}$$

The accessible fluorophore fraction (f) can be calculated by modified Stern–Volmer equation (Lakowicz 2006):

$$F_0/(F_0) = 1/f\,K[Q] + 1/f \tag{7}$$

where F_0 is the initial fluorescence intensity and F is the fluorescence intensity in the presence of quenching agent (or interacting molecule). K is the Stern–Volmer quenching constant, $[Q]$ is the molar concentration of quencher and f is the fraction of accessible fluorophore to a polar quencher, which indicates the fractional fluorescence contribution of the total emission for an interaction with a hydrophobic quencher (Lakowicz 2006). The plot of $F_0/(F_0 - F)$ *vs.* $1/[Q]$, yields f^{-1} as the intercept on y-axis and $(f\,K)^{-1}$ as the slope. Thus, the ratio of the ordinate and the slope gives K.

7.2.4   Molecular Modelling

Molecular modelling has been widely used to determine the ligand binding sites on both DNA and RNA (Privé *et al* 1991; Sussman *et al* 1978). The structures of DNA and tRNA were obtained from the Protein Data Bank (PDB; ID: 6TNA) and the retinoid three-dimensional structures were generated from PM3 semi-empirical calculations, using Chem3D Ultra 6.0. The docking study was carried out to locate the binding sites of vitamin A components on DNA and RNA.

7.3   Structural Characterization

7.3.1   FTIR Spectra of Retinoid–DNA and Retinoid–RNA Complexes

FTIR spectroscopy is widely used to characterize the nature of ligand bindings to DNA and RNA (Ahmed Ouameur and Tajmir-Riahi 2004; Ahmed Ouameur *et al* 2010; Alex and Dupuis 1989). Figures 7.1 and 7.2 present the infrared spectra and difference spectra of DNA and RNA complexes with retinol and retinoic acid. The assignments of typical IR bands related to DNA and tRNA are given in Tables 7.1 and 7.2. Marker DNA infrared bands were selected at 1717 (guanine), 1663 (thymine), 1609 (adenine), 1492 (cytosine), 1222 (PO_2) and 1088 cm^{-1} (PO_2) (Table 7.1) to detect retinoid–base and retinoid–phosphate bindings for both retinol and retinoic acid–DNA complexes (Figure 7.1). The difference spectra showed spectral changes (shifting and intensity changes) for the marked bands, related to DNA bases and the backbone phosphate group, indicating retinoid interaction with both DNA bases and the PO_2 group (Figure 7.1). Based on spectral changes observed, the major retinoid-binding sites were guanine and adenine N7 and thymine O2 sites, as well as the backbone PO_2 group (Figure 7.1).

Similarly, RNA marker bands at 1698 (guanine), 1660 (uracil), 1609 (adenine), 1485 (cytosine), 1240 (PO_2) and 1085 cm^{-1} (PO_2) (Table 7.2) were examined to detect the binding of retinoid to RNA bases and the backbone phosphate group (Figure 7.2). The difference spectra showed major spectral changes (shifting and intensity increase) for RNA marker bands related to RNA bases and the backbone phosphate group (Figure 7.2). Based on the spectral changes, retinoid interaction was mainly with guanine (N7 site), uracil (O2) and adenine (N7 site) as well as the backbone PO_2 group (Figure 7.2). The intensity increases of the RNA marker bands were similar for retinoid–RNA complexes (Figure 7.2).

7.3.2   CD Spectra of Retinoid–DNA and Retinoid–RNA Complexes

CD spectroscopy is a powerful tool used to analyse the conformational aspects of DNA, RNA and proteins. The CD spectrum of the free DNA is composed

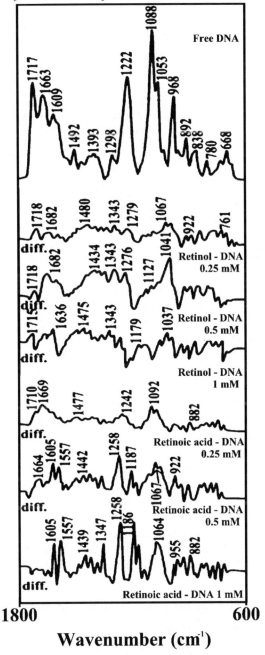

Figure 7.1 FTIR spectra and difference spectra [(DNA solution + retinoid solution) –(DNA solution)] (diff.) in the region of 1800–600 cm^{-1} for the free calf-thymus DNA and its retinol and retinoic acid complexes in aqueous solution at pH 7.4 with various retinoid concentrations and constant DNA concentration (12.5 mM).

Table 7.1   Principal infrared absorption bands, relative intensities and
                 assignments for calf-thymus DNA.

Wavenumber (cm^{-1})	Intensity[a]	Assignment
1717	vs	Guanine (C=O stretching)
1663	vs	Thymine (C2=O stretching)
1609	s	Adenine (C7=N stretching)
1578	sh	Purine stretching (N7)
1529	w	In-plane vibration of cytosine and guanine
1492	m	In-plane vibration of cytosine
1222	vs	Asymmetric PO_2^- stretch
1088	vs	Symmetric PO_2^- stretch
1053	s	C–O deoxyribose stretch
968	s	C–C deoxyribose stretch
893	m	Deoxyribose, B-marker
838	m	Deoxyribose, B-marker

[a]Relative intensities: m, medium; s, strong; sh, shoulder; vs, very strong; w, weak.

of four major peaks at 211 (negative), 221 (positive), 247 (negative) and 280 nm
(positive) (Figure 7.3). This is consistent with the CD spectrum of double-
helical DNA in B conformation (Kypr and Vorlickova 2002; Vorlickova 1995).
No major shifting of CD bands were observed (Figure 7.3) upon addition of
retinoid. The band at 280 nm showed no shifting upon retinoid complexation
(Figure 7.3). This was due to no DNA conformational changes upon retinol
and retinoic acid interactions. However, minor alterations of the intensity of
the CD band at 211, 221, 247 and 280 nm were observed due to DNA
aggregation and condensation at high retinoid contents (Figures 3A and 3B).

The CD spectrum of free tRNA is composed of four major peaks at 210
(negative), 222 (positive), 236 (negative) and 267 nm (positive) (Figure 7.3). This
is consistent with the CD spectrum of double-helical RNA in A conformation
(Kypr and Vorlickova 2002; Vorlickova 1995). No major shifting of the bands
was observed (Figure 7.3) upon addition of retinoid. However, as retinoid
concentration increased, major decrease in molar ellipticity of the band at 210,
222, 236 and 267 nm was observed due to RNA aggregation and particle
formation (Figures 7.3C and 7.3D). Since there was no major shifting of the
band at 267 nm, tRNA remains in the A-conformation.

7.3.3   Fluorescence Spectra and Stability of Retinoid-DNA and Retinoid-RNA Complexes

Since DNA or RNA is a weak fluorophore, the titration of retinoids was done
against various polynucleotide concentrations, using retinol and retinoic
excitation at 330–350 nm and emission at 450–500 nm (Gross *et al* 2000;
Kennedy *et al* 1995). When retinoid interacts with DNA and RNA, the
fluorescence may change depending on the impact of such an interaction on
the retinoid conformation, or *via* the direct quenching effect. The decrease in

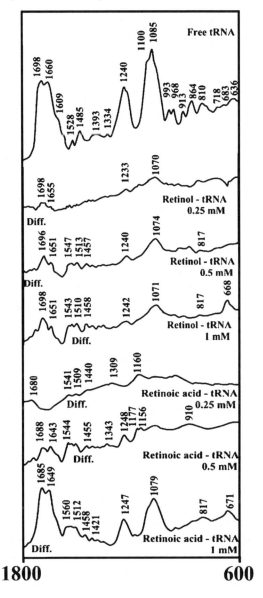

Wavenumber (cm⁻¹)

Figure 7.2 FTIR spectra and difference spectra [(RNA solution + retinoid solution) – (RNA solution)] (Diff.) in the region of 1800–600 cm⁻¹ for the free tRNA and its retinol and retinoic acid complexes in aqueous solution at pH 7.4 with various retinoid concentrations and constant DNA concentration (12.5 mM).

Table 7.2 Principal infrared absorption bands, relative intensities and assignments for Baker's yeast tRNA.

Wavenumber *(cm^{-1})*	*Intensity*[a]	*Assignment*[b]
1698	vs	Guanine (C=O, C=N stretching)
1660	vs	Uracil (C=O stretching)
1609	s	Adenine (C=N stretching)
1528	w	In-plane ring vibration of cytosine and guanine
1485	m	In-plane ring vibration of cytosine
1393	s	In-plane ring vibration of guanine in *anti*-conformation
1240	vs	Asymmetric PO$_2^-$ stretch
1085	vs	Symmetric PO$_2^-$ stretch
1063	s	C–O ribose stretch
968	m	C–C ribose stretch
913	m	C–C ribose stretch
864	m	Ribose–phosphodiester, A-marker
810	m	Ribose–phosphodiester, A-marker

[a]Relative intensities: m, medium; s, strong; sh, shoulder; vs, very strong; w, weak. [b]Assignments have been taken from the literature and relevant references are given in FTIR discussion.

fluorescence intensity of retinoid has been monitored at 485 nm (retinol) and 525 nm (retinoic acid) for retinoid–polynucleotide systems. The plot of $F_0/(F_0 - F)$ *vs.* 1/[polynucleotides] is shown in Figure 7.4. Assuming that the observed changes in fluorescence come from the interaction between retinoid and polynucleotides, the quenching constant can be taken as the binding constant of the complex formation. The K values given here are averages of four-replicate and six-replicate runs for retinoid–polynucleotide systems, each run involving several different concentrations of DNA or RNA (Figure 7.4). The binding constants obtained were $K_{\text{retinol (ret)–DNA}} = 3.0$ (± 0.50) $\times$ 10^3 M^{-1}, $K_{\text{retinoic acid (retac)–DNA}} = 1.0$ (± 0.20) $\times$ 10^4 M^{-1}, $K_{\text{ret–RNA}} = 2.0$ (± 0.50) $\times$ 10^4 M^{-1} and $K_{\text{retac–RNA}} = 6.0$ (± 1) $\times$ 10^4 M^{-1} (Figure 7.4). The association constants calculated for the retinoid–polynucleotide complexes suggest low-affinity retinoid–polynucleotide binding, compared to the other strong ligand–DNA and ligand–RNA complexes (Froehlich *et al* 2011; Marty *et al* 2009a; 2009b; Nafisi *et al* 2007). The *f* values obtained in Figure 7.4, suggest that DNA and RNA also interact with the fluorophore *via* hydrophobic interactions.

The number of retinoid molecules bound per polynucleotides (*n*) is calculated from log [($F_0 - F$)/F] = logK_S + *n* log [polynucleotides] for the static quenching (Charbonneau *et al* 2009; Mandeville *et al* 2010). The linear plot of log [($F_0 - F$)/F] as a function of log [polynucleotides] is shown in Figure 7.5. The *n* values from the slope of the straight line are 0.84 for retinol–DNA and 1.3 for retinoic acid–DNA, and 1.35 for retinol–RNA and 1.66 for retinoic acid–RNA (Figure 7.5).

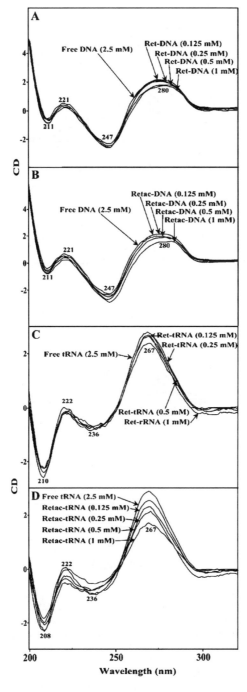

Figure 7.3 CD of the free calf-thymus DNA and tRNA and their retinoid complexes in aqueous solution with 2.5 mM DNA concentration and 0.125 to 1 mM retinoid concentrations at pH 7.4.

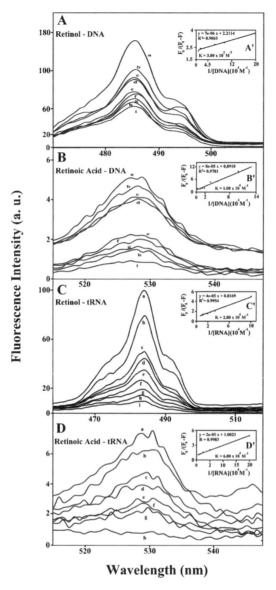

Figure 7.4 Fluorescence emission spectra of retinoid–DNA and retinoid–tRNA systems in 10 mM Tris/HCl buffer pH 7.4 at 25°C. (**A**) Retinol–DNA: (a) free retinol (30 μM), (b–h) with retinol–DNA complexes at 5, 10, 15, 70, 90, 100 and 140 μM with (i) free DNA (140 μM). (**B**) Retinoic acid–DNA: (a) free retinoic acid (30 μM), (b–h) with acid–DNA complexes at 5, 10, 15, 70, 90, 100 and 140 μM with (i) free tRNA (140 μM). (**C**) Retinol–tRNA: (a) free retinol (30 μM), (b–h) with retinol–RNA complexes at 5, 10, 15, 70, 90, 100 and 140 μM with (i) free tRNA (140 μM). (**D**) Retinoic acid–RNA: (a) free retinoic acid (30 μM), (b–g) with retinoic acid–RNA complexes at 10, 15, 70, 90, 100 and 140 μM with (h) free tRNA (140 μM).

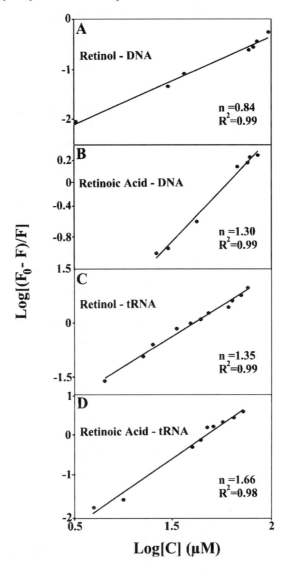

Figure 7.5 The plot of log $(F_0 - F)/F$ as a function of log [polynucleotides] for calculation of number of bound retinoid molecule (n) in retinoid–polynucleotide complexes: (**A**) retinol–DNA, (**B**) retinoic acid–DNA, (**C**) retinol–tRNA and (**D**) retinoic acid–tRNA complexes.

7.3.4 Docking Studies

Molecular modelling is reported based on the spectroscopic data for retinoid–DNA and retinoid–tRNA complexes. The docking results presented in Figure 7.6 and Table 7.3 showed that retinol is surrounded by G7, T8, C9, T10, C11, C12, A15, G16, A17, C18 and C19 with the binding energy of −4.52

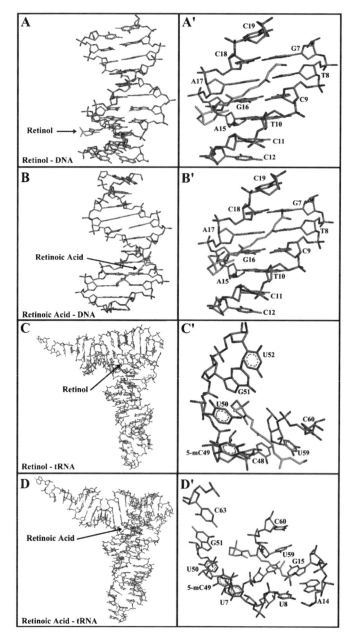

Figure 7.6   The retinoids are shown in green color. (**A**) DNA (PDB ID: 2K0V) in sphere-filling model with the retinol binding site in sticks (**A'**). (**B**) DNA in sphere-filling model with the retinoic acid binding site in sticks (**B'**). For tRNA (PDB ID: 6TNA), (**C**) shows RNA in sphere-filling model with the retinol binding sites represented in sticks (**C'**). For retinoic acid–tRNA shows RNA in sphere-filling model (**D**) with the retinoic acid binding sites represented in sticks (**D'**).

Table 7.3 Ribonucleotides in the vicinity of retinol and retinoic acid with DNA (PDB ID: 2K0V) and tRNA (PDB ID: 6TNA) with the free binding energies of the docked complexes.

Complex	Nucleobases involved in retinoid interactions	$\Delta G_{binding}$ $(kcal\ mol^{-1})$
Retinol– DNA	G7, T8, C9, T10, C11, C12, A15, G16, A17, C18, C19	−4.52
Retinoic acid– DNA	G7, T8, C9, T10, C11, C12, A15, G16, A17, C18, C19	−4.11
Retinol– tRNA	C48, 5-mC49, U50, G51, U52, U59, C60	−4.36
Retinoic Acid–tRNA	U7, U8, A14, G15, 5-mC49, U50, G51, U59, C60, C63	−4.53

kcal mol^{-1} and average binding distances of retinol–DNA bases are 1.5–2.5 Å (Figure 7.6A and Table 7.3). Retinoic acid is located in the vicinity of G7, T8, C9, T10, C11, C1, A15, G16, A17, C18 and C19 with the binding energy of − 4.11 kcal mol^{-1} and average binding distances of retinoic acid–DNA bases are 1.5–2.5 Å (Figure 7.6B and Table 7.3). Similarly, docking study was carried out for tRNA and retinoids. The models showed that retinol is surrounded by C48, 5-mC49, U50, G51, U52, U59 and C60 with the binding energy of −4.36 kcal mol^{-1} and average binding distances of retinol–RNA bases are 1.5–2.5 Å (Figure 7.6C and Table 7.3). Retinoic acid was located in the vicinity of U7, U8, A14, G15, 5-mC49, U50, G51, U59, C60 and C63 with the binding energy of −4.53 kcal mol^{-1} and average binding distances of retinoic acid–RNA bases are 1.5–2.5 Å (Figure 7.6D and Table 7.3). It is worth mentioning that the binding sites of retinol and retinoic acid with DNA and RNA are different.

Summary Points

In conclusion,

- Retinol and retinoic acid interact with guanine and adenine N7 sites and thymine O2 atom in the major and minor grooves, as well as the backbone phosphate group of DNA duplex.
- Retinoids bind tRNA *via* guanine and adenine N7 sites and uracil O2 atom as well as the backbone phosphate group.
- Hydrophobic contacts occurred between retinoid tails and DNA and RNA bases. Stronger affinity was observed for retinoid-RNA adducts than those of retinoid-DNA complexes.
- At high retinoid concentration, major DNA and RNA aggregation occurred, in the presence of both retinol and retinoic acid while DNA remained in B-family structure and RNA in A-form.
- FTIR spectroscopy was used to locate the binding mode of retinoids to DNA and RNA, while CD spectroscopy was used for conformational analysis of retinoid-polynucleotide complexes.

- Molecular modeling located the positions of each retinoid in the vicinity of DNA and RNA bases. These analytical methods were essential in elucidating the nature of vitamin A complexes with DNA and RNA.
- The structural information generated can be correlated to the biological implication of vitamin A adduct formation and its role in preventing DNA damage.

Key Facts

- Retinol and retinoic acid binding sites include guanine and adenine N7 and thymine (uracil) O2 atoms in the major and minor grooves of DNA and RNA duplexes.
- Retinoids form stronger complexes with RNA than DNA.
- Hydrophobic contacts occur between retinoid tails and DNA and RNA bases.
- DNA and RNA aggregation and particle formation occur in the presence of retinol and retinoic acid.
- DNA remains in the B-family structure and RNA in the A-form upon retinoid complexation.
- The structural information generated here can be correlated to the biological implication of vitamin A adduct formation and its role in preventing DNA damage.

List of Abbreviations

CD   circular dichroism
FTIR   Fourier-transform infrared
PDB   Protein Data Bank
ret   retinol
retac   retinoic acid

Acknowledgements

This work is supported by grant from Natural Sciences and Engineering Research Council of Canada (NSERC).

References

Ahmed Ouameur, A. and Tajmir-Riahi, H. A., 2004. Structural analysis of DNA interactions with biogenic polyamines and cobalt (III) hexamine studied by Fourier transform infrared and capillary electrophoresis. Journal of Biological Chemistry. 279: 42041–42054.

Ahmed Ouameur, A., Bourassa, P. and Tajmir-Riahi, H. A., 2010. Probing tRNA interaction with biogenic polyamines. RNA. 16: 1968–1976.

Alex, S. and Dupuis, P., 1989. FTIR and Raman investigation of cadmium binding by DNA. Inorganica Chimica Acta. 157: 271–281.

Block, G., 1992. The data support a role for antioxidants in reducing cancer risk. Nutrition Reviews 50: 207–213.

Charbonneau, D., Beauregard, M. and Tajmir-Riahi, H. A., 2009. Structural analysis of human serum albumin complexes with cationic lipids. Journal of Physical Chemistry B. 113: 1777–1784.

Froehlich, E., Mandeville, J. S., Kreplak, L. and Tajmir-Riahi, H. A., 2011. Bundling and aggregation of DNA by cationic dendrimers. Biomacromolecules. 12: 511–517.

Gross, E. A., Li, G. R., Ruuska, S. E., Boatright, J. H. and Nickerson, J. M., 2000. G239T mutation in repeat 1 of human IRBP: possible implications for more than one binding site in a single repeat. Molecular Vision. 6: 51–62.

Kennedy, M. W., Britton, C., Price, N. C., Kelly, S. M. and Cooper, A., 1995. The DvA-1 polyprotein of parasitic nematode *Dictyocaulus viviparous*. Journal of Biological Chemistry. 270: 19277–19281.

Kleinjans, J. C. S., van Herwijnen, M. H. M., van Maanen, J. M. S., Maas, L. M., Herlad, T. M. C., Moonen, J. and Briede, J. J., 2004. *In vitro* investigations into the interaction of β-carotene with DNA: evidence for the role of carbon-centered free radicals. Carcinogenesis. 25: 1249–1256.

Kypr, J. and Vorlickova, M., 2002. Circular dichroism spectroscopy reveals invariant conformation of guanine runs in DNA. Biopolymers (Biospectroscopy) 67: 275–277.

Lakowicz, J. R., 1999. Principles of Fluorescence Spectroscopy. 2nd edn, Kluwer/Plenum, New York, USA.

Mandeville, J.S., N'soukpoé-Kossi, C. N., Neault, J. F. and Tajmir-Riahi, H. A., 2010. Structural analysis of DNA interaction with retinol and retinoic acid. Biochemistry and Cell Biology. 88: 469–477.

Marty, R., N'soukpoe-Kossi, C. N., Charbonneau, D., Weinert, C. M., Kreplak, L. and Tajmir-Riahi, H. A., 2009a. Structural analysis of DNA complexation with cationic lipids. Nucleic Acids Research. 37: 749–757.

Marty, R., N'soukpoe-Kossi, C. N., Charbonneau, D., Kreplak, L. and Tajmir-Riahi, H.A., 2009b. Structural characterization of cationic lipid–tRNA complexes. Nucleic Acids Research. 37: 5197–5207.

Murata, M. and Kawanishi, S., 2000. Oxidative DNA damage by vitamin A and its derivative *via* superoxide generation. Journal of Biological Chemistry. 275: 2003–2008.

Nafisi, Sh., Saboury, A. A., Keramat, N., Neault, J.F., and Tajmir-Riahi, H. A., 2007. Stability and structural features of DNA intercalation with ethidium bromide, acridine orange and methylene blue. Journal of Molecular Structure. 827: 35–43.

N'soukpoé-Kossi, C. N., Ahmed Ouameur, A., Thomas, T., Shirahata, A., Thomas, T. J. and Tajmir-Riahi, H. A., 2008. DNA interaction with polyamine analaogues: a comparison with biogenic polyamines. Biomacromolecules. 9: 2712–2718.

N'soukpoé-Kossi, C. N., Ahmed Quameur, A., Thomas, T., Thomas T. J., and Tajmir-Riahi, H. A., 2009. Interaction of tRNA with antitumore polyamine analogues. Biochemistry and Cell Biology. 87: 621–630.

Omenn, G. S., Goodman, G. E., Thornquist, M. D., Balmes, J., Cullen, M. R., Glass, A., Keogh, J. P., Meyskens, F. L., Valanis, B., Williams, J. H., Barnhart, S. and Hammar, S., 1996. Effects of a combination of β-caroteine and vitamin A on lung cancer and cardiovascular disease. New England Journal of Medicine. 334: 1150–1155.

Palli, D., Masala, G.,Vineis, P., Garte, S., Saieva, C., Krogh, V., Panico, S., Tumino, R., Munnia, A., Riboli, E. and Marco, P., 2003. Biomarkers of dietery intake of micronutrients modulate DNA adduct levels in healthy adults. Carcinogenesis. 24: 739–746.

Privé, G. G.,Yanagi, K. and Dickerson, R. E., 1991. Structure of the B-DNA decamer C-C-A-A-C-G-T-T-G-G and comparaison with isomorphous decamers C-C-A-A-A-G-A-T-T-G-G and C-C-A-G-G-C-C-T-G-G. Journal of Molecular Biology. 217: 177–199.

Slaga, T. J., 1995. Nutrition and biotechnology in heart disease and cancer. Longenecker, J. B., Kritchevesky, D. and Drezner, M. K. (eds). Plenum Publishing Corporation, New York, USA, pp. 167–174.

Sussman, J. L., Holbrook, S. R., Warrant, R. W., Church, G. M. and Kim, S.-H., 1978. Crystal structure of yeast phenylalanine transfer RNA: I. Crystallographic refinement. Journal of Molecular Biology. 123: 607–630.

Szekely, J. and Gates, K. S., 2006. Noncovalent DNA binding and the mechanism of oxidative DNA damage by Fecapentaene-12. Chemical Research in Toxicology. 19: 117–121.

Vorlickova, M., 1995. Conformational transitions of alternating purine-pyrimidine DNAs in the perchlorate ethanol solutions. Biophysical Journal. 69: 2033–2043.

Wang, Y., Ichiba, M., Oishi, H., Iyadomi, M., Shono, N. and Tomokuni, K., 1997. Relationship between plasma concentrations and β-carotene and α-tocopherol and life-style factors and levels of DNA adducts in lymphocytes. Nutrition and Cancer. 27: 69–73.

Wargoviche, M. J., 1997. Experimental evidence for cancer preventive elements in foods. Cancer Letters. 114: 11–17.

Willet, W. C., 1994. Diet and health: what should we eat? Science. 264: 532–537.

Zheng, X., Chang, R. L., Cui, X.X., Avila, G. E., Lee, S., Lu, Y. P., Lou, Y. R., Shih, W.J., Lin, Y., Reutl, K., Newmark, H., Rabson, A. and Conney, A. H., 2004. Inhibitory effect of 12-O-tetradecanoylphorbol-13-acetate alone or in combination with all-*trans*-retinoic acid on the growth of LNCaP prostate tumors in immunodeficient mice. Cancer Research. 64: 1811–1820.

CHAPTER 8

Encapsulation of Vitamin A: A Current Review on Technologies and Applications

BEATRICE ALBERTINI*, MARCELLO DI SABATINO AND NADIA PASSERINI

Department of Pharmaceutical Sciences, Faculty of Pharmacy, University of Bologna, Via San Donato 19/2, 40127 Bologna, Italy
*E-mail: beatrice.albertini@unibo.it

8.1   Introduction

Vitamin A belongs to the group of fat-soluble vitamins and *in vivo* is found as the free alcohol [all-(*E*)-retinol] form or esterified with a fatty acid (European Pharmacopoeia, 2005), commonly indicated as retinoids. The dietary source of vitamin A is represented by many animal-based foods and commonly is referred to as preformed vitamin A. Plant-based foods do not contain vitamin A, but many fruits and vegetables contain carotenoids that serve as a pro-vitamin A source because they are converted into retinol during absorption. Among the carotenoids, β-carotene is the most biologically active, even if its adsorption rate is very low if compared to pre-formed vitamin A (Nagao 2009).

Figure 8.1 shows the chemical structure of different retinoids and β-carotene.

Food and Nutritional Components in Focus No. 1
Vitamin A and Carotenoids: Chemistry, Analysis, Function and Effects
Edited by Victor R Preedy
© The Royal Society of Chemistry 2012
Published by the Royal Society of Chemistry, www.rsc.org

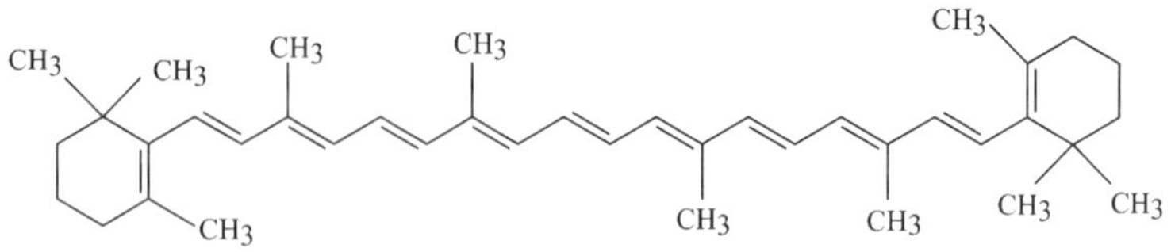

Figure 8.1 Chemical structures of natural sources of vitamin A and of β-carotene. R
= H, all-*trans*-retinol; R = CO–CH₃, retinyl acetate; R = CO–C₂H₅,
retinyl propionate; R = CO–C₁₅H₃₁, retinyl palmitate.

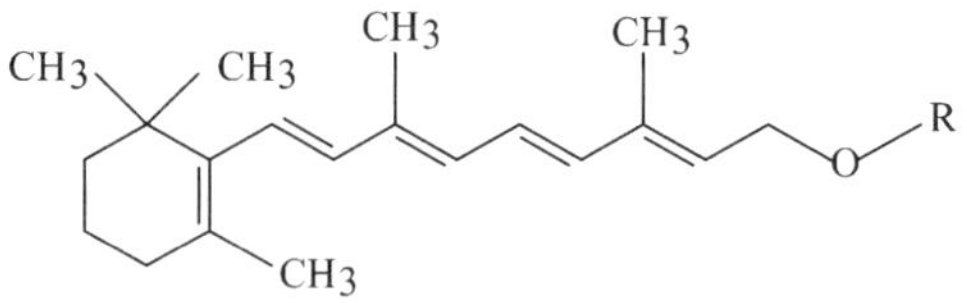

8.1.1 Vitamin A Supplementation: Aims and Problems

Vitamin A plays an important role to support several physiological functions.
The main scope of human vitamin A supplementation is the prevention or
treatment of vitamin A deficiency (VAD). VAD may cause blindness,
abnormal bone growth, modification of epithelial cell functionality and
increased risk of mortality for infections (Bates 1995). To treat VAD, vitamin
A is orally administered through several pharmaceutical preparations (soft
capsules, tablets, solutions or emulsions). On the other hand, VAD is quite
rare in rich countries, where the growing interest in wellness and healthy
products has favoured the development of fortified foods with vitamins and
where nutrient-rich foods are intended as the key for a healthy life. To this aim,
besides the previously described physiological effects, food fortification with
vitamin A is of great deal of interest due to its antioxidant properties and
consequently to its anticancer activity.

Vitamin A supplementation is also required in a properly balanced animal
feeding program (Bondi and Sklan 1984).

Therefore, although for different reasons, there is a real requirement of
vitamin A dietary supplementation both in human and in animals. The major
concern about vitamin A supplementation is related to the chemical instability.
Both retinoids and carotenoids are very sensitive to oxidation due to light,
oxygen, heat and humidity. To extend their half-lives, these compounds have
to be handled in dark containers under an inert atmosphere and kept at a
temperature between 8 and 15 °C (European Pharmacopoeia, 2005).

8.1.2 Why Encapsulate Vitamin A?

To achieve an effective food fortification, the nutrient should be stable under
proper conditions of storage and use. As a consequence, the preservation of

vitamin A chemical structure in medicines as well as in food/feedstuff represents the main objective for the agro/food, pharmaceutical and cosmetic industry.

Since several years encapsulation has emerged as a promising approach to preserve the integrity of vitamin A and β-carotene over time (Gonnet *et al* 2010). The potential use of this technology in the pharmaceutical industry has been considered since the 1960s (Bakan 1994), while the first encapsulation processes in the food industry were developed approximately in the 1970s (Gibbs *et al* 1999). The potential benefits of encapsulating active substances are significant and are all summarised in the Key Facts of Encapsulation Purposes (Bakan 1994). In the case of vitamin A, encapsulation has mainly been used to protect it from the surrounding environment, especially moisture, free radicals and UV. Some food constituents can also affect retinoids integrity.

In this Chapter the encapsulation strategies used to formulate and stabilize vitamin A or β-carotene are reviewed. Concerning food applications, it implies the use of food-grade or GRAS (Generally Recognised As Safe) materials by the Food and Drug Administration (FDA) for the formulation development (Loveday and Singh 2008). Therefore particular attention has been paid to the techniques and the approaches useful for dietary supplementation. Drug delivery systems (DDS) with pharmaceutical and/or cosmetic applications obtained with food-unsafe manufacturing processes are not taken into consideration.

8.2   Encapsulation Technologies

Encapsulation is a technological procedure that enable the formation of solid particles, which are called microparticles when the dimensions range in the μm size (10^{-6} m) and nanoparticles if they are in the submicron size (10^{-9} m). Microparticles with dimensions higher than 800 μm are better identified as beads. Based on their structure, micro/nanoparticles are usually divided in two groups: micro/nanocapsules and micro/nanospheres (Figure 8.2).

Other types of particles are biphasic or triphasic systems (Figure 8.3), such as oil in water (O/W) or oil in water in oil (O/W/O) micro/nanoemulsions, respectively, and liposomes. Further DDS are inclusion complexes with cyclodextrins (CDs) (Figure 8.4). Liposomes and inclusion complexes can be solid or liquid, whereas O/W or O/W/O emulsions are in the liquid form.

Table 8.1 shows the range of techniques used for vitamin A and β-carotene encapsulation.

The materials used for vitamin A encapsulation depend on the selected manufacturing procedure. Hydrophilic polymers include proteins such as gelatin and albumin, polysaccharides such as alginate, chitosan, starch and its derivatives (*i.e.* maltodextrines, CDs), cellulose and its derivatives, gums and carragenaan. Lipid-based carriers such as fatty acids, fatty alcohols, waxes, vegetable oils and phospholipids are commonly used in liquid emulsions, liposomal systems or in solid lipid nanoparticles (SLNs) prepared by hot homogenization with the aid of surfactants and co-surfactants. Lipids are also

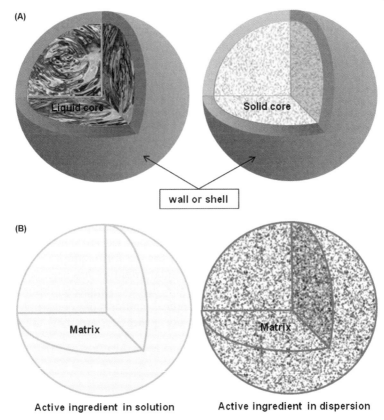

Figure 8.2   Schematic diagrams of: (A) monolayer micro/nanocapsules and (B) micro/nanospheres. The size and morphology of the particles can be tailored to achieve the desired product characteristics.

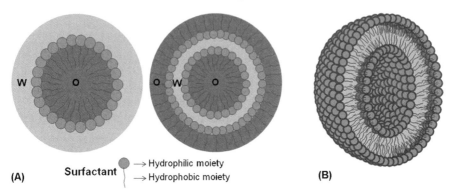

Figure 8.3   Schematic diagrams of: (A) O/W emulsion and O/W/O double emulsion and (B) liposome. The O/W and O/W/O emulsions are stabilised with the surfactant that stays at the interface of the phases, arranging the hydrophilic moiety toward the water phase and the hydrophobic one to the oil. In the O/W emulsion, the oily phase is dispersed within water. Liposomes are made up of at least a bilayer of the surfactant.

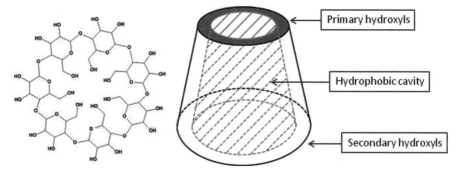

Primary hydroxyls

Hydrophobic cavity

Secondary hydroxyls

Figure 8.4 β-CD chemical structure and its schematic three-dimensional configuration. The three-dimensional configuration shows the hydrophobic cavity that can host the vitamin.

Table 8.1 Encapsulation methods used for vitamin A (1) and β-carotene (2) in dietary supplementation. The preparation methods of micro/nanoparticles are separated in physico-chemical and mechanical processes.

Physic-chemical methods	Mechanical methods
O/W (1, 2) and O/W/O (1) emulsification	Spray drying (1,2)
Coacervation or phase separation (1)	Fluid bed coating (1,2)
Ionical cross-linking or ionotropic gelation (1, 2)	Freeze drying (2)
Complexation with cyclodextrins (1,2)	High pressure homogenization or melt emulsification (1,2)
Inclusion in self-nanoemulsified drug delivery systems (SNEDDS) (1)	Extrusion (1)
Liposome entrapment (1)	

used to prepare self-emulsified drug delivery systems (SEDDS) and to coat preformed particles containing the nutrient in order to create an impermeable barrier to moisture.

8.3   Development of Dietary Supplements

The strategies reported in literature for the encapsulation of vitamin A and carotenoids can serve two main purposes: (i) stabilization and protection, and (ii) bioavailability enhancement. Table 8.2 summarises all of the characteristics of the encapsulated systems.

8.3.1   Stabilization and Protection

The majority of the methods listed in Table 8.1 involve as the first step the formation of O/W emulsion due to the lipophilic nature of the active

Table 8.2 Characteristics and stability of the micro/nano particulate dietary supplements. The data reported are taken from literature.

Manufacturing method	System	Active component	Amount (%, w/w)	Encapsulation efficiency (%)	Vitamin retention during storage	Reference
Coacervation	Polymeric beads with high amylose corn starch	Retinyl palmitate	–	–	–	Wurtburg et al (1970)
	Gelatin–acacia beads	Retinyl palmitate	5–12	83	–	Junyaprasert et al (2001)
Ionotropic gelation	Calcium alginate beads with dextran (alginate conc. 0.5–2%, w/w)	Vitamin A	10	–	–	Lim and Moss (1983)
	Beads of chitosan–alginate functionalized with palmitoyl chloride	β-Carotene	–	–	88% after 4 months at 23 °C and 56% RH	Han et al (2008)
	Chitosan–calcium alginate beads with BHT, EDTA, TiO_2 (alginate conc. 2.5%, w/w)	Retinyl palmitate	42	94	61% after 1 year at RT	Albertini et al (2010)
Ionotropic gelation plus fluid bed coating	100–450 μm calcium–alginate microcapsules (alginate conc. 1.5% w/w) coated with waxes, fats and polysaccharides	Vitamin A carotenoids	–	–	97% after 40 days at RT	Blatt et al (2003)

Table 8.2 (*Continued*)

Manufacturing method	System	Active component	Amount (%, w/w)	Encapsulation efficiency (%)	Vitamin retention during storage	Reference
Spray drying	Maltodextrin microspheres	β-Carotene	–	–	<50% after 6 weeks at RT and low RH level	Desobry et al (1997)
	Maltodextrin–glucose microspheres (20–40 μm)	β-Carotene	–	–	50% after 4 months at RT and low RH level	Desobry et al (1999)
	Starch-modified microcapsules with a sugar and vitamin E acetate at different concentrations	β-Carotene	–	–	80–88% after 14 days at 55 °C	Gellenbeck (1999)
	Trehalose–gelatin microspheres	β-Carotene	–	–	80% after 6 months at RT and 11% RH	Elizalde et al (2002)
	Starch-modified microcapsules	Retinyl acetate	–	–	95%, 87% and 75% after 1 month at RT and at 33%, 60% and 86% RH	Xie et al (2007)
Emulsification	O/W/O with antioxidants	Retinol	–	–	77% after 1 month at 50 °C	Yoshida et al (1999)
High-pressure homogenization	SLN	Retinyl palmitate	5	–	50% after 3 months	Jenning and Gohla (2001)
High-pressure homogenization	NLC with tocopherol	β-Carotene	0.045	–	Stable for 19 days at 20 °C	Hentschel et al (2008)
Thin film dehydration–rehydration	Liposomes	Retinol	–	97–99	80% after 8 days	Lee et al (2002)
Complexation with CDs plus coating	Microcapsules	Carotenoids	–	–	94% after 14 days at 55 °C	Reuscher et al (2004)

ingredient. Hence, vitamin A (or β-carotene) is solubilised in the oil-based carrier, which is then incorporated into an aqueous polymer solution forming an O/W emulsion that is stabilised by means of a suitable surfactant. The stability of the encapsulated retinoid depends on the droplet physical characteristics (*e.g.* solid/liquid state, interfacial layer thickness) and chemical characteristics (*e.g.* degree of unsaturation and length of fatty acids, presence of antioxidants) (McClements *et al* 2007). The O/W emulsion is then processed by means of one procedure and the impact of the selected technology on the possible degradation of the nutrient has to be considered.

8.3.1.1 *Polymer-based Micro/Nanoparticles*

Coacervation was the first technological procedure applied to vitamin A encapsulation and the protective shell of the microcapsules comprised different polymers such as gelatin, alginate, arabic gum, starch and especially high amylose starches (Bakan 1994; Wurtburg *et al* 1970). The toughness, strength, appearance of cracks and impermeability to oxygen of the particle shell are critical variables for the microcapsule stability and the drying method has to be as mild as possible. Gelatin–acacia complex coacervation has been investigated (Junyaprasert *et al* 2001) to prepare microcapsules of vitamin A palmitate. Freeze drying produced the best microcapsule morphology and a high encapsulation efficiency. However, the stability of these systems has not been investigated. Moreover, formaldehyde or glutaraldehyde used as a hardening agent give rise to a significant toxicity problem.

Ionotropic gelation was used to produce microcapsules or beads by dropping or spraying the vitamin A O/W emulsion. Gelled calcium alginate droplets were then formed, hardened for a suitable time and finally dried. Lim and Moss (1983) prepared beads using concentrations of alginate from 0.5 to 2% (w/w) with the addition of polysaccharides. A similar procedure was patented by Blatt *et al* (2003). Before drying, microcapsules were rinsed with an aqueous acidic solution causing shrinkage of the particles. Particles were then dried in a fluidized bed apparatus and subsequently *in situ* coated with a mixture of waxes, fats and polysaccharides. Stability was improved when microcapsules were washed in acidic solution. The microcapsules are suitable for tablet preparation, hard shell capsule filling and incorporation in different foods. Recently, Han and co-workers (2008) produced beads loaded with 0.5% (w/v) β-carotene, and chitosan and alginate functionalized with palmitoyl chloride increased the shell impermeabilty. Stability data are reported Table 8.2. With respect to beads formed without alginate functionalization, the novel DDS was more resistant in acidic stomach pH. The latter encapsulation approach was developed by Albertini *et al* (2010). Biodegradable polyelectrolyte complexes loaded with high amount of retinyl palmitate were prepared. Double-layer microcapsules with high vitamin loading (>40%, w/w) and high encapsulation efficiency were formulated. Moreover, the high concentration of alginate (2.5%, w/w) provided a compact microcapsule external layer, thus decreasing surface permeability. This system showed increased protection against degradation of the vitamin during 1 year of storage. This is a very important result, considering the poor shelf-life of the vitamin stabilized with butylated hydroxytoluene (BHT) (45 days).

Microencapsulation by means of spray drying may provide a protective network around the carotenoid and allows the drying of the product in the same equipment. The microcapsules produced by spray drying may have very small size (1–5 μm); therefore they present a large surface of exchange between oxygen and the labile moiety. An homogenized suspension of β-carotene in a maltodextrin aqueous solution was processed using different techniques and the extent of the carotenoid loss depended on the severity of the drying process (Desobry *et al* 1997). A further study demonstrated that the addition of glucose to maltodextrin extended the half-life of the encapsulated β-carotene, due to the permeability reduction of the membrane induced by glucose (Desobry *et al* 1999). A similar procedure, with a higher carotenoid concentration, was performed: first a 30% (w/w) β-carotene in oil dispersion was obtained, then milled and emulsified with the encapsulating solution containing starch, a sugar and an antioxidant. The emulsion was homogenized at high pressure and spray dried (Gellenbeck 1999). Probably, the sugar contributed to the formation of a glass-like crystalline coating with good oxidative protection. On the contrary, the encapsulation of β-carotene in a mixture of gelatin and trehalose, a disaccharide, showed a significant improvement in the stability due to the matrix amorphous form. However, the exposure of these particles to different relative humidity (RH) levels hardly influenced the recrystallization and thus the degree of β-carotene retention (Elizalde *et al* 2002). In a further study, the influence of RH on the retention of vitamin A acetate in spray-dried microcapsules was evaluated during storage (Xie *et al* 2007). Starch octenylsuccinate was employed as a wall microcapsule material; it was dissolved in warm water solution and emulsified with a 30% (w/w) solution of the vitamin dissolved in coconut oil. After high-pressure homogenization the emulsion was spray dried; the microcapsules were then stored at different RH levels for 1 month, showing a good vitamin retention (Table 8.2).

Microcapsules with a high carotenoid content were obtained by fluid bed coating in the bottom configuration (DeFreitas *et al* 2003). The carotenoid powder can be used as starting material and fluidized with the aid of hot air and then sprayed on with the coating material; otherwise, sugar spheres, as the inert material, are fluidized in the hot bed air and coated by the encapsulating solution containing carotenoids. The process provided a coating which protects the carotenoid from losses during the processing of the dietary supplement.

8.3.1.2 Lipid-based Micro/Nanoparticles

Yoshida *et al* (1999) studied the stability of retinol O/W, W/O and O/W/O emulsions, showing that the stability of retinol was highest in the latter system.

Moreover, the addition of antioxidants such as BHT or sodium ascorbate improved the stability of retinol from 57% up to 77% after 1 month at 50 °C. However, the low drug-loading capacity and the limited commercial applications of liquid multiple emulsions are the main drawbacks of such systems.

SLNs represent an innovative attempt to formulate retinoids as well as carotenoids. The interest in SLNs for food applications is related to the biocompatibility of the carrier system and to the absence of organic solvents during manufacturing. Different retinoids were encapsulated into SLNs (Jenning and Gohla 2001): a mixture of solid lipid was melted and the drug was added. This mixture was then dispersed in a hot surfactant solution, homogenized at high pressure and then cooled to room temperature (RT). The results showed that the loading capacity decreased with the polarity of molecules, whereas it increased by adding liquid lipids in the starting phase. The recovery of retinyl palmitate demonstrated a certain level of degradation (Table 8.2). More recently, nanostructured lipid carriers (NLCs) were prepared containing β-carotene in formulation with tocopherol (Hentschel *et al* 2008). NLCs differ from SLNs in the physical state of the lipids: they are solid at RT but are prepared with liquid and solid lipids. The particle dispersions stored at 20 °C showed aggregation due to coalescence after 9 weeks, whereas those stored in the fridge did not aggregate for at least 30 weeks. With regards to the stabilization of β-carotene, formulations were stable at least 11 days in dilution with demineralized water and 19 days as pure formulation at 20 °C. These nanoparticles can be applied as dyes in beverages.

Another attempt pursued to improve retinol stability is represented by liposomes (Lee *et al* 2002). 20% of retinol degraded after 8 days of storage, whereas free aqueous retinol completely degraded within 2 days. Their low entrapment efficiency and their reduced physical stability are further drawbacks. Moreover, for food applications, the choice of solvents in the liposome manufacturing is restricted. They are actually used in dairy products to encapsulate proteinases, lipases and vitamin D (Taylor *et al* 2005).

8.3.2   Bioavailability Enhancement

In addition to their chemical instability, retinoids are characterized by poor solubility in water, which leads to unfavourable pharmacokinetics and thus poor bioavailability. Therefore the physiological benefits are negligible.

8.3.2.1   *Polymer-based Micro/Nanoparticles*

It is well known that inclusion of poorly water soluble actives into CDs can improve their solubility in aqueous systems and thereby their bioavailability. With regards to food applications, complexation with CDs was employed to protect carotenoids from oxidation and to improve cellular uptake of carotenoids into intestinal cells (Reuscher *et al* 2004). In particular, complexes of carotenoids and CDs, preferably 2-hydroxypropyl-β-CD and γ-CD were prepared in water, to which an emulsion, dispersion or solution of the coating component was then added. Several coating methods can be useful, including ionic gelation, spray drying and fluid bed coating according to the employed materials. These complexes in oil were stable against oxidation and mostly exhibited higher bio-uptake than oil-based and micellular carotenoid compositions. Recently, an innovative method for enhanced absorption of carotenoids was patented (Bartlett *et al* 2010). This method involves the complexation of γ-CD with a carotenoid and incorporation of the inclusion complex into other nutritional supplements (*i.e.* fish oil and vitamin E). The final suspension is then encapsulated in soft gelatin capsules.

8.3.2.2 *Lipid-based Micro/Nanoparticles*

SEDDS are able to form a fine emulsion on dilution with physiological fluid and gentle blending. The drug, therefore, remains in solution in the gastrointestinal tract, avoiding the dissolution step that frequently limits the absorption rate of hydrophobic drugs from the crystalline state. Moreover, as vitamin A adsorption is related to the adsorption of lipids, SEDDS have a greater chance of improving vitamin oral bioavailability. Nanosize SEDDS containing retinyl acetate at 25% (w/w) were prepared using different concentrations of soya-bean oil as solvent and of a surfactant and a co-surfactant. This system was then filled in hydroxypropyl methyl cellulose capsules, showing an effective increase of the dissolution rate, especially when the surfactant to co-surfactant ratio was 2:1 (Taha *et al* 2004). The results obtained *in vivo* in rats (Taha *et al* 2007) confirmed the bioavailability enhancement.

8.4   Final Remarks and Future Perspectives

A large variety of encapsulation methods, matrix polymers and coating materials have been applied to provide the integrity and the benefits of vitamin A, obtaining great advances in technologies as well as in the design of novel DDS.

    With regards to the processing of vitamin A, the main limitation is represented by the high costs of production; indeed, the manufacturing methods must be easily moved to a larger scale. Spray drying is the most widespread technological procedure of encapsulation; it is not very expensive, it is scalable and exploitable for different materials and particle production (Gibbs *et al* 1999). The encapsulation strategies reported here highlight that spray drying of a dispersion or an emulsion system containing one polymer as the wall material is not enough to effectively protect vitamin A against degradation for a long time. Undoubtedly, the coating thickness plays an important role as a moisture barrier. Owing to their chemical affinity to vitamin A, liposomes or SLNs/NLCs could have the potential to increase the health benefits of food products. However, the high costs of production, the difficult scalability and the use of processes that involve heat are significant drawbacks.

In our opinion, a multilayer stratification process and a combination of chemically different molecules to build up both a physical and chemical barrier would favourable. The later studies show the great potential of the ionotropic gelation method due to the use of food grade, biocompatible and low cost polymers. It is also an easy, economical and safe method for the industrial production of vitamin A microcapsules. Moreover, the loading capacity of these systems can be high, even if the degradation increases by raising the retinoid content. From the industrial point of view, the prilling technology may be considered the best choice of obtaining relatively stable vitamin A microcapsules.

Finally, with regards β-carotene as source of vitamin A in food fortification, the cost factor is considered too high. Therefore encapsulated β-carotene is used mainly to maintain colour in foods and to inhibit the light-catalysed oxidation of fats and oils.

Summary Points

- This Chapter focuses on vitamin A encapsulation. An overview of the available technologies and of their application in the food/feed field is presented.
- Vitamin A and its chemically related substances are very unstable compounds because they are sensitive to the action of air, oxidising agents, acids and light. It is important to preserve vitamin A and β-carotene to ensure a stable amount of these nutraceuticals in foods and animal feeds during storage.
- Many encapsulation methods are available to protect vitamin A from oxidation and are divided into physico-chemical or mechanical methods. All these methods form particles, which dimensions can be tailored to obtain micro- or nanoparticles as a function of their final destination in food/feedstuff.
- Concerning food applications, food-grade materials have to be used and procedures that involve organic solvents or other unsafe substance are not permitted.
- Encapsulation approaches that have been developed in the scientific and patent literature mainly concern the preservation and stability of vitamin A. Strategies to enhance its poor bioavailability have been also assessed.
- Micro/nanoparticles can be polymer-based or lipid-based, according to the selected procedure. To gain a better vitamin A protection two encapsulation methods can be employed consecutively. Thus a multilayer stratification and a combination of chemically different molecules to build up both a physical and chemical barrier would favourable.

- The selection of the manufacturing method must be a good compromise between costs and effectiveness in terms of vitamin A stability over time, loading capacity and bioavailability.

Key Facts of Encapsulation Purposes

- Conversion of oils and other liquids into solids for ease of handling.
- Taste masking of bitter drugs.
- Environmental protection of labile moieties.
- Delay of volatilization.
- Separation of incompatible materials (for example, choline salts from vitamin A).
- Improvement of the technological properties (mixability and flowability) of the final products.
- Safe handling of toxic substances.
- Protection against gastrointestinal drug degradation or the reduction in mucosal gastro-irritation caused by some substances.
- Production of controlled-release and targeted DDS.

Definition of Words and Terms

Coacervation: The separation of a polymer solution in two liquid immiscible phases: a dense coacervate phase and a dilute liquid phase. When two or more polymers of opposite charge are present, it is referred to as complex coacervation. The separation is the result of the polymer dehydration after the addition of salts or non-solvents, temperature change or chemical interactions.

Drug Delivery System (DDS): A physical platform designed to deliver an active pharmaceutical ingredient to humans and animals in order to achieve a therapeutic effect.

Fluid bed coating: Solid core particles are fluidized by air pressure in a fluid-bed granulator/coater and a spray of dissolved wall material is applied from the perforated bottom of the fluidization chamber parallel to the heated air stream and on to the solid core particles (bottom spray or Wurster process).

Freeze drying: A single-step drying method that involves the freezing of a solution followed by solvent sublimation at low temperature and pressure. The freeze drying of an emulsion and/or dispersion leads to polymer–drug particle formation.

Inclusion complexes: Formed between two compounds called host and guest. The host has an inner hydrophobic cavity in which the guest is located, due to the chemical affinity.

Ionotropic gelation: A polymer in the ionized form is used to react with an oppositely charged poly-ion (*e.g.* sodium alginate with $CaCl_2$) or two opposite charged polymers linked each other to form gelled microcapsules/beads, which after a suitable linking time becomes solid.

Liposomes: Nanosize lipid vesicles formed from aqueous dispersion of polar lipids that produce bilayer structures. They may consist of one or several bilayer membranes.

Microlnanocapsules: Constituted by a membrane or shell, in a single or multi-layer, containing the fill material that can be liquid or solid. This structure is also called reservoir system.

Microlnanospheres: Matrix particles in which the active or particles are dispersed within the carrier material.

Self-emulsified drug delivery systems (SEDDS): Isotropic mixtures of drug, oil/lipid, surfactant and/or co-surfactant, which form fine emulsion on dilution with physiological fluid and gentle blending.

Spray drying: A single-step drying process that produces solid microparticles, starting from a liquid phase by means of heated air stream and an atomization device. If the liquid phase is a solution or dispersion, microspheres are obtained. Starting from an emulsion, microcapsules are formed.

List of Abbreviations

BHT	butylated hydroxytoluene
CD	cyclodextrin
DDS	drug delivery system(s)
EDTA	ethylendiaminetetra-acetic acid
HPMC	hydroxypropyl methyl cellulose
NLC	nanostructured lipid carriers
O/W	oil in water
O/W/O	oil in water in oil
RH	relative humidity
RT	room temperature
SEDDS	self-emulsified drug delivery system(s)
SLN	solid lipid nanoparticle
VAD	vitamin A deficiency

References

Albertini, B., Di Sabatino, M., Calogerà, G., Passerini N. and Rodriguez, L., 2010. Encapsulation of vitamin A palmitate for animal supplementation: formulation, manufacturing and stability implications. Journal of Microencapsulation. 27: 150–161.

Bakan J. A., 1994. Microencapsulation. In: Boylan, J.C. and Swarbrick J. (ed.) Encyclopedia of Pharmaceutical Technology. Marcel Dekker, New York, USA, vol. 9, pp. 423–441.

Bartlett, M. R., Mastaloudis, A., Smidt, C. R. and Poole, S. J., 2010. Nanosized carotenoid cyclodextrin complexes. NSE Product, U.S. Patent, 7781572 B2.

Bates, C.J., 1995. Vitamin A. Lancet. 345: 31–35.

Blatt, Y., Pinto, R., Safronchik, O., Sedlov, T. and Zelkha, M., 2003. Stable coated microcapsules. Bio-Dar Ltd, U.S. Patent, 0064133 A1.

Bondi, A. and Sklan, D., 1984. Vitamin A and carotene in animal nutrition. Progress in Food and Nutrition Science. 8: 165–191.

DeFreitas, Z., Hall, H., Newman, J. and Gordon, M., 2003. Microcapsules having high carotenoid content. Kemin Foods, U.S. Patent, 6663900 B2.

Desobry, S. A., Netto, F. M. and Labuza, T. P., 1997. Comaprison of spray-drying, drum-drying and freeze-drying for β-carotene encapsulation and preservation. Journal of Food Science. 62: 1158–1162.

Desobry, S. A., Netto, F. M. and Labuza, T. P., 1999. Influence of maltodextrin systems at an equivalent 25DE on encapsulated β-carotene loss during storage. Journal of Food Processing and Preservation. 23: 39–55.

Elizalde, B. E., Herrera, M. I. and Buera, M. P., 2002. Retention of β-carotene encapsulated in trehalose-based matrix as affected by water content and sugar crystallization. Journal of Food Science. 67: 3039–3045.

European Pharmacopoeia, 2005. Vitamin A. 5th edn. Vol. 2, pp. 2682–2684.

Gellenbeck, K. W., 1999. Dry carotenoid-oil powder and process for making same. Amway Corporation, U.S. Patent, 5976575.

Gibbs, B. F., Kermasha, S., Alli, I. and Mulligan, C. N., 1999. Encapsulation in the food industry. International Journal of Food Science and Nutrition. 50: 213–224.

Gonnet, M., Lethuaut, L. and Boury, F., 2010. New trends in encapsulation of liposoluble vitamins. Journal of Controlled Release. 146: 276–290.

Han, J., Guenier, A. S., Salmieri, S. and Lacoix, M., 2008. Alginate and chitosan functionalization for micronutrient encapsulation. Journal of Agricultural and Food Chemistry. 56: 2528–2535.

Hentschel, A., Gramdorf, S., Müller, R. H. and Kurz, T., 2008. β-Carotene-loaded nanostrucured lipid carriers. Journal of Food Science. 73: 1–6.

Jenning, V. and Gohla, S., 2001. Encapsulation of retinoids in solid lipid nanoparticles. Journal of Microencapsulation. 18: 149–158.

Junyaprasert, V. B., Mitrevej, A., Sinchaipanid, N., Boonme, P. and Wurster, D. E., 2001. Effect of process variables on the microencapsulation of vitamin A palmitate by gelatin–acacia coacervation. Drug Development and Industrial Pharmacy. 27: 561–566.

Lee, S. C., Yik, H. G., Lee, D. H., Hwang, Y. I. and Ludescher, R. D., 2002. Stabilization of retinol through incorporation into liposomes. Journal of Biochemistry and Molecular Biology. 35: 358–363.

Lim, F. and Moss, R. D., 1983. Vitamin encapsulation. Damon Corporation, U.S. Patent, 4389419.

Loveday, S. M. and Singh, H., 2008. Recent advances in technologies for vitamin A protection in food. Trends in Food Science and Technology. 19: 657–668.

McClements, D. J. and Decker, E. A., 2007. Emulsion-based delivery systems for lipophilic bioactive components. Journal of Food Science. 72: R109–R124.

Nagao, A., 2009. Absorption and function of dietary carotenoids. Forum of Nutrition. 61: 55–63

Reuscher, H., Kagan, D. I. and Madhavi, D. L., 2004. Coated carotenoid cyclodextrin complexes. Bioactives LLC, Wacker Biochem Coorporation, U.S. Patent, 0109920 A1.

Taha, E. I., Al-Saidan, S., Samy, A. M. and Khan, M. A., 2004. Preparation and *in vitro* characterization of self-nanoemulsified drug delivery system (SNEDDS) of all-*trans* retinol acetate. International Journal of Pharmaceutics. 285: 189–119.

Taha, E. I., Ghorab, D. and Zaghloul, A., 2007. Bioavailability assessment of vitamin A self-nanoemulsified drug delivery systems in rats: a comparative study. Medical Principles and Practice. 16: 355–359.

Taylor, T. M., Bruce, B. D. and Weiss, J., 2005. Liposomal nanocapsules in food science and agriculture. Critical Reviews in Food Science and Nutrition. 45: 587–605.

Wurtburg, O. B., Trubiano P. C. and Herbst, W., 1970. Encapsulation of water insoluble materials. National Starch and Chemical Corporation, U.S. Patent, 3499962

Xie, Y. L., Zhou, H. M. and Zhang, Z. R., 2007. Effect of relative humidity on retention and stability of vitamin A microencapsulated by spray draying. Journal of Food Biochemistry. 31: 68–80.

Yoshida, K., Sekine, T., Matsuzaki, F., Yanaki, T. and Yamaguchi, M., 1999. Stability of vitamin A in an oil-in-water-in-oil-type multiple emulsion. Journal of the American Oil Chemists' Society. 76: 195–200.

CHAPTER 9

Thermal Degradation of β-Carotene in Food Oils

ALAM ZEB

Department of Biotechnology, University of Malakand, Chakdara, Pakistan
E-mail: alamzeb01@yahoo.com

9.1 Introduction

β-Carotene (β,β-carotene) is one of the most important and widely studied carotenoids with a strong red–orange colour. Chemically, β-carotene consists of eight isoprenoid units, which are joined in such a fashion that the arrangement of isoprenoid units is reversed at the centre of the molecule. The two central methyl groups are in a 1,6-position, whereas the remaining non-terminal methyl groups are in a 1,5-position. The central chain ends on both sides by two cyclohexene rings known as β-rings (Britton 1995). β-Carotene produces vitamin A upon enzymatic cleavage during digestion, and thus has been assigned 100% activity of vitamin A. In foods and vegetables, β-carotene is a good source of vitamin A for humans. β-Carotene also serves as a biological antioxidant and is helpful in maintaining human health. In food industries it is used as a colorant and also as a source of pro-vitamin A.

Generally, β-carotene is commercially produced by chemical synthesis at a large scale and also from different plant sources. Red palm oil and sea buckthorn oil are now the main sources of the plant oils containing large amount of β-carotene (Zeb and Malook 2009). β-Carotene can also be produced by fermentation and from microalgae (Dufosse 2009). In fruits and vegetables β-carotene may occur as a mixture of cis/trans (E/Z) geometrical isomers

Food and Nutritional Components in Focus No. 1
Vitamin A and Carotenoids: Chemistry, Analysis, Function and Effects
Edited by Victor R Preedy
© The Royal Society of Chemistry 2012
Published by the Royal Society of Chemistry, www.rsc.org

(Figure 9.1). During processing *E/Z* isomerization is quite common and more frequent. Therefore in order to stop the isomerization and oxidation of carotenoids, antioxidants, such as butylated hydroxytoluene (BHT) or ascorbic acid, are added. The Z-β-carotene isomers are present naturally in mango fruits (Godoy and Rodriguez-Amaya 1994; Pott *et al* 2003). The exposure of mangos to heat and light may result in to *E–Z* isomerization. It has been reported that 13-Z- and 9-Z-β-carotene are commonly formed during vegetable preparations due to the effects of heat and illumination, respectively. The formation of 13-Z-β-carotene was enhanced in carrot juice in the presence of lipids at temperatures of 80–100 °C for 30–60 min, and also using the addition of grape-seed oil prior to thermal preservation (Marx *et al* 2003). Thus thermal stress produces significant oxidative degradation of β-carotene. This consequently decreases the nutritional value of vitamin A and its activity as antioxidant, and causes significant loss of the natural flavour and chromophores in foods. Ultimately, the food products are less acceptable or unacceptable to consumers. In food industries β-carotene can also be added to some commercial fruits and vegetable beverages (Rodríguez-Comesana *et al* 2002). In order to maintain the high quality of nutrition and flavour of the food products, it is particularly important to protect β-carotene during processing and storage. Thus the study of chemical reactions, antioxidant functions and degradation or oxidation of β-carotene are important tasks for the present day food chemist.

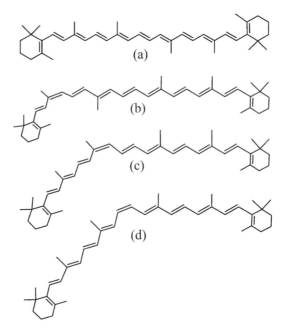

Figure 9.1   Structures of the β-carotene and its isomers. (a) All-*E*-β-carotene, (b) 9-*Z*-β-carotene, (c) 13-*Z*-β-carotene and (d) 15-*Z*-β-carotene.

9.2 Thermal Degradation of β-Carotene in Oils

Degradation usually occurs by the oxidation of the substrate. β-Carotene is degraded by thermal treatment at different temperatures. Increasing the temperature has been found to increase the oxidation rate. The first step in the oxidation of β-carotene is its conversion into *Z*-isomers. There is no change in the overall colour during isomerization. Other reactions are: intra-molecular cyclization to form volatile compounds, and degradation to form aldehydes and ketones with a low molecular weight. Generally, isomerization is reversible, whereas degradation or oxidation are non-reversible reactions. Carotenoid degradation in food or similar model systems is a highly complex phenomenon. However, some authors proposed a first-order kinetic model for the degradation (Chen *et al* 1994; Takahashi *et al* 2003). Borsarelli and Mercadante (2010) showed a simplified mechanism of overall changes occurring in carotenoids during heating. The mechanism, however, does not provide information about the susceptibility of *Z*-isomers towards thermal oxidation or degradation, or about the formation of the new oxygenated species. The mechanism fits to the degradation of carotenoids in solution. In solutions, the *E*- and *Z*-isomers are in equilibrium. However, the solvent also affects the isomerization induced by heating. For example, the isomerization rate of β-carotene is higher in non-polar solvents (petroleum ether and toluene) than in polar solvents.

9.2.1 Kinetics of β-Carotene Degradation

β-Carotene plays an important role during the oxidation of oil and fats. The outcomes from early investigations showed that the behaviour of β-carotene in the oxidation process depends strongly on its concentration and on the lipid medium, as well as whether the process proceeds in the dark or in the light and in the presence and absence of other antioxidants and pro-oxidants. For example, Hunter and Krackenberger (1947) oxidized β-carotene at 50 °C in benzene and in peanut oil. They found that the rate of the oxidation process was much higher in benzene than in the oil. They attributed this fact to the presence of a natural antioxidant in the peanut oil. Henry *et al* (1998) studied the thermal and oxidative degradation of all-*E*-β-carotene, 9-*Z*-β-carotene, lycopene and lutein in safflower seed oil at 75, 85 and 95 °C for 24, 12 and 5 h, respectively. All the carotenoids were found to degrade following a first-order kinetic model. The rates of degradation were: lycopene > all-*E*-β-carotene ≈ 9-*Z*-β-carotene > lutein. The authors also found that there was no significant difference in the rates of degradation for 9-*E*- and all-*E*-β-carotene under the experimental conditions used.

In another study a heat-induced degradation of carotenoids was studied in two crude paprika oleoresins diluted with high oleic or high linoleic oil (Perez-Galvez and Minguez-Mosquera 2004). The degradation rate constants were not significantly different. During thermal treatment the *E*-to-*Z* isomer conversion was initially higher than degradation of *Z*-isomers. β-Carotene was degraded more than the β-cryptoxanthin. Under the reaction conditions, oleic and linoleic fatty acids showed the same reactivity and induced degradative effects. Similarly, the stability of carotenoids was studied in marigold oil extracts prepared from Myritol 312®, paraffin oil, almond oil, olive oil, sunflower oil, grape-seed oil and soya-bean oil. Carotenoids were degraded with a first-order kinetics model, which was dependent on concentration. The highest degradation rates were observed in extracts prepared with linoleic acid containing oils such as sunflower oil, soybean oil and grape-seed oil (Bezbradica *et al* 2005). The connection between the degradation of carotenoids and lipid auto-oxidation were confirmed. The authors also suggest that the effect of the oil solvents on the stability of oil extracts of *Calendula officinalis* is a factor that must be considered when selecting a solvent for the production of marigold oil extracts. Achir *et al* (2010) studied the kinetics and degradation at 120–180 °C of all-*E*-β-carotene and all-*E*-lutein in palm olein and Vegetaline®. The variations in the *E*- and *Z*-isomers of carotenoids were investigated using high-performance liquid chromatography (HPLC)-diode array detection (DAD). The authors found that initial all-*E*-β-carotene and all-*E*-lutein degradation rates increased with temperature in both oils. All-*E*-lutein was found to have more thermal resistance to degradation than all-*E*-β-carotene. The isomers identified were 13-*Z*- and 9-*Z*-β-carotene, and 13-*Z*-, 9-*Z*-, 13'-*Z*-, and 9'-*Z*-lutein. They found that *E–Z* isomerization of carotenoids was involved in the degradation reactions at rates that increased with temperature.

9.2.2   Thermal Degradation Products of β-Carotene

The study carried out by Ouyang *et al.* (1980) showed the effects of deodorization of palm oil at high temperatures (170–250 °C) with respect to the formation of short-chain apo-carotenals. Later, Onyewu *et al* (1986) identified apo-carotenone and apo-carotenals formed by heating of palm oil at 210 °C. Long-chain apo-carotenals such as apo-8'-, apo-10'-β-carotenals have been observed at lower-to-medium temperatures during extrusion cooking, whereas at higher treatment temperatures only the short-chain apo-carotenals were detected (Marty and Berset 1990). However, little is known about the effect of these oxygenated species on the oil in these studies.

Liquid chromatography coupled to DAD and mass spectrometry (MS) are commonly used for the determination of β-carotene degradation products. Recently, the degradation products of β-carotene were identified in corn oil (Zeb and Murkovic 2011a). Typical chromatograms of β-carotene and its oxidation products in corn oil thermally oxidized at 110 °C in a Rancimat are shown in Figure 9.2. Only eight oxidation products (peaks 1–8) were identified. These were all-*E*-5,8-epoxy-5,8-dihydro-β,β-carotene, all-*E*-5,6-epoxy-5,6-dihydro-β,β-carotene and its isomer 15-*Z*-5,6-epoxy-5,6-dihydro-β,β-carotene, 13-*Z*-5,6,5',6'-diepoxy-5,6,5',6'-tetrahydro-β,β-carotene, 5,6-epoxide, 5,8-epoxides, all-*E*-8'-apo-β-caroten-8'-al, all-*E*-6'-apo-β-caroten-6'-al, all-*E*-5,6-epoxy-8'-apo-β-caroten-8'-al, β,β-caroten-2,2'-dione, 13-*Z*-β,β-carotene and all-*E*-β-carotene. With increasing the length of the heating time, the formation of *Z*-isomers of β-carotene was found to increase significantly. Similarly, the prolongation of the thermal treatment led to a complete degradation of each intermediate product.

Based on the above results, Zeb and Murkovic (2011a) hypothesize that β-carotene may first be converted into its *Z*-isomers or oxidized directly to 5,6-epoxide or diones without isomerization. The epoxides may also be converted into the *Z*-isomers or form *Z*-isomers of apo-carotenals, whereas the *Z*-isomer of β-carotene, upon oxidation, may produce *Z*-isomers of epoxides or apo-carotenals (Figure 9.3). Apo-carotenals can also be produced directly upon oxidative breakdown of β-carotene. This mechanism is in accordance with the scheme reported by Caris-Veyrat (2001) for the oxidation of β-carotene in organic solution. The final degradation products are short-chain products which are volatile in nature and thus evaporated from the food oils during thermal treatment.

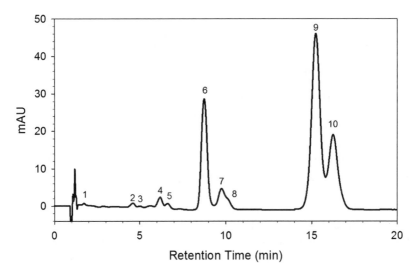

Figure 9.2 Typical chromatogram of HPLC separation profile of the thermal oxidation products of β-carotene in corn oil at 110 °C. Separation was achieved using reversed phase HPLC and identification was carried out using DAD and atmospheric pressure chemical ionization (APCI)-MS system. The gradient mobile phase consists of methanol, methanol/methyl *tert*-butyl ether (MTBE)/water and acetone. Peaks: (1) 8'apo-β-carotenal, (2) 6'-apo-β-carotenal, (3) all-*E*-5,6-epoxy-β-caroten-8'-al, (4) β-carotene-2,2'-dione, (5) 13-*Z*-5,6,5',6'-diepoxy-β-carotene, (6) all-*E*-5,8-epoxy-β-carotene, (7) all-*E*-5,6-epoxy-β-carotene, (8) 15-*Z*-5,6-epoxy-β-carotene, (9) all-*E*-β-carotene and (10) 13-*Z*-β-carotene. Data are from Zeb and Murkovic (2010a), with permission from Publisher.

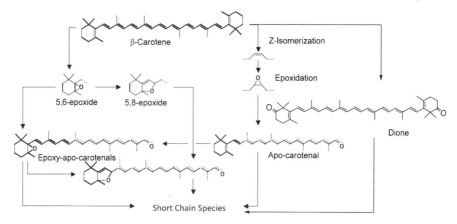

Figure 9.3   Proposed mechanism of the thermal degradation of β-carotene in corn oil. Schematic representation of the possible mechanism during thermal oxidation of β-carotene in corn oil based on the detected compounds. β-Carotene may be converted directly into dione, which may be degraded to short-chain species, or isomerize and epoxidize in the centre to form apo-carotenals. Epoxidation can occur near the ring, which may degrade to epoxy-apo-carotenals. The dotted line represents the varied chain length.

9.3   Role of β-Carotene during Thermal Degradation

9.3.1   Antioxidant Action

The antioxidant role of β-carotene is well known during auto-oxidation at lower temperatures. Warner and Frankel (1987) showed that β-carotene had a significant effect in protecting soybean oil against light-induced oxidation. Similarly, the effects of 0, 5, 10 and 20 ppm of β-carotene on the oxidation of the soybean oil/methylene chloride model system containing 4 ppm chlorophyll in light (4000 lux) was found to reduce the oxidation of soybean oil (Lee and Min 1988). Most of the earlier studies of carotenoid auto-oxidation described the effect of lipids or other antioxidants on the auto-oxidation of carotenoids (Budowski and Bondi 1960; Lisle 1951). However, the pro-oxidant action of β-carotene during thermal oxidation of food oils has the consequence of lowering the stability.

9.3.2   Pro-oxidant Action

The pro-oxidant action of β-carotene is quite common during thermal oxidation. It is believed that degradation compounds as well as β-carotene are promoting the oxidation of food oils. The relative oxidative stability of soybean oil containing either thermally degraded β-carotene or lycopene was determined at every 4 h for 24 h (Steenson and Min 2000). Soybean oil containing 50 ppm degraded β-carotene stored in the dark at 60 °C displayed significantly ($P < 0.01$) headspace oxygen depletion (HOD) values compared with controls. Lycopene degradation products significantly ($P < 0.05$) decreased HOD of samples when stored in the dark. The results indicated that, during auto-oxidation of soybean oil in the dark, β-carotene thermal degradation products acted as pro-oxidants, whereas thermally degraded lycopene displayed antioxidant activity in similar soybean oil systems.

Similarly, the effects of β-carotene, α-tocopherol and ascorbyl palmitate (AP) were studied by Karabulut (2010) on the oxidation of butter oil triacylglycerols (BO-TAGs). An accelerated oven-oxidation test was carried out at 60 °C to determine the most effective dosages of the antioxidants. α-Tocopherol was found to be the most effective at a concentration of 50 μg g^{-1}. A better stability was achieved with a combination of all three antioxidants. However, the author reported pro-oxidant effects of β-carotene and ascorbyl palmitate when used individually or in combination.

In model triacylglycerols (TAGs), β-carotene was found to degrade faster than TAGs (Zeb and Murkovic 2010). Most of the β-carotene was found to degrade in the first few hours of thermal treatment. β-Carotene was much stable in olive oil (Zeb and Murkovic 2011b), as shown in Table 9.1 than model TAGs, where the TAGs composition of olive oil is closely related to the model TAGs. This suggests that food oils containing other antioxidants such as tocopherols, various phenolic compounds are responsible for the stability of β-carotene. However, β-carotene was pro-oxidant and has lower stability than astaxanthin during thermal degradation in olive oil. In case of corn oil, β-carotene was found more stable than olive oils and model TAGs samples. The complete degradation of β-carotene was observed before 12 h of thermal treatment in a Rancimat at 110 °C (Zeb and Murkovic 2011b). In all of these

Table 9.1   Thermal degradation of β-carotene in food oils and model TAGs from food oils. A specific amount (300 μg g^{-1}) of standard β-carotene was added to the food oils or food oil TAGs. The samples were oxidized in a Rancimat at 110 °C at different time intervals. Data are from Zeb and Murkovic (2010; 2011a; 2011b), with permission from Publishers.

	β-Carotene (μg g^{-1})		
Time (h)	*Corn oil*	*Olive oil*	*Model TAGs*
0	293.55	302.34	300
1	269.44	294.25	296.34
2	199.09	253.56	153.88
3	157.19	215.93	137.18
4	137.07	212.00	96.14
5	109.25	180.26	53.68
6	57.18	138.98	0
8	27.47	111.42	0
10	19.00	24.16	0
12	0	0	0

cases β-carotene acts as pro-oxidant by promoting the oxidation of food oils, and thus increasing the chances of deterioration and ultimately unacceptability of food products. The pro-oxidant action was confirmed by the status of oxidation of food oils.

9.4 Fate of Food Oils

During thermal oxidation significant changes can be observed in the chemistry of selected food oils. The oxidation can be attributed to the formation of hydroperoxides, which can be measured in terms of analytical parameter called peroxide value (PV). Jia *et al* (2007) revealed that unoxidized soybean oil could increase the stability of β-carotene, whereas oxidized oils act as pro-oxidant for β-carotene. PVs were significantly increased with exposure time in the presence of β-carotene than in control samples. Figure 9.4 shows the formation of more hydroperoxides in the presence of β-carotene in model TAGs than in olive oil containing β-carotene under similar conditions of experiments. β-Carotene was stable and protected during the thermal treatment of olive oil. This could be attributed to the presence of other antioxidant compounds present in olive oils, which are not present in model TAGs of similar composition (Zeb and Murkovic 2010; 2011b). In corn oil (Figure 9.5) under similar conditions, β-carotene produced more hydroperoxides than control samples, suggesting a

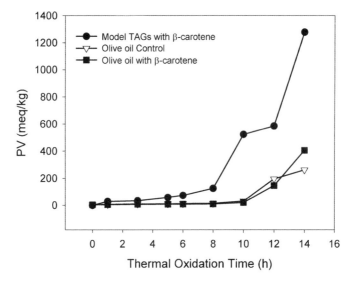

Figure 9.4   The effects of β-carotene on the formation of hydroperoxides in model TAGs and refined olive oils. The samples were oxidized in a Rancimat at 110 °C from 1 to 14 h. The effect of β-carotene (300 μg g) on the formation of hydroperoxides shows that model TAGs having similar TAG composition to olive oil are producing more peroxides than the control and β-carotene-containing olive oils. Data are from Zeb and Murkovic (2010b), with permission from Publisher.

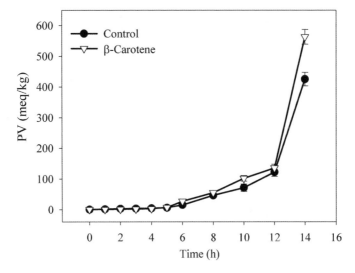

Figure 9.5 The effects of β-carotene on the formation of hydroperoxides in corn oils. The corn oil samples were oxidized in a Rancimat at 110 °C from 1 to 14 h. The effect of β-carotene (300 μg g) on the formation of hydroperoxides shows that more peroxides are produced in β-carotene-containing corn oils than control. Data are from Zeb and Murkovic (2010b), with permission from Publisher.

pro-oxidant role of β-carotene (Zeb and Murkovic 2011a). In all these cases, β-carotene does not produce any effects on the change in PV during the initial hours of thermal treatment. However, at higher exposure times the oxidation of food oils was enhanced in the presence of β-carotene.

The detailed studies of the reactions by liquid chromatography–mass spectrometry (LC-MS) reveals the structures of these primary oxidation products of TAG formed during food oil thermal treatments (Zeb and Murkovic 2010; 2011a; 2011b). The interactions of individual hydroperoxides were also determined. It was found that β-carotene or its degradation products produce different effects on the promotion or suppression of individual hydroperoxide.

Summary Points

- β-Carotene is the most important carotenoids present in or added into various food oils.
- β-Carotene is one of major source of vitamin A in general, as well as in food oils.
- During thermal treatment, β-carotene is either converted into *cis*-isomers or directly oxidized.
- The degradation of β-carotene in food oils generally follows first-order kinetics.

- The β-carotene degradation products are usually mono-epoxides, di-epoxides, dione and apo-carotenals.
- β-Carotene acts as pro-oxidant during thermal treatment at higher exposure time and high temperature.
- The oxidation and degradation of β-carotene started earlier than the oxidation of food oils.
- β-Carotene or its degradation products enhance the oxidation of TAGs in food oils.

Key Facts

Key Fact for Food Oils

- The oils used in or for foods preparation. It is also known as edible or vegetable oils.
- The major composition of food oils are TAGs, tocopherols and phenolic compounds.
- The common examples are olive, sunflower, soybean and rapeseed oils.
- The more unsaturated the food oils, the more they will thermally oxidized.
- Red palm oil is one of the carotenoid-rich food oils, containing more saturated fatty acids than unsaturated fatty acids.
- Carotenoids are more stable in saturated food oils than in unsaturated food oils.

Key Fact for Oxidative Degradation

- The kind of reaction in which a molecule is initially oxidized and then broken down into two or more compounds.
- Oxygen is one of the main reactant in the initial phase of degradation.
- Different oxygen-containing compounds are produced.
- In case of oxidative degradation of carotenoids, an epoxide is formed in the central polyene chain that results in to the formation of apo-carotenals.
- Lipo-oxygenase action on β-carotene also involves oxidative degradation.

Definitions of Words and Terms

Apo-carotenals: The short-chain oxygen-containing derivative compound of carotenoid produced upon the oxidation of the original carotenoid.

Calendula officinalis: A pot marigold used as ornamental and medicinal plant.

Chromophore: The part of a compound which imparts a characteristics colour. In case of carotenoids, the central polyene chain is responsible for colour.

Diode array detection (DAD): A type of detector which detects light on a number of photosensitive diodes placed side-by-side forming a sandwich. The output is scanned, stored and subsequently processed by a computer. This kind of detector is very useful in carotenoid analysis by liquid chromatography.

First-order kinetics: The reaction which depends on the concentration of the one reactant only. It is also known as a uni-molecular reaction. β-Carotene usually follows first-order kinetics during thermal degradation.

Isomerization: The process of formation of different compounds with the same molecular formula but different structure.

Isoprenoid: An iso-pentene unit, which is the precursor of many metabolic compounds, especially carotenoids and sterols.

LC-MS: Liquid chromatography coupled to mass spectrometric detector. The most reliable instrument now used for carotenoid analysis.

lux: The SI unit used in photometry for luminance. It measures the luminous power per area.

Peroxide value (PV): The analytical parameters used to quantify the amount of hydroperoxides present in a biological sample.

Rancimat: The instrument that analyses the oxidation of oil or fat. An oil sample is oxidized in the presence of air and heat, from which the volatile compounds are collected in deionized water and the conductivity are measured.

Thermal degradation: The breakdown of a molecule using heat is known as thermal degradation. It may be oxidative or none oxidative.

Triacylglycerols (TAGs): The neutral fats made from 3 mol of fatty acids and 1 mol of glycerol. TAGs are the most common and major form of food oils.

List of Abbreviations

DAD	diode array detection
E/Z	*cis/trans*
HOD	headspace oxygen depletion
MS	mass spectrometry
PV	peroxide value
TAG	triacylglycerol

References

Achir, N., Randrianatoandro, V. A., Bohuon, P., Laffargue, A. and Avallone, S., 2010. Kinetic study of β-carotene and lutein degradation in oils during heat treatment. European Journal of Lipid Science and Technology. 112: 349–361.

Bezbradica, D., Milik-Askrabic, J., Petrovic, S. D. and Siler-Marinkovic, S., 2005. An investigation of influence of solvent on the degradation kinetics of carotenoids in oil extracts of *Calendula officinalis*. Journal of the Serbian Chemical Society. 70: 115–124.

Borsarelli, C. D. and Mercadante, A. Z., 2010. Thermal and photochemical degradation of carotenoids. In: Landrum, J. T. (ed.) Carotenoids: Physical, Chemical, and Biological Functions and Properties. CRC Press, Boca Raton, USA, pp. 229–253.

Britton, G., 1995. UV/visible spectroscopy. In: Britton, G., Liaaen-Jensen, S., Pfander, H. (eds) Carotenoids: Spectroscopy, vol. 1B. Birkhäuser Verlag, Basel, Switzerland, pp. 13–62.

Budowski, P. and Bondi, A., 1960. Autoxidation of carotene and vitamin A. Influence of fat and antioxidants. Archives of Biochemistry and Biophysics. 89: 66–73.

Caris-Veyrat, C., Amiot, M. J., Ramasseul, R. and Marchon, J. C., 2001. Mild oxidative cleavage of β,β-carotene by dioxygen induced by a ruthenium porphyrin catalyst: characterization of products and of some possible intermediates. New Journal of Chemistry. 25: 203–206.

Chen, B. H., Chen, T. M. and Chien, J. T., 1994. Kinetic model for studying the isomerisation of α- and β-carotene during heating and illumination. Journal of Agriculture and Food Chemistry. 42: 2391–2397.

Duffosse, L., 2009. Microbial and microalgal carotenoids as colorants and supplements. In: Britton, G., Liaaen-Jensen, S., Pfander, H. (ed.) Carotenoids: Nutrition and Health, vol. 5. Birkhäuser Verlag, Basel, pp. 83–98.

Godoy, H. T. and Rodriguez-Amaya, D. B., 1994. Occurrence of *cis*-isomers of pro-vitamin A in Brazilian fruits. Journal of Agriculture and Food Chemistry. 42: 1306–1313.

Henry, L. K., Catignani, G. L. and Schwartz, S. J., 1998. Oxidative degradation kinetics of lycopene, lutein, and 9-*cis* and all-*trans*-β-carotene. Journal of the American Oil Chemist's Society. 75: 823–829.

Hunter, R. F. and Krakenberger, R. M., 1947. The oxidation of β-carotene in solution by oxygen. Journal of the Chemical Society. January: 1–4.

Jia, M., Kim, H. J. and Min, D. B., 2007. Effects of soybean oil and oxidized soybean oil on the stability of β-carotene. Food Chemistry. 103: 695–700.

Karabulut, I., 2010. Effects of α-tocopherol, β-carotene and ascorbyl palmitate on oxidative stability of butter oil triacylglycerols. Food Chemistry. 123: 622–627.

Lee, E. C. and Min, D. B., 1988. Quenching mechanism of β-carotene on the chlorophyll-sensitized photooxidation of soybean oil. Journal of Food Science. 53: 1894–1895.

Lisle, E. B., 1951. The effect of carcinogenic and other related compounds on the autoxidation of carotene and other autoxidizable systems. Cancer Research. 11: 153–156.

Marty, C. and Berset, C., 1990. Factors affecting the thermal degradation of all-*trans*-carotene. Journal of Agriculture and Food Chemistry. 38: 1063–1067.

Marx, M., Stuparić, M., Schieber, A. and Carle, R., 2003. Effects of thermal processing on *trans–cis*-isomerisation of β-carotene in carrot juices and carotene-containing preparations. Food Chemistry. 83: 609–617.

Onyewu, P. N., Ho, C.-T. and Daun, H., 1986. Characterisation of β-carotene thermal degradation products in a model food system. Journal of the American Oil Chemist's Society. 63: 1437–1441.

Ouyang, J. M., Daun, H., Chang, S. S. and Ho, C.-T., 1980. Formation of carbonyl compounds from β-carotene during palm oil deodorization. Journal of Food Science. 43: 1214–1222.

Perez-Galvez, A. and Minguez-Mosquera, M. I., 2004. Degradation, under non-oxygen-mediated autooxidation, of carotenoid profile present in paprika oleoresins with lipid substrates of different fatty acid composition. Journal of Agriculture and Food Chemistry. 52: 632–637.

Pott, I., Marx, M., Neidhart, S., Mühlbauer, W. and Carle, R., 2003. Quantitative determination of β-carotene stereoisomers in fresh, dried, and solar-dried mangoes (*Mangifera indica* L.). Journal of Agriculture and Food Chemistry. 51: 4527–4531.

Rodríguez-Comesaña, M., García-Falcón, M. S. and Simal-Gándara, J., 2002. Control of nutritional labels in beverages with added vitamins: screening of β-carotene and ascorbic acid contents. Food Chemistry. 79: 141–144.

Steenson, D. F. and Min, D. B., 2000. Effects of β-carotene and lycopene thermal degradation products on the oxidative stability of soybean oil. Journal of the American Oil Chemist's Society. 77: 1153–1160.

Takahashi, A., Shibasaki-Kitakawa, N. and Yonemoto, T., 2003. Kinetic analysis of β-carotene oxidation in a lipid solvent with or without an antioxidant. In: Kamal-Eldin, A. (ed.) Lipid Oxidation Pathways. AOCS Press, Champaign, IL, USA, pp. 110–135.

Warner, E. and Frankel, E. N., 1987. Effects of β-carotene on light stability of soybean oil. Journal of the American Oil Chemist's Society. 64: 213–218.

Zeb, A. and Malook, I., 2009. Biochemical characterization of sea buckthorn (*Hippophae rhamnoides* L. spp. turkestanica) seed. African Journal of Biotechnology. 8: 1625–1629.

Zeb, A. and Murkovic, M., 2010. Characterization of the effects of β-carotene on the thermal oxidation of triacylglycerols using HPLC-ESI-MS. European Journal of Lipid Science and Technology. 112: 1218–1228.

Zeb, A. and Murkovic, M., 2011a. Determination of β-carotene and triacylglycerols thermal oxidation in corn oil. Food Research International. doi: 10.1016/j.foodres.2011.02.039.

Zeb, A. and Murkovic, M., 2011b. Carotenoids and triacylglycerols interactions during thermal oxidation of refined olive oil. Food Chemistry. 127: 1584–1593.

CHAPTER 10

Provitamin A Carotenoids: Occurrence, Intake and Bioavailability

TORSTEN BOHN

Centre de Recherche Public Gabriel Lippmann, Environment and Agro-Biotechnologies Department, Nutrition and Toxicology Unit, 41, rue du Brill, L-4422, Luxembourg
E-mail: bohn@lippmann.lu or torsten.bohn@gmx.ch

10.1 Introduction

Carotenoids range, together with terpenoids, among the most prevalent, lipid-soluble secondary plant metabolites. In plants, carotenoids are often affiliated with the light-harvesting complexes, preventing the photo-oxidation of chlorophyll, and also absorbing light in the 450–500 nm region, closing the "green-gap" of chlorophyll absorption. In addition, they can be found in chromoplasts, adding to the typical yellow-to-red coloration of many plants.

This prevention of photo-oxidation has added to the reputation of carotenoids as antioxidants, *i.e.* compounds that can react and quench reactive oxygen species (ROS). In plants, this is possibly their most important function, either reacting with singlet oxygen ($^1O_2$), or acting as electron donors and acceptors for free radicals such as peroxyl (ROO·), hydroxyl (OH·) or alkoxyl (RO·) radicals. Usually, the longer the extended conjugated π-electron system that may stabilize the radical, the higher the antioxidant capacity.

Food and Nutritional Components in Focus No. 1
Vitamin A and Carotenoids: Chemistry, Analysis, Function and Effects
Edited by Victor R Preedy
© The Royal Society of Chemistry 2012
Published by the Royal Society of Chemistry, www.rsc.org

Carotenoid biosynthesis in plants is achieved *via* the condensation of isopentyl diphosphate (IPP) units, resulting in the formation of phytoene possessing nine double bonds, which is then further desaturated to lycopene (13 double bonds), following then further cyclization to α- or β-ionone rings, and/or hydroxylation processes, resulting in the formation of oxocarotenoids or xanthophylls such as lutein or zeaxanthin. Finally, expoxidation reactions can occur, with the formation of, for example, capsanthin, violaxanthin, neoxanthin, and fucoxanthin (Bohn 2008).

Carotenoids can also be further classified into provitamin A and non-provitamin A carotenoids. Although it is assumed that provitamin A carotenoids contribute only to a third of vitamin A supply in the typical Western diet, this proportion is possibly much higher in developing countries with low meat intake or for vegetarians. Among the provitamin A carotenoids that can be cleaved by human enzymes and transferred into retinol are β-

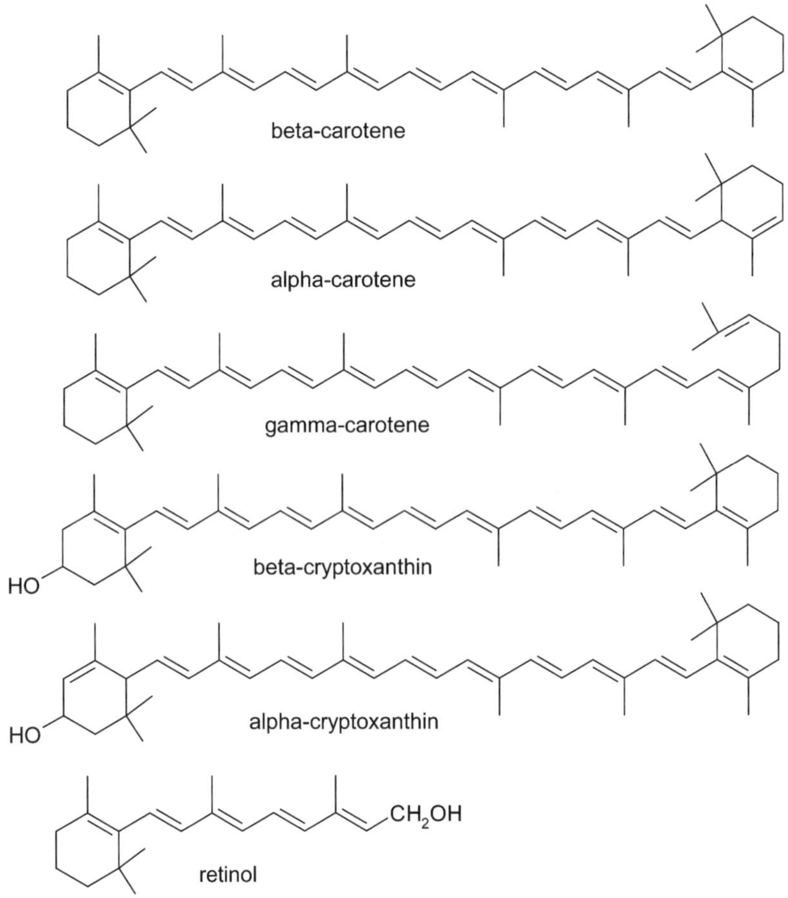

Figure 10.1 Carotenoids with provitamin A activity and retinol. Structures of provitamin A carotenoids frequently found in the diet.

carotene, α-carotene, γ-carotene, and α- and β-cryptoxanthin (Figure 10.1). While central enzymatic cleavage of β-carotene eventually results in the formation of two vitamin A active compounds, all other provitamin A carotenoids can only yield one vitamin A active molecule.

The biological activity of provitamin A carotenoids depends on many factors, including dietary and host-related factors. Provitamin A carotenoid absorption is usually low, in the area of 10–20% (Bohn 2008), and not all absorbed provitamin A carotenoids are finally cleaved by β-carotene 15,15′-mono-oxygenase (BCMO1) into vitamin A active compounds. Thus, the

Table 10.1   Major mechanisms for anticipated beneficial health effects of provitamin A carotenoids. Despite their main activities relying on their vitamin A activity and acting as antioxidants, a number of additional effects could contribute to their proclaimed health benefits. COX-2, cyclo-oxygenase 2; EGF endothelial growth factor; IGF, insulin-like growth factor; IL, interleukin; MDA, malondialdehyde; NF-κB, nuclear factor κB.

Effect	*Description*	*Reference*
Antioxidant	Protection of DNA against peroxidation and strand break	Bouayed and Bohn (2010)
	Prevention of lipid peroxidation	Bouayed and Bohn (2010)
	Reduction of MDA and other reactive products of peroxidation	Reifen *et al* (2004)
	Improving endothelial function *via* impact on NO formation	Fang *et al* (2009)
	Inhibition of LDL oxidation, thrombosis, platelet aggregation	Halliwell (2000)
Cell proliferation	Tumour suppression activity	Reddy *et al* (2005)
	Increased intracellular oxidative stress in tumor cells	Palozza *et al* (2004)
	EGF and IGF receptor protein down-regulation	
Apoptosis	Several pathways:	Palozza *et al* (2004)
	- up-regulation of the anti-apoptotic Bcl-2 protein	
	- increased mitochondria pore transition permeability	
	- induced mitochondrial cytochrome *c*	
	- activation of caspases	
	- increasing NF-κB-binding sites and mediating apoptosis	
Anti-inflammatory	COX-2 down-regulation	Palozza *et al* (2004)
	Reduction of pro-inflammatory cytokines IL-1β and IL-6 mRNA in macrophages	Konishi *et al* (2008)
Cellular communication	Improved gap junction communication	Stahl and Sies (2001)

efficiency of cleavage of provitamin A carotenoids into vitamin A, also referred to as bioefficacy, has been the topic of many controversial discussions. Some organizations such as the Food and Agriculture Organization (FAO)/World Health Organization (WHO) claim an activity of 1:6 compared with pure vitamin A, whereas others suggest 1:12 (Institute of Medicine 2001). Thus, given a dietary reference intake (DRI), *e.g.* the recommended dietary allowance (RDA) for vitamin A of 900 µg day^{-1} for an adult male (Institute of Medicine 2001), an intake of 11 mg of β-carotene would ensure sufficient vitamin A supply, assuming a conversion rate of 1:12. The average dietary β-carotene consumption is typically lower, approx. 5–10 mg, suggesting that additional vitamin A or other provitamin A carotenoids have to be consumed to meet the dietary recommendations for vitamin A intake.

However, the potential biological activity of provitamin A carotenoids, also including non-vitamin-A-related activities, have recently received increased attention due to epidemiological studies associating their consumption within fruits and vegetables with several beneficial health effects, especially for their potential to reduce the incidence of chronic diseases such as cancer and cardiovascular diseases (CVDs). Although often predominantly attributed to their antioxidant potential, carotenoids may further impact on gap junction communication between cells, regulate cell differentiation, growth and apoptosis, and modulate gene expression and interact with drug-metabolizing enzymes (Table 10.1).

10.2 Occurrence of Provitamin A Carotenoids in the Diet

In most native plant food items, both provitamin A carotenoids and non-provitamin A carotenoids occur concurrently. Due to their affiliation with the chlorophylls and the light-harvesting complex, and their presence in chromoplasts, coloured fruits and vegetables, especially leafy vegetables, are rich sources of carotenoids, up to approx. 10–15 mg (100 g)$^{-1}$. Some root and tuber vegetables, such as carrots and sweet potato, respectively, also constitute provitamin A carotenoid-rich sources (Table 10.2). Also plant species residing in the sea, such as green algae, can be rich in provitamin A carotenoids, however, these sources are rarely consumed in Western countries.

Although animals cannot produce carotenoids, some species accumulate carotenoids in their tissues, such as some fish and seafood. For example, salmon is comparatively rich in the non- provitamin A astaxanthin. Eggs can also be a source of secondary carotenoids, especially lutein and zeaxanthin, originating from maize fodder, whereas their β-carotene content is very low (Table 10.2). However, in general, animal food products are poor sources of carotenoids. A number of foods are further fortified with carotenoids, especially β-carotene. Predominant examples include butter, dairy products, such as cheese, and a number of beverages. In addition to a potential prevention of oxidation and the formation of off-flavours or rancidity, such as in margarine, β-carotene is also added for colour preservation, resulting in a

Table 10.2 Dietary sources rich in carotenoids. List of common food items rich in carotenoids, listing both provitamin A carotenoids and total carotenoids. n.d., no data.

Source/food item	Total carotenoids[a] [mg (100 g)$^{-1}$]	Provitamin A carotenoids[b] [mg (100 g)$^{-1}$]	Reference
Carrots, raw	10.4	10.2	O'Neill *et al* (2001)
Sweet potato	n.d.	8.6	Max Rubner-Institut (2011)
Kale, raw	n.d.	5.2	Max Rubner-Institut (2011)
Salmon	n.d.	5	Max Rubner-Institut (2011)
Spinach, cooked	10.8	4.5	O'Neill *et al* (2001)
Grapefruit, pink	4.7	1.3	O'Neill *et al* (2001)
Apricot, fresh	1.1	1.0	O'Neill *et al* (2001)
Lettuce, butterhead	2.1	0.9	O'Neill *et al* (2001)
Chicken eggs	<0.1	0.6	O'Neill *et al* (2001)
Tomato, raw	3.4	0.6	O'Neill *et al* (2001)
Green bell-pepper, raw	n.d.	0.5	Max Rubner-Institut (2011)
Pumpkin, cooked	1.1	0.5	O'Neill *et al* (2001)
Butter	0.4	0.4	O'Neill *et al* (2001)
Green beans, cooked	1.1	0.4	O'Neill *et al* (2001)
Watermelon, peeled, ripe	3.7	0.2	O'Neill *et al* (2001)
Plum, raw, unpeeled	0.1	0.1	O'Neill *et al* (2001)
Sweet corn	0.9	0.1	O'Neill *et al* (2001)

[a]In edible portion. [b]Sum of α- and β-carotene.

yellow hue. Another source are dietary supplements, either in form of multimineral/multivitamin supplements, or as individual supplements, with typical doses of up to 15 mg of β-carotene unit^{-1}, most typically a capsule.

Of the carotenoids present in the diet, the predominant ones include β-carotene, lycopene, lutein, β-cryptoxanthin and α-carotene (O'Neill *et al* 2001). Some databases containing information on carotenoid content include the US carotenoid database (Mangels *et al* 1993), and a European database (O'Neill *et al* 2001), and more general databases such as the German Bundeslebensmittelschlüssel (Max Rubner-Institut 2011). It is interesting to note that many carotenoids have so far not been systematically included in these databases. This is especially true for a number of non-provitamin A carotenoids, such as phytoene or phytofluene, or expoxycarotenoids including violaxanthin and neoxanthin (Biehler *et al* 2011a).

It is further noteworthy that carotenoids are present in different forms in various food items. Although in many vegetables such as tomato and carrots, carotenoids are mostly present as agglomerates in crystalline form in the chromoplasts and/or chloroplasts; carotenoids in some fruits such as oranges and watermelon exist dissolved in oil. These differences are of potential importance, as the presence in crystalline form could correspond with reduced release and solubilization from the food matrix and therefore reduced bioavailability. Another distinguishing form includes esters of hydroxycarotenoids and fatty acids, *i.e.* α- and β-cryptoxanthin esters. Some fruits and vegetables may contain the majority of β-cryptoxanthin in form of various esters (Breithaupt and Bamedi 2001). Due to their increased apolarity, these are also assumed to be of poorer bioavailability compared with the free form.

10.3   Dietary Intake of Provitamin A Carotenoids

In contrast to vitamin A, no dietary intake recommendations exist for provitamin A carotenoids. Nevertheless, with respect to the associated health benefits, dietary carotenoid consumption has been investigated in several countries. Two common ways to study dietary patterns include studying food disappearance, which is based usually on sales data and suffers from inaccuracy due to the fact that not all purchased food is eventually consumed by humans; or, preferably, food consumption studies where data from individual subjects are compiled and averaged. However, even these estimates are easily biased due to: (1) source of data with respect to carotenoid content of food items, (2) the population studied, *e.g.* rural *vs.* metropolitan, (3) assessment methods such as 24 h recall or diet history, and (4) neglecting certain dietary sources such as supplements.

Despite these difficulties, a number of studies have aimed at assessing carotenoid intake, showing that average carotenoid consumption is in the region of 10–20 mg day^{-1}, and approx. 30–50% of this is usually attributed to provitamin A carotenoids (O'Neill *et al* 2001). The majority of the carotenoids consumed are in form of α- plus β-carotene, followed by lycopene, lutein and β-cryptoxanthin. Similar as for the food databases, consumption of carotenoid precursors (phytoene, phytofluene) and several minor abundant carotenoids, such as the expoxycarotenoids, are usually, but not always (Biehler *et al* 2011a) neglected, albeit they could contribute to approx. 25% of total carotenoid intake.

Despite differences between countries in their carotenoid consumption, with, for example, northern countries typically consuming less carotenoids than, for example, Mediterranean countries, carotenoid intake depends on an individual's diet. For example, vegetarians consume more fruits and vegetables than omnivorous people, which was found to translate into approx. 25% higher plasma β-carotene levels (Johnson *et al* 1995).

10.4   Detection of Provitamin A Carotenoids in Food Items and Body Tissues

Following extraction methods, typically employing liquid/liquid extraction such as with hexane/acetone or clean-up by solid phase extraction (SPE), the detection of carotenoids including provitamin A carotenoids most often relies on their strong ultraviolet–visible (UV-Vis) activity, making use of their specific, often three-finger-type absorption bands between 450–500 nm. However, absorption wavelengths could differ, with shorter wavelengths for carotenoids with a less extended conjugated electron system such as phytoene, absorbing in the area of approx. 285 nm. A reduction of the conjugated π-electron system also results in decreased absorption sensitivity, typically expressed by their specific absorption coefficients (Table 10.3).

The easiest quantification of carotenoids is *via* simple spectrophotometric methods. However, carotenoids mostly occur in mixtures, and their detection can be further impeded by the presence of other spectrophotometric active compounds such as chlorophylls. If the aim is to only detect the sum of carotenoids, chlorophylls could be degraded *via*, for example, saponification, and total carotenoid content can be estimated based on the mean of their absorption coefficients (Biehler *et al* 2009). Another method is to subtract chlorophylls mathematically *via* measuring at several wavelengths to distinguish between chlorophylls and carotenoids (Lichtenthaler 1987). However, individual carotenoids cannot be determined.

Thus, prior to their UV-Vis detection, carotenoids are usually separated chromatographically. Although carotenoids are too fragile to be determined by gas chromatography, high-pressure liquid chromatography (HPLC) is able to allow for individual separation. Often as extracts from plants are analyzed which could contain residues of water, reversed phase (RP) chromatography has become the method of choice for carotenoid detection. Due to their long chain length, in addition to the standard RP-18 configuration, special carotenoid columns with RP-30 materials have been developed, which even allow the separation of various isomers, such as of 9-*cis* and all-*trans* β-

Table 10.3   Specific molecular absorption coefficients of carotenoids. Selected, dietary predominant carotenoids, their specific molecular absorption coefficients and their maximum absorption wavelengths in the UV-Vis area.

Carotenoid	Solvent	*Absorption maxima (nm)*	*Specific molecular absorption coefficient [L mol^{-1} cm^{-1}]*	*Reference*
β-Carotene	Acetone	452	140663	Britton *et al* (2004)
Lutein	Ethanol	445	144900	Britton *et al* (2004)
Lycopene	Acetone	448	120600	Britton *et al* (2004)
Zeaxanthin	Acetone	452	133118	Britton *et al* (2004)
β-Cryptoxanthin	Petrol Ether	449	131915	Britton *et al* (2004)
α-Carotene	Hexane	445	145472	Britton *et al* (2004)
Phytoene	Hexane	286	73567	Campbell *et al* (2007)
Phytofluene	Hexane	348	67863	Campbell *et al* (2007)

carotene, and also allow for detecting different lycopene isomers (Hadley *et al* 2003). However, to date, no ultra-pressure/performance liquid chromatography (UPLC) column with this material exists. When employing RP-30 HPLC, the time for a complete carotenoid profile detection is approx. 25–40 min. The detector of choice coupled to HPLC analysis is the diode array detector (DAD), allowing the detection of each individual cartotenoid at its maximum sensitive wavelength.

In addition to UV-Vis based methods, two other techniques are worth mentioning: electrochemical detection (ECD) and mass spectrometry (MS), both of which are usually coupled to HPLC. ECD oxidizes carotenoids at specifically applied electrochemical potentials, thus another dimension is added for separating carotenoids in addition to retention time. ECD equipment, such as with the Coularray detector (ESA, Chelmsford, MA, USA), has been claimed to be 10–1000 times more sensitive compared with UV-Vis detectors, albeit due to the high UV-Vis activity of carotenoids, the gain in detection is possibly on the lower side. However, several methods have reported on the use of ECD for carotenoid analyses (Bohn *et al* 2011; Ferruzzi *et al* 1998), with detection limits as low as 0.1 ng mL^{-1} for standards. This may be the method of choice for studying carotenoid breakdown products or carotenoids with shorter conjugated π-electron systems and reduced UV-Vis activity.

MS is possibly the priciest choice to detect carotenoids. Both HPLC-MS and HPLC-tandem MS (MS-MS) methods, especially in combination with atmospheric pressure chemical ionization (APCI) have been suggested, with the latter technique reaching sensitivities of 0.1 ng mL^{-1} in human plasma (Gundersen *et al* 2007).

10.5 Aspects of Bioavailability of Provitamin A Carotenoids

10.5.1 Overview of Provitamin A Carotenoid Absorption

Factors that impact on provitamin A carotenoids have been summarized by the mnemonic term SLAMENGHI: species of carotenoids, molecular linkage, amount of carotenoids ingested, food matrix effects, dietary factors effecting absorption and bioconversion, nutrient status of the host, genetic factors of the host, host factors such as intestinal passage time, and the interaction of all these factors. The absorption of carotenoids can be thought of as the following sequence (Failla and Chitchumroonchokchai 2005) (see Figure 10.2):

1. Release from food matrix
2. Solubilization and micellarization for cellular uptake
3. Uptake into the epithelial cells of the small intestine
4. Transport through the epithelial cells, partly central cleavage, reaction to retinal, then retinol and re-esterification to, for example, retinyl palmitate

5. Sequestration for further absorption into chylomicrons and transport *via* lymph to the bloodstream
6. Remodelling in the liver and further transport in various lipoprotein fractions
7. Transport to various organs, incorporation into, for example, lipid bilayers
8. Excretion *via* urine and feces, following breakdown into smaller apo-carotenals and unknown compounds.

Thus, all factors impacting any of these processes can affect the bioavailability of provitamin A carotenoids (Table 10.4). In the following, the above stages are outlined in further detail.

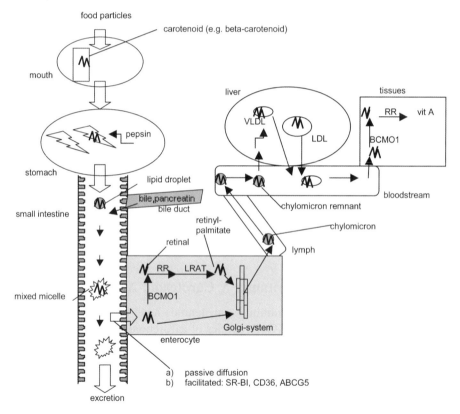

Figure 10.2 Various stages of biodistribution of provitamin A carotenoids following ingestion *via* the diet. The Figure shows carotenoid release from the matrix, micellarization, cellular uptake, cleavage into retinol and re-esterification, sequestration into chylomicrons, remodelling in the liver and further transport in lipoproteins and tissue uptake. RR, retinal reductase; vitA, vitamin A; LRAT, lecithin:retinol acyltransferase; SR-BI, scavenger receptor class B type I; ABCG5, ATB-binding cassette G5; BCMO1, beta-carotene 15,15′-mono.oxygenase 1.

10.5.1.1   Release from Food Matrix

Following dietary intake, carotenoids have first to be released from the food matrix by chewing and enzymatic digestion. The more intact the plant cell matrix, the harder the plant material and cell walls, and the more viscous the food item during gastro-intestinal digestion, the less available the carotenoid fraction is for further uptake. In addition, carotenoids in larger, more crystalline-like agglomerations, such as in the chloroplasts of leafy vegetables, or in chromoplasts, such as in tomato or carrots, could be hypothesized to be less soluble compared with carotenoids present in lipid droplets, such as in orange and watermelon.

Processing, especially the heating of fruits and vegetables, has shown to considerably improve carotenoid absorption, especially that of lycopene, in humans and animals, most likely due to maceration of the plant cell walls, destruction of the crystalline agglomerates or release of the protein-bound carotenoids. However, heating has also been reported to foster *trans–cis* isomerization of carotenoids, *i.e.* formation of 9-*cis* β-carotene from all-*trans* β-carotene. While this form has been reported to undergo micellarization to a larger extent compared with all-*trans* β-carotene, it also appeared to be of lower cellular uptake (Canene-Adams and Erdman 2009). Carotenoids in supplements are usually extracted from carotenoid-rich plants, such as spinach, and can be assumed to be at least equally well released and absorbed compared with dietary sources.

10.5.1.2   Solubilization/Micellarization for Cellular Uptake

Due to their low solubility in water, carotenoids have to undergo micellarization, *i.e.* emulsification prior to their uptake by the epithelium in the small intestine. This micellarization usually requires the presence of some dietary-derived lipids such as triacylglycerols and phospholipids which are later broken down enzymatically, such as into free fatty acids and mono- and diglycerides. During digestion, carotenoids are first present in lipid droplets, which then, during gastro-intestinal digestion, and together with free fatty acids, phospholipids, cholesterol and the bile salts, result in the formation of mixed micelles of approx. 8 nm size (Parker 1996). The more polar carotenoids, such as xanthophylls including β-cryptoxanthin, are located in the outer layers of this micelle, whereas the more apolar ones, such as β-carotene, are believed to be located in the core (Borel *et al* 1996). Due to their improved water solubility, the more polar carotenoids typically show higher micellarization (Biehler *et al* 2011b). The same has been found for the *cis*-forms as opposed to the normally present all-*trans*-forms, perhaps due to lower apparent chain length and easier incorporation into mixed micelles (Ferruzzi *et al* 2006).

A number of factors have been reported to interact with the formation of mixed micelles. Several studies have shown that the consumption of dietary lipids together with a carotenoid-rich meal improves the availability of carotenoids, presumably due to improved micellarization (Biehler *et al* 2011b;

Unlu *et al* 2005). A negative impact on the micellarization process has been suggested by compounds competing with emulsification, such as higher amounts of other carotenoids, *e.g.* lutein, or other lipid-soluble vitamins such as vitamin E, vitamin D or phytosterols, albeit there exists no conclusive data (reviewed by Bohn 2008). Furthermore, a lack of digestion enzymes, such as pancreatin, and bile acids would result in impaired micellarization (Biehler *et al* 2011b). Recently, it has been suggested that high concentrations of divalent minerals could reduce the availability of fatty acids and bile salts in the gut, due to the formation of insoluble soaps, also reducing carotenoid micellarization (Biehler *et al* 2011b).

10.5.1.3   *Uptake into the Epithelial Cells of the Intestine*

Uptake into the epithelial cells of the small intestine is believed to occur following the diffusion of the micelles through the unstirred water layer to the apical site of the mucosal cells. Earlier, it was believed that cellular uptake is based on solely passive diffusion. In recent years, however, based on Caco-2 cellular trials, it was suggested that also facilitated uptake mechanisms for several carotenoids, such as *via* the scavenging receptor type B class I (SR-BI), do play a role. This protein is normally responsible for cholesterol uptake, and blocking this receptor (with ezetimibe) also inhibited uptake of carotenoids, especially of β-carotene and α-carotene, but to a lesser extent that of lycopene and lutein (During *et al* 2005). The fact that β-carotene uptake has been shown to be saturable (Failla and Chitchumroonchokchai 2005) is in further agreement with this assumption. It therefore appears likely that the uptake of the more essential provitamin A carotenoids is additionally regulated at this stage of absorption. Other transport proteins, including cluster determinant 36 (CD36) and the ATB-binding cassette G5 (ABCG5), have also been assumed to play a role during carotenoid uptake, albeit their role remains to be elucidated further.

10.5.1.4   *Transport through the Epithelial Cells, Central Cleavage and Re-Esterification*

Following their uptake into the enterocytes, provitamin A carotenoids may then be cleaved by central enzymatic cleavage, by the enzyme BCMO1, possibly in the cytosol. Retinal is then further reduced by retinal reductase into retinol which is then esterified by lecithin:retinol acyltransferase (LRAT) with C-16 or C-18 fatty acids such as palmitate for further transport. Also, an asymmetric or eccentric cleavage by, for example, β-carotene 9′,10′-dioxygenase (BCDO2) may occur (Lindqvist *et al* 2005), resulting in the formation of different apo-carotenals, compounds that could still possess vitamin A activity, as chain reduction and retinal formation could still occur. Both enzymes appear to be active in a variety of epithelial cells, including the parenchymal cells of the liver, the adrenal gland, kidney tubules, among others. Also non-enzymatic oxidation of carotenoids to apo-carotenals could occur, but this is unlikely to be the major pathway of biotransformation, as other animal species, such as the cat, lack the ability to produce significant amounts of vitamin A from β-carotene. 9-*cis* β-Carotene, which may be formed during food processing or gastro-intestinal digestion, is believed to be isomerized into its *trans*-form within the enterocytes, either prior to or after cleavage, as their plasma levels are very low.

Based on human studies with isotopically labelled β-carotene, it has been estimated that 35–71% of absorbed β-carotene can be cleaved into vitamin A active compounds (van Vliet *et al* 1995), albeit lower conversions of approx. 20% have also been discussed (reviewed by Parker 1996). Factors affecting the intracellular cleavage process are only poorly understood, but it can be speculated that re-esterification could depend on the presence of recently absorbed dietary-derived fatty acids.

10.5.1.5 Sequestration for Further Absorption into Chylomicrons and Transport in Lymph and Blood

The non-cleaved carotenoids and the re-esterified retinyl palmitate are then sequestered *via* the Golgi apparatus of the enterocyte into chylomicrons, which then reach via the lymphatic system and the thoracic duct the bloodstream, where they are transformed under the cleaving influence of lipoprotein lipase into chylomicron remnants. As only newly absorbed cartotenoids can be found in this fraction, this is a preferred analytical target to study carotenoid absorption by investigating the so-called triacylglycerol-rich lipoprotein fraction (TRL), mostly consisting of chylomicrons (Parker 1996), especially its area under a curve (AUC) (concentration *vs.* time).

The sequestration of chylomicrons by the Golgi apparatus is assumed to be impacted on by following test meals containing dietary lipids, as carotenoid absorption following a single carotenoid meal is usually biphasic, with a second peak in the plasma-TRL fraction following a carotenoid-free but fat-containing lunch (Unlu *et al* 2005). The additional lipids are assumed to foster the sequestration of additional chylomicrons into the lymph and therefore facilitate carotenoid absorption.

10.5.1.6 Remodelling in the Liver and Further Transport in Lipoproteins

In the liver, chylomicron remnants are endocytosed by parenchymal hepatocytes and may be stored or packaged into various lipoproteins, considered to be the only transport vehicle of carotenoids in the bloodstream. While the more apolar carotenes are transported preferably by very-low-density lipoproteins (VLDLs) and low-density lipoproteins (LDLs), approx. 10–16% and 58–73%, respectively, the dihydroxy-xanthophylls are rather associated with high-density lipoproteins (HDLs; approx. 53%, *vs.* 31% and 16% in LDLs

Table 10.4   Factors impacting aspects of provitamin A cartotenoid bioavailability. The impact of positive and negative host and external factors on provitamin A carotenoid availability at various steps of biodistribution. LCFAs, long-chain fatty acids; N/A, not available; TGs, triacylglycerols; ROS, reactive oxygen species; SR-B1, scavenger receptor class B type 1.

Stage of biodistribution	Factor influencing bioavailability	Positive/negative effect	Literature
Digestion, release	Dietary fibre	−	Failla and Chitchumroonchokchai (2005)
	Viscous matrix	−	Biehler *et al* (2011b)
	Heat processing of food	+	Unlu *et al* (2007)
Micellarization	Competing compounds (*e.g.* other carotenoids)	−	Bohn (2008)
	Minerals	−	Biehler *et al* (2011b)
	Dietary lipids	+	Unlu *et al* (2005)
	Presence as carotenoid ester	−	Bohn (2008)
	Lack of digestion enzymes or bile salts	−	Biehler *et al* (2011b)
Cellular uptake	Drugs blocking SR-B1 (*e.g.* ezetemibe)	−	During *et al* (2005)
	High viscosity of chyme	−	Failla and Chitchumroonchokchai (2005)
Cleavage into vitamin A	Presence of vitamin A in diet	−	Failla and Chitchumroonchokchai (2005)
Chylomicron sequestration	Following test meals containing TGs composed of LCFAs	+	Bohn *et al* (2011)
Biodistribution	Unknown	N/A	N/A
Metabolism/ excretion	Presence of ROS, UV-Vis damage of skin	+	Speculative

and VLDLs, respectively). β-Cryptoxanthin was found to equally partition between LDLs and HDLs, approx. 40% each (Furr 2004). Overall, LDLs are believed to account for >50% of total carotenoid transport in the human body (Canene-Adams and Erdman 2009). Peak appearance of β-carotene in LDLs were found approx. 24–48 h following dietary intake (reviewed by Parker 1996), which is much delayed compared with the plasma-TRL fraction, where carotenoids usually peak at approx. 4–6 h following absorption.

In general, the most prevalent carotenoids found in plasma are lutein and lycopene, followed by β-carotene and ζ-carotene, β-cryptoxanthin and α-carotene (Khachik 2006), which together possibly contribute to approx. 90% of the carotenoids in the human body (Rao and Rao 2007). However, the profile of carotenoids in plasma can vary largely and is thought to reflect the general dietary patterns. On carotenoid-depleted diets, significant reductions in carotenoid blood levels by over 50% were detected, with half-lives between 26–76 days. The normal average concentrations of carotenoids in blood plasma are in the range of approx. 0.3–0.7 μmol L (Canene-Adams and Erdman 2009).

10.5.1.7 Target Organs in the Human Body

There exists no predominant single target organ for provitamin A carotenoids, therefore, these can be found in all tissues. Instead, there are several organs with increased affinity, depending presumably on the number of LDL receptors and SR-BIs. Adipose tissue, prostate, liver, testis and adrenal tissue possess a high expression of LDL receptors (reviewed by Biehler and Bohn 2010) and therefore constitute preferential targets for provitamin A carotenoids. Adipose tissue may thus constitute a useful marker of long-term carotenoid intake (Canene-Adams and Erdman 2009). In addition, comparatively high concentrations of β-carotene and degradation products were found also in lung, breast and colonic tissue (Khachik *et al* 2002). A number of carotenoids including β-carotene can also be detected in the skin where they could offer protection from UV-induced damage. In addition, the liver remains the major storage tissue for provitamin A carotenoids, and returning lipoproteins including their carotenoids could be remodelled in this organ. In all organs, additional provitamin A carotenoid conversion into vitamin A can continue.

10.5.1.8 Aspects of Excretion of Provitamin A Carotenoids and Vitamin A

Not much is known on the further biological fate of provitamin A carotenoids following uptake by body tissue cells. In lipid bilayers, provitamin A carotenoids have been reported to be present in the core of the phospholipid membrane. This is in contrast to the xanthophylls which reside, due to increased hydrophilicity, more in the outer parts (Parker 1996). Thus, a high fraction of carotenoids is usually affiliated with the membrane fraction of cells, protecting these from ROS and lipid peroxidation.

Excretion of non-absorbed provitamin A caroteoids occurs mostly *via* the feces, whereas absorbed provitamin A carotenoids are partly excreted *via* the bile or pancreas into the feces. Interestingly, large amounts of isotopically labelled β-carotene were recovered in the urine (approx. 35% of those absorbed) (Lemke *et al* 2003), possibly following breakdown *via* shorter apo-carotenals and their hydroxylated products.

Summary Points

- This Chapter focuses on provitamin A carotenoids.
- The main sources of provitamin A carotenoids include leafy vegetables and coloured fruits, containing up to approx. 15 mg/100 g of provitamin A carotenoids.
- The main dietary provitamin A source is β-carotene, followed by β-cryptoxanthin and α-carotene.
- Predominant detection and quantification methods include spectrophotometry (UV-Vis) and HPLC, especially reversed phase HPLC coupled to a diode array detector.
- Bioavailability is low, mainly due to low bioaccessibility and limited cellular uptake.
- Health aspects include not only vitamin A aspects, but also anti-inflammatory, anti-proliferative and antioxidant effects.

Key Facts

Key Facts of General Properties of Provitamin A Carotenoids

- Among the primary function of provitamin A carotenoids is their central cleavage in various cells of the human body by 15,15′-monooxygenase and their subsequent transformation into retinoic acid.
- β-Carotene has the highest provitamin A activity of all carotenoids, as its central cleavage will eventually produce two vitamin A molecules. A key feature of possessing vitamin A activity is the intact β-ionone ring.
- α-Carotene, γ-carotene, α-cryptoxanthin and β-cryptoxanthin can result in only one vitamin A molecule following central cleavage.
- The bioefficacy, *i.e.* the efficiency of provitamin A carotenoids to result in the formation of vitamin A in the human body, is controversial, and estimates include both a biological activity of 1:6 and 1:12 compared with vitamin A.
- In addition to their potential vitamin A activity, provitamin A carotenoids have, especially in *in vitro* and animal studies, further shown to possess anti-inflammatory, anti-proliferative and pro-apoptotic properties. Their interaction with gene expression has also been highlighted.

Key Facts of Dietary Intake and Bioavailability of Provitamin A Carotenoids

- Rich dietary sources are leafy vegetables and some coloured fruits. Dietary carotenoids are either bound to proteins, *i.e.* to the light-harvesting complex in the chloroplasts of plants, occur in rather crystalline form within chromoplasts, appear dissolved in lipid droplets as in some fruits.

- The general dietary intake of provitamin A carotenoids is approx. 5–10 mg day^{-1}, which may not be enough to result by itself in sufficient vitamin A supply to meet the recommended dietary intake (approx. 0.9 mg day^{-1}).
- Even if high amounts of provitamin A carotenoids are consumed, they are typically of low bioaccessibility and absorption. Co-ingested dietary lipids can enhance their absorption, whereas dietary fibre will possibly reduce bioavailability.
- Earlier thought to occur by passive diffusion, cellular uptake of provitamin A carotenoids could also take place *via* facilitated uptake *via* proteins (e.g. SR-BI).
- Absorption of newly consumed carotenoids in the body can be followed by examining the plasma chylomicron fraction, while plasma and especially LDL-affiliated carotenoids represent the major circulating fraction.

Definitions of Words and Terms

Antioxidant capacity: The potential of a nutrient or phytochemical to act as an antioxidant and to prevent the formation or negative activities of ROS.

Bioaccessibility: The amount of a nutrient or phytochemical that can be released from the food matrix, solubilized and be available for further uptake and absorption.

Bioavailability: The fraction of a nutrient or phytochemical that can be absorbed and is available for its physiological functions and/or storage.

Bioefficacy: In this Chapter, the efficiency or extent to which provitamin A carotenoids are cleaved by central cleavage and result eventually in the formation of vitamin A.

DRI-RDA: The dietary reference intake (DRI) of the US Food and Nutrition Board is an umbrella term for intake recommendations of certain nutrients, especially vitamins and minerals. The recommended dietary allowance (RDA) is the intake that supplies the majority of the population (>97%) with a sufficient amount to ensure optional body health and functions.

Phytochemical: A dietary compound with no strict essentiality for humans, but which may nevertheless an impact on the human body. Examples include polyphenols and carotenoids.

Provitamin A carotenoids: Carotenoids that can result, following enzymatic central cleavage in human cells, in the formation of vitamin A. Only carotenoids with an intact β-ionone ring possess provitamin A activity.

Provitamin A carotenoid status: A measure of the overall supply of the human body with provitamin A carotenoids. The best measure of a status of a nutrient is the measurement of its concentration in the primary target organ. As such a target tissue does not exist for carotenoids, measurements usually include blood plasma or adipose tissue.

Reactive oxygen species (ROS): These can be classified into radicals and non-radicals, but both types have the potential to result in cellular damage, such as of the lipid bilayer membrane, or could result in protein and DNA damage. Examples include the hydroxyl radical (OH·) or hydrogen peroxide (H_2O_2).

Uptake: Transport of a nutrient or phytochemical from an extracellular space (*e.g.* the intestinal lumen) through the cellular membrane (*e.g.* of the enterocyte) into the cytosol of the cell.

List of Abbreviations

ABCG5	ATB-binding cassette G5
BCMO1	β-carotene 15,15′-mono-oxygenase 1
CD36	cluster determinant 36
ECD	electrochemical detection
HDL	high-density lipoprotein
HPLC	high-pressure liquid chromatography
LDL	low-density lipoprotein
LRAT	lecithin:retinol acyltransferase
MS	mass spectrometry
ROS	reactive oxygen species
RP	reversed phase
RR	retinol reductase
SR-BI	scavenger receptor class B type I
TRL	triacylglycerol-rich lipoprotein fraction
UV-Vis	ultraviolet or visible light
VLDL	very-low-density lipoprotein

References

Biehler, E. and Bohn, T., 2010. Methods for assessing aspects of carotenoid bioavailability. Current Nutrition and Food Science. 6: 44–69.

Biehler, E., Mayer, F., Hoffmann, L., Krause, E. and Bohn, T., 2009. Comparison of 3 spectrophotometric methods for carotenoid determination in frequently consumed fruits and vegetables. Journal of Food Science. 75: C55–C61.

Biehler, E., Alkerwi, A., Hoffman, L., Krause, E., Guillaumec, M., Lair, M. L. and Bohn, T., 2011a. Contribution of violaxanthin, neoxanthin, phytoene and phytofluene to total carotenoid intake – assessment in Luxembourg. Journal of Food Composition and Analysis. 25: 56–65. (submitted April 2011).

Biehler, E., Kaulmann, A., Krause, E., Hoffman, L. and Bohn, T., 2011b. Dietary and host-related factors influencing carotenoid bioaccessibility from spinach (*Spinacia oleracea*). Food Chemistry. 125: 1328–1334.

Bohn, T., 2008. Bioavailability of non-provitamin A carotenoids. Current Nutrition and Food Science. 4: 240–258.

Bohn, T., Blackwood, M., Francis, D., Schwartz, S. J. and Clinton, S. K., 2011. Bioavailability of phytochemical constituents from a novel soy fortified lycopene rich tomato juice developed for targeted cancer prevention trials. Nutrition and Cancer (in press). DOI: 10.1080/01635581.2011.630156, ahead of print.

Borel, P., Grolier, P., Armand, M., Partier, A., Lafont, H., Lairon, D. and Azais-Braesco, V., 1996. Carotenoids in biological emulsions: solubility, surface-to-core distribution, and release from lipid droplets. Journal of Lipid Research. 37: 250–261.

Bouayed, J. and Bohn, T., 2010. Exogenous antioxidants – double-edged swords in cellular redox state: health beneficial effects at physiologic doses versus deleterious effects at high doses. Oxidative Medicine and Cellular Longevity. 3: 228–237.

Breithaupt, D. E. and Bamedi, A., 2001. Carotenoid esters in vegetables and fruits: a screening with emphasis on β-cryptoxanthin esters. Journal of Agricultural and Food Chemistry. 49: 2064–2070.

Britton, G., Liaaen-Jensen, S. and Pfander, H., 2004. Carotenoid Handbook. Birkhäuser, Basel, Switzerland.

Campbell, J. K., Engelmann, N. J., Lila, M. A. and Erdman, Jr, J. W., 2007. Phytoene, phytofluene, and lycopene from tomato powder differentially accumulate in tissues of male Fisher 344 rats. Nutrition Research (New York). 27: 794–801.

Canene-Adams, K. and Erdman, J. W., 2009. Absorption, transport, distribution in tissues and bioavailability. In: Britton, G., Liaaen-Jensen, S. and Pfander, H. (ed.) Carotenoids: Volume 5: Nutrition and Health. Birkhäuser, Basel, Switzerland, pp. 115–148.

During, A., Dawson, H. D. and Harrison, E. H., 2005. Carotenoid transport is decreased and expression of the lipid transporters SR-BI, NPC1L1, and ABCA1 is downregulated in Caco-2 cells treated with ezetimibe. Journal of Nutrition. 135: 2305–2312.

Failla, M. and Chitchumroonchokchai, C., 2005. *In vitro* models as tools for screening the relative bioavailabilities of provitamin A carotenoids. Harvest Plus, Washington DC, USA.

Fang, J., Seki, T. and Maeda, H., 2009. Therapeutic strategies by modulating oxygen stress in cancer and inflammation. Advanced Drug Delivery Reviews. 61: 290–302.

Ferruzzi, M. G., Sander, L. C., Rock, C. L. and Schwartz, S. J., 1998. Carotenoid determination in biological microsamples using liquid chromatography with a coulometric electrochemical array detector. Analytical Biochemistry. 256: 74–81.

Ferruzzi, M. G., Lumpkin, J. L., Schwartz, S. J. and Failla, M., 2006. Digestive stability, micellarization, and uptake of β-carotene isomers by Caco-2 human intestinal cells. Journal of Agricultural and Food Chemistry. 54: 2780–2785.

Furr, H. C., 2004. Analysis of retinoids and carotenoids: problems resolved and unsolved. Journal of Nutrition. 134: 281S–285S.

Gundersen, T. E., Bastani, N. E. and Blomhoff, R., 2007. Quantitative high-throughput determination of endogenous retinoids in human plasma using triple-stage liquid chromatography/tandem mass spectrometry. Rapid Communications in Mass Spectrometry. 21: 1176–1186.

Hadley, C. W., Clinton, S. K. and Schwartz, S. J., 2003. The consumption of processed tomato products enhances plasma lycopene concentrations in association with a reduced lipoprotein sensitivity to oxidative damage. Journal of Nutrition. 133: 727–732.

Halliwell, B., 2000. Lipid peroxidation, antioxidants and cardiovascular disease: how should we move forward? Cardiovascular Research. 47: 410–418.

Institute of Medicine, 2001. Dietary reference intakes for vitamin A, vitamin K, arsenic, boron, chromium, copper, iodine, iron, manganese, molybdenum, nickel, silicon, vanadium, and zinc. National Academy Press, Washington, D.C., USA.

Johnson, E. J., Suter, P. M., Sahyoun, N., Ribaya-Mercado, J. D. and Russell, R. M., 1995. Relation between β-carotene intake and plasma and adipose tissue concentrations of carotenoids and retinoids. American Journal of Clinical Nutrition. 62: 598–603.

Khachik, F., 2006. Distribution and metabolism of dietary carotenoids in humans as a criterion for development of nutritional supplements. Pure and Applied Chemistry. 78: 1551–1557.

Khachik, F., de Moura, F. F., Zhao, D. Y., Aebischer, C. P. and Bernstein, P. S., 2002. Transformations of selected carotenoids in plasma, liver, and ocular tissues of humans and in nonprimate animal models. Investigative Ophthalmology and Visual Science. 43: 3383–3392.

Konishi, I., Hosokawa, M., Sashima, T., Maoka, T. and Miyashita, K., 2008. Suppressive effects of alloxanthin and diatoxanthin from *Halocynthia roretzi* on LPS-induced expression of pro-inflammatory genes in RAW264.7 cells. Journal of Oleo Science. 57: 181–189.

Lemke, S. L., Dueker, S. R., Follett, J. R., Lin, Y., Carkeet, C., Buchholz, B. A., Vogel, J. S. and Clifford, A. J., 2003. Absorption and retinol equivalence of β-carotene in humans is influenced by dietary vitamin A intake. Journal of Lipid Research. 44: 1591–1600.

Lichtenthaler, H. K., 1987. Chlorophylls and carotenoids, the pigments of photosynthetic biomembranes. In: Douce, R. and Packer, L. (ed.) Methods in Enzmology. Academic Press, New York, USA, pp. 350–382.

Lindqvist, A., He, Y. G. and Andersson, S., 2005. Cell type-specific expression of β-carotene 9′,10′-monooxygenase in human tissues. Journal of Histochemistry and Cytochemistry. 53: 1403–1412.

Mangels, A. R., Holden, J. M., Beecher, G. R., Forman, M. R. and Lanza, E., 1993. Carotenoid content of fruits and vegetables: an evaluation of analytic data. Journal of the American Dietetic Association. 93: 284–296.

Max Rubner-Institut, 2011. Bundeslebensmittelschlüssel. Available at: http://www.bls.nvs2.de/. Accessed 20 May 2011.

O'Neill, M. E., Carroll, Y., Corridan, B., Olmedilla, B., Granado, F., Blanco, I., Van den Berg, H., Hininger, I., Rousell, A. M., Chopra, M., Southon, S. and Thurnham, D. I., 2001. A European carotenoid database to assess carotenoid intakes and its use in a five-country comparative study. British Journal of Nutrition. 85: 499–507.

Palozza, P., Serini, S., Di Nicuolo, F. and Calviello, G., 2004. Modulation of apoptotic signalling by carotenoids in cancer cells. Archives of Biochemistry and Biophysics. 430: 104–109.

Parker, R. S., 1996. Absorption, metabolism, and transport of carotenoids. FASEB Journal. 10: 542–551.

Rao, A. V. and Rao, L. G., 2007. Carotenoids and human health. Pharmacological Research. 55: 207–216.

Reddy, M. K., Alexander-Lindo, R. L. and Nair, M. G., 2005. Relative inhibition of lipid peroxidation, cyclooxygenase enzymes, and human tumor cell proliferation by natural food colors. Journal of Agricultural and Food Chemistry. 53: 9268–9273.

Reifen, R., Nissenkorn, A., Matas, Z. and Bujanover, Y., 2004. 5-ASA and lycopene decrease the oxidative stress and inflammation induced by iron in rats with colitis. Journal of Gastroenterology. 39: 514–519.

Stahl, W. and Sies, H., 2001. Effects of carotenoids and retinoids on gap junctional communication. Biofactors. 15: 95–98.

Unlu, N. Z., Bohn, T., Clinton, S. K. and Schwartz, S. J., 2005. Carotenoid absorption from salad and salsa by humans is enhanced by the addition of avocado or avocado oil. Journal of Nutrition. 135: 431–436.

Unlu, N. Z., Bohn, T., Francis, D. M., Nagaraja, H. N., Clinton, S. K. and Schwartz, S. J., 2007. Lycopene from heat-induced *cis*-isomer-rich tomato sauce is more bioavailable than from all-*trans*-rich tomato sauce in human subjects. British Journal of Nutrition. 98: 1–7.

van Vliet, T., Schreurs, W. and van den Berg, H., 1995. Intestinal β-carotene absorption and cleavage in men: response of β-carotene and retinyl esters in the triglyceride-rich lipoprotein fraction after a single oral dose of β-carotene. American Journal of Clinical Nutrition. 62: 110–116.

CHAPTER 11

Vitamin A – Serum Vitamin A Analysis

RONDA F. GREAVES[a,b]

[a] Clinical Biochemistry, School of Medical Sciences, RMIT University, PO Box 71, Bundoora, Victoria 3083, Australia; [b] Centre for Hormone Research, Murdoch Children's Research Institute, Flemington Rd, Parkville, Victoria 3052, Australia
E-mail: ronda.greaves@rmit.edu.au

11.1 Introduction

11.1.1 Preamble

The clinical importance of the first fat soluble vital amine, vitamin A, (Funk 1912; McCollum and Kennedy 1916) was described in 1938 by Dr Dorothy Andersen in association with cystic fibrosis (CF) of the pancreas (Andersen 1938). As the mechanism and therefore the understanding of the disease CF evolved so did the association of the deficiency of other fat soluble vitamins, including D, E and K. Today, we appreciate that pathological processes are associated with both deficiency and excess absorption of these fat soluble vitamins. However, vitamin A still holds centre stage, as its deficiency has been listed by the World Health Organization (WHO) to be in epidemic proportions in developing countries (World Health Organization 2009). To predict deficient from adequate or excessive vitamin intake, it is essential that the analytical method employed to measure serum/plasma vitamin concentrations is robust and fit for purpose.

This Chapter describes the methodology employed, including its advantages and limitations, for serum/plasma vitamin A in a clinical setting.

Food and Nutritional Components in Focus No. 1
Vitamin A and Carotenoids: Chemistry, Analysis, Function and Effects
Edited by Victor R Preedy
Published by the Royal Society of Chemistry, www.rsc.org

11.1.2 Definitions, Nomenclature and Terminology

The term "vital amine", which soon became known as "vitamine" was coined in 1912 by Casimir Funk in his ground-breaking work "The Vitamines" in which he described a growth factor present in food which was essential for life (Funk 1912). This was followed quickly by McCollum's insight in 1916 where it was established that there was more than one growth factor and divided them into two classes: fat soluble A and water soluble B (McCollum and Kennedy 1916). Four years later, in 1920, the proposal was put forward to delete the "e" from "vitamine" in order to remove the incorrect assumption that the structures were related to amines (Drummond 1920). This coincided with the alphabetical expansion of the vitamin series which progressed into the 13 currently recognised vitamins, four fat-soluble (A, D, E and K) and nine water-soluble (the eight B group vitamins and vitamin C) vitamins. The inclusion of these 13 compounds as vitamins is based on the definition of a vitamin as an essential organic compound required as a nutrient in minute quantities which cannot be synthesised in adequate amounts and therefore must be obtained in the diet.

In 1960 vitamin A was recognised to exist in three parent forms; retinal, retinol and retinoic acid (International Union of Pure and Applied Chemistry 1960). The term vitamin A is now defined as "the generic descriptor for retinoids exhibiting qualitatively the biological activity of retinol" (International Union of Pure and Applied Chemistry and International Union of Biochemistry 1982). Retinol is the compound routinely measured in serum or plasma vitamin A assays; the structure of which is provided in Figure 11.1.

For chromatographic analysis, as presented in this Chapter, the term "vitamin A" may be considered to be equivalent to the term "retinol".

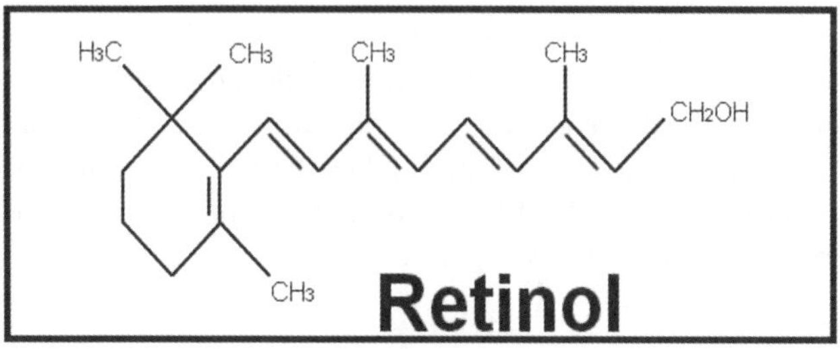

Figure 11.1 Chemical structure of retinol. Chemical structure of (2*E*,4*E*,6*E*,8*E*)-3,7-dimethyl-9-(2,6,6-trimethylcyclohex-1-en-1-yl)nona-2,4,6,8-tetraen-1-ol, which is commonly known as retinol (also known as vitamin A and vitamin A1). The structure of retinol exhibits the characteristics of a monocyclic ring with an acyclic carbon arm ending with the CH_2OH functional group and includes a total of five carbon-carbon double bonds (International Union of Pure and Applied Chemistry and International Union of Biochemistry 1982).

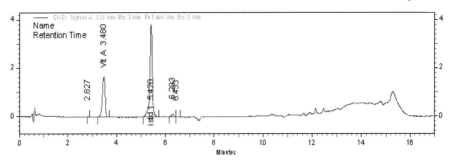

Figure 11.2   Vitamin A chromatogram. Example of a HPLC chromatogram demonstrating the separation of vitamin A from its internal standard (retinol acetate) with an octadecylsilane column with ultraviolet (UV) detection at 325 nm.

11.1.3   Role of Vitamin A in the Body

The intake of vitamin A is either in the form of retinyl esters (from animal products) or carotenoids especially β-carotene (from pigmented fruit and vegetables). Vitamin A is transported in the plasma bound to a complex of retinol-binding protein and prealbumin (transthyretin) (World Health Organization 2011). Normally, the liver stores most of the vitamin A for the body. The levels of vitamin A in plasma are thought to be homeostatically regulated, with stores from the liver released into plasma in response to tissue demand (Penniston and Tanumihardjo 2006). This infers that minor changes in storage levels will not influence plasma vitamin A levels and only pathological extremes will result in significant changes to vitamin A levels in plasma.

   The most renowned role of vitamin A is with the eye and the formation of retinal which complexes to opsin to form rhodopsin (a photoreceptor pigment in the rods of the retina) to allow for dim light vision. In addition to this important role in the chemical events associated with visual excitation, vitamin A has also been implicated in protein synthesis, development of skin tissue and mucous membranes, electron transfer reactions and maintenance of the integrity of cell membranes or cell organelles. In short, vitamin A is required for eye sight, normal growth and development, reproduction, immunity and health of epithelial tissues including skin.

11.1.4   Pathophysiology

11.1.4.1   Hypovitaminosis A

The earliest symptom of vitamin A deficiency (VAD) is impaired vision. This impaired vision starts as night blindness but can progress to total and irreversible blindness. Normally, vitamin A is oxidised in the rods of the eyes to retinal which then complexes to opsin to form rhodopsin, which allows for the dim light vision. However, in VAD, the depletion of retinal in the retina causes opsin to be destabilised and catabolized, causing the permanent destruction of the rod cells which results in night blindness. As impaired vision is one of the earliest clinical manifestations of VAD, it is often used as part of the initial clinical assessment of vitamin A status.

11.1.4.1.1 WHO

VAD is listed by the WHO as a major health issue in developing countries and is the main cause of preventable childhood blindness. In 1987, the WHO estimated that VAD was endemic in 39 countries, based on the ocular manifestations of xerophthalmia or deficient serum (or plasma) retinol concentrations (<0.35 μmol L^{-1}). This estimate was updated in 1995, recognising VAD to be of public health significance in 60 countries, and likely to be a problem in an additional 13 countries. The current estimates released in 2009, based on 1995 to 2005 data, indicates VAD is of public health significance in preschool children in 45 countries based on the prevalence of night blindness and 122 countries based on biochemical vitamin A deficiency (serum retinol < 0.70 μmol L^{-1}) (World Health Organization 2009).

The WHO also recognises the relationship between VAD and increased risk of morbidity and mortality, which affects the most vulnerable – pre-school children and pregnant women. Based on data from the year 2000, 60% of deaths among pre-school children in non-industrialised countries were related to malnutrition (WHO 2009). Improvement in the vitamin A status in populations with VAD is recognised to decrease the risk of mortality associated with malnutrition and other causes of mortality by 23% (Beaton 1993).

11.1.4.1.2 CF and Pancreatic Insufficiency

In developed countries the prime objectives for measuring vitamin A are to assess recent immigrants from developing countries for VAD and to monitor patients with malabsorption of fat. The impaired absorption of fat may accompany a variety of pathologies including: pancreatic dysfunction, *e.g.* CF; obstructive liver disease, *e.g.* biliary atresia; intestinal diseases, *e.g.* coeliac disease; and following surgical removal of a large part of the intestine.

Acute vitamin A deficiency is commonly associated with CF due to pancreatic insufficiency. As vitamin A is fat soluble, exocrine pancreatic enzymes are required for absorption of vitamin A in association with triglycerides. Classical CF includes deficiency of the relevant pancreatic enzymes which results in both fatty stools and a deficiency of fat-soluble vitamins including vitamin A. These CF patients routinely receive enzyme replacement therapy in association with vitamin supplements such as VitABDECK (Technipro-PulmoMed, Sydney, Australia). Plasma levels of vitamin A and the other fat-soluble vitamins are monitored to verify the adequacy of such replacement therapy.

11.1.4.2   Hypervitaminosis A

Hypervitaminosis A can lead to toxic effects such as loss of hair, joint pains, drowsiness, headaches, vomiting, skin defects, increased intracranial pressure, abdominal pains, excessive sweating and brittle nails.

The concentration of serum vitamin A that indicates toxicity is not well defined and varies between acute and chronic toxicity. Acute toxicity is associated with an intake of 20–100 times the recommended daily intake. Chronic toxicity is associated with a daily intake of 25 000 international units (IU) for 6 years or 100 000 IU for 6 months and serum levels may be within the reference intervals. Fasting retinyl ester concentrations >10% of total circulating vitamin A has been proposed as a biomarker for toxicity (Penniston and Tanumihardjo 2006).

Vitamin A toxicity can occur from over-dosage of the vitamin. The symptoms of over-dosage include hepatomegaly, nausea, jaundice, irritability, blurred vision, headaches, muscle and abdominal pain, drowsiness and an altered mental state. Acute toxicity will be associated with serum vitamin A levels above the reference interval.

Long-term (chronic) over-dosing can lead to hair loss, drying of the mucous membranes, insomnia, weight loss, bone fractures, anaemia and diarrhoea. As vitamin A is fat soluble, the tissue (storage) levels of vitamin A may be built up, however, the serum vitamin A level may still be within the reference interval.

In summary, supplementation of deficient individuals is common either directly or in association with pancreatic enzyme replacement therapy. The beneficial range of vitamin A exists within a narrow window and toxicity can be associated with both acute and chronic supplementation of vitamin A. Acute toxicity is readily assessed with the measurement of vitamin A levels in serum. However, measurement of serum vitamin A lacks sensitivity for the assessment of chronic toxicity but may still be important to monitor vitamin supplementation in association with clinical signs to prevent the effects of vitamin A toxicity (Penniston and Tanumihardjo 2006). In practice, analysis of serum vitamin A is infrequently requested to assess toxicity as the turnaround time to generate a result is relatively slow and therefore not clinically expedient.

11.2   Measurement

11.2.1   Overview of Method

Vitamin A, as retinol, is routinely measured by high-pressure liquid chromatography (HPLC) with spectrophotometric detection for clinical purposes. To prepare the serum, proteins binding vitamin A are precipitated using solvents such as ethanol which usually contains the internal standard, *e.g.* retinol acetate. The aqueous phase, which contains the freed fat-soluble components, is then extracted by the addition of a liquid organic solvent such as hexane. The organic layer is removed, evaporated and then re-dissolved in a solvent that is compatible with the mobile phase to be used for the chromatographic separation of the sample constituents. A proportion of this prepared sample is injected on to the HPLC column. This separation is usually performed with an alkane-bonded silica column and vitamin A is quantitated spectrophotometrically at its absorption maximum of 325 nm.

A 2007 external quality assurance participant questionnaire demonstrated the methods commonly employed for routine clinical measurement of vitamin A. This questionnaire serves as the basis to further describe the routine clinical methods for measurement of vitamin A in Section 11.2 (Greaves *et al* 2010).

11.2.2 Pre-analytical Considerations

11.2.2.1 Specimen Collection

Vitamin A is photo-labile, therefore samples for vitamin A analysis must be protected from fluorescent lighting and sun light from the time of collection by wrapping samples in foil (Young 1997).

Generally, 2 mL of plain or heparinised plasma is sufficient to collect by venipuncture for vitamin A analysis. Provided sufficient sample can be collected, the equivalent may be collected by capillary or arterial phlebotomy techniques. The sample tube should be wrapped in foil or equivalent immediately post collection to protect it from light.

Despite vitamin A levels increasing post-prandially, samples are often collected from non-fasting patients. Where there is doubt, vitamin A should be measured in a fasting patient to ensure there is no pre-analytical cause resulting in an artefactual increase in serum vitamin A.

11.2.2.2 Specimen Processing

Once collected, the sample should be sent to the laboratory in a timely manner as relevant studies have not been conducted on the stability over hours while vitamin A remains in whole blood.

On receipt by the laboratory, the sample requires centrifugation to separate the serum from the cells. An aliquot of serum should be frozen at –20 °C until analysis (Comstock *et al* 1993). Vitamin A remains stable with freeze–thaw cycles (Hsing *et al* 1989). During these steps, the sample should continue to be protected from light.

11.2.3 Sample Preparation for Analysis

Vitamin A is often analysed simultaneously with vitamin E and less frequently with β-carotene and other carotenoids for clinical purposes. Therefore the sample preparation described below may be generally extrapolated for these analytes.

The frozen aliquot should be thawed at room temperature, protected from light and then mixed to ensure homogeneity of the serum. Between 100 and 500 μL of this aliquot is pipetted into either borosilicate glass or plastic tubes for sample cleanup. A stabiliser (*e.g.* butylated hydroxytoluene or ascorbate) may be added to the sample at this stage but the advantage of inclusion is not clear and not used by most clinical laboratories.

Note: calibrators and controls are treated in the same manner as the patient samples throughout the method.

Typically laboratories perform a liquid extraction step for sample preparation which includes the addition of an internal standard (*i.e.* a compound that is similar to the compound of interest that is added to each sample as a constant to account for variations during sample preparation and analysis). Solid phase extraction as a preparation technique is not routinely used by clinical testing laboratories.

For analysis, retinol needs to be freed from its binding proteins; this is performed by the addition of a precipitant such as ethanol. This solution usually contains the internal standard, which is used to account for losses during processing. A common and appropriate internal standard is retinol acetate. Tocopherol acetate and δ-tocopherol are also used by some, however, the selection of these internal standards are less ideal (for vitamin A) as they do not share the same similarity in structure and absorbance maximum as retinol acetate, and δ-tocopherol is found in small amounts in serum. After addition of the precipitating solution, the samples should be mixed by vortexing or similar.

Hexane is commonly added as the organic solvent for the liquid extraction of vitamin A. Isopropanol is used by some in isolation or as part of a hexane/isopropanol mix for the extraction. After addition of the organic solvent, the sample is mixed for approx. 10 min. Vitamin A will move into the organic phase in preference to remaining in the aqueous phase due to its hydrophobicity.

The ratio of the precipitating solution to the extraction solvent in relation to the sample varies between clinical laboratories. In general, the ratio of sample to precipitating solution is 1:1. The ratio of precipitating solution to organic solvent varies from 1:2.5–15 (*e.g.* 1 part ethanol to 2.5–15 parts hexane) for liquid extraction of vitamin A.

After extraction, the sample may be centrifuged briefly to clearly separate the aqueous (bottom) layer from the organic (upper) layer. The upper layer should then be transferred into a fresh tube, usually glass, and evaporated to dryness. A stream of nitrogen is commonly used for evaporation at temperatures, varying from room temperature to 40 °C. After evaporation the sample is commonly reconstituted in ethanol, methanol or mobile phase. The reconstituted sample can then be transferred to a light-protected vial for analysis.

11.2.4 Chromatographic Analysis

Following the sample preparation steps in 11.2.3, the prepared samples are ready for chromatographic analysis. Calibrator, control and sample vials are loaded on to the HPLC. A portion (often 50 ± 20 µL) of each vial is injected sequentially on to the HPLC column. Chromatographic separation is achieved using a unique combination of liquid mobile phase or phases and columns. Pressure and temperature need to be kept constant to ensure repeatability from injection to injection. A detector is used to ascertain the magnitude of the separated peaks. Results are calculated by comparing the area or height of the analyte to its internal standard and then comparing this to the calibration curve.

The majority of clinical laboratories employ isocratic separation with 97 to 100% methanol as the mobile phase. Gradient methods are also used which allow for an increase in the percentage of methanol over the run time; an example gradient system is provided in Table 11.1. The advantage of a gradient solvent program is that it elutes late-eluting peaks that occur in some samples and that cause interference with succeeding chromatograms when an isocratic protocol is used. Other mobile phase constitutes less frequently used include: acetonitrile, ammonium acetate, chloroform, ethanol, ethyl acetate, isopropanol, tetra-hydrofuran and triethylamine.

An alkane-bonded silica octadecylsilane (C18) column from a variety of manufacturers is typically used for the solid phase. Routinely diode array or UV/visible detectors are used for vitamin A at an absorbance of 325nm. Note: A mass spectrometer potentially could be used as a detector, however, currently clinical laboratories are not employing such technology for routine vitamin A assessment.

11.3   Standardisation

Standardisation is a fundamental concept for clinical biochemists. This is achievable when the method base is similar, which is the case with the methods routinely used for vitamin A. With standardisation common reference intervals can be applied between laboratories and results from one laboratory can be directly compared with another laboratory. This ultimately improves patient care.

As a result, in 2002 the Joint Committee for Traceability in Laboratory Medicine (JCTLM) was established to address the need for establishment of lists of higher-order reference methods and reference materials. As part of the process, proposed reference methods are examined to ensure conformity with appropriate international documented standards and reference materials are verified by measurement institutes with demonstrated competency (Joint Committee for Traceability in Laboratory Medicine 2002).

11.3.1   Reference Measurement System

There is no reference measurement system currently available for serum vitamin A (Joint Committee for Traceability in Laboratory Medicine 2012). Liquid chromatography coupled with tandem mass spectrometry would appear to be a suitable approach for the development of a candidate reference method.

11.3.2 Primary Calibrators

There is no certified reference material currently listed on the JCTLM database. Previously, the National Institute of Standards and Technology (NIST) (Gaithersburg, MD, USA) have listed standard reference material (SRM)-968c on this database. This listing was removed as the supply was depleted. The NIST have since released two replacements for SRM-968c, being SRM-968d now superseded by SRM-968e; neither of which is listed on the JCTLM database (Joint Committee for Traceability in Laboratory Medicine 2012).

The principal options for vitamin A primary reference material currently is the NIST SRM-968e or the purchase of pure retinol from Sigma–Aldrich (St Louis, MO, USA) or similar.

11.3.3 Secondary Calibrators

Secondary (*i.e.* routine calibrators) are either made within the laboratory or purchased from a commercial vendor. Such calibrators are referenced back to a primary reference material. It is recommended that primary reference materials, *i.e.* calibrators, should achieve lower error for inaccuracy compared with the equivalent secondary calibrator (Stockl *et al* 2009). From the biological variation database, desirable inaccuracy of secondary calibration material is suggested to be no greater than 5.5 to 5.8% (Ricos *et al* 1999).

Major vendors of commercial secondary serum calibrators for vitamin A align their material with the NIST SRM-968 reference material. Some vendors also supply values aligned with Sigma–Aldrich material. The target values assigned for vitamin A by these vendors differ between Sigma–Aldrich and NIST. However, even when all secondary calibrators are aligned to NIST, a relatively wide spread of vitamin A results is evident between laboratories, as demonstrated by result submissions in external quality assurance schemes (Greaves *et al* 2011; RCPA Quality Assurance Programs 2011). This highlights the need for the development of a higher-order reference method plus material to provide alignment between laboratories analysing serum vitamin A.

11.4 Interpretation of Results

11.4.1 Reference Intervals

The WHO defines vitamin A deficiency as a serum retinol level of <0.7 µmol L^{-1} (World Health Organization 2009). Children are known to have lower serum vitamin A levels than adults. Reference intervals for serum vitamin A change during pregnancy (Table 11.2). There appears to be no sex difference between vitamin A levels.

Commercial vendors for vitamin A calibrators include: Bio-Rad, Munich, Germany; Chromsystems Instruments and Chemical GmbH, Munich, Germany; and Recipe, Munich, Germany.

However, comparison of reference intervals between vitamin A testing laboratories indicates that there is a lack of agreement. Laboratories differ in the inclusion of age-related reference intervals. Some laboratories only provide one reference interval for vitamin A, irrespective of age and tend to have a lower limit of 0.7–0.8 μmol L^{-1}. Laboratories which provide age-related reference intervals appear to use a higher lower limit for vitamin A for adults. This comparison is provided in Figure 11.3. Ideally, age-stratified reference intervals should be reported.

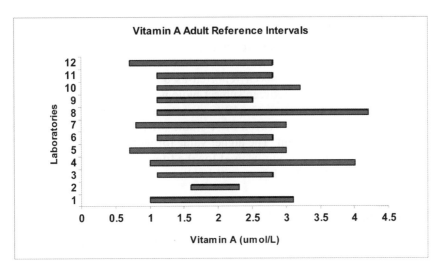

Figure 11.3 Adult vitamin A reference intervals applied by clinical laboratories. Serum vitamin A adult reference intervals across laboratories enrolled in the RCPA Quality Assurance Programs – Chemical Pathology Vitamin A program in 2007 (Greaves *et al* 2010). Authors' publication, permission inherently granted.

Table 11.1 HPLC gradient mobile phase example parameters. Example gradient mobile phase set-up with solvent A (80% methanol in water) and solvent B (100% methanol).

Time (min)	Flow rate mL min^{-1}	Solvent A (%) (80% methanol)	Solvent B (%) (100% methanol)
0	2.0	100	0
12	2.0	0	100
14	2.0	0	100
15	2.0	100	0
17	2.0	100	0

Table 11.2   Pregnancy reference intervals for vitamin A. Gestation appropriate pregnancy reference intervals for vitamin A. (Adapted from Lockitch *et al* 1993).

Gestational age (weeks)	Retinol ($\mu mol\ L^{-1}$)
<12	0.9–2.3
12–18	0.9–1.9
18–24	1.1–1.8
24–28	1.1–1.7
>28	0.9–1.9

11.4.2   Biological Variation

Estimates of biological variation for vitamin A have been conducted in both plasma and serum. The within subject biological variation data for plasma and serum vitamin A levels have been reported as 6.2% and 13.6% respectively. From the same studies the between subject biological variation for plasma and serum was reported to be 21% and 19% respectively. Although the biological variation data is consistent between these studies, the within subject data does not show the same agreement. The higher value for within subject variation, as found in the study using serum, is used for the fitness for purpose calculation example in Section 11.5.2.1 (Ricos *et al* 1999).

11.4.3   Additional Analytes

In addition to reporting Vitamin A, a minority of laboratories measure and report retinyl esters and retinol-binding protein and ratios in conjunction with vitamin A results. This is thought to be useful for the assessment of hypervitaminosis A.

11.5   Method Validation

Each method introduced into a laboratory's repertoire requires an understanding of its parameters and limitations. To do this, method validation experiments need to be conducted. Studies to validate a chromatographic vitamin A method include: linearity; sensitivity; accuracy; imprecision; recovery and interference (Westgard 1999).

11.5.1   Analytical Range

Linearity and sensitivity experiments are used to determine the reportable range of results. For vitamin A, the reportable range should span all decision points, such as the levels of hypovitaminosis and hypervitaminosis A. Ideally, sensitivity of the assay should be to 0.1 μmol L^{-1} and the upper limit of quantitation without the need for dilution should be to at least 4.0 μmol L^{-1}.

11.5.2 Imprecision

Accuracy and precision are both important characteristics of method validation studies. The desirable between run imprecision for serum vitamin A is estimated to be $\leq 6.8\%$ (Ricos *et al* 1999). This imprecision should be ascertained at the clinical decision points. In practice, between run control data generated over multiple runs (i.e. >20 runs over at least a one month or preferably longer period) is used to generate this information (Westgard 2009).

11.5.2.1 Fitness for Purpose

Total intra-individual variation (CV_t), which is the combined estimate of analytical imprecision (CV_a) and intra-individual biological variation (CV_w) can be calculated with the equation: $CV_t = \sqrt{[(Cv_a)^2 + (CV_w)^2]}$ (Fraser 2001; Koerbin *et al* 2008).

The analytical goal for fitness for purpose (White and Farrance 2004) for an assay is based on the CV_w and can be determined based on the following criteria:

Optimum: $CV_a \leq 0.25 \times CV_w$

Desirable: $CV_a \leq 0.5 \times CV_w$

Minimum: $CV_a \leq 0.75 \times CV_w$.

As the CV_w for serum vitamin A (retinol) is estimated as 13.6%, the desirable specifications for the analytical imprecision based on the above calculations must be $\leq 6.8\%$ (Ricos *et al* 1999; White and Farrance 2004).

11.5.2.2 Uncertainty of Measurement

Ascertainment of the uncertainty of measurement (MU) of the vitamin A assay is required for clinical laboratories under International Organization for Standardization (ISO) document number 15189 (International Organization for Standardization 2011). Presuming negligible bias, the assay imprecision can be used to calculate the uncertainty of measurement, which is effectively the two standard deviation range either side of a result if it were to be repeated.

11.5.3 Recovery

Recovery and interference studies should be performed prior to releasing a method for clinical samples. Recovery is important to ensure the efficiency of the sample preparation. Experimental design should be based on the criteria set by the International Union of Pure and Applied Chemists (Burns *et al* 2002). Recovery experiment results for vitamin A should ideally fall into the range of 90 to 110%.

11.5.4 Interference

Efficient chromatography with adequate retention and resolution from interferences, including early-eluting carotenoids, is attainable for vitamin A methods.

11.6   Quality Specifications

Quality refers to the features encompassing the analysis, being pre-analytical, analytical and post-analytical, which provides assurance that the result (*e.g.* serum vitamin A concentration) meets clinical expectations. In 1999 the meeting in Stockholm Sweden on "Strategies to Set Global Quality Specifications in Laboratory Medicine" developed a hierarchy for goal setting for analytical performance. This has become known as the "Stockholm consensus hierarchy" which is:

  I. Evaluation of the effect of analytical performance on clinical outcomes in specific clinical settings
 II. Evaluation of the effect of analytical performance on clinical decisions in general
      A. Data based on components of biological variation
      B. Data based on analysis of clinicians opinions
III. Published professional recommendations
      A. From national and international expert bodies
      B. From expert local groups or individuals
 IV. Performance goals set by
      A. Regulatory bodies
      B. Organisers of External Quality Assessment (EQA) schemes
  V. Goals based on current state of the art
      A. As demonstrated by data from EQA or Proficiency Testing schemes
      B. As found in current publications on methodology" (Westgard 2009).

   As part of this meeting, a substantial biological variation data base was developed by Ricos and colleagues (1999). Together, this information has provided direction for setting quality specifications for many analytes including serum vitamin A.

11.6.1   Internal Quality Control

Commercial internal quality control material for vitamin A is available from a variety of manufacturers including Bio-Rad Munich, Germany; Chromsystems Instruments and Chemical GmbH, Munich, Germany; and Recipe, Munich, Germany. Each of these manufacturers produces their control material as multi-level lyophilized serum vials.

   In line with Level II of the Stockholm consensus hierarchy, internal quality control for serum vitamin A should achieve between run imprecision of ≤6.8% (Ricos *et al* 1999) to meet the desirable specifications for fitness for purpose (White and Farrance 2004).

11.6.2   External Quality Assurance

Participation in an external quality assurance program ensures on-going proficiency of the analytical component of vitamin A. There are a number of vitamin external quality assurance programs for vitamin A. Some of these are detailed in Table 11.3.

As an example external quality assurance program, the Royal College of Pathologists Australasia (RCPA) QAP Vitamin program has been in operation since 1999 and is a worldwide program, with participants from both the southern (*e.g.* Australia, New Zealand, Singapore and Thailand) and the northern (*e.g.* South Africa, Israel, France and USA) hemispheres (RCPA Quality Assurance Programs 2012). Each month participating laboratories analyse two samples alongside their clinical samples. Twelve samples are analysed within a cycle, which runs for a period of 6 months. There are two cycles per annum. Participants are asked to describe their method in terms of analytical principle, measurement system, reagent source and calibrator source. Targets for this program have been set intermittently using NIST SRM-968 for vitamin A (Greaves *et al* 2011). The program's allowable limits of performance have been developed in line with Callum Fraser's Biological Variation recommendations (Fraser 2001) and the database for biological variation data (Ricos *et al* 1999) and fall into Level II of the Stockholm consensus hierarchy (Westgard 2009). Participation in this or similar programs (Table 11.3) allows for direct comparison of vitamin A results from one laboratory to another.

11.7 Needs and Opportunities

The routine clinical methods for vitamin A, retinol, demonstrate a significant degree of commonality as they are all chromatographically based assays with similar principles for sample preparation. There is, however, a significant degree of spread in the results obtained based on the data available from comparison of reference intervals between laboratories and external quality assurance program studies. Further standardisation of assays appears to be needed which could be achieved by comparison with both robust primary reference materials and higher order analytical methods.

There is, however, currently no reference method for clinical serum vitamin A methods. Liquid chromatography coupled with tandem mass spectrometry would appear to be a suitable approach for the development of a candidate reference method. The establishment of such a method would allow for both primary and secondary calibrators to be assigned with improved accuracy. This would effectively allow for the full commutability of results between laboratories and aid in defining upper and lower clinical decision limits for vitamin A.

Summary Points

- Retinol is the compound routinely measured in serum vitamin A assays.
- The intake of vitamin A is either in the form of retinyl esters (from animal products) or carotenoids especially β-carotene (from pigmented fruit and vegetables).

Table 11.3 External quality assurance (EQA) programs for serum vitamin A. Non-exhaustive list of external quality assurance programs offering vitamin A. CAP, College of American Pathologists; DGKL, Deutsche Vereinte Gesellschaft fur Klinische Chemie und Laboratoriumsmedizin; RCPA QAP, RCPA Quality Assurance Programs; NEQAS, National External Quality Assurance Program; SKML, Stichting Kwaliteitsbewaking Medische Laboratoriumdiagnostiek.

EQA program organiser	Country	Survey name	Vitamins offered as part of the same program	Sample frequency	Website
CAP	USA	Bone Markers and Vitamin Survey	A, E and 1,25-vitamin D	3 samples per survey, 2 surveys per year. Total 6 samples per annum	http://www.cap.org/
Centre for Disease Control and Prevention	USA	VITAL	A	9 samples to be run in duplicate over 3 days per survey, 2 surveys per year. Total 18 samples.	http://www.cdc.gov/labstandards/vitaleqa.html
DGKL	Germany	Vitamins and analgesics	A, B6, B12, D, E and folate	2 samples per survey, 2 surveys per year. Total 4 samples per annum	www.dgkl-rfb.de
NEQAS	UK	Vitamin Assays (Carotene and Vitamins A & E)	A, E, β-carotene; total carotenoids; carotenoid fractions	5 samples per survey, 6 surveys per year. Total 30 samples per annum	www.ukneqas.org.uk/

Table 11.3 (*Continued*)

EQA program organiser	Country	Survey name	Vitamins offered as part of the same program	Sample frequency	Website
NIST	USA	Micronutrients Quality Assurance Program	Total retinol, *trans* retinol, vitamin E fractions, β-carotene, carotenoid fractions	5 samples per survey; 2 surveys per year. Total of 10 samples per annum	http://www.nist.gov/mml/analytical/mmqaprogram.cfm
SKML	The Netherlands	Vitamin A/E and β carotene	A, E, β-carotene	6 samples per survey, 2 surveys per year. Total 12 samples per annum	http://www.skml.nl/en/
RCPA QAP	Australia	Vitamin	A, E, B1, B2, B6, C, β-carotene and total carotenoids	2 samples per survey; 12 surveys per year. Total of 24 samples per annum.	www.rcpaqap.com.au/chempath

- Vitamin A is required for eye sight, normal growth and development, reproduction, immunity and health of epithelial tissue including skin.
- The earliest symptom of vitamin A deficiency is impaired vision. This impaired vision starts as night blindness but can progress to total and irreversible blindness.
- Vitamin A is oxidised in the rods of the eyes to retinal which then complexes to opsin to form rhodopsin, which allows for the dim light vision.
- In vitamin A deficiency the depletion of retinal in the retina causes opsin to be destabilised and catabolised causing the permanent destruction of the rod cells, which results in night blindness.
- Vitamin A is photo-labile. Samples for vitamin A analysis must be protected from fluorescent lighting and sunlight from the time of collection by wrapping samples in foil.
- Vitamin A, as retinol, is routinely measured by high pressure liquid chromatography (HPLC) with spectrophotometric detection for clinical purposes.
- There is no reference measurement system currently available for serum vitamin A for clinical analysis.
- Accuracy and precision should be ≤5.8% and ≤6.8% respectively to meet the desirable specifications for fitness for purpose.
- Development of a liquid chromatography coupled with tandem mass spectrometry vitamin A method would appear to be a suitable approach for the development of a candidate reference method.

Key Facts

Key Facts of Cystic Fibrosis

- Cystic fibrosis (CF) is probably the oldest genetic disease known to man that is still prevalent today.
- Medieval folklore probably describes CF "woe is the child when kissed who tastes salty for they are bewitched and soon must die".
- CF as a disease entity was described as "cystic fibrosis of the pancreas" in 1938 based on the appearance of the pancreas at autopsy of young babies; the primary cause was thought to be vitamin A deficiency.
- People with cystic fibrosis have more salt in their sweat.
- This feature of salty sweat is used to diagnosis CF.
- In 1959 the Gibson and Cooke sweat collection method was developed and is still the basic method for collection of sweat.
- In the 1980's the gene defect was identified.
- This lead to us understanding that the disease caused an abnormality in the chloride channel *i.e.* the negative ion of salt.
- This loss of activity of the chloride channel causes thick mucus in the lungs and pancreatic insufficiency.

- Gene testing was introduced in the 1990's.
- In many countries CF is part of a newborn screening program, where babies have a blood collection about 2 days after birth for screening.
- This has led to a significant improvement in survival.
- Supplementation of CF patients with pancreatic enzymes allows for vitamin A and other fat-soluble vitamins to be absorbed.
- Vitamin supplementation is also given.
- Currently, there are >1850 mutations and polymorphisms recognised in the CF gene.
- A cure for CF is being vigorously sort.

Key Facts of the World Health Organization

- Commonly known by its abbreviation WHO.
- Originally known as the "League of Nations".
- Established in 1920.
- After the United Nations was formed in 1945, it was recognised that an international system for public health was needed.
- The WHO was established as the directing and co-ordinating authority for health within the United Nations system.
- The WHO has six agenda items: 1. promoting development; 2. fostering health security; 3. strengthening health systems; 4. harnessing research, information and evidence; 5. enhancing partnerships; and 6. improving performance.
- Vitamin A deficiency is listed by the WHO as a major health issue in developing countries.
- WHO estimates from 2005 indicate vitamin A deficiency is of public health significance in pre-school children in 45 countries based on the prevalence of night blindness.
- WHO estimates from 2005 indicate vitamin A deficiency is of public health significance in pre-school children in 122 countries based on biochemical vitamin A deficiency.
- Vitamin A deficiency in association with malnourishment is linked to an increase in morbidity and mortality in pre-school children and pregnant woman.
- The WHO administers vitamin A to pre-school children in association with immunisation programs in some developing countries.
- This has led to an improvement in population-based vitamin A levels and associated morbidity.

Definitions of Words and Terms

Analytical variation: The variation in results that is normally expected due to the imprecision of the analytical method. This is often expressed as coefficient of variation and can be differences within the one assay (intra or within assay imprecision) or over successive analytical assays (inter or between assay imprecision).

Biological variation: Biological variation is the natural fluctuation or change that occurs in the body around a homeostatic set point such as the change in concentration of vitamin A that is normally expected over time. This can be described within the one individual (intra-individual biological variation) or between a group or population of individuals (inter-individual biological variation).

Chromatography: An analytical technique allowing for the separation of sample components, based on their chemical and physical properties using a mobile phase (either liquid or gas) and a stationary phase (column).

Coefficient of variation: A statistical calculation used in clinical laboratories to express imprecision independent of units. Percentage coefficient of variation is calculated from the standard deviation divided by the mean and then multiplied by 100.

Cystic fibrosis (CF): A genetic disease, with >1850 mutations and polymorphisms described, resulting in a defect in the transport of chloride across membranes. This defect causes a thick mucus build up in areas such as the lungs and digestive system; which results in chronic lung disease and inability to absorb fat.

Fitness for purpose: A description used by quality organisations to express the appropriateness of a clinical method in relation to biological variation and clinical decision requirements. Fitness for purpose is usually described for quantitative methods as Optimal, Desirable and Minimum.

Internal standard: A compound that is similar to the compound of interest that is added to each sample as a constant to account for variations during sample preparation and analysis.

Quality: Quality refers to the features encompassing the analysis, being pre-analytical, analytical and post-analytical, which provides assurance that the result (*e.g.* serum vitamin A concentration) meets clinical expectations.

Retinol: Common name for the chemical (2*E*,4*E*,6*E*,8*E*)-3,7-dimethyl-9-(2,6,6-trimethylcyclohex-1-en-1-yl)nona-2,4,6,8-tetraen-1-ol, which is also known as vitamin A and vitamin A1.

Standardisation: Standardisation is a concept whereby agreement of test results is achieved by establishing traceability to higher-order reference materials and measurement procedures.

Total intra-individual variation: Total intra-individual variation is the combined estimate of analytical and biological variation within the one individual.

Uncertainty of measurement: A statistical parameter associated with the analysis of vitamin A that characterises the likely dispersion of results that could be reasonably expected due to the imprecision of the analytical method.

Vitamin: An organic compound required as a nutrient at low concentrations which cannot be synthesised in adequate amounts and therefore must be obtained in the diet.

Vitamin A: "The generic descriptor for retinoids exhibiting qualitatively the biological activity of retinol" (IUPAC-IUB Joint Commission on Biological Nomenclature 1982). For chromatographic analysis, as presented in this Chapter, the term "vitamin A" may be considered to be equivalent to the term "retinol".

List of Abbreviations

CF	cystic fibrosis
CV_a	analytical imprecision
CV_t	total intra-individual variation (combination of analytical and biological variation)
CV_w	intra-individual biological variation
HPLC	high-pressure liquid chromatography
IU	international units
JCTLM	Joint Committee for Traceability in Laboratory Measurement
NIST	National Institute of Standards and Technology
SRM	standard reference material
VAD	vitamin A deficiency
WHO	World Health Organization

Acknowledgements

I wish to extend my appreciation to all members of the Australasian Association of Clinical Biochemists (AACB) Vitamins Working Party for their expertise and teamwork over the years, and, in particular, Kirsten Hoad and Jan Gill for reviewing this manuscript.

References

Andersen, D., 1938. Cystic fibrosis of the pancreas and its relationship to celiac disease: Clinical and pathologic study. American Journal of Disease in Childhood. 56: 344.

Beaton, G., Martorell, R., Aronson, K., Edmonston, B., McCabe, G., Ross, A. and Harvey, B., 1993. Effectiveness of Vitamin A Supplementation in the Control of Young Child Morbidity and Mortality in Developing Countries. ACC/SCN State-of-the-Art Series, Nutrition Policy Discussion Paper No. 13. ACC/SCN, Geneva. Available at: http://www.unscn.org/layout/modules/resources/files/Policy_paper_No_13.pdf. Accessed 4 June 2011.

Burns, D. T., Danzer, K. and Townshend, A., 2002. Use of the terms "Recovery" and "Apparent Recovery" in analytical procedures (IUPAC Recommendations 2002). Pure and Applied Chemistry. 74: 2201–2205.

Comstock, G. W., Alberg, A. J. and Helzlouer, K. J., 1993. Reported effects of long-term freezer storage on concentrations of retinol, β-carotene, and α-tocopherol in serum or plasma summarized. Clinical Chemistry. 39: 1075–1078.

Drummond, J. C., 1920. The nomenclature of the so-called accessory food factors (vitamins). Biochemistry Journal. 14: 660.

Fraser, C. G., (2001) Biological Variation: From Principles to Practice. AACC Press, Washington DC, USA, pp. 151.

Funk, C., 1912. The Vitamines. Journal of State Medicine. 20: 341–368.

Gallagher, S. K., Johnson, A. K. and Milne, D. B., 1992. Short and long term variability of selected indices related to nutritional status. II. Vitamins, lipids, and protein indices. Clinical Chemistry. 38: 1449–1453.

Greaves, R., Jolly, L., Woollard, G. and Hoad, K., 2010. Serum vitamin A and E analysis: comparison of methods between laboratories enrolled in an external quality assurance programme. Annals of Clinical Biochemistry. 47: 78–80.

Greaves, R. F., Hoad, K. E., Woollard, G. A., Walmsley, T. A., Briscoe, S. Johnson, L. A., Carter, W. D. and Gill, J. P., 2011. External Quality Assurance target setting with NIST SRM 968d material: Performance in the 2010 Royal College of Pathologists of Australasia Quality Assurance Program with retinol, α-tocopherol and β-carotene. Annals of Clinical Biochemistry. 48: 480–482.

Hsing, A.W., Comstock, G.W. and Polk, B.F., 1989. Effect of repeated freezing and thawing on vitamins and hormones in serum. Clinical Chemistry. 35: 2145.

International Organization for Standardization (ISO)., 2011. Available at: http://www.iso.org/iso/catalogue_detail?csnumber=42641. Accessed 20 May 2011.

International Union of Pure and Applied Chemistry (IUPAC) Commission on the Nomenclature of Biological Chemistry., 1960. Definitive rules for the nomenclature of vitamins. Journal of the American Chemical Society. 82: 5581–5583.

IUPAC-IUB Joint Commission on Biological Nomenclature (JCBN)., 1982. Nomenclature of Retinoids. 129: 1–5.

Joint Committee for Traceability in Laboratory Medicine., 2002. Appendix III. The JCTLM Framework: A framework for the international recognition of available higher-order reference materials, available higher-order reference measurement procedures and reference measurement laboratories for laboratory medicine. Available at: www.bipm.org/utils/en/pdf/jctlm_framework.pdf. Accessed 6 May 2012.

Joint Committee for Traceability in Laboratory Medicine., 2011. Database of higher-order reference materials, measurement methods/procedures and services. Available at: www.bipm.org/jctlm/. Accessed 6 May 2012.

Koerbin, G., Greaves, R., Robins, H., Farquhar, J. and Hickman, P. E., 2008. Total intra-individual variation in sweat sodium and chloride concentrations for the diagnosis of cystic fibrosis. Clinica Chimica Acta. 393: 128–129.

Lockitch, G., 1993. Handbook of Diagnostic Biochemistry and Hematology in Normal Pregnancy. CRC Press, Baco Raton, FL, USA, pp. 537.

McCollum, E. V. and Kennedy, C., 1916. The dietary factors operating in the production of polyneuritis. Journal of Biological Chemistry. 24: 491–502.

Olmedilla, B., Granado, F., Blanco, I. and Rojas-Hidalgo, E., 1994. Seasonal and sex-related variations in six serum carotenoids, retinol, and α-tocopherol. American Journal of Clinical Nutrition. 60: 106–110.

Penniston, K. L. and Tanumihardjo, S. A., 2006. The acute and chronic toxic effects of vitamin A. American Journal of Clinical Nutrition. 83: 191–201.

RCPA Quality Assurance Programs Pty Ltd - Chemical Pathology. 2011. Available at: http://www.rcpaqap.com.au/chempath/. Accessed 6 May 2012.

Ricos, C., Alvarez, V., Cava, F., Garcia-Lario, J. V., Hernandez, A., Jimenez, C. V., Minchinela, J., Perich, C. and Simon, M., 1999. Current databases on biologic variation: pros, cons and progress. Scandinavian Journal of Clinical Laboratory and Investigation. 59: 491–500. Available at: http://www.westgard.com/biodatabase1.htm. Accessed 4 June 2011.

Stockl, D., Sluss, P. M. and Thienpont, L. M., 2009. Specifications for trueness and precision of a reference measurement system for serum/plasma 25-hydroxyvitamin D analysis. Clinica Chimica Acta. 408: 8–13.

Talwar, D. K., Azharuddin, M. K., Williamson, C., Teoh, Y. P., McMillan, D. C. and O'Reilly, D. 2005. Biological variation of vitamins in blood of healthy individuals. Clinical Chemistry. 51: 2145–2150.

Westgard, J. O., 2008. Basic Method Validation 3rd Edn. Westgard QC Inc, Madison WI, USA, pp. 320.

Westgard, J. O.. 2009. Westgard QC. Available at: http://www.westgard.com/right-quality-goal.htm. Accessed 4 June 2011.

White, G. H. and Farrance, I., 2004. Uncertainty of measurement in quantitative medical testing – a laboratory implementation guide. Clinical Biochemist Reviews. 25: S1–S4.

World Health Organization. 2009. Global prevalence of vitamin A deficiency in populations at risk. 1995–2005 WHO global database on vitamin A deficiency. pp. 55. ISBN 9789241598019. Available at: whqlibdoc.who.int/publications/2009/9789241598019_eng.pdf. Accessed 4 June 2011.

World Health Organization., 2011. Serum retinol concentrations for determining the prevalence of vitamin A deficiency in populations. Vitamin and Mineral Nutrition Information System., World Health Organization, Geneva. (WHO/NHM/NHD/MNM/11.3). Available at: http://www.who.int/vmnis/indicators/retinol.pdf. Accessed 27th May 2011.

Young, D. S., 1997. Effects of Preanalytical Variables on Clinical Laboratory Tests, 2nd edn. AACC Press, Washington DC, USA, pp. 1283.

CHAPTER 12

Liquid Chromatography-based Assay for Carotenoids in Human Blood

TAIKI MIYAZAWA, KIYOTAKA NAKAGAWA AND
TERUO MIYAZAWA*

Department of Food and Biodynamic Chemistry, Graduate School of
Agricultural Science, Tohoku University, Sendai 981-8555, Japan
*E-mail: miyazawa@biochem.tohoku.ac.jp

12.1 Introduction

Membrane phospholipid peroxidation has received attention in relation to
oxidative stress occurring during pathophysiological changes such as
atherogenesis and aging (Kinoshita *et al* 2000; Miyazawa *et al* 1988; 1992a;
Moriya *et al* 2001; Stocker *et al* 2004). We previously confirmed that
significantly higher concentrations of oxidized phospholipids [*i.e.*, phospholipid hydroperoxides (PLOOH)] are accumulated in the red blood cells (RBC)
of patients with senile dementia (Figure 12.1) (Miyazawa *et al* 1992b).
Compounds that can minimize such accumulation may be used therapeutically
as effective drugs or functional foods to prevent the disease. Therefore, we
carried out animal studies and found carotenoids as potential compounds to
inhibit phospholipid peroxidation of RBC (Figure 12.2 and Figure 12.3)
(Nakagawa *et al* 1996).

From these results, we are particularly interested in the effects of carotenoids
on human RBC, but the information is very limited. This is in contradiction to

Food and Nutritional Components in Focus No. 1
Vitamin A and Carotenoids: Chemistry, Analysis, Function and Effects
Edited by Victor R Preedy
© The Royal Society of Chemistry 2012
Published by the Royal Society of Chemistry, www.rsc.org

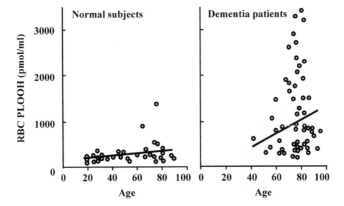

Figure 12.1   Phospholipid hydroperoxide concentrations in RBC of normal subjects and dementia patients. The RBC of dementia patients have a high concentration of PLOOH. In contrast, the RBC of normal subjects have a low concentration of PLOOH.

plasma carotenoids, which have been investigated thoroughly for analytical methods as well as biological significance and metabolism (Aust *et al* 2001; Su *et al* 2002). To the best of our knowledge, there have been at least ten reports to date on carotenoids in human RBC (Bjornson *et al* 1976; Futouhi *et al* 1996; Leo *et al* 1995; Morinobu *et al* 1994; Murata *et al* 1992; 1994; Nakagawa *et al* 2008; Norkus *et al* 1990; Shah *et al* 1989). Some studies have successfully detected carotenoids in RBC (mainly β-carotene) (Bjornson *et al* 1976; Futouhi *et al* 1996; Leo *et al* 1995; Morinobu *et al* 1994; Murata *et al* 1992; 1994), whereas others have failed to detect them (Norkus *et al* 1990; Shah *et al* 1989). Incorporation of carotenoid (β-carotene) into RBC after oral supplementation

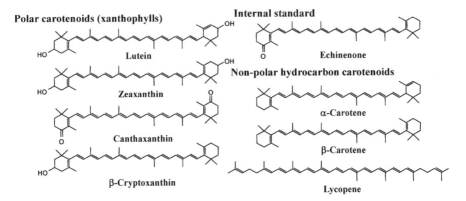

Figure 12.2   Chemical structures of all-*trans* carotenoids, including xanthophylls, an internal standard, and hydrocarbon carotenoids. The xanthophylls are lutein, zeaxanthin, canthaxanthin, and β-cryptoxanthin. The internal standard is echinenone. The hydrocarbon carotenoids are α-carotene, β-carotene, and lycopene.

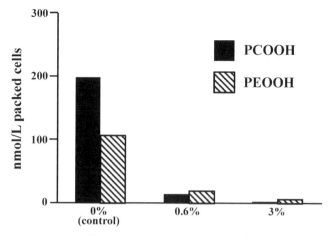

β-carotene concentration in diet

Figure 12.3 β-Carotene supplementation prevents phospholipid hydroperoxide accumulation in the RBC of mice. Three groups of 24 mice each were fed for 1 week on a semi-synthetic diet supplemented with either 0.6 or 3.0% β-carotene per diet or maintained on a control (β-carotene-unsupplemented) diet. RBC PLOOH showed a significant decrease followed by an increase of β-carotene intake; *i.e.*, 201, 16 and 4 pmol of PCOOH mL of packed cells, and 108, 22 and 8 pmol of PEOOH mL of packed cells in the mice given the control diet, 0.6% β-carotene diet, and 3.0% β-carotene diet, respectively.

has been described in some reports (Futouhi *et al* 1996; Murata *et al* 1992). However, there has been no study evaluating whether administered carotenoids other than β-carotene are distributed to the RBC. Therefore, the actual occurrence and roles of carotenoids in human RBC have never been elucidated, and this has been due, in part, to a lack of suitable analytical methods.

To address this need, we first tried to develop a quantitative method to analyze human RBC carotenoids using high-performance liquid chromatography (HPLC) coupled to ultraviolet–diode array detection (UV-DAD) and atmospheric pressure chemical ionization/mass spectrometry (APCI/MS) (Nakagawa *et al* 2008).

12.2 Occurrence of Carotenoids in Human RBC

12.2.1 HPLC Analysis of Standard Carotenoids

It is well known that the major carotenoids in human fluid samples (*i.e.*, plasma) are xanthophylls (lutein, zeaxanthin, canthaxanthin, and β-cryptoxanthin) and hydrocarbon carotenoids (α-carotene, β-carotene, and lycopene) (Aust *et al* 2001; Futouhi *et al* 1996; Goulinet and Chapman 1997; Leo *et al* 1995; Su *et al* 2002; Yeum *et al* 1996). Therefore, we focused on these seven carotenoids because their presence was expected in RBC. C18 reversed-phase columns have been frequently used for carotenoid analysis, but separation of lutein and zeaxanthin is still difficult. In contrast, C30 columns offer another type of stationary phase that permits a separation of lutein and zeaxanthin. Therefore, we decided to use C30 column, and developed HPLC conditions (Nakagawa *et al* 2008) as follows.

HPLC consisted of two Shimadzu LC-10ADvp pumps (Kyoto, Japan), a Shimadzu DGU-12AM degasser, and Rheodyne 7125 injector (Cotati, CA, USA) with a 100 μL sample loop. A C30 carotenoid column (5 μm, 4.6 × 250 mm; YMC, Kyoto, Japan) was used. The column was eluted using a binary gradient consisting of the following HPLC solvents: (A) methanol/methyl *tert*-butyl ether (MTBE)/water (83:15:2, v/v/v) containing 3.9 mmol L^{-1} ammonium acetate; (B) methanol/MTBE/water (8:90:2, v/v/v) containing 2.6 mmol L^{-1} ammonium acetate. The gradient profile was as follows: 0 to 12 min, 10 to 45% B linear; 12 to 24 min, 45 to 100% B linear; 24 to 30 min, 100% B liner. The flow rate was adjusted to 1 mL min, and the column temperature was maintained at 20 °C. The column eluent was monitored by a Shimadzu SPD-M10Avp DAD detector for carotenoids at 463 nm. UV spectra were recorded from 200 to 700 nm. After the column eluent was passed through the DAD detector, the eluent was analyzed on a Shimadzu LCMS-2010 quadrupole mass spectrometer equipped with an APCI interface. MS was carried out in the positive ion measurement mode with a detection voltage of 1.6 kV, an APCI temperature of 400 °C, a curved desolvation line of 250 °C, and a block temperature of 200 °C. The flow rate of the nebulizer gas was 2.5 mL min. Full scan spectra were obtained by scanning masses between *m/z* 200 and 800.

When standard carotenoids were subjected to HPLC–DAD using a C30 column, they were well separated, eluting as follows: lutein (6.6 min), zeaxanthin (7.3 min), canthaxanthin (8.3 min), β-cryptoxanthin (11.0 min), α-carotene (13.9 min), β-carotene (15.4 min), and lycopene (25.6 and 25.9 min) (Figure 12.4). Two peaks of lycopene represent the all-*trans* (25.6 min) and 5,6-

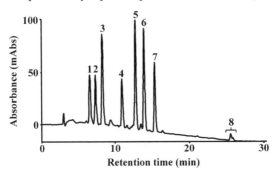

Figure 12.4 A typical HPLC–DAD chromatogram of standard carotenoids. Standard carotenoids (50–100 pmol of each) were analyzed by HPLC–DAD (at 463 nm) equipped with a C30 column. Detailed analytical conditions are described in the text. Peak identifications are as follows: (1) lutein, (2) zeaxanthin, (3) canthaxanthin, (4) β-cryptoxanthin, (5) echinenone, (6) α-carotene, (7) β-carotene, and (8) lycopene.

Table 12.1   External standard curves of carotenoids. The standard curve was linear for each carotenoids.

Carotenoid	Standard curve	R^2
Lutein	y = 10126x + 11201	0.9999
Zeaxanthin	y = 10372x + 10251	0.9999
Canthaxanthin	y = 9705x + 8059	0.9999
β-Cryptoxanthin	y = 9963x + 9215	0.9999
Echinenon	y = 9488x + 508	0.9999
α-Carotene	y = 9887x + 9643	0.9999
β-Carotene	y = 9996x + 11441	0.9999
Lycopene	y = 11976x + 8656	0.9997

cis (25.9 min) isomers (Britton 1995), and lycopene quantification was carried out based on total peak area of both isomer peaks. DAD limits of carotenoids were in the range of 0.1 to 0.25 pmol at a signal-to-noise ratio of 3. The external standard curve was linear for each carotenoid in the concentration range from 0.5 to 300 pmol (Table 12.1). On the other hand, the use of an internal standard is necessary in the quantitation of RBC carotenoids because extraction loss would be an unavoidable matter. We used echinenone as an internal standard due to its structural and chemical similarity to other carotenoids. On the DAD chromatogram, the internal standard echinenone appeared at 12.8 min (between β-cryptoxanthin and α-carotene).

12.2.2   Extraction of Carotenoids from Human RBC

According to previous extraction procedures of RBC carotenoids, some researchers treated or did not treat RBC with potassium hydroxide (KOH) in the presence or absence of pyrogallol as an antioxidant. We investigated these extraction conditions, and the optimal condition (Nakagawa *et al* 2008) is as follows.

Human RBC were washed three times with phosphate-buffered saline (pH 7.4) to prepare packed cells. Packed cells (2.5 ml) were diluted with 2.5 mL of water and were mixed with 5 mL of 80 mmol L^{-1} ethanolic pyrogallol, 1.0 mL of 1.8 mol L^{-1} aqueous KOH, and 40 μL of 1 μmol L^{-1} ethanolic echinenone (internal standard). The mixture was directly extracted without incubation; that is, after the addition of 1.25 mL of 0.1 mol L^{-1} aqueous sodium dodesylsulfate (SDS), RBC carotenoids were extracted with 15 mL of hexane/dichloromethane (5:1, v/v). Under this optimal condition, recovery of echinenone (internal standard) was calculated as greater than 70%. To minimize light- and thermo-induced isomerization and degradation of carotenoids, all procedures were conducted in the dark room and under the ice (–4 °C).

In these procedures, because RBC possess high concentrations of molecular oxygen as well as ferrous ions, prevention by pyrogallol of possible oxidative degradation of RBC carotenoids during extraction is very effective (Figure 12.5A). In addition, KOH increased the extraction rate of RBC

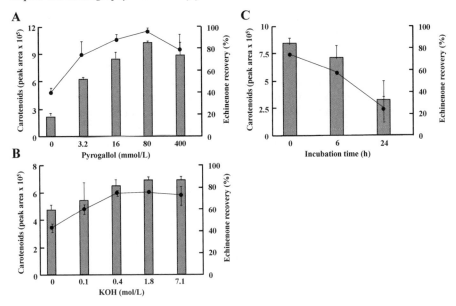

Figure 12.5 Effects of pyrogallol or KOH on extraction efficacy of RBC carotenoids. To investigate the effect of pyrogallol (A), packed cells (2.5 ml) were diluted with 2.5 mL of water. The cell suspension (5 mL) was mixed with 5 mL of ethanolic pyrogallol at various concentrations (0–400 mmol L), 1 ml of 1.8 mol L KOH aqueous solution, and 40 μl of 1 μmol L ethanolic echinenone (internal standard). To evaluate the effect of KOH (B), the RBC cell suspension (5 mL) was mixed with 5 mL of 16 mmol L ethanolic pyrogallol, 1 ml of KOH aqueous solution at various concentrations (0–7.1 mol L), and 40 μl of 1 μmol L ethanolic echinenone (internal standard). To investigate the effect of incubation (C), the cell suspension (5 mL) was mixed with 5 ml of 16 mmol L ethanolic pyrogallol, 1 ml of 1.8 mol L KOH aqueous solution, and 40 μl of 1 μmol L ethanolic echinenone (internal standard) and then incubated at 37 °C for 24 h. RBC carotenoids were then extracted and quantified by HPLC–DAD. Bars indicate the total peak areas of carotenoids, whereas circles indicate the recovery of the internal standard echinenone. Values are means ± S.D. ($n = 3$).

carotenoids (Figure 12.5C), indicating that carotenoid extraction in alkaline conditions is favorable, probably due to increased decomposition of background contaminants of RBC (*e.g.*, RBC lipids). Considering the fact that incubation of both pyrogallol- and KOH-treated packed cells caused lower extraction rates (Figure 12.5B), saponifiable carotenoids such as esterified forms of xanthophylls would be negligible in RBC.

12.2.3 HPLC Analysis of RBC Carotenoids

When an RBC extract was directly subjected to HPLC–DAD, several unknown peaks appeared, especially around echinenone and α-carotene, hindering RBC carotenoid determination. To remove these impurities, the RBC carotenoid extract was applied to a solid-phase extraction with Sep-Pak silica cartridge prior to HPLC analysis [Sep-Pak procedures are referred to in Nakagawa *et al* (2008)]. On the UV chromatogram (Figure 12.6A), RBC carotenoids were eluted as defined peaks: lutein (6.6 min), zeaxanthin (7.3 min), β-cryptoxanthin (11.0 min), α-carotene (13.9 min), β-carotene (15.4 min), and lycopene (25.8 and 26.1 min). These carotenoid peaks were identified based on their MS and UV spectrum profile. For instance, a peak component with a retention time of 6.6 min gave ions $[M+H]^+$ at *m/z* 569.7, $[M+H–H_2O]^+$ at *m/z* 551.7, and $[M+H–H_2O–92]^+$ at *m/z* 459.6, which were identical to lutein (Figure 12.6C). The peak (6.6 min) showed a UV maximal absorption at 444.7 nm (Figure 12.6B), which coincided with standard lutein. In addition, a peak at 7.3 min gave ion $[M+H]^+$ at *m/z* 569.7 (Figure 12.6C) and UV maximal absorption at 449.8 nm (Figure 12.6B), which were identical to zeaxanthin. β-Cryptoxanthin, α-carotene, β-carotene, and lycopene were characterized similarly (Table 12.2). Canthaxanthin was below the detection limit in all measurements. These results documented that six carotenoids are actually present in RBC.

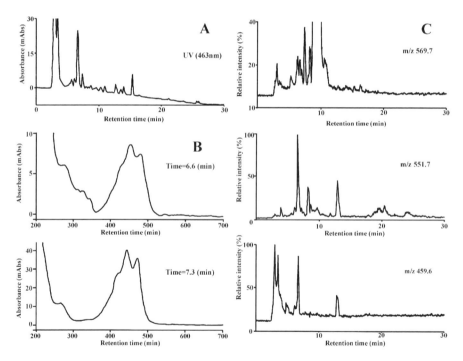

Figure 12.6   HPLC–DAD–APCI/MS analysis of human RBC carotenoids. Detailed analytical conditions are given in the text. (A) Typical DAD (at 463 nm) chromatogram. (B) UV spectra of peak 1 (6.6 min) and peak 2 (7.3 min) detected in the chromatogram in (A). Peak identities are the same as those provided in Figure 12.4. (C) Single ion plots of the mass corresponding to *m/z* 569.7, 551.7, and 459.6.

Table 12.2 UV and MS spectra profiles of carotenoids detected in human RBCs. Canthaxanthin was below the detection limit, therefore it is not in this Table.

Carotenoid	Retention time (min)	λ_{max} (nm)	Major ion(s) (m/z)
Lutein	6.6	444.7,472.3	$[M + H]^+$ 569.7
			$[M + II - H2O]^+$ 551.7
			$[M + H - H2O - 98]^+$ 459.6
Zeaxanthin	7.3	449.8, 477.3	$[M + H]^+$ 569.7
β-Cryptoxanthin	11	449.8	$[M + H]^+$ 553.6
α-Carotene	13.9	446	$[M + H]^+$ 537.6
β-Carotene	15.4	451	$[M + H]^+$ 537.6
Lycopene	25.8,26.1	472	$[M + H]^+$ 537.6

Next, to ascertain the recovery of carotenoids from RBC, we spiked standard carotenoids into packed cells and then the concentrations of carotenoids were calculated using the equation corresponding to the external standard curve (Table 12.1) and were adjusted by the percentage recovery of the added echinenone (the internal standard). The spiked amounts were equivalent to half and twice the amounts of endogenous RBC carotenoids. Recoveries of carotenoids were very close to 100% (Table 12.3). The results show how useful the internal standard is and indicate the high accuracy of the current determination method.

By using the established method, we analyzed carotenoids in the RBC of healthy human volunteers. Lutein, zeaxanthin, β-cryptoxanthin, α-carotene, β-carotene, and lycopene were found in all RBC samples (Table 12.4), with the relative amounts of each carotenoid consistently as follows: lutein > β-cryptoxanthin > zeaxanthin > β-carotene > α-carotene > lycopene. Therefore, it is apparent that in human RBC, xanthophylls (polar carotenoids) are the prevalent ones. In contrast, non-polar carotenoids were more abundant

Table 12.3 Recoveries of carotenoids from human RBC. Values are mean $\pm$ S.D. ($n = 5$).

Carotenoid	Recovery of spiked carotenoid (%)	
	Half[a]	Twice[a]
Lutein	102.0 ± 4.6	100.7 ± 7.7
Zeaxanthin	97.8 ± 6.7	89.0 ± 6.7
β-Cryptoxanthin	100.5 ± 9.3	90.2 ± 3.6
α-Carotene	99.1 ± 4.4	94.0 ± 1.7
β-Carotene	97.2 ± 4.4	93.2 ± 1.4
Lycopene	96.4 ± 9.8	98.4 ± 9.0

[a]The spiked amounts were equivalent to half and twice the amount of endogeneous RBC carotenoids.

Table 12.4 Carotenoid concentrations in RBC and plasma of healthy human subjects. Values are mean ± S.D. (n = 5).

Carotenoid	RBC			Plasma		
	Total (n = 20)	Men (n = 11)	Women (n = 9)	Total (n = 20)	Men (n = 11)	Women (n = 9)
	pmol/mL packed cells			pmol/mL plasma		
Lutein	70.2 ± 24.4	53.9 ± 11.0	90.2 ± 21.0[b]	468 ± 213	338 ± 88	626 ± 217[f]
Zeaxanthin	24.2 ± 6.7	21.1 ± 6.3	27.9 ± 5.3[c]	130 ± 57	101 ± 36	165 ± 60[g]
β-Cryptoxanthin	31.9 ± 12.3	25.8 ± 8.1	39.5 ± 12.6[d]	487 ± 330	332 ± 224	675 ± 352[c]
α-Carotene	2.1 ± 1.6	1.6 ± 0.9	2.7 ± 2.1	290 ± 192	181 ± 113	423 ± 188[f]
β-Carotene	5.1 ± 3.7	2.3 ± 1.1	8.4 ± 2.8[e]	981 ± 723	475 ± 229	1600 ± 626[e]
Lycopene	1.3 ± 0.7	1.3 ± 0.7	1.2 ± 0.8	336 ± 181	312 ± 147	366 ± 220
	μg/g total lipids[a]			μg/g total lipids[a]		
Lutein	10.3 ± 2.7	9.0 ± 2.1	12.0 ± 2.4[g]	57 ± 23	44 ± 9	73 ± 24[h]
Zeaxanthin	3.6 ± 0.9	3.5 ± 1.1	3.7 ± 0.7	16 ± 5	13 ± 4	19 ± 5[f]
β-Cryptoxanthin	4.6 ± 1.5	4.1 ± 1.4	5.1 ± 1.6	58 ± 40	43 ± 31	77 ± 42[f]
α-Carotene	0.3 ± 0.2	0.2 ± 0.2	0.3 ± 0.3	34 ± 23	22 ± 13	48 ± 25[d]
β-Carotene	0.7 ± 0.4	0.4 ± 0.2	1.0 ± 0.3[b]	111 ± 75	58 ± 28	175 ± 63[d]
Lycopene	0.2 ± 0.1	0.2 ± 0.1	0.2 ± 0.1	40 ± 21	39 ± 19	41 ± 25

[a]Total lipids are the combined value of phospholipids, cholesterol, and triacylglycerol. [b,c,g]Significantly different from men (Student's t test); [b]$P<0.001$, [c]$P<0.05$, [g]$P<0.01$. [d,f]Significantly different from men (Mann–Whitney U test); [d]$P<0.05$, [f]$P<0.01$. [e,h]Significantly different from men (Welch's t test); [e]$P<0.001$, [h]$P<0.01$.

Table 12.5 Correlations of carotenoids between RBC and plasma.

Carotenoid	r (n=20)	P
Lutein[a]	0.89	<0.001
Zeaxanthin[a]	0.70	<0.001
β-Cryptoxanthin[b]	0.40	<0.1
α-Carotene[b]	0.58	<0.05
β-Carotene[a]	0.95	<0.001
Lycopene[a]	0.59	<0.01

[a]Analyzed by Pearson's correlation coefficient test. [b]Analyzed by Spearman's rank correlation coefficient test.

in human plasma. Total RBC carotenoid concentration (as expressed in µg g of total lipids) was approximately 16 times lower than total plasma carotenoids.

A positive correlation between RBC and plasma carotenoids (Table 12.5) implies that carotenoids in plasma lipoprotein particles may be transferred into RBC. It is plausible that xanthophylls and non-polar carotenoids are located in the outer and inner regions of plasma lipoprotein, respectively, facilitating the transfer of outer xanthophylls from lipoproteins to RBC. The predominant presence of xanthophylls in RBC may be related to the characteristic distribution of carotenoids in the retina, where xanthophylls (lutein and zeaxanthin) are the only carotenoids detected (Bone *et al* 1993). Given that a xanthophyll-binding protein in the retina has been recently reported (Bhosale *et al* 2004), RBC may have such a protein. This possibility is still speculative, and further studies are needed. On the other hand, differences in RBC and plasma carotenoid levels between men and women were observed (Table 12.4). This may reflect dietary differences, given that carotenoid intake is generally higher in women than in men (El-Sohemy *et al* 2002). Both RBC and plasma canthaxanthin were present at trace levels (Table 12.4). This seems to be because the volunteers did not ingest foods enriched in canthaxanthin.

12.2.4 Xanthophylls are a Potential Antioxidant in RBC

It is well recognized that one of the major physiological activities of carotenoids is their antioxidant action and their protective effect against lipid peroxidation in biological membranes. Xanthophylls have recently gained increasing scientific interest due to their antioxidative (Chitchumroonchokchai *et al* 2004), anti-obesity (Maeda *et al* 2005), and anti-inflammatory (Izumi-Nagai *et al* 2007) activities, all of which differ somewhat from those of non-polar carotenoids. As mentioned in the introductory paragraphs, we previously found a higher accumulation of PLOOH in RBC of dementia patients (Figure 12.1) (Miyazawa *et al* 1992b). In animal trials, the inhibitory effect on PLOOH formation in RBC membranes was confirmed in mice by dietary supplementation with carotenoids (Figure 12.3) (Nakagawa *et al* 1996). As described here, a pronounced distribution of polar carotenoids (xanthophylls) to RBC was observed. Therefore, it seems that carotenoids, especially xanthophylls, have a potential to act as important antioxidant molecules in RBC and, thereby, may contribute to the prevention of dementia.

12.3   Antioxidant Effect of Lutein towards Phospholipid Oxidation in RBC

12.3.1   Lutein Supplementation Study

As mentioned above, xanthophylls such as lutein have gained attention as potential inhibitors against RBC phospholipid hydroperoxidation, thereby making them plausible candidates for preventing dementia. To evaluate the hypothesis, we next investigated whether orally administered lutein is distributed to human RBC, and inhibits RBC phospholipid hydroperoxidation (Nakagawa *et al* 2009). To date, although many *in vitro* studies on the antioxidative property of food constituents have been reported, little has been known about the biological functions of dietary antioxidants *in vivo* (especially in humans), except for major antioxidants (for example, tocopherols and ascorbic acid). Since the bioavailability of food constituents is limited by their digestibility and metabolic fate, human oral administration trials are favored to evaluate their biological function.

Six healthy subjects (aged 21–28 years) participated in this study. The subjects gave their written informed consent to the experimental protocol. Subjects took one capsule of Flora-Glo Lutein (containing 9.67 mg of lutein, 0.73 mg of zeaxanthin and 0.12 mg of α-tocopherol; Kemin Foods, Des Moines, IA, USA) once per day (after breakfast) for 4 weeks. During the experimental period, the subjects were instructed to avoid foods rich in carotenoids (for example, spinach, broccoli, carrots, and tomatoes). Before, 2 and 4 weeks after the ingestion, blood was collected. Packed cells were prepared from the blood. RBC carotenoids were determined by HPLC–DAD–APCI/MS as described above (Nakagawa *et al* 2008).

For determination of RBC PLOOH (Miyazawa *et al* 1988; 1992b), total lipids were extracted from packed cells with a mixture of 2-propanol and chloroform containing butylated hydroxytoluene. PLOOH [*i.e.*, phosphatidylcholine hydroperoxide (PCOOH) and phosphatidylethanolamine hydroperoxide (PEOOH)] in the total lipids were measured by HPLC with chemiluminescence (CL) detection. The column was a 4.6 × 250 mm, 5 μm Finepak SIL NH2-5 (Japan Spectroscopic Co., Tokyo, Japan), the eluent was 2-propanol/methanol/water (135:45:20, v/v/v), and the flow rate was 1 ml min. Post-column CL detection was carried out using a CLD-100 detector (Tohoku Electronic Industries Co., Sendai, Japan). A mixture of luminol and cytochrome *c* in 50 mmol L^{-1} borate buffer (pH 10.0) was used as a hydroperoxide-specific post-column CL reagent. Calibration was carried out using PCOOH or PEOOH standards (Ibusuki *et al* 2008).

12.3.2 Lutein Inhibits RBC Phospholipid Oxidation

In a typical DAD chromatogram of the RBC extract taken before the ingestion of Flora-Glo Lutein, six endogenous carotenoids (lutein, zeaxanthin,β-cryptoxanthin, α-carotene, β-carotene and lycopene) were separated, detected by DAD (Figure 12.7), and concurrently identified based on APCI/MS and UV spectra profiles. The relative amounts of each carotenoid were consistently as follows: lutein > β-cryptoxanthin > zeaxanthin > β-carotene > lycopene > α-carotene (Table 12.6). Therefore, these data support well the above finding (Table 12.4) that, in human RBC, xanthophylls (especially lutein) are the most prevalent carotenoids. In contrast, non-polar carotenoids (for example, β-carotene) were more abundant in plasma. In a typical CL chromatogram of RBC total lipids taken before the ingestion, PCOOH and PEOOH were the predominant forms of PLOOH. On the other hand, only PCOOH was detected in plasma.

After ingestion of Flora-Glo Lutein for 4 weeks, RBC lutein concentration increased from baseline (160 pmol g of hemoglobin) to 449 pmol g ofhemoglobin, whereas RBC PLOOH (sum of PCOOH and PEOOH) decreased from 4.9 μmol/mol phospholipids to 1.5 μmol/mol phospholipids (Table 12.6). Flora-Glo Lutein ingestion did not affect the levels of zeaxanthin, β-carotene, lycopene, and α-carotene in RBC and plasma, PLOOH in plasma, and tocopherols in RBC and plasma. Also, RBC hemoglobin and RBC phospholipid, as well as plasma phospholipid, were not affected by Flora-Glo Lutein ingestion. Based on these results, it is suggested that when humans ingest lutein, lutein is absorbed, distributed, and accumulated in RBC, where it acts as an antioxidant molecule, thereby reducing PLOOH as an index of oxidative stress. To date, although there have been many reports about the health benefits of xanthophylls (Izumi-Nagai et al 2007; Nakagawa et al 2009; 2011), they have never provided any information about the distribution and

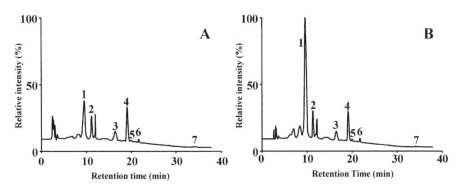

Figure 12.7 Typical photodiode array detection chromatograms of carotenoids in RBC: (A) before lutein supplementation, and (B) after 4 weeks lutein supplementation. Peak identifications are as follows: (1) lutein, (2) zeaxanthin, (3) β-cryptoxanthin, (4) echinenone (internal standard), (5) α-carotene, (6) β-carotene, and (7) lycopene.

Table 12.6 Carotenoids, phospholipid hydroperoxides, and tocopherols in RBC and plasma. Values are mean ± S.D. ($n = 6$).

		Ingestion periods		
		0 week	2 weeks	4 weeks
Erythrocytes	Carotenoids	pmol/g hemoglobin		
	Lutein	160 ± 42[a]	364 ± 113[b]	449 ± 120[b]
	Zeaxanthin	52 ± 18	66 ± 21	71 ± 18
	β-Cryptoxanthin	66 ± 19	64 ± 17	68 ± 21
	α-Carotene	8 ± 5	9 ± 6	5 ± 4
	β-Carotene	20 ± 9	22 ± 14	11 ± 4
	Lycopene	12 ± 6	14 ± 10	5 ± 2
	Xanthophylls[1]	278 ± 47[a]	495 ± 116[b]	589 ± 118[b]
	Non-polar carotenoids[2]	40 ± 18	46 ± 28	21 ± 10
	Total carotenoids	318 ± 57[a]	540 ± 110[b]	610 ± 110[b]
PLOOH		μmol/mol phospholipids		
	PCOOH	2.7 ± 1.4[a]	1.7 ± 0.6[a]	0.7 ± 0.5[b]
	PEOOH	2.1 ± 1.1[a]	0.7 ± 0.2[b]	0.8 ± 0.8[b]
	Total PLOOH	4.9 ± 2.5[a]	2.4 ± 0.8[a]	1.5 ± 1.2[b]
	Tocopherols	nmol/g hemoglobin		
	α-Tocopherol	20.2 ± 3.1	21.2 ± 2.9	20.7 ± 0.8
	γ-Tocopherol	3.0 ± 0.6	3.8 ± 1.1	3.2 ± 0.5
	Total tocopherols	23.2 ± 3.3	25.0 ± 3.2	23.9 ± 0.9

Table 12.6 (*Continued*)

		Ingestion periods		
		0 week	2 weeks	4 weeks
Plasma	Carotenoids		pmol/mL plasma	
	Lutein	341 ± 119a	682 ± 280ab	768 ± 295b
	Zeaxanthin	115 ± 33	122 ± 46	120 ± 38
	β-Cryptoxanthin	411 ± 111b	284 ± 77a	224 ± 31a
	α-Carotene	184 ± 94	147 ± 56	135 ± 48
	β-Carotene	475 ± 189	409 ± 168	337 ± 175
	Lycopene	303 ± 140	249 ± 119	301 ± 112
	Xanthophylls[1]	868 ± 151	1083 ± 361	1112 ± 317
	Non-polar carotenoids[2]	961 ± 377	805 ± 303	772 ± 255
	Total carotenoids	1829 ± 452	1894 ± 445	1885 ± 279
PLOOH	PCOOH	μmol/mol phospholipids		
		9.5 ± 5.1	7.7 ± 2.2	11.3 ± 3.9
	Tocopherols	nmol/mL plasma		
	α-Tocopherol	24.4 ± 4.3	22.7 ± 4.0	23.7 ± 3.6
	γ-Tocopherol	2.7 ± 0.6	3.1 ± 0.5	3.0 ± 0.4
	Total tocopherols	27.1 ± 4.6	25.7 ± 3.6	26.6 ± 3.8

[a,b,c] Mean values within a column with unlike superscript letters were significantly different at $P<0.05$. [1] Xanthophylls are the sum of lutein, zeaxanthin, and β-cryptoxanthin. [2] Non-polar carotenoids are the sum of α-carotene, β-carotene, and lycopene.

the antioxidant effect of xanthophylls in RBC. Therefore, our findings (the inhibitory effect of lutein on RBC PLOOH) would provide a new insight into the application of lutein as a possible anti-dementia agent.

To evaluate the peroxidisability, we measured PLOOH. Because PLOOH are the primary oxidation products of phospholipid, an increase in PLOOH directly reflects *in vivo* oxidative stress (Miyazawa *et al* 1988; 1992b). As shown here, we demonstrates that when human subjects ingest food-grade lutein (Flora-Glo Lutein), lutein is absorbed, distributed and accumulated in RBC, where it exhibits antioxidative effects *in vivo* (inhibition of RBC phospholipid hydroperoxidation) (Table 12.6). It is interesting that the antioxidative effect observed in the present study was brought about by relatively short-term supplementation with lutein (2–4 weeks). Each volunteer received 9.67 mg lutein per day. The dosage was approx. 2-fold compared with that of human daily intake. The concentrations of endogenous antioxidants (*i.e.*, carotenoids and tocopherols) other than RBC and plasma lutein and plasma β-cryptoxanthin showed no changes before and after Flora-Glo Lutein ingestion. This is advantageous for elucidation of the antioxidative contribution of lutein. The antioxidative effect of lutein was confirmed on the RBC membrane, but not in the plasma. RBC are rich in polyunsaturated fatty acids (PUFAs) in the phospholipid bilayer, and contain high concentrations of molecular oxygen and ferrous ions as constituents of oxyhemoglobin. The oxidation of hemoglobin accompanies the formation of superoxides, the source of more reactive oxygen species. Therefore, RBC membrane phospholipid would be more susceptible to peroxidation than other organelle membranes, although being protected by several antioxidant systems such as superoxide dismutase, catalase and glutathione peroxidase.

12.4   Conclusions

We previously found that the higher accumulation of PLOOH in RBC of dementia patients (Miyazawa *et al* 1992b) and the inhibitory effect on PLOOH formation in RBC membranes was confirmed in mice by dietary supplementation with carotenoids (Asai *et al* 1999; Nakagawa *et al* 1996). As described here, a determination method for RBC carotenoid was developed, and a pronounced distribution of xanthophylls (for example, lutein) to RBC was confirmed (Nakagawa *et al* 2008). In addition, orally administered lutein was incorporated into RBC, and RBC PLOOH levels decreased (Nakagawa *et al* 2009). Similar results were observed by another xanthophyll study, in which the middle-aged and senior volunteers daily received astaxanthin (Nakagawa *et al* 2011). According to our on-going study, RBC PLOOH increased, and RBC xanthophylls (for example, lutein) decreased, in correlation with the severity of dementia. RBC high in lipid hydroperoxides have been suggested to have a decreased ability to transport oxygen to brain, and they may impair blood rheology, thus facilitating dementia. On the basis of these points, it seems that xanthophylls, especially lutein, have the potential to act as important antioxidant molecules in RBC, and they thereby may contribute to the prevention of dementia. This possibility warrants their testing in other models of dementia with a realistic prospect of their use in human therapy.

Summary Points

- This Chapter focuses on analysis and antioxidant function of carotenoids in human red blood cells.
- Carotenoids were determined by high-performance liquid chromatography coupled to ultraviolet photodiode array detection and atmospheric pressure chemical ionization mass spectrometry.
- Lutein is a major antioxidant carotenoid present in red blood cells.
- In patients with senile dementia, low lutein and high lipid hydroperoxide concentrations are found in their red blood cells.
- Dietary supplementation of lutein increases lutein and decreases the lipid hydroperoxide concentrations of red blood cells in normal subjects.

Key Facts of Lutein

- Lutein is the major carotenoid responsible for the prevention of membrane lipid peroxidation in human red blood cells. In contrast, the major carotenoids responsible for the prevention of plasma are non-polar hydrocarbon carotenoids.
- Lutein is present in green vegetables such as spinach.
- Xanthophylls, such as lutein, is the major carotenoids in human brain.
- Lutein is thought of as antioxidant in human brain, because it is thought that protects important brain component such as docosahexaenoic acid.
- Lutein reduces the risk of degenerative eye disease.

Definitions of Words and Terms

APCI/MS: APCI/MS is an abbreviation for atmospheric pressure chemical ionization/mass spectrometry that is employed for the determination of carotenoids in human red blood cells.

Atherogenesis: One of the blood vessel disorders that causes atherosclerotic artery lesions especially in cardiac infarction.

Carotenoids: The most widespread group of hydrophobic pigments in food, and indispensable cellular components in both animals and humans. Human blood contains β-carotene, lycopene, and lutein as representative carotenoids.

DAD: DAD is the abbreviation for diode array detection in the wavelength region of 200 nm and 700 nm, and is employed for the detection of carotenoids coupled with high-performance liquid chromatography.

Dark room: Carotenoids are easily degraded by sunlight, therefore, if extracted in this room, protect the carotenoid.

HPLC: HPLC is an abbreviation for high-performance liquid chromatography, and is used for the separation and determination of carotenoids and other biological molecules.

Lipid hydroperoxides: Lipid hydroperoxides are primary oxidation products of unsaturated lipids, and has been linked to cause cellular oxidative damage and ageing.

Oxidized phospholipids: Typically found in aged red blood cells in patients with senile dementia at higher concentrations than those of normal subjects. Membrane phospholipids yield oxidized phospholipids after exposure to oxygen radicals and to lipoxygenases.

Senile dementia: One of the age-related diseases of the brain that shows a decrease in recognition and memory. Alzheimer's disease is the representative disease.

Xanthophylls: Xanthophylls are polar carotenoids. Lutein, zeaxanthin, canthaxanthin, and β-cryptoxanthin are the representatives that are present in foods.

List of Abbreviations

APCI	atmospheric pressure chemical ionization
CL	chemiluminescence
DAD	diode array detection
HPLC	high-performance liquid chromatography
KOH	potassium hydroxide
MS	mass spectrometry
PCOOH	phosphatidylcholine hydroperoxide
PEOOH	phosphatidylethanolamine hydroperoxide
PLOOH	phospholipid hydroperoxides
RBC	red blood cell(s)
MTBE	methyl *tert*-butyl ether
UV	ultraviolet

References

Aust, O., Sies, H., Stahl, W. and Polidori, M. C., 2001. Analysis of lipophilic antioxidants in human serum and tissues: tocopherols and carotenoids. Journal of Chromatography A. 936: 83–93.

Bhosale, P., Larson, A. J., Frederick, J. M., Southwick, K., Thulin, C. D. and Bernstein, P. S., 2004. Identification and characterization of an isoform of glutathione S-transferase (GSTP1) as a zeaxanthin-binding protein in the macula of the human eye. Journal of Biological Chemistry. 279: 49447–49454.

Bjornson, L. K., Kayden, H. J., Miller, E. and Moshell, A. N., 1976. The transport of α-tocopherol and β-carotene in human blood. Journal of Lipid Research. 17: 343–352.

Bone, R. A., Landrum, J. T., Hime, G. W., Cains, A. and Zamor, J., 1993. Stereochemistry of the human macular carotenoids, Investigation of Ophthalmology and Visible Science. 34: 2033–2040.

Britton, G., 1995. UV/Visible spectroscopy. In: Britton, G., Liaaen-Jensen, S. and Pfander, H. (ed.) Carotenoids, vol. 1B: Spectroscopy. Birkhauser, Basel, Switzerland, pp. 13–62.

Chitchumroonchokchai, C., Bomser, J. A., Glamm, J. E. and Failla, M. L., 2004. Xanthophylls and α-tocopherol decrease UVB-induced lipid peroxidation and stress signaling in human lens epithelial cells. Journal of Nutrition. 134: 3225–3232.

El-Sohemy, A., Baylin, A., Kabagambe, E., Ascherio, A., Spiegelman, D. and Campos, H., 2002. Individual carotenoid concentrations in adipose tissue and plasma as biomarkers of dietary intake. American Journal of Clinical Nutrition. 76: 172–179.

Fotouhi, N., Meydani, M., Santos, M. S., Meydani, S. N., Hennekens, C. H. and Gaziano, J. M., 1996. Carotenoid and tocopherol concentrations in plasma, peripheral blood mononuclear cells, and red blood cells after long-term-carotene supplementation in men. American Journal of Clinical Nutrition. 63: 553–558.

Goulinet, S. and Chapman, M. J., 1997. Plasma LDL and HDL subspecies are heterogeneous in particle content of tocopherols and oxygenated and hydrocarbon carotenoids. Arteriosclerosis, Thrombosis, and Vascular Biology. 17: 786–796.

Ibusuki, D., Nakagawa, K., Asai, A., Oikawa, S., Masuda, Y., Suzuki, T. and Miyazawa, T., 2008. Preparation of pure lipid hydroperoxides. Journal of Lipid Research. 49: 2668–2677.

Izumi-Nagai, K., Nagai, N., Ohgami, K., Satofuka, S., Ozawa, Y., Tsubota, K., Umezawa, K., Ohno, S., Oike, Y. and Ishida, S., 2007. Macular pigment lutein is antiinflammatory in preventing choroidal neovascularization. Arteriosclerosis, Thrombosis, and Vascular Biology. 27: 2555–2562.

Kinoshita, M., Oikawa, S., Hayasaka, K., Sekikawa, A., Nagashima, T., Toyota, T. and Miyazawa, T., 2000. Age-related increases in plasma phosphatidylcholine hydroperoxide concentrations in control subjects and patients with hyperlipidemia. Clinical Chemistry. 46: 822–828.

Leo, M. A., Ahmed, S., Aleynik, S. I., Siegel, J. H., Kasmin, F. and Lieber, C. S., 1995. Carotenoids and tocopherols in various hepatobiliary conditions. Journal of Hepatology. 23: 550–556.

Maeda, H., Hosokawa, M., Sashima, T., Funayama, K. and Miyashita, K., 2005. Fucoxanthin from edible seaweed, *Undaria pinnatifida*, shows antiobesity effect through UCP1 expression in white adipose tissues. Biochemistry and Biophysics Research Communications. 332: 392–397.

Miyazawa, T., Yasuda, K., Fujimoto, K. and Kaneda, T., 1988. Presence of phosphatidylcholine hydroperoxide in human plasma. Journal of Biochemistry. 103: 744–746.

Miyazawa, T., Suzuki, T., Fujimoto, K. and Yasuda, K., 1992a. Chemiluminescent simultaneous determination of phosphatidylcholine hydroperoxide and phosphatidylethanolamine hydroperoxide in the liver and brain of the rat. Journal of Lipid Research. 33: 1051–1059.

Miyazawa, T., Suzuki, T., Yasuda, K., Fujimoto, K., Meguro, K. and Sasaki, H., 1992b. Accumulation of phospholipid hydroperoxides in red blood cell membranes in Alzheimer disease. In: Yagi, K., Kondo, M., Niki, E. and Yoshikawa, T. (ed.) Oxygen Radicals. Elsevier Science, Amsterdam, The Netherlands, pp. 327–330.

Morinobu, T., Tamai, H., Murata, T., Manago, M., Takenaka, H., Hayashi, K. and Mino, M., 1994. Changes in β-carotene levels by long-term administration of natural β-carotene derived from *Dunaliella bardawil* in humans. Journal of Nutritional Science and Vitaminology. 40: 421–430.

Moriya, K., Nakagawa, K., Santa, T., Shintani, Y., Fujie, H., Miyoshi, H., Tsutsumi, T., Miyazawa, T., Ishibashi, K., Horie, T., Imai, K., Todoroki, T., Kimura, S. and Koike, K., 2001. Oxidative stress in the absence of inflammation in a mouse model for hepatitis C virus-associated hepatocarcinogenesis. Cancer Research. 61: 4365–4370.

Murata, T., Tamai, H., Morinobu, T., Manago, M., Takenaka, A., Takenaka, H. and Mino, M., 1992. Determination of β-carotene in plasma, blood cells, and buccal mucosa by electrochemical detection. Lipids. 27: 840–843.

Murata, T., Tamai, H., Morinobu, T., Manago, M., Takenaka, H., Hayashi, K. and Mino, M., 1994. Effect of long term administration of β-carotene on lymphocyte subsets in humans. American Journal of Clinical Nutrition. 60: 597–602.

Nakagawa, K., Fujimoto, K. and Miyazawa, T., 1996. β-Carotene as a high-potency antioxidant to prevent the formation of phospholipid hydroperoxides in red blood cells of mice. Biochimica et Biophysica Acta. 1299: 110–116.

Nakagawa, K., Kiko, T., Hatade, K., Asai, A., Kimura, F., Sookwong, P., Tsuduki, T., Arai, H. and Miyazawa, T., 2008. Development of a high-performance liquid chromatography-based assay for carotenoids in human red blood cells: application to clinical studies. Analytical Biochemistry. 381: 129–134.

Nakagawa, K., Kiko, T., Hatade, K., Sookwong, P., Arai, H. and Miyazawa, T., 2009. Antioxidant effect of lutein towards phospholipid hydroperoxidation in human erythrocytes. British Journal of Nutrition. 102: 1280–1284.

Nakagawa, K., Kiko, T., Miyazawa, T., Burdeos, G. C., Kimura, F., Satoh, A. and Miyazawa, T., 2011. Antioxidant effect of astaxanthin on phospholipid peroxidation in human erythrocytes. British Journal of Nutrition. 105: 1563–1571.

Norkus, E. P., Bhagavan, H. N. and Nair, P. P., 1990. Relationship between individual carotenoids in plasma, platelets, and red blood cells (RBC) of adult subjects. FASEB Journal. 4: A1774.

Shah, S. N., Johnson, R. C. and Singh, V. N., 1989. Antioxidant vitamin (A and E) status of Down's syndrome subjects. Nutrition Research. 9: 709–715.

Stocker, R. and Keaney, J.F., 2004. Role of oxidative modifications in atherosclerosis. Physiological Review. 84: 1381–1478.

Su, Q., Rowley, K. G. and Balazs, N. D., 2002. Carotenoids: Separation methods applicable to biological samples. Journal of Chromatography B. 781: 393–418.

Yeum, K. J., Booth, S. L., Sadowski, J. A., Liu, C., Tang, G., Krinsky, N. I. and Russell, R. M., 1996. Human plasma carotenoid response to the ingestion of controlled diets in fruits and vegetables. American Journal of Clinical Nutrition. 64: 594–602.

CHAPTER 13

Capillary Liquid Chromatographic Analysis of Fat-soluble Vitamins and β-Carotene

SHENG ZHANG AND LI JIA*

Ministry of Education, Key Laboratory of Laser Life Science and Institute of Laser Life Science, College of Biophotonics, South China Normal University, Guangzhou 510631, China
*E-mail: jiali@scnu.edu.cn

13.1 Introduction

High-performance liquid chromatography (HPLC) has been widely used in routine analytical work, method development, process monitoring, and quality control since its introduction in the late 1960s. In comparison with conventional liquid chromatography (LC) in columns with diameters of 4.6 mm, the down-sizing of separation columns has become a mainstream of LC development. This is due to the advantages of narrow-bore column LC, including significant reduction of solvent consumption, the small amounts of sample required and easy coupling with other techniques. In terms of the size of separation column, there is a tendency to define 0.50–1.0 mm internal diameter (I.D.) columns as micro LC, 100–500 μm I.D. columns as capillary liquid chromatography (CLC), and 10–100 μm I.D. columns as nanoscale LC (Chervet *et al* 1996). The initial development towards miniaturization in HPLC

Food and Nutritional Components in Focus No. 1
Vitamin A and Carotenoids: Chemistry, Analysis, Function and Effects
Edited by Victor R Preedy
Published by the Royal Society of Chemistry, www.rsc.org

is attributed to Horváth and co-workers (Horváth *et al* 1967; Horváth and Lipsky 1969), and Ishii and co-workers (Ishii *et al* 1977). Horváth and co-workers examined 1 mm I.D. stainless steel columns packed with 50 μm pellicular particles for the separation of nucleotides. Ishii and co-workers demonstrated the use of 0.5 mm polytetrafluoroethylene (PTFE) columns packed with 30 μm pellicular particles. Since then, LC in small-bore columns has shown considerable progress both in instrumentation and applications. This Chapter will focus on the features and instrumentation of CLC, and its applications for the analysis of fat-soluble vitamins and β-carotene.

13.2   CLC

13.2.1   Features

The attractive features of CLC are attributed to the use of columns with diameters smaller than 0.5 mm:

1. The small column diameter leads to the operation at low flow rates when the linear velocity independent of the column diameter is kept constant, which in turn leads to the low consumption of the mobile phase, and allows us to use exotic or expensive mobile phases or mobile phase additives. This is environmentally friendly.
2. The small column diameter leads to the low consumption of the stationary phase, which facilitates the use of valuable and expensive packing materials or long columns.
3. The small column diameter leads to the requirement of small amounts of sample. This is especially of benefit when the sample amounts available are restricted, as in the case of biological samples.
4. The extracolumn band broadening effect has considerable contribution to the loss of efficiency of a column with small diameter. In order to achieve the maximal performance of a column with small diameter, the extracolumn band broadening effect must be reduced accordingly.
5. A minute sample injection volume and limited optical path length for on-column photometric detection in CLC result in its low detection sensitivity. In order to enhance the detection sensitivity in CLC, hyphenation with sample pre-concentration techniques is needed.

13.2.2   Instrumentation

A CLC equipment consists of a pump, an injector, a capillary column, a detector, and a data acquisition system. The basic insrumental setup to accomplish CLC is depicted in Figure 13.1. A capillary column is heart of a CLC system.

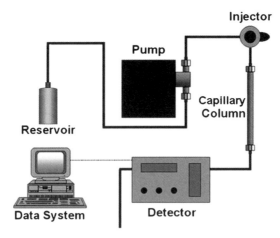

Figure 13.1   Basic instrumental setup for a CLC system. The basic CLC system consists of five modules (including pump, injector, column, detector, and data acquisition system) connected by appropriate tubing and fittings.

13.2.2.1   Capillary Column

A column in a CLC system enables the separation of compounds based on the interactions between the compounds and the stationary phases in a column. Three types of columns are used in CLC, including open tubular columns (Hibi *et al* 1978), packed columns (Crescentini *et al* 1988), and monolithic columns (Chen *et al* 2009):

1. Open tubular columns in CLC usually have an internal diameter of 10 μm or less. The stationary phases are adsorbed or covalently coated on the inner walls of a fused-silica capillary. The advantages of the open tubular columns are their easy preparation, the variety, and availability of suface modification. However, they exhibit some limitations due to the low sample-loading ability. To overcome the intrinsic drawback, several approaches including etched capillary and porous layer coated capillary have been proposed to increase the surface area.
2. Packed columns filled with high surface area particles have an improved sample capacity over open tubular columns. The major disadvantage of packed columns over open tubular columns is its relatively high flow impedance.
3. Monolithic columns are prepared in capillaries, where the skeleton are bonded to the inner walls of a capillary. The stationary phase in a monolithic column is a well-ordered array of skeleton with through-pores and meso-pores. They possess permeability between that of open-tubular columns and densely packed columns. In comparison with packed columns, monolithic columns offer some improvements, including low back pressure, fast mass transfer, absence of end frits, and so on.

13.2.2.2   Pumping System

In CLC, the small column diameter leads to the operation at low flow rates. The flow rates are usually in the order of a few nL up to μL. There are three ways to achieve the low flow rates. The first one is to use a syringe pump. This option provides stable flows, but the system needs to be taken off-line periodically to refill the syringe. The second one is to use a standard style reciprocating HPLC pump with a small piston. These pumps may have large pressure fluctuations and reliability problems. The third one is to use a conventional LC pump with a flow splitter. The split ratio on a flow splitter is pressure dependent, therefore setting the desired flow rate is not easy. The disadvantage of any flow-splitting scheme is that you can not achieve the solvent consumption savings, since most of the mobile phase goes to a waste reservoir attached to the pump, and only a small fraction of the mobile phase goes through the column and detector.

13.2.2.3   Injector

The large injection volume has a great influence on the column efficiency in CLC due to the effect of the extracolumn band broadening from the injection part. The injection volumes are usually in the order of a few nL for 50–100 μm I.D. columns, up to approximately 1 μL for 1.0 mm I.D. columns. Injection volumes ranging from a few μL down to 5 nL can be routinely performed with injection valves equipped with an internal loop. The injection volumes below 5 nL can be performed by positioning a split vent between the injector and the column.

13.2.2.4   Detector

The detection system in CLC includes on-column and post-column detectors. Ultraviolet/visible (UV/Vis) and fluorescence spectrometry detectors can be used as on-column or post-column in CLC. In order to prevent the extacolumn band broadening in the detection part from deteriorating the column efficiency, the detection cell volume must be reduced. When UV/Vis and fluorescence spectrometry detectors are used as on-column optical detectors, a portion of the separation column itself can be employed as the detection cell. The attracting benefit of on-column detection is that there is no extracolumn band broadening caused by column–detector connections, and detector cell void volume. In addition, the detection cell volume is a function of the column internal diameter and the illuminated length. When UV/Vis and fluorescence spectrometry detectors are used as post-column detectors, a cylindrical quartz tube is usually used as the detection cell. In this case, the dead volume of the connection between the separation column and the detection cell must be minimized.

Electrochemical detectors and mass spectrometry (MS) are placed post-column in CLC. The dead volume from the connection between the column and detection cell needs to be decreased as possible. The electrochemical detector just responds to substances that are either oxidizable or reducible, and the electrical output is an electron flow generated by a reaction that takes place at the surface of the electrodes. The electrochemical detector is extremely sensitive but it demands that the mobile phase must be extremely pure, oxygen-free and devoid of metal ions (Erickson 2000). A MS detector coupled with CLC is capable of providing the molecular identity of a wide range of components.

13.2.3   On-line Sample Pre-concentration

In CLC, a minute sample injection volume and limited optical path length for on-column photometric detection result in its low detection sensitivity. In order to enhance the detection sensitivity in CLC, hyphenation with sample pre-concentration techniques is needed. One technique is on-column focusing, which dissolves analytes in a weaker solvent than the mobile phase and allows the injection of a large sample volume without considerable effect on band broadening (Šlais *et al* 1983). In this case, the analyte dissolved in a weak solvent undergoes on-column concentration at the column inlet during its injection. Other sample enrichment techniques are on-line or off-line solid-phase microextraction (SPME) (Pawliszyn 1997). On-line in-tube SPME, introduced by Eisert and Pawliszyn (1997), enables continuous extraction, concentration, desorption, and injection using an autosampler, which not only shortens the total analysis time but also provides better accuracy and precision relative to manual techniques, as reviewed by Kataoka (2002) and Saito and Jinno (2002). In this technique, an open tubular fused silica capillary column was usually used as an extraction device. The ratio of the surface area of the coated layer contacted with sample solution to the volume of the capillary column is insufficient for mass transfer. Repetition of the draw/eject cycles of the sample solution is needed to improve extraction efficiency. To overcome the problem, Shintani *et al* (2003) first introduced a monolithic silica-*n*-octadecyl silica (ODS) column used for in-tube SPME, which demonstrated better pre-concentration efficiency compared with the conventional in-tube SPME due to the merits of higher permeability and porosity possessed by a monolithic column. Jia and co-workers developed an in-tube SPME coupled to a CLC system for analysis of cellular flavins (Jia *et al* 2004) and fat-soluble vitamins (Xu and Jia 2009), in which a monolithic silica-ODS column was used as an extraction device.

13.3   Applications of CLC

13.3.1   Fat-soluble Vitamins

Vitamins are a group of organic compounds, which are essential for the normal growth and self-maintenance of humans and animals. Lack of vitamins can lead to serious diseases, however, only small concentrations are required to maintain health. Humans and animals cannot synthesize vitamins and have to assimilate them through their diets or pharmaceutical vitamin preparations. According to their solubility, vitamins can be categorized into two groups: fat-soluble vitamins and water-soluble vitamins. Fat-soluble vitamins include vitamins A, D, E, and K. Their structures are shown in Figure 13.2.

The concentrations of fat-soluble vitamins in most food, blood, and serum samples are low. Also, these vitamins are usually found in the presence of larger amounts of other substances (*e.g.* sterols, triglycerides, phospholipids) that may cause interference. Therefore, a separation technique is needed for the analysis of fat-soluble vitamins in complicated samples. Over the past 10 years, HPLC has been the most commonly used technique for the determination of fat-soluble vitamins. Due to the advantages of CLC, few papers also reported the analysis of fat-soluble vitamins.

The first paper on the separation of fat-soluble vitamins by CLC was reported in 1987 (Wahyuni and Jinno 1987). The fat-soluble vitamins (including vitamins A, E, D_2, D_3, K_1, and vitamin E acetate) were separated on a fused-silica column of 500 × 0.53 mm I.D. packed with polymer-based octadecyl-bonded silica with a binary mobile phase system of 75% acetonitrile and 25% methanol, as shown in Figure 13.3. The injection volume was 0.1 μL and the flow rate was 4 μL min^{-1}.

Gomis *et al* (1995) reported the separation of fat-soluble vitamins (including vitamins D_3, E, K_1, and vitamin A palmitate) on a packed ODS fused-silica capillary column (300 × 0.32 mm I.D.) using methanol/tetrahydrofuran as the mobile phase, as shown in Figure 13.4. The detection limits of these vitamins were 2.9 pg (vitamin A palmitate), 1 pg (vitamin D_3), 29 pg (vitamin E), and 2.6 pg (vitamin K_1). The proposed method allows the simultaneous and sensitive determination of trace amounts of fat-soluble vitamins in complex plasma sample with the minimum amount of sample availability (60 nL). In 2000, Gomis and co-workers successfully separated the fat-soluble vitamins A, D_2, D_3, E and K, retinyl acetate, retinyl palmitate, tocopherol acetate, ergosterol, and 7-dehydrocholesterol in milk on a packed ODS fused-silica capillary column (150 × 0.3 mm I.D.) in gradient mode with detection limits between 0.02 ng mL^{-1} for retinol and 2 ng mL^{-1} for vitamin E, as shown in Figure 13.5 (Gomis *et al* 2000). In the method, on-column focusing was used as a pre-concentration method to improve the detection limits of the vitamins. The detection limits of the vitamins are of the order of ng mL^{-1}.

In 2009, our research group developed a CLC method in combination with in-tube SPME for analysis of fat-soluble vitamins on a monolithic silica-ODS column (270 × 0.1 mm I.D.) (Xu and Jia 2009). The in-tube SPME system comprised a high-pressure microflow pump, a six-port valve, and an extraction column (monolithic silica–ODS column, 100 × 0.2 mm I.D.). The schematic diagram of the in-tube SPME/CLC system is illustrated in Figure 13.6. The system allows for a large injection volume without deteriorating the column efficiency. Figure 13.7 shows the chromatogram of the fat-soluble vitamins

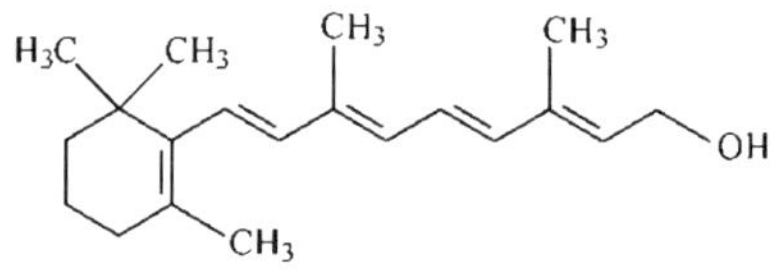

Vitamin A

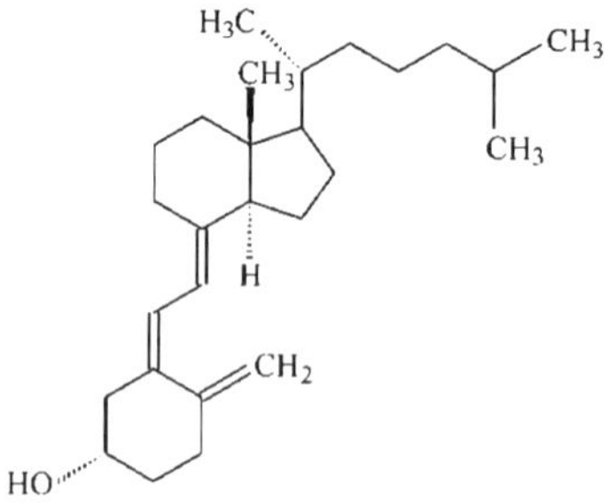

Vitamin D

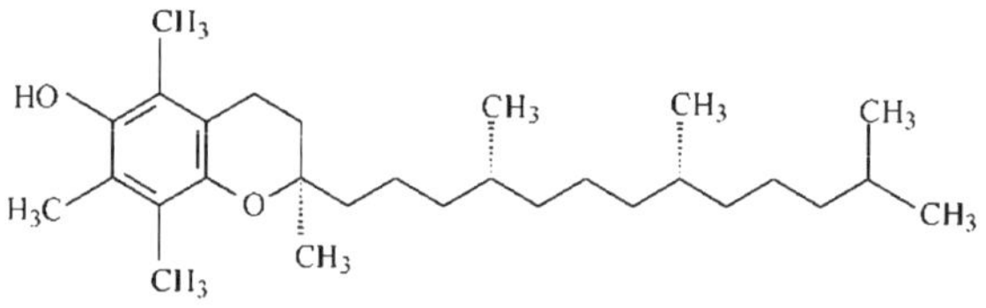

Vitamin E

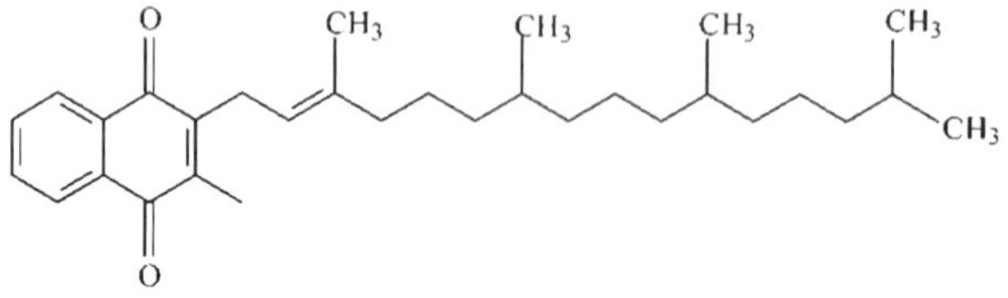

Vitamin K

Figure 13.2 Structures of vitamins A, D, E, and K. The diagram indicates the arrangement of atoms and the chemical bonds that hold the atoms together in each fat-soluble vitamin. The differences in the chemical structures of these fat-soluble vitamins are shown.

analyzed by the system at an injection volume of 10 μL. The detection limits of the vitamins are 3.6 (vitamin A), 16.4 (vitamin K_1), and 173 (vitamin E) ng mL^{-1}. The developed method was applied for the determination of vitamin E in corns.

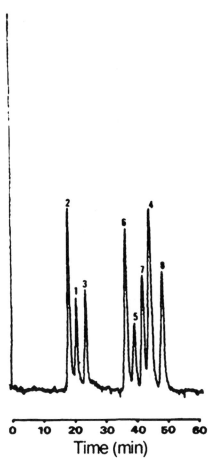

Figure 13.3 Chromatogram of fat-soluble vitamins and vitamin B_{12}. The chromatogram shows the separation of fat-soluble vitamins and vitamin B_{12} by
CLC. The conditions were as follows: column, 500×0.53 mm I.D.,
packed with polymeric octadecyl-bonded silica; mobile phase, 75%
acetonitrile/25% methanol; flow rate; 4 μL min^{-1}; injection volume, 0.1
μL; detection wavelength, 280 nm; column temperature, 20 °C. Peak
assignment: 1, vitamin K_3; 2, vitamin B_{12}; 3; vitamin A; 4, vitamin E; 5,
vitamin D_2; 6, vitamin E acetate; 7, vitamin D_3; 8, vitamin K_1.
Reproduced with permission from Wahyuni and Jinno (1987).

13.3.2 β-Carotene

Carotenoids are a class of fat-soluble compounds, which are synthesized by
plants and many micro-organisms. They have many physiological and
biological functions, such as provitamin A activity, antioxidants, enhancers
of the immune response, and free radical scavengers (Van den Berg *et al* 2000).
The term "provitamin A" is accepted to differentiate carotenoid precursors of
vitamin A from carotenoids without vitamin A activity. Provitamin A active

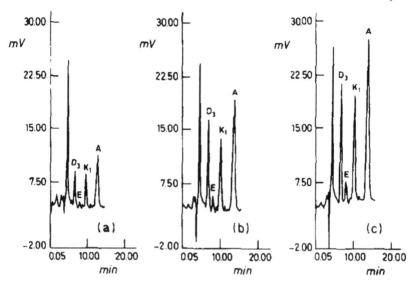

Figure 13.4   Chromatogram of fat-soluble vitamins by CLC. The chromatogram shows the separation of fat-soluble vitamins by CLC. The conditions were as follows: column, 300 × 0.32 mm I.D., fused silica, packed with 5 μm Spherisorb ODS-2; mobile phase, methanol/tetrahydrofuran (80:20, v/v); injection volume, (a) 100 nL; (b) 200 nL; (c) 300 nL; flow rate, 5 μL min^{-1}; detection wavelength, λ_A = 328 nm, λ_D = 265 nm, λ_E = 284 nm, λ_K = 250 nm. Reproduced with permission from Gomis *et al* (1995).

carotenoids include α-, β-, γ-carotene, and cryptoxanthin. Their structures are shown in Figure 13.8. The most often used method for the separation of carotenoids is conventional reversed phase HPLC.

In 2009, our research group first prepared a triacontyl-functionalized monolithic silica capillary column (300 × 0.1 mm I.D.) and applied the column in CLC for the separation of α- and β-carotene in isocratic mode with methanol as the mobile phase (Chen *et al* 2009). More recently, we reported the separation of α- and β-carotene on a monolithic silica-ODS column (350 × 0.2 mm I.D.) by in-tube SPME/CLC system (Zhang *et al* 2010), in which a monolithic silica-ODS column (100 × 0.2 mm I.D.) was used as the extraction medium. The chromatogram for the separation of the standard carotene mixture is shown in Figure 13.9. The detection limits for α- and β-carotene were 15 and 20 ng mL^{-1}, respectively. The proposed method was successfully used to determine trace amount of β-carotene in corns.

13.4   Future Prospects of CLC

Development of capillary columns with higher efficiency than conventional LC columns will shift users from conventional to CLC due to the benefits of CLC, as seen in the development of capillary gas chromatography (GC). Capillary

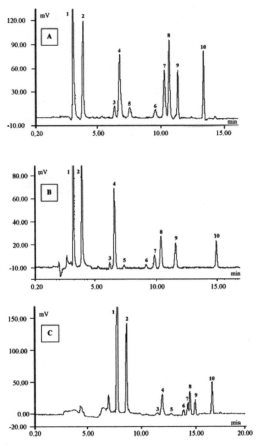

Figure 13.5 Chromatogram of fat-soluble vitamins by CLC. The chromatogram shows the separation of fat-soluble vitamins by CLC. The conditions were as follows: column, 150 × 0.30 mm I.D., fused silica, 3 μm Hypersil C_{18} BDS; flow rate, 6 μL min^{-1}. (A) Injected volume, 60 nL; concentrations are 50 mg mL^{-1} in methanol for all of the vitamins; gradient I conditions were programmed as follows: 0–4 min, 0% B; 4–10 min, 100% B; 10–15 min, 100% B; 15–17 min, 0% B [A: methanol/water (99:1); B: methanol/tetrahydrofuran (70:30)]. (B) Injected volume, 500 nL; concentrations are 1 mg mL^{-1} for all of the vitamins in methanol/ water (50:50); gradient I was used. (C) Injected volume, 5 μL; concentrations between 0.1 and 0.25 mg mL^{-1} in methanol/water (50:50); gradient II conditions were programmed as follows: 0–6 min, 0% B; 6–12 min, 100% B; 10–14 min, 100% B; 14–16 min, 0% B [A: methanol/water (99:1); B: methanol/tetrahydrofuran (70:30)]. The detection wavelength was set at 325 nm for retinol, retinyl acetate, and retinyl palmitate; 264 nm for vitamin D_2 and vitamin D_3, 280 nm for vitamin K_1, tocopherol, tocopherol acetate, ergosterol, and 7-dehydrocholesterol. Peak assignments: 1, retinol; 2, retinyl acetate; 3, vitamin D_2; 4, vitamin D_3; 5, vitamin E; 6, provitamin D_2; 7, provitamin D_3; 8, tocopherol acetate; 9, vitamin K_1; 10, retinyl palmitate. Reproduced with permission from Gomis *et al* (2000).

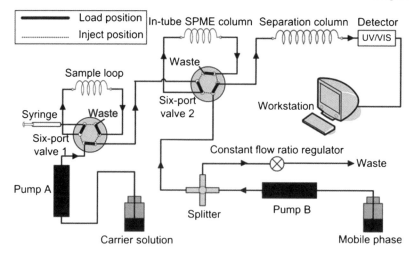

Figure 13.6   Schematic diagram of the in-tube SPME/CLC system. The schematic
diagram shows that an in-tube SPME device was on-line coupled to a
CLC instrument. The in-tube SPME system comprised a high-pressure
microflow pump, a six-port valve, and an extraction column.
Reproduced with permission from Zhang *et al* (2010).

columns with higher column efficiency must have higher permeability in
comparison with common densely packed columns. Open tubular capillary
columns and monolithic columns are expected to be the most hopeful
candidates for a highly permeable separation system. On the other hand, when
the pressure limit can be increased up to much higher levels, densely packed
columns will come back to the candidate. Ultra-high-speed and high-efficiency
separation, ultra-micro pumping systems, and the down-sizing of a whole CLC
system are to be a trend in future.

Summary Points

- This Chapter focuses on capillary liquid chromatography (CLC) and its
  applications for the analysis of fat-soluble vitamins and β-carotene.
- The trends toward miniaturization lead to the advent of CLC.
- CLC offers several advantages including significant reduction of solvent
  consumption, the small amounts of sample required, and easy coupling with
  other techniques.
- Open tubular capillary columns and monolithic columns are expected to be
  the most hopeful candidates for a CLC system.
- In order to enhance the detection sensitivity in CLC, hyphenation with
  sample pre-concentration techniques is needed.
- The application of CLC for the analysis of fat-soluble vitamins and β-
  carotene has demonstrated the benefits of CLC.

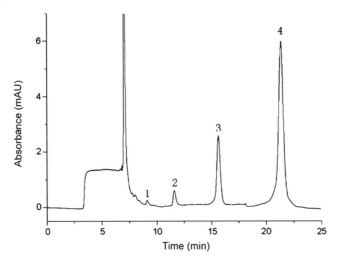

Figure 13.7 Chromatogram of the fat-soluble vitamins by the SPME/CLC at an
injection volume of 10 μL. The chromatogram shows the separation of
fat-soluble vitamins by in-tube SPME/CLC at an injection volume of 10
μL. In-tube SPME/CLC conditions were as follows: monolithic silica-
ODS columns (270 × 0.1 mm I.D.) and (100 × 0.2 mm I.D.) were used
as the separation column and extraction column, respectively. Isocratic
elution was performed with CH_3OH as the mobile phase at a flow rate
of 1.0 μL min^{-1}. Sample matrix was 50% CH_3OH. The detection
wavelength was set at 290 nm for vitamin E, 250 nm for vitamin K_1, 325
nm for vitamin A, and 452 nm for β-carotene. The concentration of
each analyte was 0.2 μg mL^{-1} except for vitamin E (1μg mL^{-1}).
Analytes: 1, vitamin E; 2, vitamin K_1; 3, vitamin A; 4, β-carotene.
Reproduced with permission from Xu and Jia (2009).

Key Facts

Key Facts of In-Tube Solid Phase Microextraction (SPME)

* In-tube SPME is an on-line sample enrichment technique.
* In-tube SPME was introduced by Eisert and Pawliszyn.
* An in-tube SPME system comprises a high-pressure microflow pump, a six-
  port valve, and an extraction column.
* In-tube SPME enables continuous extraction, concentration, desorption,
  and injection of sample.
* In-tube SPME not only shortens the total analysis time but also provides
  better accuracy and precision.

Key Facts of Mass Spectrometry (MS)

* MS is an analytical technique that measures the mass-to-charge ratio of
  charged molecules.

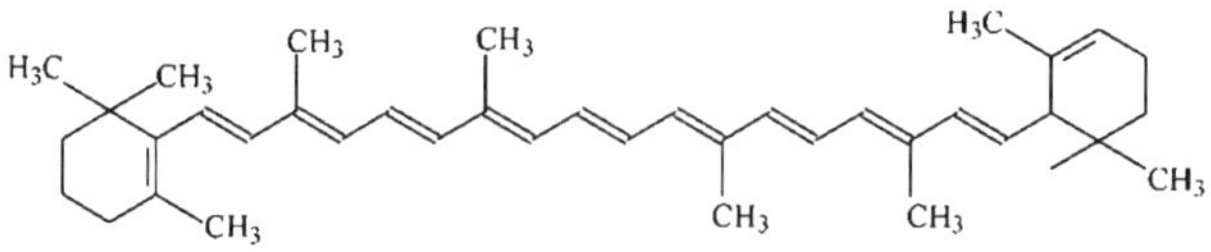

α-Carotene

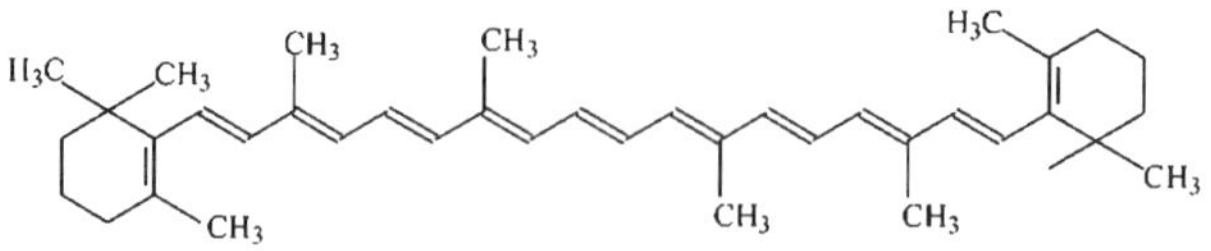

β-Carotene

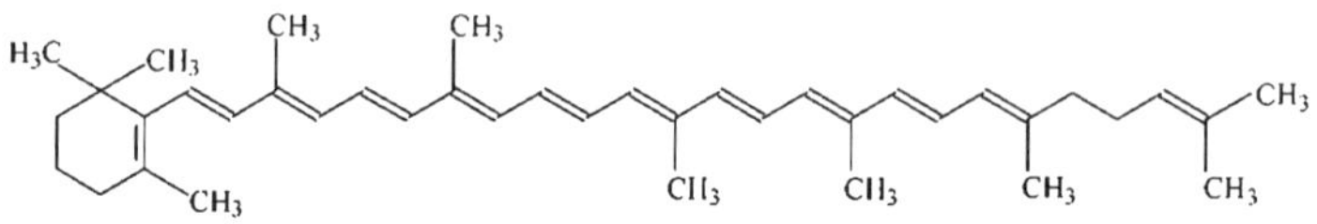

γ-Carotene

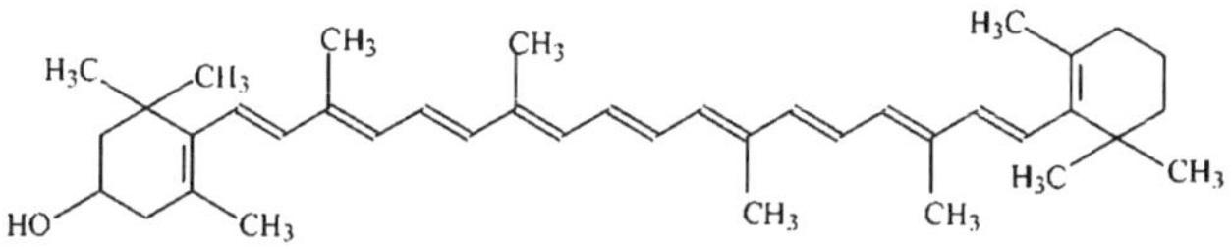

Cryptoxanthin

Figure 13.8   Structures of α-, β-, γ-carotene, and cryptoxanthin. The diagram shows the difference in the chemical structures of these carotenes, which was useful for comprehending the natures of these compounds and optimizing the separation conditions.

• MS is used for determining the masses and chemical structures of molecules based on measuring the mass-to-charge ratios of charged molecules and their distinctive fragments.
• MS instruments consist of three fundamental parts, namely the ionization source, the analyzer, and the detector.
• MS can be used for qualitative and quantitative analysis, including identifying unknown compounds and quantifying the amount of a compound in a sample.
• MS can be coupled to chromatographic separation techniques owing to its mass resolving and mass determining capabilities.

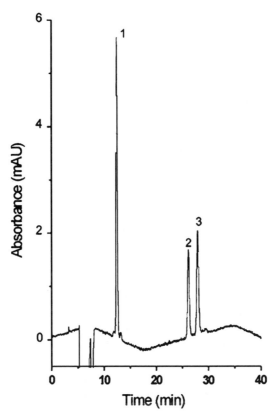

Figure 13.9 Chromatogram of the carotenes analyzed by in-tube SPME/CLC with the injection volume at 3 μL. The in-tube SPME/CLC conditions were as follows: monolithic silica-ODS columns of 350 × 0.2 mm I.D. and 100 × 0.2 mm I.D. were used as the separation column and extraction column, respectively. Isocratic elution was performed with acetonitrile containing 0.1% butylated hydroxytoluene (BHT) as the mobile phase at a flow rate of 60 μL min^{-1} before splitting. Detection wavelength was set at 450 nm. The sample matrix was 90% acetonitrile. The concentration of lutein was 0.5 μg mL^{-1}, α-carotene 0.33 μg mL^{-1}, and β-carotene 0.67 μg mL^{-1}. Analytes: 1, lutein; 2, α-carotene; 3, β-carotene. Reproduced with permission from Zhang *et al* (2010).

Definitions of Words and Terms

Capillary liquid chromatography (CLC): CLC is a high-performance liquid chromatography (HPLC) technique, in which the inner diameter of the separation column is 100–500 μm.

Column efficiency: Column efficiency is defined as the number of theoretical plates in a column, which is an important characteristic of a column. It can determine the ability of a column to produce sharp and narrow peaks.

Extracolumn band broadening effect: It refers to the contributions to the bandwidth of a peak from the injector, the connections between the column and injector or detector, and the detection cell.

Gradient elution: Gradient elution is that the composition of the mobile phase changes during the elution process. Usually, the solvent strength of the mobile phase increases during the run.

Isocratic elution: Isocratic elution is that the composition of the mobile phase is unchanged during the entire elution process.

Monolithic columns: The stationary phase in a monolithic column is a single piece of solid made of either porous cross-linked polymer or porous silica with interconnected skeletons and interconnected flow paths that go through these skeletons.

On-line sample pre-concentration: On-line sample pre-concentration is an effective approach for enhancement of concentration sensitivity, which allows the injection of large volume diluted sample without deteriorating the column efficiency.

Open tubular columns: The stationary phase in an open tubular column is adsorbed or covalently coated on the inner walls of a fused-silica capllary.

Packed columns: The stationary phase in a packed column is a packing particle, which is stacked in a hollow tube and sealed with frits.

Solid-phase microextraction (SPME): SPME is a sample preparation technique involving the use of an extraction device coated with an extracting phase, which can be performed without solvent. It can be coupled to gas chromatography (GC) or liquid chromatography (LC).

List of Abbreviations

CLC	capillary liquid chromatography
HPLC	high-performance liquid chromatography
I.D.	internal diameter
LC	liquid chromatography
MS	mass spectrometry
ODS	*n*-octadecyl silica
SPME	solid-phase microextraction
UV/Vis	ultraviolet/visible

References

Chen, Y. S., Chen, J. and Jia, L., 2009. Study of triacontyl-functionalized monolithic silica capillary column for reversed-phase capillary liquid chromatography. Journal of Chromatography A. 1216: 2597–2600.

Chervet, J. P., Ursem, M. and Salzmann, J. P., 1996. Instrumental requirements for nanoscale liquid chromatography. Analytical Chemistry. 68: 1507–1512.

Crescentini, G., Bruner, F., Mangani, F. and Guang, Y., 1988. Preparation and evaluation of dry-packed capillary columns for high-performance liquid chromatography. Analytical Chemistry. 60: 1659–1662.

Eisert, R. and Pawliszyn, J., 1997. Automated in-tube solid-phase microextraction coupled to high-performance liquid chromatography. Analytical Chemistry. 69: 3140–3147.

Erickson, B. E., 2000. Electrochemical detectors for liquid chromatography. Analytical Chemistry. 72: 353A–357A.

Gomis, D. B., Arias, V. E., Fidalgo Alvarez, L. E. and Gutikrez Alvarez, M. D., 1995. Determination of fat-soluble vitamins by capillary liquid chromatography in bovine blood plasma. Analytica Chimica Acta. 315: 177–181.

Gomis, D. B., Fernández, M. P. and Gutiérrez Alvarez, M. D., 2000. Simultaneous determination of fat-soluble vitamins and provitamins in milk by microcolumn liquid chromatography. Journal of Chromatography A. 891: 109–114.

Hibi, K., Ishii, D., Fujishima, I., Takeuchi, T. and Nakanishi, T., 1978. Studies of open tubular micro capillary liquid chromatography. 1. The development of open tubular micro capillary liquid chromatography. Journal of High Resolution Chromatography and Chromatography Communications. 1: 21–27.

Horváth, C. and Lipsky, S. R., 1969. Rapid analysis of ribonucleosides and bases at the picomole level using pellicular cation exchange resin in narrow bore columns. Analytical Chemistry. 39: 1227–1234.

Horváth, C. G., Preiss, B. A. and Lipsky S. R., 1967. Fast liquid chromatography: an investigation of operating parameters and the separation of nucleotides on pellicular ion exchangers. Analytical Chemistry. 39: 1422–1428.

Ishii, D., Asai, K., Hibi, K., Jonokuchi, T. and Nagaya, M., 1977. A study of micro-high-performance liquid chromatography: I. Development of technique for miniaturization of high-performance liquid chromatography. Journal of Chromatography A. 144: 157–168.

Jia, L., Tanaka, N. and Terabe, S., 2004. Capillary liquid chromatographic determination of cellular flavins. Journal of Chromatography A. 1053: 71–78.

Kataoka, H., 2002. Automated sample preparation using in-tube solid-phase microextraction and its application – a review. Analytical and Bioanalytical Chemistry. 373: 31–45.

Pawliszyn, J., 1997. Solid Phase Microextraction: Theory and Practice. Wiley-VCH, New York, , NY, USA.

Saito, Y. and Jinno, K., 2002. On-line coupling of miniaturized solid-phase extraction and microcolumn liquid-phase separations. Analytical and Bioanalytical Chemistry. 373: 325–331.

Shintani, Y., Zhou, X. J., Furuno, M., Minakuchi, H. and Nakanishi, K., 2003. Monolithic silica column for in-tube solid-phase microextraction coupled to high-performance liquid chromatography. Journal of Chromatography A. 985: 351–357.

Šlais, K., Kouřilová, D. and Krejčí, M., 1983. Trace analysis by peak compression sampling of a large sample volume on microbore columns in liquid chromatography. Journal of Chromatography. 282: 363–370.

Van den Berg, H., Faulks, R., Granado, H. F., Hirschberg, J., Olmedilla, B., Sandmann, G., Southon, S. and Stahl, W., 2000. The potential for the improvement of carotenoid levels in foods and the likely systemic effects. Journal of the Science of Food and Agriculture. 80: 880–912.

Wahyuni, W. T. and Jinno, K., 1987. Optimization of fat soluble vitamin separation in reversed-phase microcolumn liquid chromatography. Chromatographia. 23: 320–324.

Xu, H. and Jia, L., 2009. Capillary liquid chromatographic analysis of fat-soluble vitamins and β-carotene in combination with in-tube solid-phase microextraction. Journal of Chromatography B. 877: 13–16.

Zhang, S., Jia, L. and Wang, S. J., 2010. Determination of β-carotene in corn by in-tube SPME coupled to micro-LC. Chromatographia. 72: 1231–1233.

CHAPTER 14
Assay of Carotenoid Composition and Retinol Equivalents in Plants

SANGEETHA RAVI KUMAR AND
VALLIKANNAN BASKARAN*

Department of Biochemistry and Nutrition, Central Food Technological
Research Institute, CSIR, Mysore - 570 020, India
*E-mail: baskaranv@cftri.org

14.1 Introduction

Carotenoids are orange/red/yellow pigments of plant origin, present in the
chloroplasts where they protect the chlorophylls from the deleterious effects of
exposure to sunlight. More than 600 carotenoids have been identified in
nature; of these the most common provitamin A carotenoids, that is, they can
be enzymatically cleaved to retinol, are β-, α-, γ-carotene and β-cryptoxanthin.
Some of the common dietary carotenoids are shown in Figure 14.1.
Carotenoids share structural similarity with the presence of a polyene chain
with alternating double bonds, and/or β-ionone ring(s). In addition, several
carotenoids also possess oxygen-containing functional groups (such as
hydroxyl, keto and epoxide) which render them more polar in nature. Thus,
carotenoids are primarily divided into two types: the highly non-polar
carotenes and the xanthophylls that contain polar functional groups that
result in their varied chemical natures (Britton 1995). Due to these chemical
peculiarities, carotenoids are potent antioxidants and have many health
benefits (Kotake-Nara *et al* 2001; Shiratori *et al* 2005). Carotenoids are known
to be beneficial in chronic degenerative disorders such as cancer, age-related
macular degeneration, cardiovascular disorders and diabetes by virtue of their

Food and Nutritional Components in Focus No. 1
Vitamin A and Carotenoids: Chemistry, Analysis, Function and Effects
Edited by Victor R Preedy
© The Royal Society of Chemistry 2012
Published by the Royal Society of Chemistry, www.rsc.org

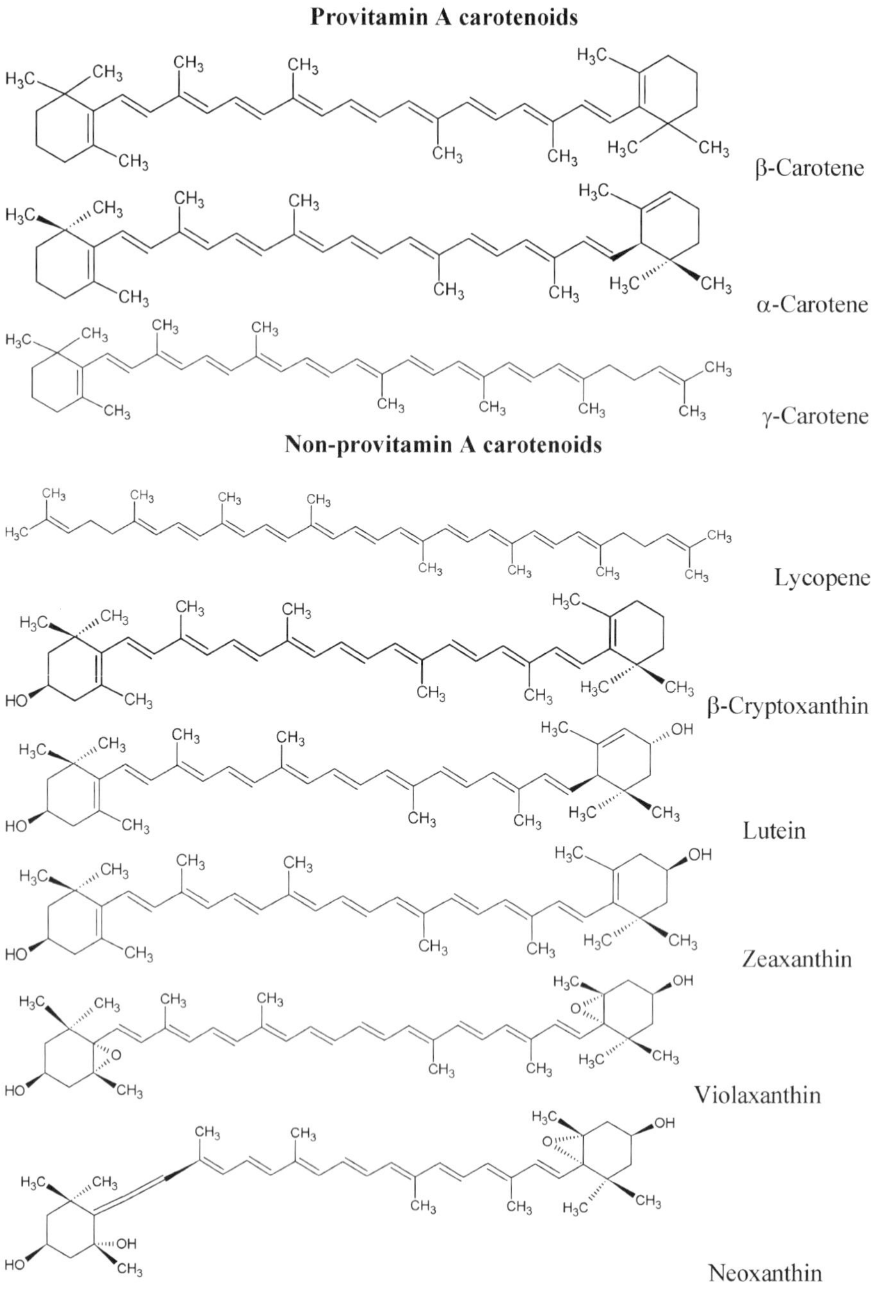

Provitamin A carotenoids

β-Carotene

α-Carotene

γ-Carotene

Non-provitamin A carotenoids

Lycopene

β-Cryptoxanthin

Lutein

Zeaxanthin

Violaxanthin

Neoxanthin

Figure 14.1   Provitamin A and non-provitamin A carotenoids commonly found in plants.

anti-proliferative, anti-inflammatory and anti-oxidative effects (Kotake-Nara *et al* 2001; Menotti *et al* 1999, Shiratori *et al* 2005). Clinical and epidemiological studies have established that consumption of fruits and vegetables (that are naturally rich in carotenoids) are inversely associated with the prevalence of degenerative disorders.

The plant kingdom abounds in pigments, which are divided into four main categories, *viz.* chlorophylls, carotenoids, anthocyanins and betalains. Although many health benefits of carotenoids are known, their bioavailability is poor (15–30%) and influenced by many factors. Increased awareness of the beneficial effects of carotenoids has led to their inclusion *via* dietary sources and as diet supplements. Provitamin A carotenoids are of particular interest due to their role as precursors of retinol in addition to their other health benefits. Of the provitamin A carotenoids, β-carotene has a unique structure, which when cleaved at exactly the centre (15–15' bond), can form two molecules of retinol. Thus, β-carotene is considered to have the highest provitamin A activity. The vitamin A activity of plants is often expressed in terms of retinol equivalents (RE), which is calculated based on the β-, α-carotene and other provitamin A carotenoid content of the plants and is useful in determining plants that are good sources of retinol precursors (National Research Council 1989; World Health Organization 1982). Determination of RE requires separation and quantitation of the provitamin A carotenoids from the food source. Thus, carotenoids, being lipid-soluble, are extracted by organic solvents along with physical destruction of the plant matrix for release of the pigments. The carotenoids can be purified by open column chromatography, confirmed by thin-layer chromatography (TLC) and roughly quantified by a spectrophotometer. The extract containing the carotenoids can be subjected to high-performance liquid chromatography (HPLC) analysis for accurate quantitation and liquid chromatography-mass spectrometry (LC-MS) analysis for their characterization. It is recommended that HPLC analysis is used for the quantitation of carotenoids in plants as colorimetric and spectrophotometric methods are often subject to over estimation of the vitamin A activity, as biologically inactive carotenoids are difficult to completely separate from the provitamin A carotenoids.

14.2 Assay Methods for Carotenoids

14.2.1 Extraction of Carotenoids

All procedures and analyses are conducted under a dim yellow light and on ice to prevent photo-isomerization and degradation of carotenoids. Edible portions of plants are washed separately with deionized water and dried on blotting paper at room temperature. Cleaned edible portion of plant materials (20–30 g) is ground well using mortar and pestle, along with 2–3 g of anhydrous sodium sulphate and 2 mM α-tocopherol (antioxidant). Carotenoids (total) are extracted using ice-cold acetone (50 ml). The extraction is repeated three times, or until the residue was colourless, indicating complete extraction of pigments (crude extract). The pooled crude extract (150–200 ml) is filtered through a Whatman No. 1 filter paper containing 10 g of anhydrous sodium sulphate. The various steps involved in the extraction and purification of carotenoids from the plant materials are given in Figure 14.2. Other solvent systems that can be employed for the extraction of carotenoids are given in Table 14.1.

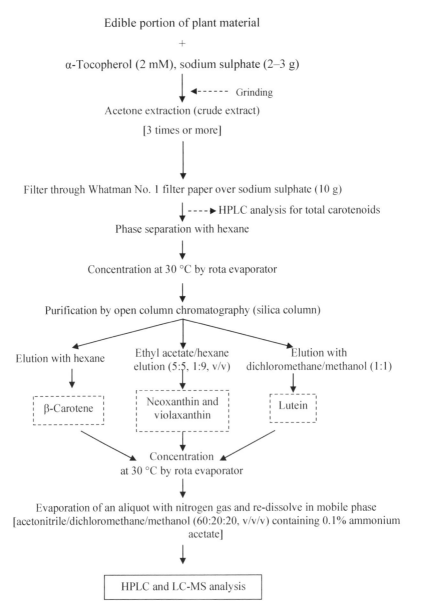

Figure 14.2 Flow diagram showing the various steps involved in extraction, purification and analysis of carotenoids from plant materials.

14.2.2 Purification by Open Column Chromatography

The crude carotenoid extract (100 ml) is evaporated using a flash evaporator and the residue is re-dissolved in hexane (1 ml). Pre-activated silica gel (60–120 mesh size) is made into a slurry using chloroform and packed into a column (30 cm length × 3 cm diameter). The column is equilibrated with hexane (2 ml min^{-1}) until all the chloroform is eluted (Figure 14.3). The hexane extract containing carotenoids is applied on to the column. The β-carotene is eluted with hexane. After the complete elution of the β-carotene, the column is eluted with ethyl acetate: hexane (5:5, v/v) followed by ethyl acetate: hexane (1:9, v/v) for the separation of neoxanthin and violaxanthin respectively. To purify lutein from the crude extract, the crude acetone extract is subjected to saponification with 30% methanolic KOH for 3 h in the dark, following which phase separation is carried out with hexane in a separating funnel. The lower aqueous layer containing chlorophylls is discarded and the hexane layer containing the carotenoids is collected and evaporated to dryness using a flash evaporator. The residue is dissolved in hexane (1 ml) and applied on to a hexane-stabilized column. The β-carotene present is eluted with hexane as mentioned earlier and the lutein is eluted with dichloromethane/methanol (1:1, v/v). The purity and quantity of the β-carotene and lutein thus obtained is checked by TLC, spectrophotometrically and HPLC analysis.

14.2.3 TLC for Separation and Isolation of Carotenoids

The purity of the carotenoids purified by open column chromatography is confirmed using TLC. Pre-coated aluminium TLC plates (20 × 20 cm) with silica gel (mesh size 60) are used for separating the purified carotenoids. Along with the samples, corresponding standards are also spotted for comparison and confirmation. Hexane/acetone (70:30, v/v) and dichloromethane/methanol (1:1, v/v) are used as mobile phase for the separation of β-carotene and lutein respectively. The distance travelled by the solvent and carotenoids is measured and used for calculation of the retardation factor (R_f), as per the formula:

R_f = (distance travelled by compound)/(distance travelled by mobile phase)

14.2.4 Spectrophotometric Estimation

Carotenoids purified by open column chromatography can be confirmed and quantified by recording their spectra and absorbance at 430–460 nm. Standard graphs are plotted with known concentrations and employed for quantitation of the respective carotenoids. In the absence of standards, absorption coefficients of the carotenoids can be used to calculate the concentration. The absorption coefficients of some of the carotenoids and their absorption maxima (λ_{max}) are given in Table 14.2. The formula that can be used for the calculation of concentration is as follows:

Table 14.1 Methods for extraction and separation of carotenoids by HPLC. ACN, acetonitrile; DCM, dichloromethane; MeOH, ethanol; MTBE, *tert*-butyl methyl ether; THF, tetrahydrofuran.

Extraction method	Mobile phase	Saponification	Column	Conditions	Run time	Reference
Acetone	ACN/DCM/MeOH (60:20:20)	No	RP-18	Isocratic 1 ml min^{-1}	18–20 min	Aruna and Baskaran (2010); Raju *et al* (2007); Sangeetha and Baskaran (2010)
Ethyl acetate	1.Acetone/water (82:18) 2A.MeOH/MTBE/water (92:4:4) 2B.MTBE/MeOH/water (90:6:4)	Yes	RP-30	Linear gradient 1 ml min^{-1} 100%A to 6%B in 80 min	80 min	Aman *et al* (2005)
Acetone, 10% ethyl ether	ACN/MeOH/ethyl acetate	No	RP-18	Concave gradient 95:5:0 to 60:20:20 in 20 min 0.5 ml min^{-1}	40 min	Niizu and Rodriguez-Amaya (2005)
Ethanol, hexane	A. ACN/THF/MeOH/1% aqueous ammonium sulphate (85:5:5:5) B. ACN/THF/MeOH/1% aqueous ammonium sulphate (55:35:5:5)	No	RP-18	Gradient 5% to 95% solvent B, as follows: 5% B for 10 min, increase linearly to 95% solvent B for 19 min, maintain 95% B for 6.9 min, decrease to 60% B for 0.1 min; maintain 60% B for 8.9 min, decrease to 5% B at 45 min. Column re-equilibrated with 5% B for 3 min. 1 ml min^{-1}	48 min	Seo *et al* (2005)

Table 14.1 (*Continued*)

Extraction method	Mobile phase	Saponification	Column	Conditions	Run time	Reference
Methanol, diethyl ether	A. MeOH B. 36% DCM in MeOH	Yes	RP-18	Gradient 0–20 min, 100% A to 35% B 1.25 ml min⁻¹	45 min	Molnar et al (2004)
Ethanol, n-hexane	ACN/MeOH/ethyl acetate (88: 10: 2)	Yes	RP-18	Isocratic 1 ml min⁻¹	–	Ismail and Fun (2003)
Acetone, petroleum ether	ACN/MeOH/ethyl acetate	No	RP-18	Concave gradient 95:5:0 to 60:20:20 in 20 min 0.5 ml min⁻¹	45 min	Kimura and Rodriguez-Amaya (2003)
MeOH, ethyl acetate, light petroleum	MTBE/MeOH/water A. 15:81:4 B. 90:6:4	Yes	RP-30	Isocratic 100% A for 10 min, Gradient 50% B at 40 min, 100% A at 55 min, Isocratic 100% A 55–60 min. 1 ml min⁻¹	40 min	Weller and Breithaupt (2003)
MeOH, THF	MeOH/MTBE/water A. 83:15:2 B. 8:90:2	No	RP-18	Gradient 100% A for 1 min, 70% A for 7 min, 70% A, 13 to 22 min, 45% A for 9 min, 5% A for 14 min, 100% A for 12 min 1 ml min⁻¹	50 min	Lienau et al (2003)
Acetone MeOH, chloroform	ACN/DCM (7:3) A. MeOH B. Water/MeOH (20:80) C. MTBE	No Yes	RP-18 RP-18	Isocratic, 1 ml min⁻¹ Gradient 95%A, 5%B for 12 min, a step to 80% A, 5% B, 15% C at 12 min, followed by a linear gradient to 30% A, 5% B, 65% C by 30 min. Conditioning for 30–60 min. 1 ml min⁻¹	40 min 85–115 min	Barua (2001) Fraser et al (2000)
MeOH, THF	MeOH/THF (95:5)	Yes	RP-18	Isocratic 1 ml min⁻¹	25 min	Konings and Roomans (1997)

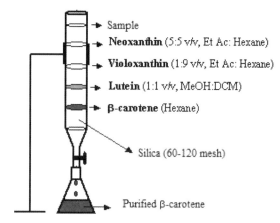

Figure 14.3 Open silica column chromatography for purification of carotenoids.

% (g/100 ml) = (optical density × dilution factor × 1)/(absorption coefficient)

14.2.5 HPLC Analysis

An aliquot of crude extract or purified carotenoid eluted from the open colum chromatography is evaporated under a stream of nitrogen and re-dissolved (100 μl) in acetonitrile/dichloromethane/methanol (60:20:20, v/v/v) containing

Table 14.2 Absorption coefficients and absorption maxima (λ_{max}) of carotenoids estimated in plants. a = Britton (1995); b = Sangeetha and Baskaran (2010); c = Eitenmiller and Lander (1999).

Carotenoid	Solvent/ absorption coefficient[a]		λ_{max} a	b	c
Neoxanthin	Ethanol	2243	439	440	439
Violaxanthin	Ethanol	2550	440	449	443
	Acetone	2400	442		
Lutein	Ethanol	2550	445	447	445
	Diethyl ether	2480	445		
Zeaxanthin	Petroleum ether	2348	449	454	452
	Ethanol	2480	450		
	Acetone	2340	452		
α-Carotene	Petroleum ether	2800	444	448	444
	Hexane	2710	445		
β-Carotene	Petroleum ether	2592	450	454	453
	Ethanol	2620	450		
	Chloroform	2396	465		

a = Britton (1995)

0.1% ammonium acetate (mobile phase) for the analysis of β-carotene, α-carotene, lutein, zeaxanthin, neoxanthin and violaxanthin by reverse-phase HPLC analysis. The extract (20 μl) is injected to HPLC system (LC-10Avp; Shimadzu, Kyoto, Japan) equipped with Shimadzu photodiode array (PDA) detector (SPD-M20A). Carotenoids are separated on a Phenomenex RP-18 column (250 mm × 4.6 mm; 5 μm) isocratically eluting with 1 ml min^{-1} of mobile phase and were monitored at 450 nm (Shimadzu Class-VP, version 6.14SP1 software). The peak identity of each carotenoid is confirmed by their ultraviolet/visible (UV-Vis) spectra recorded with the PDA detector. The details of other HPLC methods that are employed for the separation and quantitation of carotenoids are described in Table 14.1.

14.2.6 Liquid Chromatography-Mass Spectrometric Analysis of Carotenoids [LC-MS, Atmospheric Pressure Chemical Ionization (APCI)]

LC-MS is used to confirm the identity and characterize the carotenoids. The positive ions of the carotenoids are recorded with an HPLC system (Alliance 2695, Waters, USA) connected to a LC-Q mass spectrometer (Waters 2996 modular HPLC system, UK) equipped with APCI module. The APCI source is heated to 130 °C and the probe is maintained at 500 °C. The corona (5 kV), HV lens (0.5 kV) and cone (30 V) voltages are optimized. Nitrogen is used as sheath and drying gas at 100 and 300 L h, respectively. The spectrometer is calibrated in the positive ion mode, and $[M+H]^+$ ion signals are recorded and confirmed with respective standards. Quantification of individual compounds is evaluated by comparing their peak area with the authentic standards.

14.3 Carotenoid Composition in Plants

The major carotenoids in plants are β- and α-carotene (provitamin A carotenoids) and lutein, zeaxanthin, neoxanthin and violaxanthin (xanthophylls). The HPLC elution profile of carotenoids under the conditions mentioned earlier in plants is almost similar (Figure 14.4) with the xanthophylls eluting first (within 4.5 min), followed by the chlorophylls (6.5–7.5 min) and then the carotenes (17–20 min). The carotenoids are confirmed from their spectra (Figure 14.5) and further characterized by LC-MS analysis (Figure 14.6). Under the HPLC conditions described earlier, the absorption maxima (λ_{max}) of β-carotene are 426, 454 and 480; α-carotene are 421, 448 and 475; lutein are 421, 447 and 475; neoxanthin are 415, 440 and 468; and violaxanthin are 425, 449 and 476. The carotenoid profile obtained from different plant materials is similar, with differences only in the concentration of the constituent carotenoids. Carotenoid composition of various leafy greens, vegetables, spices and medicinal plants quantified by the authors are given in Table 14.3. From Table 14.3, it is seen that the β- and α-carotene content (mg/100 g of dry weight) in leafy greens is in the range of 0.04 (green cabbage)

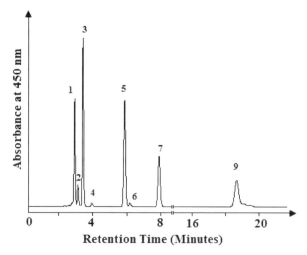

Figure 14.4   Representative HPLC chromatogram of carotenoids extracted from *Chenopodium album*. 1 = neoxanthin, 2 = violaxanthin, 3 = lutein, 4 = zeaxanthin 5 = chlorophyll b, 6 = chlorophyll a, 7 = α-tocopherol, 8 = α-carotene and 9 = β-carotene.

to 120 (lamb's quarters) and 0.31 (hog weed, chilli leaf) to 36 (jio). In vegetables, β- and α-carotene content (mg/100 g of dry weight) ranges from 0.01 (onion, turnip, bitter orange, white colocasia, red radish) to 159 (green capsicum) and <0.01 (bitter orange, white colocasia, white cucumber, green

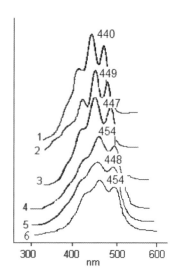

Figure 14.5   Typical UV-Vis spectra of carotenoids along with absorption maxima (λ_{max}) extracted from plants. 1 = neoxanthin, 2 = violaxanthin, 3 = lutein, 4 = zeaxanthin, 5 = α-carotene and 6 = β-carotene.

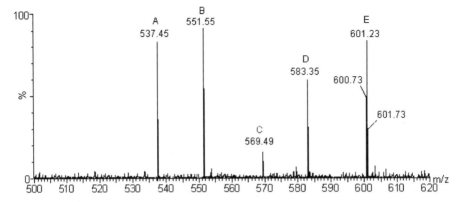

Figure 14.6 Representative LC-MS (APCI) profile of carotenoids extracted from *C. album*. A = β-carotene (537 $[M+H]^+$), B = lutein (551 $[M+H-H_2O]^+$), C = zeaxanthin (569 $[M+H]^+$), D = neoxanthin, violaxanthin (583 $[M+H-H_2O]^+$), E = neoxanthin, violaxanthin (601 $[M+H]^+$).

zucchini, cluster beans, broad bean seed, tender red gram, cowpea) to 110 (carrot). Green chilli contains highest β- and α-carotene levels (9.1, 0.5 mg/100 g of dry weight) among the spices and they are lowest for spillanthes (0.03) and black cardamom (0.02) respectively. β-Carotene (mg/100 g of dry weight) content ranges between 2 (Australian pea, chickpea) and 68 (red parboiled unpolished rice) in cereals and pulses, whereas α-carotene (mg/100 g of dry weight) is in the range of 5 (finger millet) to 53 (whole black gram). Among medicinal plants, β- and α-carotene (mg/100 g of dry weight) is in the range 3 (bitter gourd leaf) to 35 (thyme leaved gratiola) and 0.05 (Tanner's cassia) to 16 (margosa).

14.4 Calculation of RE in Plants

Vitamin A activity of β- and α-carotene is calculated in terms of RE based on the *in vivo* conversion factor proposed by the World Health Organization (WHO) and National Research Council (NRC) (National Research Council 1989; World Health Organization 1982), where 1 RE = 1 μg of retinol = 6 μg of β-carotene or 12 μg of other provitamin A carotenoids such as α-carotene. The RE values of leafy greens, vegetables, spices and medicinal plants are given in Table 14.3. RE is a convenient means of gauging and determining plant-based foods that would be better sources of retinol (vitamin A). Recently, the United States Institute of Medicine (Food and Nutrition Board 2001) proposed that the retinol activity equivalents (RAE) may be a better tool for their dietary reference intakes where 1 RAE = 1 μg of retinol = 2 μg of all-*trans*-β-carotene as a supplement = 12 μg of all-*trans*-β-carotene in a food matrix = 24 μg of other provitamin A carotenes in a food matrix. However, RE continues to be an accepted measure of vitamin A activity for plants. In addition, 1 RE = 3.33 international units (IU) of vitamin A activity from

Table 14.3 Carotenoid composition (mg/100 g dry weight) and RE of provitamin A carotenoids in plants. Values are mean of duplicate analysis. ND, not detected; NA, not applicable.

Botanical name[d,e]	Common name	Xanthophylls					Provitamin A			RE[c]
		Neo-xanthin	Viola-xanthin	Lutein	Zea-xanthin	Total[a] xanthophylls	α-Carotene	β-Carotene	Total provitamin[b] A carotenoids	
Leafy greens										
Allium cepa	Onion stalks	6.61	1.83	30.28	0.21	38.93	ND	16.9	16.9	2816
Allium schoenoprasum	Chives	23.53	3.15	73.4	44.4	144.47	ND	48.14	48.14	8020
Almania nodiflora	Celosia	7.74	8.9	23.1	0.18	39.92	ND	9.63	9.63	1605
Alternanthera pungens Kunth	Khaki weed	ND	42.03	71.86	0.67	114.56	ND	34.66	34.66	5776
Alternanthera sessilis (*L.*) Dc.	Joy weed	13.89	23.93	32.47	0.26	70.55	ND	27.07	27.07	4511
Amaranthus gangeticus	Amaranth leaves	1.46	26.5	32.02	0.34	60.32	ND	18.67	18.67	3111
Amaranthus polygonoides	Amaranth leaves	12.51	2.41	25.73	1.19	41.84	ND	20.61	20.61	3430
Amaranthus sp.	Amaranth leaves (Keerai)	38.08	4.92	71.92	1.3	116.21	1.03	66.39	67.42	11150
Amaranthus sp.	Amaranth leaves (Yelavare)	16.42	2.74	29.19	0.95	49.3	1.38	23.2	24.57	3980
Amaranthus spinosus	Prickly amaranth	54.07	3.27	116.86	2.78	176.98	ND	97.88	97.88	16130
Amaranthus tristis	Arai keerai	1.15	19.15	30.3	0.23	50.83	ND	16.76	16.76	2793

Table 14.3 (Continued)

| Botanical name | Common name | Xanthophylls | | | | | Provitamin A | | | RE[c] |
		Neo-xanthin	Viola-xanthin	Lutein	Zea-xanthin	Total[a] xanthopylls	α-Carotene	β-Carotene	Total provitamin[b] A carotenoids	
Amaranthus viridis	Slender amaranth	12.63	84.06	90.43	1.04	188.16	6.75	58.95	65.7	10387
Apium graveolus var. dulce	Celery	30.79	2.87	56.52	1.45	91.63	ND	49.93	49.93	8320
Basella alba	Indian spinach	7.74	6.27	113.82	1.76	129.59	18.23	43.82	62.05	8822
Basella rubra	Red spinach	12.01	6.94	67.94	2.25	89.14	ND	32.42	32.42	5403
Beta vulgaris	Beat greens	6.39	3.97	26.86	0.14	38.36	1.54	12.5	14.04	2211
Boerhavia diffusa	Hog Weed	1.64	0.16	54.13	0.32	56.24	0.31	2.43	2.75	430
Brassica chinesis	Cabbage, Chinese	0.69	3.07	22.42	0.63	26.24	0	1.5	1.5	250
Brassica oleracea var. Capitata	Cabbage, green	0.09	0.06	0.37	0.06	0.58	ND	0.04	0.04	10
Brassica oleracea var. Capitata	Cabbage, red	0.3	ND	1.34	ND	1.64	0.36	ND	0.36	30
Brassica oleracea var. caulorapa	Knol khol greens	18.5	4.73	42.97	1.57	67.76	0.67	27.11	27.78	4570
Brassica oleracea var.botrytis L.	Broccoli	0.52	1.45	4.5	ND	6.47	ND	3.85	3.85	641
Brassica rapa	Turnip Greens	0.54	5.41	11.73	1.54	19.21	ND	8.98	8.98	1500
Capsicum annuum	Chilli leaf	2.73	0.27	136.24	0.39	139.63	0.31	3.48	3.79	610
Chenopodium album	Lamb's Quarters	0.03	140.5	185.2	5	330.73	ND	120.2	120.2	20030
Colocasia anti-quorum	Colocasia Leaves	44.14	1.33	104.68	2.43	152.58	6.7	57.28	63.98	10110

Table 14.3 (Continued)

Botanical name	Common name	Xanthophylls					Provitamin A			RE[c]
		Neo-xanthin	Viola-xanthin	Lutein	Zea-xanthin	Total[a] xanthopylls	α-Carotene	β-Carotene	Total provitamin[b] A carotenoids	
Commelina benghalensis	Jio	0.03	100.3	175.6	2.06	277.99	35.6	95.7	131.3	18920
Coriandrum sativum L.	Coriander leaves	5.47	83.43	9.92	ND	98.82	ND	67.5	67.5	11250
Cucurbita maxima Duchesne	Winter squash	ND	15.58	27.18	0.25	43.02	ND	10.27	10.27	1712
Daucus carota L.	Carrot greens	2.09	7	40.17	0.59	49.85	21.53	12.09	33.62	3809
Gynandropsis pentaphylla L.	Spider wisp	49.31	ND	42.65	1.28	93.24	ND	37.04	37.04	6173
Hibiscus cannabinus L.	Kenaf	5.95	ND	33.97	0.14	40.06	ND	26.02	26.02	4338
Hydrocotyle asiatica L.	Indian pennywort	0.89	0.65	15.93	ND	17.47	ND	9.02	9.02	1503
Ipimoea pes-tigridis	- (I.pestigridis)	2.71	0.13	134.78	0.31	137.93	ND	4.31	4.31	720
Lactuca sativa	Lettuce, green	1.39	0.19	2.4	0.05	4.03	ND	1.57	1.57	260
Lactuca sativa	Lettuce iceberg	2.53	0.15	26.48	0.04	29.19	ND	2.87	2.87	480
Lactuca sativa	Lettuce, red	2.19	0.17	2.58	0.05	4.99	ND	2.48	2.48	410
Mentha spicata L.	Spear mint	2.11	5.62	17.74	0.26	25.73	ND	7.48	7.48	1246
Moringa oleifera Lam.	Drumstick	9.6	18.3	50.4	4.13	82.43	ND	22.89	22.89	3815
Murraya koenigii	Curry leaf tree	4.39	6.68	27.2	0.16	38.43	2.87	8.95	11.82	1730

Table 14.3 (Continued)

Botanical name	Common name	Xanthophylls					Provitamin A			RE[c]
		Neo-xanthin	Viola-xanthin	Lutein	Zea-xanthin	Total[a] xanthopylls	α-Carotene	β-Carotene	Total provitamin[b] A carotenoids	
Petroselinum crispum	Parsley	10.4	2.56	29.71	0.81	43.48	ND	18.43	18.43	3070
Peucedanum sowa	Indian Dill	13.5	7.3	170.1	3.1	194	ND	61.5	61.5	10250
Phyllanthus niruri	Chanca piedre	33.6	3.67	77.55	1.63	116.45	ND	60.88	60.88	10146
Piper betle	Betel leaf	0.82	0.89	36.43	0.47	38.62	18.42	13.35	31.77	3760
Portulacae oleracae	Common purslane	2.37	0.09	36.19	0.2	38.84	ND	4.2	4.2	700
Raphanus sativus	Radish	3.3	5.4	22.3	0.75	31.75	ND	11.2	11.2	1866
Rumex acetosella	Sheep sorrel	7.7	1.45	144.3	ND	153.45	ND	70.83	70.83	11805
Sesbania grandiflora	Agathi	1.85	4.33	16.9	0.57	23.65	ND	13.28	13.28	2213
Solanum nigrum	Black night shade	2.79	22.17	84.86	ND	109.82	ND	50.11	50.11	8351
Spinacia oleracea	Spinach	58	65	77.58	1.51	202.09	ND	36.53	36.53	6088
Talinum cuniefolium Willd.	Ceylon spinach	7.95	10.51	89.79	1.22	109.47	18.77	42.44	61.21	8637
Tamarindus indica	Tamarind	2.43	0.03	16.4	0.55	19.41	0.76	4.77	5.53	860
Trianthema portulacastrum	Desert horse purslane	2.5	5	41.51	0.44	49.45	ND	37.76	37.76	6293
Tribulus terrestris	Puncture vine	1.34	8.84	56.39	0.04	66.61	6.1	30.81	36.91	5643
Trigonella foenum-graecum	Fenugreek	3.32	33.26	59.6	0.95	97.13	ND	12.13	12.13	2021

Table 14.3 (Continued)

Botanical name	Common name	Xanthophylls					Provitamin A			RE[c]
		Neo-xanthin	Viola-xanthin	Lutein	Zea-xanthin	Total[a] xanthopylls	α-Carotene	β-Carotene	Total provitamin[b] A carotenoids	
Vegetables[f]										
Allium cepa	Onion	0.03	0.02	0.02	ND	0.07	ND	0.01	0.01	NA
Allium cepa (small)	Onion	ND	0.03	0.2	<0.01	0.23	ND	0.13	0.13	0.02
Allium porrum	Leek	3.98	0.26	16.53	ND	20.77	8.3	ND	8.3	0.69
Artocarpus heterophyllus	Jackfruit raw	0.53	0.22	3.37	0.14	4.26	0.02	0.29	0.31	0.05
Beta vulgaris	Beetroot	ND	ND	0.31	0.02	0.33	ND	0.02	0.02	NA
Brassica oleracea var. acephala	Kohlrabi	0.04	0.09	0.48	0.01	0.62	ND	0.37	0.37	0.06
Brassica oleracea var. botrytis	Brocolli	0.31	3.43	27.26	ND	31	16.2	0.65	0.65	1.46
Brassica oleracea var. botrytis	Cauliflower	0.01	0.04	0.65	<0.01	0.7	ND	0.33	0.33	0.06
Brassica rapa	Turnip	0.01	<0.01	0.01	ND	0.02	ND	0.01	0.01	NA
Cajanus cajan	Pigeon pea	0.12	0.3	0.05	0.01	0.48	0.04	0.14	0.18	0.03
Capsicum annuum	Chilli long	<0.01	0.07	0.53	<0.01	0.6	ND	0.38	0.38	0.06
Capsicum annuum var. Grossa	Capsicum green	25.56	ND	419.37	ND	444.93	21.62	158.78	180.4	28.27
Capsicum annuum var. Grossa	Capsicum red	9.44	45.52	275.25	ND	330.21	ND	37.69	37.69	6.28

Table 14.3 (*Continued*)

Botanical name	Common name	Xanthophylls					Provitamin A			RE^c
		Neo-xanthin	Viola-xanthin	Lutein	Zea-xanthin	$Total^a$ xanthopylls	α-Carotene	β-Carotene	Total provitamin[b] A carotenoids	
Capsicum annuum var. Grossa	Capsicum yellow	11.75	38.26	60.75	ND	110.76	15.05	14.36	29.41	3.65
Citrus aurantium[g]	Bitter orange	<0.01	ND	0.02	<0.01	0.02	<0.01	0.01	0.01	NA
Coccinia cordifolia	Ivy gourd	10.37	1.99	33.66	ND	46.02	25.04	1.32	26.36	2.31
Colocasia esculenta var. Schott	Colocasia white	0.01	<0.01	0.04	<0.01	0.05	<0.01	0.01	0.01	NA
Colocasia esculenta	Colocasia	<0.01	ND	0.08	<0.01	0.08	ND	0.08	0.08	0.01
Cucumis anguria var. anguria	Gherkin	0.01	ND	0.08	<0.01	0.09	<0.01	0.02	0.02	NA
Cucumis sativa	Cucumber green	ND	0.29	1.74	<0.01	2.03	ND	0.87	0.87	0.15
Cucumis sativa	Cucumber white	0.03	0.02	0.31	0.01	0.37	<0.01	0.12	0.12	0.02
Cucurbita maxima	Pumpkin (small, red)	21.04	ND	18.94	ND	39.98	0.49	12.06	12.55	2.05
Cucurbita pepo	Zucchini green	0.01	0.01	3.27	0.02	3.31	<0.01	0.05	0.05	0.01
Cucurbita pepo	Zucchini yellow	29.87	8.66	209.79	ND	248.32	10.9	1.02	11.92	1.08
Curcuma amada	Mango ginger	ND	ND	ND	ND	0	ND	0.08	0	0.01
Cyamopsis tetragonoloba	Cluster bean	0.01	0.1	0.49	<0.01	0.6	<0.01	0.43	0.43	0.07

Table 14.3 (Continued)

Botanical name	Common name	Xanthophylls					Provitamin A			RE[c]
		Neo-xanthin	Viola-xanthin	Lutein	Zea-xanthin	Total[a] xanthophylls	α-Carotene	β-Carotene	Total provitamin A[b] carotenoids	
Daucus carota var sativa	Carrot	ND	ND	14.46	ND	14.46	110.41	50.31	160.72	17.59
Decalepis hamiltonii	Decalpis	0.09	0.07	0.08	<0.01	0.24	0.01	0.05	0.06	0.01
Diosurea alata	Greater yam	0.4	0.03	0.3	0.03	0.76	ND	0.26	0.26	0.04
Dolichos lablab	Australian pea	0.3	0.58	1	0.01	1.89	0.05	0.45	0.5	0.08
Ipomoea batatas	Sweet potato	<0.01	0.01	0.02	<0.01	0.03	ND	0.1	0.1	0.02
Lycopersicon esculentum[h]	Tomato	ND	ND	3.89	ND	3.89	8.8	ND	8.8	0.73
Mangifera indica	Mango raw	0.01	0.01	0.08	<0.01	0.1	ND	0.13	0.13	0.02
Manihot esculenta	Tapioca	0.01	ND	0.06	<0.01	0.07	ND	0.03	0.03	0.01
Moringa oleifera	Drumstick	0.22	0.02	0.58	0.01	0.83	ND	0.27	0.27	0.05
Musa paradisiacal	Banana flower	<0.01	0.03	0.28	<0.01	0.31	0.05	0.29	0.34	0.05
Musa paradisiacal	Banana raw	ND	ND	0.1	ND	0.1	ND	0.82	0.82	0.14
Musa paradisiacal	Banana stem	0.04	0.01	0.67	0.02	0.74	0.01	0.16	0.17	0.03
Nelumbo nucifera	Lotus stem	0.01	<0.01	0.25	<0.01	0.26	ND	0.32	0.32	0.05
Phaseolus vulgaris	Romano bush bean	0.01	0.1	0.31	<0.01	0.42	0.06	0.19	0.25	0.04
Piper nigrum	Green pepper	0.01	0.13	0.51	<0.01	0.65	0.04	0.23	0.27	0.04
Plectranthus rotundifolius	Plectranthus	0.04	ND	1.48	0.07	1.59	0.08	0.01	0.09	0.01

Table 14.3 (Continued)

Botanical name	Common name	Xanthophylls					Provitamin A			RE^c
		Neo-xanthin	Viola-xanthin	Lutein	Zea-xanthin	Total^a xanthopylls	α-Carotene	β-Carotene	Total provitamin^b A carotenoids	
Raphanus sativus	Radish red	0.01	ND	0.05	<0.01	0.06	ND	0.01	0.01	NA
Sechium edule	Cho-cho	0.09	0.01	0.48	0.02	0.6	0.02	0.21	0.23	0.04
Solanum tuberosum	Potato	0.01	0.03	0.38	<0.01	0.42	ND	0.04	0.04	0.01
Trichosanthes dioic	Pointed gourd	0.01	0.15	1.07	<0.01	1.23	ND	0.6	0.6	0.1
Vicia faba	Broad bean (tender)	2.61	1.81	1.09	0.39	5.9	1.96	6.19	8.15	1.2
Vicia faba	Broad bean seed	0.04	0.02	0.04	<0.01	0.1	<0.01	0.02	0.02	NA
Vigna sinensis	Red gram (tender)	0.01	0.07	0.24	<0.01	0.32	<0.01	0.25	0.25	0.04
Vigna unguiculata	Cowpea	<0.01	0.04	0.23	<0.01	0.27	<0.01	0.24	0.24	0.04
Zea mays	Baby corn	0.11	0.08	0.24	0.01	0.44	ND	0.05	0.05	0.01
Spices^i										
Allium sativum	Garlic	ND	ND	0.05	ND	0.05	ND	ND	0.00	NA
Amomum subulatum	Black cardamom	ND	0.14	0.14	0.01	0.29	0.02	0.22	0.24	NA
Brassica nigra	Mustard	0.01	0.06	1.20	0.01	1.28	ND	ND	0.00	NA
Capsicum annum	Green chilli	1.03	0.13	13.30	0.44	14.90	0.45	9.06	9.51	1.5
Cinnamomum zeylanicum	Cinnamon	ND	0.23	0.15	0.01	0.39	ND	ND	0.00	NA
Coriander sativum	Coriander seeds	0.01	0.17	0.45	0.03	0.66	0.12	0.22	0.34	NA

Table 14.3 (*Continued*)

| Botanical name | Common name | Xanthophylls | | | | | Provitamin A | | | |
		Neo-xanthin	Viola-xanthin	Lutein	Zea-xanthin	Total[a] xanthopylls	α-Carotene	β-Carotene	Total provitamin[b] A carotenoids	RE[c]
Cuminum cyminum	Black pepper	0.17	0.26	0.31	0.04	0.78	ND	ND	0.00	NA
Curcuma longa	Turmeric	ND	ND	ND	ND	0.00	ND	ND	0.00	NA
Elattaria cardamomum	Green cardamom	0.02	0.34	0.44	ND	0.80	ND	ND	0.00	NA
Illicium verum	Star anise	ND	0.11	0.02	ND	0.13	0.07	0.04	0.11	NA
Mentha sicata	Spearmint	2.1	5.62	17.74	0.26	25.72	ND	7.48	7.48	1.2
Myrstica fragrens	Mace	0.19	0.86	0.46	0.16	1.67	ND	ND	0.00	NA
Myrstica fragrens	Nutmeg	ND	ND	0.01	ND	0.01	ND	ND	0.00	NA
Papaver somniferum	Popper seeds	ND	ND	0.01	ND	0.01	ND	ND	0.00	NA
Pimpinella aninsum	Anise seeds	0.01	0.60	0.68	0.02	1.31	ND	0.06	0.06	NA
Piper nigrum	Cumin seeds	0.01	0.02	0.32	ND	0.35	0.04	0.09	0.13	NA
Piper nigrum	White pepper	0.18	ND	0.02	ND	0.20	ND	ND	0.00	NA
Spilanthes mauritiana	Spillanthes	ND	0.17	0.13	ND	0.30	ND	0.03	0.03	NA
Synzyngium aromaticum	Cloves	ND	0.09	0.06	ND	0.15	ND	ND	0.00	NA
Tamarindus indica	Tamarind	ND	ND	ND	ND	0.00	ND	ND	0.00	NA
Tarchyspermum ammi	Carum seeds	0.02	0.16	0.84	0.01	1.03	ND	0.17	0.17	NA
Trigonella foenum-graceum	Fenugreek	ND	ND	0.79	ND	0.79	ND	ND	0.00	NA

Table 14.3 (*Continued*)

Botanical name	Common name	Xanthophylls					Provitamin A			RE^c
		Neo-xanthin	Viola-xanthin	Lutein	Zea-xanthin	$Total^a$ xanthopylls	α-Carotene	β-Carotene	Total provitamin^b A carotenoids	
Zingiber officinale	Ginger	ND	ND	0.10	ND	0.10	ND	0.30	0.30	0.1
Cereals^f										
Eleusine coracana	Finger millet	ND	ND	87.50	0.20	87.70	4.80	3.20	8.00	0.93
Hordeum vulgare	Barley	ND	0.01	21.70	ND	21.71	11.50	ND	11.50	0.96
Linum usitatissimum	Flaxseed	0.01	0.01	185.10	0.50	185.62	ND	ND	0.00	NA
Oryza sativa	Rice (polished)	ND	ND	28.70	0.50	29.20	ND	ND	0.00	NA
Oryza sativa	Rice (red parboiled polished)	ND	ND	7.80	ND	7.80	ND	ND	0.00	NA
Oryza sativa	Rice (red parboiled unpolished)	ND	ND	22.40	0.20	22.60	ND	67.60	67.60	11.27
Setaria italic	Foxtail	ND	ND	100.50	0.20	100.70	ND	2.80	2.80	0.47
Sorghum vulgare	Sorghum	ND	ND	28.10	ND	28.10	ND	ND	0.00	NA
Triticum aestivum	Wheat (Punjab)	ND	ND	77.30	0.20	77.51	ND	6.30	6.30	1.05
Triticum aestivum	Wheat (soft)	ND	ND	72.50	0.30	72.80	ND	12.70	12.70	2.12
Pulses^f										
Dolichos lablab	Australian pea	ND	ND	63.8	0.3	64.1	7.8	2.0	9.8	1.0
Cicer arietinum	Chickpea (split)	ND	ND	199.4	0.5	199.9	ND	2.0	2.0	0.3
Cicer arietinum	Chickpea (whole)	ND	ND	200.1	0.1	200.2	ND	ND	0.0	NA

Table 14.3 (Continued)

Botanical name	Common name	Xanthophylls					Provitamin A			RE[c]
		Neo-xanthin	Viola-xanthin	Lutein	Zea-xanthin	Total[a] xanthophylls	α-Carotene	β-Carotene	Total provitamin[b] A carotenoids	
Cicer arietinum	Chickpea, (puffed)	ND	ND	34.0	1.8	35.9	ND	ND	0.0	NA
Phaseolus mungo	Black gram (split)	ND	ND	36.6	0.2	36.8	ND	ND	0.0	NA
Phaseolus mungo	Black gram (whole)	ND	ND	27.2	ND	27.3	52.7	9.7	62.4	6.0
Vigna catjang	Cowpea	ND	ND	91.4	0.2	91.6	ND	3.8	3.8	0.6
Phaseolus aureus	Green gram (split)	ND	ND	ND	ND	0.0	ND	8.9	8.9	1.5
Phaseolus aureus	Green Gram (whole)	ND	ND	88.0	0.1	88.1	ND	6.6	6.6	1.1
Medicinal Plants										
Acalypha indica	Acalypha	7.57	1.28	419.8	0.75	429.4	0.33	14.31	14.64	2410
Aegle marmelos Correa	Bel	2.89	0.42	193.17	0.27	196.75	0.6	5.28	5.88	930
Aloe vera	Aloe	2.27	0.15	14.85	0.39	17.67	0.87	6.59	7.46	1170
Aristolochia indica	Indian birthwort	11.05	1.19	391.53	0.79	404.57	0.69	14.23	14.91	2430
Azadirachta indica	Margosa, Neem	3.33	0.4	11.75	0.48	15.96	0.27	6.93	7.19	1180
Bacopa monnieri	Thyme leaved gratiola	18.47	6.37	54.69	1.52	81.05	ND	34.7	34.7	5780
Calotropis gigantean	Gigantic swallow wort	4.47	1.75	134.66	0.48	141.36	0.57	11.16	11.73	1910

Table 14.3 (*Continued*)

Botanical name	Common name	Xanthophylls					Provitamin A			RE^c
		Neo-xanthin	Viola-xanthin	Lutein	Zea-xanthin	$Total^a$ xanthopylls	α-Carotene	β-Carotene	Total provitamin^b A carotenoids	
Cassia auriculata	Tanner's Cassia	3.54	0.63	195.84	0.19	200.2	0.05	6.96	7.01	1170
Centella asiatica	Indian pennywort	4.83	1.22	234.88	0.55	241.48	ND	8.88	8.88	1480
Clitoria ternatea	Butterfly pea	15.33	2.39	678.99	1.44	698.15	1.5	31.16	32.66	5320
Coleus aromaticus	Indian Borage	4.85	1.08	28.64	1.01	35.59	15.73	0.35	16.08	1370
Convolvulus sepium	Bindweed	11.83	0.92	394.78	1.07	408.6	0.86	20.22	21.08	3440
Curcuma domestica (longa)	Turmeric leaf	5.29	0.96	292.88	0.77	299.91	3.06	7.64	10.7	1530
Cymbopogon citratus	Lemongrass	4.55	0.76	291.98	0.26	297.55	ND	7.85	7.85	1310
Cynodon dactylon	Conch/dog grass	8.79	0.69	484.08	0.7	494.26	0.22	16.48	16.7	2770
Datura stramonium	Jimson weed	2.83	9.7	307.07	0.62	320.23	0.16	13.79	13.95	2310
Ficus glomerata	Country fig	3.85	0.21	184.94	0.25	189.25	4.42	3.27	7.69	910
Lawsonia inermis.	Henna	2.52	0.11	360.67	0.38	363.68	0.67	8.93	9.6	1540
Leucas aspera	- (L. aspera)	4.31	0.68	134.41	0.12	139.52	0.07	8.11	8.17	1360
Melissa officinalis	Lemon balm	6.81	0.46	225.18	0.64	233.08	3.32	9.44	12.75	1850
Momordica charantia	Bittergourd leaf	2.15	0.1	108.49	0.23	110.98	0.09	3.03	3.13	510
Ocimum canum	Hoary Basil	4.66	1.29	458.3	0.92	465.18	0.87	12.72	13.59	2190

Table 14.3 (*Continued*)

Botanical name	Common name	Xanthophylls					Provitamin A			RE[c]
		Neo-xanthin	Viola-xanthin	Lutein	Zea-xanthin	Total[a] xanthopylls	α-Carotene	β-Carotene	Total provitamin[b] A carotenoids	
Ocimum sanctum	Holy Basil	11.36	2.75	48.22	1.58	63.91	0.72	31.23	31.96	5270
Physalis alkekengi	Chinese lantern	3.49	0.68	184.65	0.48	189.29	ND	4.47	4.47	740
Plumbago zeylanica	-	4.46	0.14	244.53	0.46	249.6	0.79	5.38	6.17	960
Prunella vulgaris	Allheal	5.61	0.92	349.28	0.72	356.53	0.19	8.67	8.86	1460
Ricinus communis	Castor	7.84	2.46	431.81	0.61	442.73	ND	20.83	20.83	3470
Salvia officinalis	Sage	3.18	1.34	300.46	0.35	305.33	0.43	7.85	8.28	1350
Solanum dulcamara	Bittersweet Nightshade	6.27	0.9	325.37	0.59	333.12	3.11	7.18	10.29	1460
Solanum surattense	Wild brinjal	5.58	0.87	260.48	0.27	267.2	ND	11.56	11.56	1930
Tridax procumbens	Mexican daisy	5.48	0.26	284.95	0.26	290.94	ND	8.72	8.72	1450
Vitex negundo	Indian privet	3	0.27	100.55	0.16	103.98	ND	9.33	9.33	1560
Zingiber officinale	Ginger leaf	1.84	0.75	100.98	0.22	103.79	2.59	1.21	3.8	420

[a]Total xanthophylls = neoxanthin + violaxanthin + lutein + zeaxanthin. [b]Total provitamin A carotenoids = α-carotene + β-carotene. [c]RE = 1 RE = 6 mg of β-carotene or 12 mg of α-carotene. [d]Raju et al (2007). [e]Sangeetha and Baskaran (2010). [f]Mamatha et al (2011). [g]β-Cryptoxanthin identified but not quantified. [h]Lycopene identified but not quantified. [i]Aruna and Baskaran (2010).

retinol = 10 IU of vitamin A activity from β-carotene. From Table 14.3, it is seen that among the leafy greens, RE (mg%) is highest in lamb's quarters (20030) and lowest in green cabbage (10), whereas in vegetables it is in the range of 10 (colocasia, green zucchini, mango ginger, decalpis, tapioca, plectranthus, potato, baby corn) to 2800 (green capsicum). RE (mg%) in spices is in range of 0.1 (ginger) to 1.5 (green chilli) and in cereals and pulses it varies from 0.3 (chickpea) to 11.3 (red parboiled unpolished rice). RE levels (mg%) are highest in thyme leaved gratiola (5780) and lowest in ginger leaf (420) in medicinal plants.

14.5 Conclusions

Carotenoid content is known to vary in plant foods due to the differences in the characteristics of cultivars, climate, growing conditions and geographical origins. Differences in carotenoid levels have been reported even within the species and these can be attributed to the cultivar, climate, growing conditions, seasonal changes, variety and stage of maturity of the samples used for analysis (Kimura and Rodriguez-Amaya, 2003). This could have been the reason for the differences in carotenoid composition of similar plants analyzed in different studies. However, information on carotenoid composition and their RE is important, as it will help to compile information on the carotenoid profile of the plants. In addition, these results will be of use to health and community workers to recommend and improve awareness about the consumption of natural and locally available plant materials to meet the requirements for β-carotene/retinol (vitamin A) and lutein (macular pigment) for the prevention of deficiency-related diseases such as night blindness and other vitamin A deficiency-related disorders, as well as age-related macular degeneration, in the populace.

Summary Points

- Carotenoids are found in leafy greens, orange and yellow vegetables and fruits. They are classified into carotenes and xanthophylls on the basis of their chemical structure and properties.
- Common carotenes in plants include β-, α-, γ-carotene and lycopene, whereas xanthophylls include lutein, zeaxanthin, neoxanthin, violaxanthin and β-cryptoxanthin.
- Carotenoids can be extracted from the plant matrix by organic solvents and separated/purified using different mobile phases by open column chromatography.
- Individual carotenoids can be quantified by spectrophotometeric and HPLC methods, whereas they can be characterized by LC-MS. HPLC is the preferred method for quantitation of carotenoids due to its accuracy and sensitivity to low concentrations.

- Various solvent systems can be used for extraction of carotenoids and different HPLC methods are available for their separation and estimation.
- Levels of individual carotenes and xanthophylls and RE of provitamin A carotenes are calculated for various plant materials.
- In general, provitamin A and xanthophyll carotenoid content and consequently RE in leafy greens are greater than other plant food sources.
- Carotenoid content in plant foods may vary due to difference in characteristics of cultivars, climate, growing conditions and geographical origins.

Key Facts

Key Facts of Vitamin A and its Deficiency

- In addition to its many functions in the body, including vision, growth, reproduction, immunity, *etc.*, vitamin A (retinol) is present in all cells and is localized in the membranes of cells where it acts as an antioxidant and is essential to maintain the integrity of the membranes.
- Thus, its deficiency leads to alteration in all the above-mentioned physiological processes, leading to various disorders.
- The most common symptom and the initial stage of vitamin A deficiency is night blindness, followed by various stages as the severity of deficiency increases and may even result in permanent vision loss.
- Children and pregnant/lactating mothers are vulnerable groups.
- The WHO has declared vitamin A deficiency as a public health nutrition problem in many countries of the developing world.
- Prophylaxis programmes for administration of synthetic vitamin A have achieved varying degrees of success. Thus, consumption of foods rich in precursors of vitamin A is recommended to prevent and cure vitamin A deficiency.
- Provitamin A carotenoids are present in plant foods such as leafy greens and vegetables.
- Thus determination of the RE in commonly consumed food ingredients would help the health officers and populace to choose foods rich in provitamin A carotenoids.

Key Facts of High Performance Liquid Chromatography (HPLC)

- HPLC is a chromatographic technique used for separation and quantitation of a mixture of compounds (such as carotenoids).
- The mobile phase is in liquid phase [consisting of a single or mixture of solution(s)/solvent(s)] and carries the sample mixture from the injection port

through the column for separation of the compounds based on their retention due to reactions with the stationary phase present in the column.

- As each component is eluted out, it is detected by the detector that converts the signal and presents it as a chromatogram with peaks, using a data processor.
- HPLC is a sophisticated technique that is preferred for its accuracy and sensitivity.
- HPLC can be of two types: normal phase (non-polar mobile phase, polar stationary phase) and reverse phase (polar mobile phase, non-polar stationary phase).
- Based on the nature of the compound of interest, different HPLC methods and columns can be selected to achieve good results.

Definitions of Words and Terms

Anti-inflammatory: Prevent/ cure inflammation in the body.

Antioxidant: A molecule that can scavenge the free radicals and break the chain reactions of free radicals.

Anti-proliferative: Prevention of growth and metastasis of cancer/tumour cells.

Bioavailability: The fractions of carotenoids that are freed from the complex plant matrix, absorbed by the enterocytes of the small intestine and enter the systemic circulation for biological activity and storage.

Carotenes: Carotenes are non-polar carotenoids with no oxygen-containing functional group in their chemical structure.

Gradient elution: Elution of the mobile phase in HPLC in which there is change composition of the constituent solutions/solvents over a period of time based on the programme set for analysis. *Isocratic elution,* Elution of the mobile phase in HPLC in which the composition of the mobile phase remains constant.

Mobile phase: Liquid phase used for separation of carotenoids by HPLC.

Provitamin A carotenoids: Carotenoids that are cleaved by the β,β-carotene-15,15′-monooxygenase to form retinol (vitamin A) *in vivo*.

Saponification: Treatment of esterified carotenoids with an alkali (methanolic potassium hydroxide) to obtain carotenoids in their free state.

Xanthophylls: Xanthophylls are polar carotenoids with oxygen-containing functional groups such as hydroxyl, keto and/or epoxides in their chemical structure.

List of Abbreviations

APCI	atmospheric pressure chemical ionization
HPLC	high-performance liquid chromatography
IU	international unit
LC-MS	liquid chromatography-mass spectrometry
PDA	photodiode array

RAE   retinol activity equivalent(s)
RE   retinol equivalent(s)
TLC   thin-layer chromatography
UV-Vis   ultraviolet/visible
WHO   World Health Organization

Acknowledgements

Sangeetha Ravi Kumar acknowledges the award of Senior Research Fellowship by University Grants Commission, Government of India.

References

Aman, R., Biehl, J., Carle, R., Conrad, J., Beifuss, U. and Schieber, A., 2005. Application of HPLC coupled with DAD, APCI-MS and NMR to the analysis of lutein and zeaxanthin stereoisomers in thermally processed vegetables. Food Chemistry. 92: 753–763.

Aruna, G. and Baskaran, V., 2010. Comparative study on the levels of carotenoids lutein, zeaxanthin and β-carotene in Indian spices of nutritional and medicinal importance. Food Chemistry. 123: 404–409.

Barua, A. B., 2001. Improved normal-phase and reversed-phase gradient high-performance liquid chromatography procedures for the analysis of retinoids and carotenoids in human serum, plant and animal tissues. Journal of Chromatography A. 936: 71–82.

Britton, G., 1995. Structure and properties of carotenoids in relation to function. FASEB Journal. 9: 1551–1558.

Eitenmiller, R. R. and Lander, Jr, W. O., 1999. Vitamin A and carotenoids. In Vitamin Analysis for the Health and Food Science. CRC Press, Boca Raton, FL, USA, pp. 3–76.

Food and Nutrition Board, Institute of Medicine., 2001. Vitamin A. In: Dietary Reference Intakes for Vitamin A, Vitamin K, Arsenic, Boron, Chromium, Copper, Iodine, Iron, Manganese, Molybdenum, Nickel, Silicon, Vanadium, and Zinc. National Academy Press, Washington, D.C., USA, pp. 65–126.

Fraser, P. D., Elisabete, M., Pinto, S., Holloway, D. E. and Bramley, P. M., 2000. Application of high-performance liquid chromatography with photodiode array detection to the metabolic profiling of plant isoprenoids. The Plant Journal. 24: 551–558.

Ismail, A. and Fun, C. S,. 2003. Determination of vitamin C, β-carotene and riboflavin contents in five green vegetables organically and conventionally grown. Malaysian Journal of Nutrition. 9: 31–39.

Kimura, M. and Rodriguez-Amaya, D. B., 2003. A scheme for obtaining standards and HPLC quantification of leafy vegetable carotenoids. Food Chemistry. 78: 389–398.

Konings, E. J. M. and Roomans, H. H. S., 1997. Evaluation and validation of an LC method for analysis of carotenoids in vegetables and fruit. Food Chemistry. 59: 599–603.

Kotake-Nara, E., Kushiro, M., Zhang, H., Sugawara, T., Miyashita, K. and Nagao, A., 2001. Carotenoids affect proliferation of human prostate cancer cells. Journal of Nutrition. 131: 3303–3306.

Lienau, A., Glaser, T., Tang, G., Dolnikowski, G. G., Grusak, M. A. and Albert, K., 2003. Bioavailability of lutein in humans from intrinsically labeled vegetables determined by LC-APCI-MS. Journal of Nutritional Biochemistry. 11: 663–670.

Mamatha, B. S., Sangeetha, R. K. and Baskaran, V., 2011. Provitamin-A and xanthophyll carotenoids in vegetables and food grains of nutritional and medicinal importance. International Journal of Food Science and Technology. 46: 305–323.

Menotti, A., Kromhout, D., Blackburn, H., Fidanza, F., Buzina, R. and Nissinen, A., 1999. Food intake patterns and 25-year mortality from coronary heart disease: cross-cultural correlations in the Seven Countries Study. European Journal of Epidemiology. 15: 507–515.

Molnar, P., Szabo, Z., Osz, E., Olah, P., Toth, G. and Deli, J., 2004. Separation and identification of lutein derivatives in processed foods. Chromatographia Supplement. 60:: S101–S105.

National Research Council. 1989. Recommended Dietary Allowances, 10th edn. National Academy Press, Washington, D.C., USA.

Niizu, P. Y. and Rodriguez-Amaya, D. B., 2005. New data on the carotenoid composition of raw salad vegetables. Journal of Food Composition and Analysis. 18: 739–749.

Raju, M., Varakumar, S., Lakshminarayana, R., Krishnakantha, T. P. and Baskaran, V., 2007. Carotenoid composition and vitamin A activity of medicinally important green leafy vegetables. Food Chemistry. 101, 1598–1605.

Sangeetha, R. K. and Baskaran, V., 2010. Carotenoid composition and their retinol equivalent in plants of nutritional and medicinal importance: efficacy of β-carotene from *Chenopodium album* in retinol deficient rats. Food Chemistry. 119: 1584–1590.

Seo, J. S., Burri, B. J., Quan, Z. and Neidlinger, T. R., 2005. Extraction and chromatography of carotenoids from pumpkin. Journal of Chromatography A. 1073: 371–375.

Shiratori, K., Ohgami, K., Llieva, I., Jin, X. H., Koyama, Y., Miyashita, K., Kase, S. and Ohno, S., 2005. Effects of fucoxanthin on lipopolysaccharide-induced inflammation *in vitro* and *in vivo*. Experimental Eye Research. 81: 422–428.

Weller, P. and Breithaupt, D. E., 2003. Identification and quantification of zeaxanthin esters in plants using liquid chromatography-mass spectrometry. Journal of Agricultural and Food Chemistry. 51: 7044–7049.

World Health Organization., 1982. Control of vitamin A deficiency and xerophthalmia. World Health Organization, Technical Report Series No. 672: Report of a Joint WHO/UNICEF/Helen Keller International/IVACG meeting.

CHAPTER 15

LC-NMR for the Analysis of Carotenoids in Foods

CHISATO TODE* AND MAKIKO SUGIURA

Central Analytical Laboratory, Kobe Pharmaceutical University, 4-19-1
Motoyamakita-machi, Higashinada-ku, Kobe 658-8558, Japan
*E-mail: c-tode@kobepharma-u.ac.jp

15.1 Introduction

Liquid chromatography-nuclear magnetic resonance (LC-NMR) is an innovative technique that connects nuclear magnetic resonance (NMR) with high-performance liquid chromatography (HPLC) online, and can offer not only one-dimensional (1-D) but also two dimensional (2-D) NMR spectra for the components separated by HPLC. Recently, LC-NMR has come into wide use because of superconductive magnets and advanced techniques, especially the solvent suppression method (Smallcombe et al 1995). For example, LC-NMR has been applied to the analysis of medicinal metabolites (Spraul et al 2003), impurities in medicinal specialties, and metabolites of natural products (Iwasa et al 2008). NMR provides information about conformational geometry and is thus a powerful tool for structural analysis.

In this work, we applied LC-NMR to the componental analysis of carotenoids in several foods. Carotenoids are red and yellow pigments, and are used as food additives. However, carotenoids isolated by extraction are easily isomerized and/or decomposed by light and air because they have a long conjugated side chain as a structural feature (Morais et al 2001). It is important to analyze carotenoids quickly to confirm their presence or determine the kind of isomers. Consequently, LC-NMR appears to be a suitable tool for analyzing

Food and Nutritional Components in Focus No. 1
Vitamin A and Carotenoids: Chemistry, Analysis, Function and Effects
Edited by Victor R Preedy

carotenoids. If LC-NMR can be established as a method of analyzing carotenoids, the time required to examine the unknown compounds and determine the component ratio in foodstuffs will be shortened.

The foodstuffs analyzed in this study were tomato juice, palm oil, and satsuma mandarin orange juice.

15.2 Sample Preparation Techniques for Carotcnoids

Although we measured ^{1}H-NMR spectra for all peaks eluted by HPLC, we show only several carotenoids for which we were able to determine their chemical structures. The structural formulas of the carotenoids analyzed in this research are shown in Figure 15.1. We have employed the standard numbering system used for carotenoids.

15.2.1 Sample Preparation

15.2.1.1 Tomato Juice

Twenty milliliters of acetone was added to 10 mL of commercial tomato juice, and the mixture was filtered through filter paper. The filtered solution was then partitioned with *n*-hexane and water. The *n*-hexane layer was dried over Na_2SO_4 and concentrated. The resulting reddish solution was loaded on to a

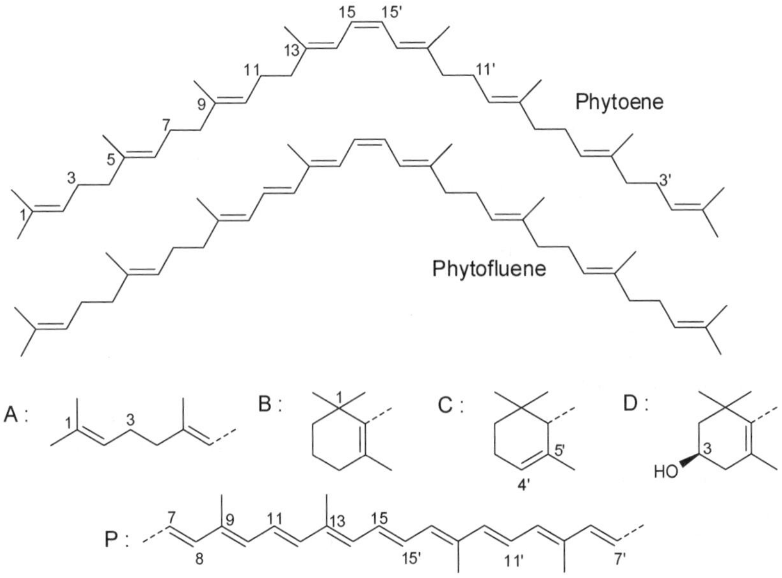

Figure 15.1 Chemical structures of the carotenoids analyzed in this work. Lycopene: A + P + A; α-carotene: B + P + C; β-carotene: B + P + B; β-cryptoxanthin: B + P + D; zeaxanthin: D + P + D.

Sep-Pak® silica cartridge. The fraction eluted with *n*-hexane (sample **1**), which contained carotenes and wax, was subjected to LC-NMR analysis.

15.2.1.2   Palm Oil

One gram crude palm oil was saponified with 5 mL of 5% KOH/methanol at room temperature for 2 h in order to remove a large amount of glycerides. Then, the unsaponifiable matter was extracted with diethyl ether (Et$_2$O)/*n*-hexane (1:1) and washed with water. The Et$_2$O/*n*-hexane extract was dried over Na$_2$SO$_4$ and allowed to evaporate. The resulting reddish residue was dissolved in *n*-hexane and was loaded on to a Sep-Pak® silica cartridge. The fraction eluted with *n*-hexane (sample **2**), which contained carotenes and wax, was subjected to LC-NMR analysis.

15.2.1.3   Satsuma Mandarin Orange Juice

Twenty milliliters of acetone was added to 10 mL of commercial satsuma mandarin juice, and the mixture was filtered through filter paper. The mixture was then partitioned with Et$_2$O/*n*-hexane (1:1) and washed with water. The organic layer was dried over Na$_2$SO$_4$ and concentrated. The resulting reddish solution was loaded on to a Sep-Pak® silica cartridge. The reddish residue was dissolved in *n*-hexane and was loaded on to a Sep-Pak® silica cartridge. The fraction eluted with Et$_2$O/*n*-hexane (1:1) (sample **3**), which contained carotenoids and terpenoids, was subjected to LC-NMR analysis.

15.2.2   LC-NMR

LC-NMR experiments were performed on a Varian UNITY INOVA-500 spectrometer (^{1}H: 499.83 MHz) equipped with a 60 μL microflow NMR probe at room temperature. H-1 1-D and 2-D spectra were obtained in the stopped-flow mode. Varian water suppression enhanced through T$_1$ effect (WET) solvent suppression and related sequences were used to suppress the peak of the residual CH$_3$CN in deuterium acetonitrile (CD$_3$CN), and the residual CHCl$_3$ in deuterium chloroform (CDCl$_3$). COSY spectra were obtained using water suppression enhanced through a T$_1$ effect gradient correlation spectroscopy (WETgCOSY) pulse sequence in which the WET element was incorporated into the gradient correlation spectroscopy (gCOSY) sequence.

The HPLC system consisted of a ternary Varian ProStar 230 pump, a Varian ProStar 335 photodiode array (PDA) detector, and a Waters model 1122 column oven (40 °C). Chromatographic separation was performed on a COSMOSIL 5C18-AR-II reversed-phase column (150 mm × 4.6 mm; particle size, 4.5 μm). The following three mobile phase conditions were used: an isocratic elution, 6% CDCl$_3$ (A) and 94% CD$_3$CN (B) (condition I); two linear gradient elutions, initial, 5% (A) and 95% (B), 15 min, 10% (A) and 90% (B), maintain the ratio for 30 min (condition II); and initial, 5% (A) and

95% (B), 15 min, 10% (A) and 90% (B), 30 min, 20% (A) and 80% (B), maintain the ratio for 40 min (condition III). The flow rate was 1 mL min for all conditions.

15.3 Measurement of Extracted Samples

15.3.1 Analysis of Sample 1 (Tomato Juice)

Figure 15.2 shows the HPLC chromatogram (extracted at 300 nm) of sample **1** extracted from tomato juice containing some carotenoids. ^{1}H-NMR spectra were obtained for each of the peaks eluted, *i.e.*, three carotenoids detected as **1a–1c**. The ultraviolet (UV) spectra of **1a–1c** showed absorption maxima at

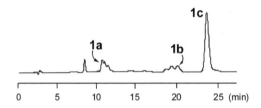

Figure 15.2   HPLC-PDA chromatogram of tomato juice (extracted at 300 nm, isocratic elution as described in Section 15.2, under condition I).

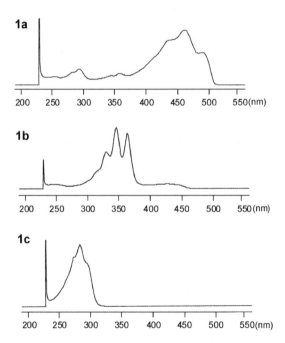

Figure 15.3   UV spectra of compounds **1a–1c** detected in tomato juice.

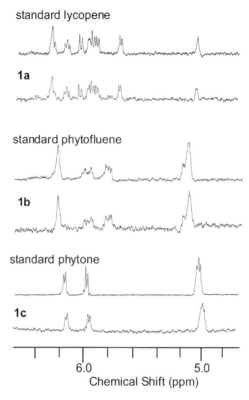

Figure 15.4   1-D ^{1}H-NMR spectra of compounds **1a–1c** detected in tomato juice and their pure standard samples.

470, 350, and 280 nm, respectively (Figure 15.3). It is known that the absorption maximum of lycopene of hendecaene is 473 nm, that of phytofluene of pentaene is 346 nm, and that of phytoene of triene is 284 nm (Takaichi and Shimada 1992). Accordingly, the UV spectra suggested that **1a–1c** were these three carotenoids. In addition, from the ^{1}H-NMR spectrum patterns in the low-field region (Figure 15.4), these carotenoids were presumed to be lycopene (**1a**), phytofluene (**1b**), and phytoene (**1c**) by comparing their spectra with those of the literature (Granger *et al* 1973; Hengartner *et al* 1992; Mercadante *et al* 1999), although the chemical shifts were slightly different because of the different solvents used. Structural assignment was further confirmed by comparing their spectra with those of the pure standards. It is well known that tomato contains these three carotenoids (Tiziani *et al* 2006) and our results agree.

15.3.2   Analysis of Sample 2 (Palm Oil)

Sample **2** was extracted from palm oil, which contained several carotenoids, and was analyzed by HPLC (extracted at 450 nm). The chromatogram is shown in Figure 15.5; ^{1}H-NMR spectra of each of the detected peaks

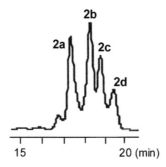

Figure 15.5 HPLC-PDA chromatogram of palm oil (extracted at 450 nm, linear gradient elution, as described in Section 15.2, under condition II).

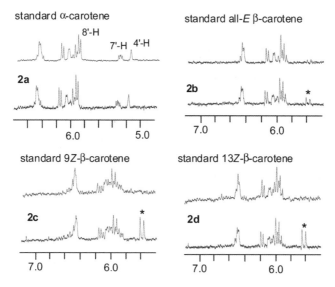

Figure 15.6 1-D ^{1}H-NMR spectra of compounds **2a–2d** detected in palm oil and their pure standard samples.

(Figure 15.6) were obtained. In the ^{1}H-NMR spectrum of **2a**, the characteristic 4' and 7'-Hs of α-carotene were obtained at δ 5.18 and 5.35 ppm, respectively. The signals of other protons were comparable with those in the spectra of the pure standard, thus supporting its identification as α-carotene. In addition, the COSY spectrum of **2a** showed a correlation of 7'-H and 8'-H (Figure 15.7). Some correlations of other olefinic protons were also observed.

Compounds **2b–2d** were identified as all-*E* β-carotene, 9-*Z* β-carotene, and 13-*Z* β-carotene, respectively, on the basis of their corresponding spectrum features. Structural assignment was further confirmed by comparing their spectra with those of the pure standard, although some minor signals appeared from impurities in the real sample. From previous research (Craft 1992), it is

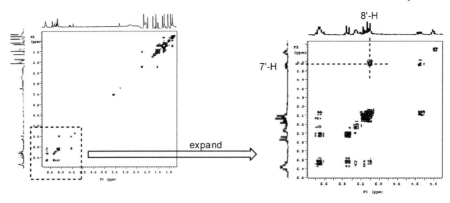

Figure 15.7   WETgCOSY spectra of **2a**.

known that these four carotenoids elute in the order of the retention times shown in Figure 15.5, and our results agree. It was possible to use LC-NMR to confirm the presence of the non-isolated carotenoids that are known to be in palm oil (Chen and Chen 1994).

15.3.3   Analysis of Sample 3 (Satsuma Mandarin Orange Juice)

Sample **3**, which was extracted from satsuma mandarin orange juice, was also analyzed by HPLC (extracted at 420 nm), as shown in Figure 15.8, and ^{1}H-NMR spectra of each of the detected peaks (Figure 15.9) were obtained. The two compounds detected and labeled as peaks **3a** and **3b** belong to distinct classes of carotenoids. In the ^{1}H-NMR spectra of both carotenoids, multiplet signals were observed at approx. δ 3.6 ppm and assigned as CH substituted with the hydroxyl group. Moreover, the spectrum pattern in the olefinic proton region was comparable with those of standard β-carotene (**2b** of Figure 15.6). These observations suggested the existence of zeaxanthin and/or β-cryptoxanthin. Further evidence was obtained by comparing the spectra obtained with those of the standard compounds, thus enabling positive identification of peaks **3a** and **b** as zeaxanthin and β-cryptoxanthin, respectively. In contrast, it is well known that satsuma mandarin orange juice contains zeaxanthin and β-cryptoxanthin, and our results agree with previous findings (Yano *et al* 2005).

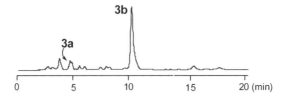

Figure 15.8   HPLC-PDA chromatogram of satsuma mandarin orange juice (extracted at 420 nm, linear gradient elution, as described in Section 15.2, under condition III).

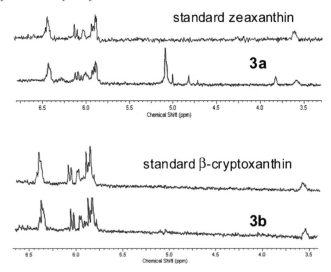

Figure 15.9   1-D ^{1}H-NMR spectra of compounds **3a** and **3b** detected in satsuma mandarin orange juice and their pure standard samples.

LC-NMR proved useful for the componential analysis of the foodstuff studied in this work, and was indisputably able to analyze content in carotenoids. Furthermore, if standard compounds are available, compounds under study can be unambiguously identified based on their chromatographic behavior and ^{1}H-NMR spectral features.

Summary Points

- Lycopene (**1a**), phytofluene (**1b**), and phytoene (**1c**) are in tomato juice.
- The presence of α-carotene (**2a**) and some isomers of β-carotene (**2b–2d**) in palm oil were confirmed.
- Zeaxanthin (**3a**) and β-cryptoxanthin (**3b**) are in satsuma mandarin orange juice.
- By using LC-NMR, we succeeded in analyzing non-isolated carotenoids, which are fragile and sensitive to light and air.
- From the observed correlations of olefinic protons, we succeeded in determining structural assignment by measuring the 2-D spectra.
- This method proved very useful for easily analyzing foodstuffs for the presence of carotenoids.

Key Facts of using LC-NMR

- LC-NMR enables searching for unknown compounds.
- LC-NMR enables separation of unstable compounds.
- LC-NMR provides rapid analysis of extracted mixtures.

- LC-NMR simplifies experimental procedures.
- LC-NMR provides direct structural analysis.

Definition of Words and Terms

α-Carotene: The form of carotene with a β-ring at one end and an ε-ring at the other. It is the second most common form of carotene. Yellow-orange vegetables and dark green vegetables are rich in α-carotene.

β-Carotene: This is a terpenoid. It is a strongly colored red-orange pigment abundant in plants and fruits. β-Carotene is also the substance in carrots that gives them their orange color. As a carotene with β-rings at both ends, it is the most common form of carotene. The structure was deduced by Karrer and colleagues in 1930. In nature, β-carotene is a precursor (inactive form) of vitamin A *via* the action of β-carotene 15,15'-monooxygenase. β-Carotene is biosynthesized from geranylgeranyl pyrophosphate.

Carotenoid: These are pigments that naturally occur in the chloroplasts and chromoplasts of plants and some other photosynthetic organisms such as algae, some types of fungus, some bacteria, and at least one species of aphid. There are over 600 known carotenoids, which are split into two classes: xanthophylls (which contain oxygen) and carotenes (which are purely hydrocarbons and contain no oxygen).

Correlation spectroscopy (COSY): One of several types of two-dimensional nuclear magnetic resonance (NMR) spectroscopy. Two-dimensional (2-D) NMR spectra provide more information about a molecule than one-dimensional (1-D) NMR spectra and are especially useful in determining the structure of a molecule, particularly for molecules that are too complicated to work with using 1-D NMR. The first 2-D experiment, COSY, was proposed by Jean Jeener, a professor at the Université Libre de Bruxelles, in 1971. This experiment was later implemented by Walter P. Aue, Enrico Bartholdi, and Richard R. Ernst, who published their work in 1976.

β-Cryptoxanthin: A natural carotenoid pigment. It has been isolated from a variety of sources including the petals and flowers of plants in the genus Physalis, orange rind, papaya, egg yolk, butter, apples, and bovine blood serum. In terms of structure, cryptoxanthin is closely related to β-carotene, with only the addition of a hydroxyl group. It is a member of the class of carotenoids known as xanthophylls. In the human body, cryptoxanthin is converted into vitamin A (retinol) and therefore, is considered a provitamin A. As with other carotenoids, cryptoxanthin is an antioxidant and may help prevent free radical damage to cells and DNA, as well as stimulate the repair of oxidative damage to DNA.

High-performance liquid chromatography (HPLC): This is liquid chromatography that generally utilizes very small packing particles and a relatively high pressure.

Liquid chromatography (LC): A separation technique in which the mobile phase is a liquid. LC can be performed either in a column or on a planar surface.

Lycopene: A bright red carotene, carotenoid pigment and phytochemical found in tomatoes and other red fruits and vegetables, such as red carrots, watermelons, and papayas (but not strawberries or cherries). Although lycopene is chemically a carotene, it has no vitamin A activity. In plants, algae, and other photosynthetic organisms, lycopene is an important intermediate in the biosynthesis of many carotenoids, including β-carotene, responsible for yellow, orange, or red pigmentation, photosynthesis, and photo-protection. Like all carotenoids, lycopene is a polyunsaturated hydrocarbon (an unsubstituted alkene). Structurally, it is a tetraterpene assembled from eight isoprene units, composed entirely of carbon and hydrogen, and is insoluble in water. The eleven conjugated double bonds of lycopene give it its deep red color and are responsible for its antioxidant activity. Due to its strong color and non-toxicity, lycopene is a useful food coloring.

Nuclear magnetic resonance (NMR): The effect whereby magnetic nuclei in a magnetic field absorb and re-emit electromagnetic (EM) energy. This energy is at a specific resonance frequency, which depends on the strength of the magnetic field and other factors. This allows the observation of specific quantum mechanical magnetic properties of an atomic nucleus. Many scientific techniques make use of NMR phenomena to study molecular physics, crystals, and non-crystalline materials through NMR spectroscopy.

Palm oils: These are edible plant oils derived from the fruits of palm trees. Palm oil is extracted from the pulp of the fruit of the oil palm *Elaeis guineensis*, palm kernel oil is derived from the kernel (seed) of the oil palm, and coconut oil is derived from the kernel of the coconut (*Cocos nucifera*). Palm oil is naturally reddish in color because it contains a high amount of β-carotene.

Zeaxanthin: One of the two primary xanthophyll carotenoids contained in the retina of the eye. In the central macula, zeaxanthin is the dominant component, whereas in the peripheral retina, lutein predominates. Zeaxanthin is important in the xanthophyll cycle.

List of Abbreviations

CDCl$_3$	deuterium chloroform
CD$_3$CN	deuterium acetonitrile
1-D	one-dimensional
2-D	two-dimensional
Et$_2$O	diethyl ether
HPLC	high-performance liquid chromatography
LC-NMR	liquid chromatography-nuclear magnetic resonance
NMR	nuclear magnetic resonance
PDA	photodiode array
UV	ultraviolet
WET	water suppression enhanced through T$_1$ effect
WETgCOSY	water suppression enhanced through T$_1$ effect gradient correlation spectroscopy

References

Chen, T. M. and Chen, B. H., 1994. Optimization of mobile phases for HPLC of *cis–trans* carotene isomers. Chromatographia. 39: 346–354.

Craft, N. E. 1992. Carotenoid reversed-phase high-performance liquid chromatography methods: reference compendium. Methods in Enzymology. 213: 185–205.

Granger, P., Maudinas, B., Herber, R. and Villoutreix, J., 1973. ^{1}H and ^{13}C NMR spectra of *cis* and *trans* phytoene isomer. Journal of Magnetic Resonance. 10: 43–50.

Hengartner, U., Bernhard, K. and Mayer, K., 1992. Synthesis, isolation, and NMR-spectroscopic characterization of fourteen (*Z*)-isomers of lycopene and some acetylenic didehydro- and tetrahydrolycopenes. Helvetica Chimica Acta. 75: 1848–1865.

Iwasa, K., Takahashi, K., Nishiyama, Y., Moriyasu, M., Sugiura, M., Takeuchi, A., Tode, C., Tokuda, H. and Takeda, K., 2008. Online structural elucidation of alkaloids and other constituents in crude extracts and cultured cells of *Nandina domestica* by combination of LC-MS/MS, LC-NMR, and LC-CD analyses. Journal of Natural Products. 71: 1376–1385.

Mercadante, A. Z., Steck, A. and Pfander, H., 1999. Carotenoids from guava (*Psidium guajava* L.): isolation and structure elucidation. Journal of Agricultural and Food Chemistry. 47: 145–151.

Morais, H., Ramos, A. C., Tibor, C. and Forgacs, E., 2001. Effects of fluorescent light and vacuum packaging on the rate of decomposition of pigments in paprika (*Capsicum annuum*) powder determined by reversed-phase high-performance liquid chromatography. Journal of Chromatography A. 936: 139–144.

Smallcombe, S. H., Patt, S. and Keifer, P., 1995. WET solvent suppression and its applications to LC-NMR and high-resolution NMR spectroscopy. Journal of Magnetic Resonance A. 117: 295–303.

Spraul, M., Freund, A. S., Nast, R. E., Withers, R. S., Mass, W. E. and Corcoran, O., 2003. Advancing NMR sensitivity for LC-NMR-MS using a cryoflow probe: application to the analysis of acetaminophen metabolites in urine. Analytical Chemistry. 75: 1536–1541.

Takaichi, S. and Shimada, K., 1992. Characterization of carotenoids in photosynthetic bacteria. Methods in Enzymology. 213: 374–385.

Tiziani, S., Schwartz, S. J., and Vodovotz, Y., 2006. Profiling of carotenoids in tomato juice by one- and two-dimensional NMR. Journal of Agricultural and Food Chemistry. 54: 6094–6100.

Yano, M., Ikoma, Y., and Sugiura, M., 2005. Recent progress in β-cryptoxanthin research. Bulletin of the National Institute of Fruit Tree Science. 4: 13–28.

CHAPTER 16

LC-DAD-tandem MS Analysis of Retinoids and Carotenoids: Applications to Bovine Milk

ALESSANDRA GENTILI* AND FULVIA CARETTI

Department of Chemistry, Faculty of Mathematical, Physical and Natural Sciences, University of Rome "La Sapienza", Piazzale Aldo Moro n. 5, 00185 Roma, Italy
*E-mail: alessandra.gentili@uniroma1.it

16.1 Introduction

Milk is the most complete single food in the human diet which constitutes the only source of feeding for newborns and which continues to be consumed as a nutritional supplement during juvenile and adult life. Nutrients of this complex physiological fluid are distributed in three chemical phases: lipids, carotenoids and fat-soluble vitamins are in form of an emulsion; casein micelles are dispersed as a colloidal suspension; lactose, minerals, water-soluble vitamins and other components are dissolved in an aqueous component (Belitz *et al* 2004).

Although raw milk contains many vitamins, it is an excellent source of vitamin A, in particular (Ball 2006). Compounds belonging to this vitamin group occur in milk mainly as retinoids (vitamin A vitamers) and to a lesser extent as their carotenoid precursors (provitamin A carotenoids). The most abundant forms are represented by the retinyl esters of saturated and unsaturated fatty acids, while only a small amount is present as free retinol (Woollard and Indyk 1989). Carotenoids, being localized in green tissues of

Food and Nutritional Components in Focus No. 1
Vitamin A and Carotenoids: Chemistry, Analysis, Function and Effects
Edited by Victor R Preedy
© The Royal Society of Chemistry 2012
Published by the Royal Society of Chemistry, www.rsc.org

plants in association with the chlorophylls, constitute the only source of vitamin A for ruminants and herbivores (Nozière *et al* 2006a). Nevertheless, only those incorporating one unsubstituted β-ionone ring with an 11-carbon polyene side chain possess provitamin A activity and may be converted into retinol. All the others are non-provitamin A carotenoids acting as antioxidant agents. Concentration of these unmetabolized dietary pigments in bovine milk depends on the nature and amount of the consumed forage (Ball 2006; Nozière *et al* 2006a).

The vitamin A percentage that the human organism is able to absorb, known as the bioavailability (Ball 2006; Bates and Heseker 1994), depends on several factors such as the kind of food and the food processing, the different chemical forms of the vitamin (retinyl esters or provitamin A carotenoids), and the physiological status of the subject.

It has been demonstrated that the availability of preformed vitamin A is much greater than that of precursor carotenoids (Bates and Heseker 1994). These forms are both present in milk, but very little is known about their qualitative and quantitative profile in bovine milk as well as in that of other animal species for human consumption. The main reason is due to a series of analytical problems such as the unavailability of standards, their cost and the complexity in development of their chromatographic separation; in fact, vitamin A vitamers are characterized by subtle differences in chemical structures, so a highly efficient and selective chromatographic system is needed for achieving their resolution.

Few works dealing with carotenoid distribution in bovine milk have been published (Chauveau-Duriot *et al* 2010; Hulshof *et al* 2006; Nozière *et al* 2006b; Ollilainen *et al* 1989). Most of them make use of high-performance liquid chromatography coupled to a diode array detector (HPLC-DAD), which is a potent tool in the carotenoid identification. Only one study (Woollard and Indyk 1989), carried out by HPLC-ultraviolet (UV) ($\lambda = 325$ nm), described the profile of retinyl esters in cow, goat and human milk. Notwithstanding the use of a tandem system of octadecylsilica (ODS or C18) chromatographic columns, the co-elution of two pairs of vitamers (retinyl–laurate/arachidonate and retinyl–miristate/palmitoleate) could not be avoided; moreover, the complexity of matrix and the poor selectivity of the detection system did not allow the analyte identification, based on the retention times exclusively.

An advanced hyphenated technique such as high-performance liquid chromatography-diode array detection-tandem mass spectrometry (HPLC-DAD-MS/MS) allowed us to establish the real occurrence and distribution of 18 retinoids (retinol, retinaldehyde, retinoic acid, retinyl propionate, retinyl caprilate, retinyl caprate, retinyl palmitoleate, retinyl laurate, retinyl miristate, retinyl pentadecanoate, retinyl arachidonate, retinyl palmitate, retinyl eptadecanoate, retinyl linoleate, retinyl oleate, retinyl stearate, retinyl linolenate, and retinyl eicosanoate) and several carotenoids (lutein, zeaxanthin, β-cryptoxanthin, and β-carotene) in raw bovine milk through the development and validation of two reliable analytical methods. Raw milk was preferred to the pasteurized one in order to analyze a matrix not submitted to thermal treatment which might be responsible for losses and artefacts creation (*e.g.* analyte isomerization). Animals, from which milk samples were obtained, were pasture-raised and fed without using forages supplemented with vitamins.

16.2 Practical Details and Techniques

16.2.1 Chemicals and Materials

The standards of carotenoids and retinoids were purchased from Aldrich-Fluka-Sigma S.r.l. (Milan, Italy) and from Chemical Research 2000 S.r.l. (Rome, Italy). Retinyl acetate and *trans*-β-apo-8′-carotenal, bought from Aldrich-Fluka-Sigma S.r.l. (Milan, Italy), were chosen as internal standards (ISs) for retinoids and carotenoids, respectively. All chemicals had a purity grade greater than 85%.

Butylated hydroxytoluene (BHT), provided by Aldrich-Fluka-Sigma S.r.l. (Milan, Italy), was used as antioxidant both in standard solutions and during the several steps of extraction procedure.

Solvents and potassium hydroxide (KOH) were purchased from Carlo Erba (Milan, Italy). Distilled water was further purified by passing it through a Milli-Q Plus apparatus (Millipore, Bedford, MA USA).

16.2.2 Standard Solutions

Solvents, concentrations and preparation frequency of the individual stock solutions of standards and ISs were established on the basis of a solubility and stability study. The working multi-standard solution and that of the ISs were prepared from the individual solutions by dilution in methanol with 0.1% (w/v) BHT. In all cases, different final concentrations were reached depending on the purpose.

All solutions were degassed with nitrogen and stored in dark glass flasks at $-18\ °C$.

16.2.3 Analytical Techniques

The liquid chromatograph consisted of a micro HPLC/autosampler/vacuum degasser system PE Series 200 (Perkin Elmer, Norwalk, CT, USA) and a ProntoSIL triacontyl silica (C30) column (3 µm; 4.6 × 250 mm; Bischoff Chromatography, Leonberg, Germany) thermostated at 19 °C and equipped with a guard column of the same type (5 µm; 4.0 × 10 mm). The mobile phase consisting of methanol (phase A) and an isopropanol/hexane (50:50, v/v) solution (phase B) was completely introduced into the DAD-triple quadrupole (QqQ) detection system at a flow rate of 1 mL min^{-1}. The mobile phase gradient profile was as follows (t in min): t_0, $B = 0\%$; t_1, $B = 0\%$; t_{21}, $B = 50\%$; $t_{21.1}$, $B = 99.5\%$; t_{30}, $B = 99.5\%$.

A Series 200 model (Perkin Elmer, Norwalk, CT, USA) DAD was coupled on-line between the chromatographic column and the mass spectrometer. The liquid chromatography (LC)-DAD chromatograms were acquired selecting the 450 nm wavelength (all-*trans* form λ_{max} of β-carotene, zeaxanthin and β-cryptoxanthin); the ultraviolet-visible (UV-Vis) spectra were recorded in the range of 200–700 nm.

The mass spectrometer was a 4000 Qtrap® (AB SCIEX, Foster City, CA, USA). Detection was performed by positive atmospheric pressure chemical ionization (APCI), setting a needle current (NC) of 3 μA and a probe temperature of 450 °C. High-purity nitrogen was used as a curtain gas (5 L min^{-1}) and collision gas (4 mTorr), while air was used as the nebulizer gas (2 L min^{-1}). Unit mass resolution was assessed by maintaining a full width at half maximum (FWHM) of approximately 0.7 ± 0.1 atomic mass unit (u).

The quantitative analysis of the target analytes was carried out in selected reaction monitoring (SRM) mode, after having studied their corresponding fragmentation spectra. The LC-MS/MS parameters used for identifying the interested compounds in bovine milk are summarized in Table 16.1 (retinol

Table 16.1 LC-SRM parameters for the identification and the quantitative analysis of the total retinol and the four target carotenoids, selected in this study. The identification of each target analyte in the raw milk samples, after hot saponification (Section 16.2.4.1), was based on its characteristic retention time, the two SRM transitions and their relative abundance.

			Ion ratio[b]
Analyte	*Retention time (min)*	*SRM transitions[a] (m/z)*	*Mean (RSD) (%)*
(Total) retinol	4.63	269.1/199.2	33 (6)
		269.1/119.1	
All-*trans*-lutein	9.27	551.4/135.2	85 (9)
		551.4/175.0	
All-*trans*-zeaxanthin	9.70	569.4/477.2	32 (8)
		551.4/135.2	
All-*trans*-β-cryptoxanthin	12.46	553.5/135.1	54 (9)
		553.5/119.1	
All-*trans*-β-carotene	16.39	537.5/119.1	78 (9)
		537.5/177.2	
Internal standard			
Retinyl acetate	5.92	269.1/199.2	–
trans-β-Apo-8'-carotenal	11.98	417.3/177.1	–

[a]The first line reports the least intense SRM transition, and the second line the most intense one.
[b]The relative abundance is calculated as the percentage ratio of the qualifier intensity/quantifier intensity; the results are reported as arithmetic average of six replicates plus the corresponding relative standard deviation (RSD).

and carotenoids extracted by hot saponification) and Table 16.2 (retinoids isolated by direct solvent extraction).

16.2.4 Sample Treatment

Two procedures have been developed to extract retinoids and carotenoids from bovine milk: (i) hot saponification, and (ii) direct extraction with solvents. The first one recovers retinol and carotenoids with high efficiency; the second isolates retinyl esters quantitatively. All operations were performed in subdued light, using BHT as an antioxidant.

16.2.4.1 Hot Saponification

Six millilitres of milk were saponified under nitrogen in a screw-capped tube (50 mL capacity) with 18 mL of absolute ethanol containing 0.1% (w/v) BHT and 1 mL of 50% (w/v) aqueous KOH. The tube was placed in a 80 °C water bath with continuous magnetic stirring for 30 min. Afterwards, the tubes were cooled in an ice bath, and the digest was diluted with 8.5 mL of Milli-Q water. The analytes were extracted twice with 12 mL of hexane with 0.1% (w/v) BHT. The organic layers were collected and spiked with the ISs. At this point, the hexane phase was evaporated up to 100 μL in a thermostated bath at 30 °C, under a gentle nitrogen flow, and diluted to a final volume of 200 μL with an isopropanol/hexane (75:25, v/v) solution containing 0.1% (w/v) BHT. Finally, 20 μL were injected into the HPLC-DAD-QqQ system.

16.2.4.2 Direct Solvent Extraction

The direct extraction of retinyl esters was carried out by deproteinization with ethanol and liquid–liquid extraction with hexane, according to the following procedure.

Six millilitres of milk were transferred into a polypropylene centrifuge tube (50 mL capacity) and 18 mL of absolute ethanol with 0.1% (w/v) BHT was added. Tube was capped and placed in an ultrasound bath for 6 min. A 3-fold extraction with 12 mL portions of hexane containing 0.1% (w/v) BHT was performed. Following the addition of each hexane aliquot, the tube was closed and the mixture was before vortex-mixed for 1 min and then placed in an ultrasound bath for 6 min; finally, after centrifugation at 6000 rpm at 0 °C for 10 min, the three organic layers were combined and stored at −18 °C for 1 h, in order to induce precipitation of colourless lipids. After centrifugation at 6000 rpm for 10 min at 0 °C, the supernatant was transferred into a glass tube with conical bottom [internal diameter (i.d.), 2 cm] and spiked with the IS (retinyl acetate). Afterwards, the extract was evaporated at 30 °C under nitrogen to a volume of 250 μL, then adjusted to 500 μL with an isopropanol/hexane (75:25, v/v) solution containing 0.1% (w/v) BHT. Finally, 20 μL was injected into the HPLC-DAD-QqQ system.

16.2.5 Identification and Quantification of Vitamin A and Carotenoids by LC-DAD-MS

The LC-DAD-MS analysis of the digests allowed the quantitation of the total retinol and the four carotenoids–lutein, zeaxantin, β-cryptoxanthin and β-carotene–whose standards were available commercially (see Table 16.1). During the same chromatographic run, the screening of other pigments occurring in bovine milk was made possible extracting the UV-Vis spectrum in correspondence of each chromatographic peak detected by DAD at $\lambda = 450$ nm. This provisional identification was based on the absorption spectrum and the expected retention time; for the unknowns being structural or geometric isomers of the target carotenoids, it was further supported by the sharing of the SRM transitions (pseudomolecular ion/product ion transitions).

A UV-Vis spectrum of good quality can easily be obtained analyzing a few ng of a carotenoid by HPLC-DAD. A typical three-peak spectrum in the region between 400 and 500 nm allows the identification of the chromophore on the basis of the position of absorption maxima and on the fine structure. The latter is defined as the percentage ratio between the height of the longest wavelength absorption peak (designated III) and that of the middle absorption peak (designated II), measured from minimum, i.e. (III/II)%.

The UV-Vis spectrum of a *cis* isomer (Britton *et al* 2008; Rodriguez-Amaya and Kimura 2004) shows the appearance of a *cis*-peak (λ_{cis} or λ_B in the near-UV range between 330–350 nm), located 142 nm below λ_{III} in the spectrum of the corresponding all-*trans* compound. The intensity of the *cis*-peak is greater as the *cis* double bond is nearer to the centre of the chromophore; in this case, a further indicator of the spectrum fine structure is the percentage ratio between of the height of the *cis* peak (B) and that of the middle main absorption peak (II) (measured from the baseline of spectrum), i.e. (II/B)%. This parameter is known as Q-ratio (Tai and Chen 2000); some other authors prefer to calculate it as (B/II)% (Rodriguez-Amaya and Kimura 2004). Other differences that are noticeable with the spectrum of the all-*trans* isomers are: (i) a small hypsochromic shift in λ_{max} (usually 2–6 nm for mono-*cis*, 10 nm for di-*cis* and 50 nm for poli-*cis* isomers); (ii) an hypochromic effect; and (iii) a reduction in fine structure.

Carotenoids show a series of ionic species ($[M+H]^+$, $[M]^{+\cdot}$, $[M–H]^+$) on their APCI full scan mass spectra; all xanthophylls, containing one hydroxyl group at least, exhibit the $[MH–H_2O]^+$ ion too. This behaviour was verified studying the ionization of the target analytes. The pseudomolecular ion $[M+H]^+$ was often detected as base peak, with the exception of the dehydrated pseudomolecular ion which resulted to be the most intense one for lutein and zeaxanthin. On the basis of fragmentation patterns, already described in previous papers (Gentili and Caretti 2011), the SRM transitions of the target carotenoids, used both for their quantification and for supporting the identification of geometric carotenoids individuated by LC-DAD, were selected.

For the quantitative analysis of total retinol, the dehydrated pseudomolecular ion at m/z 269 was selected as precursor ion; also in this case, SRM transitions were chosen on the basis of previous fragmentation studies (Capote *et al* 2007; Gundersen *et al* 2007).

16.2.6   Identification and Quantification of Retinyl Esters by LC-tandem MS

The extracts obtained by the direct solvent extraction procedure were analyzed by LC-SRM for determining the amount of free retinol and retinyl esters.

During the study of ionization, a gas-phase cleavage of retinyl esters with concurrent generation of an intense ion at m/z 269, corresponding to dehydrated retinol, was observed to occur in the APCI source. This behaviour was further verified by operating a LC-SRM analysis of retinoids: retinyl esters were characterized by capacity factor higher than the retinol one, but they were broken in the ion source producing the $[MH\text{–fatty acid–}H_2O]^+$ ion. Therefore, the extraction of the retinol characteristic SRM transitions (see Table 16.2) from the chromatogram of a standard solution, containing the 16 retinyl esters, showed as many chromatographic peaks.

Retinaldehyde and retinoic acid exhibited their characteristic SRM transitions.

16.2.7   Method Validation

Since blank samples were not available, the LC-SRM quantitative analysis was performed by means of the standard-addition method; in this way, besides estimating the unknown quantity of the analytes occurring in the sample, it was possible to valuate sensitivity and linear dynamic range. Recoveries, precision, limits of detection (LODs) and limits of quantitation (LOQs) were calculated after having determined the endogenous concentrations of each compound (see Table 16.3). Regarding the two SRM transitions, the one with the highest intensity (quantifier transition) was used to perform quantitative analysis, whereas the least intense one (qualifier transition) for identification purposes and for defining the method limits (LODs and LOQs).

16.3   Concentrations of Retinoids and Carotenoids in Bovine Milk

It is well known that the liver of food-producing animals is the richest source of vitamin A, since this organ is the main storage site of the vitamin. Other products of animal origin such as milk, dairy products and eggs are considerable dietary sources. With the exception of egg yolk, where the major retinoid is the free retinol, in most of other foods containing vitamin A, retinol is esterified with long chain fatty acids, palmitic acid in particular (Ball 2006).

Table 16.2 LC-SRM parameters for the identification and the quantitative analysis of the free retinol and retinoids in bovine milk. The identification of each target analyte in the raw milk samples, after direct solvent extraction (Section 16.2.4.2), was based on its characteristic retention time, the two SRM transitions and their relative abundance.

Analyte	Retention time (min)	SRM transitions[a] (m/z)	Ion ratio[b] Mean (RSD) (%)
(Free) retinol	4.63	269.1/199.2 269.1/119.1	92 (10)
Retinaldehyde	5.32	285.2/161.1 285.2/175.1	48 (11)
Retinyl acetate (IS)	5.92	269.1/199.2	–
Retinoic acid	6.19	300.2/285.1 300.2/185.1	15 (4)
Retinyl propionate	6.37	269.1/119.1 269.1/199.2	80 (8)
Retinyl caprylate	9.71	269.1/199.2 269.1/119.1	91 (13)
Retinyl caprate	11.85	269.1/119.1 269.1/199.2	98 (12)
Retinyl laurate	14.54	269.1/199.2 269.1/119.1	84 (9)
Retinyl linolenate	14.54	269.1/119.1 269.1/199.2	89 (19)
Retinyl arachidonate	14.94	269.1/199.2 269.1/119.1	90 (9)
Retinyl palmitoleate	15.48	269.1/119.1 269.1/199.2	98 (15)
Retinyl linoleate	15.81	269.1/119.1 269.1/199.2	95 (18)
Retinyl miristate	17.07	269.1/199.2 269.1/119.1	90 (11)
Retinyl oleate	17.64	269.1/199.2 269.1/119.1	72 (14)
Retinyl pentadecanoate	18.58	269.1/199.2 269.1/119.1	85 (15)
Retinyl palmitate	20.13	269.1/199.2 269.1/119.1	77 (17)
Retinyl eptadecanoate	21.77	269.1/119.1 269.1/199.2	100 (20)
Retinyl stearate	22.88	269.1/199.2 269.1/119.1	77 (16)
Retinyl eicosanoate	23.48	269.1/93.0 269.1/119.1	59 (14)

[a]The first line reports the least intense SRM transition, and the second line the most intense one.
[b]The relative abundance is calculated as the percentage ratio of the qualifier intensity/quantifier intensity; the results are reported as arithmetic average of six replicates plus the corresponding relative standard deviation (RSD).

Table 16.3 Validation data for performing quantitative analysis of the target analytes in bovine milk. After a preliminary screening, method validation was carried out only for the analytes actually occurring in bovine milk. Method precision was expressed as the relative standard deviation (RDS).

Analyte	Recovery[a] (%)	Precision (RSD) (%)	Method limits[b] ($\mu g\ L^{-1}$)	
			LOD	LOQ
Hot saponification				
Total retinol	88	8	14.7	44.1
All-*trans*-lutein	85	12	0.52	1.56
All-*trans*-zeaxanthin	62	15	1.75	5.25
All-*trans*-β-cryptoxanthin	77	15	0.82	2.46
All-*trans*-β-carotene	91	18	4.34	13.0
Direct liquid extraction				
Free retinol	75	9	3.00	9.00
Retinyl esters[c]	92–100	8–16	22.0–110	66.0–330

[a]The fortification levels applied for the recovery and precision assessment were chosen in order to increase the original content of vitamins and carotenoids by a factor of 2–3. [b]LODs and LOQs are reported as mean of six replicates. [c]For the detected retinyl esters, the range of obtained values is reported.

Nevertheless, there is little information about the distribution of vitamin A vitamers in nature (Woollard and Indyk 1989). For this reason our study was aimed at expanding this topic, taking advantage of advanced analytical techniques such as LC-tandem MS. Raw biological milk samples, kindly given by a farm of the Pontina Plain (Latium, Central Italy), were collected in September–October 2010 and analyzed by means of the above described method; the results of the quantitative analysis are reported in Table 16.4, while those of the screening analysis are resumed in Table 16.5.

Milk showed a pale yellow colour associable with pasture; this production system based on a "natural" feeding was also confirmed by the measured high concentrations of retinoids and carotenoids.

The vitamin A content in whole milk usually ranges between 300–400 $\mu g\ L^{-1}$ (Ball 2006; Belitz *et al* 2004), depending on several factors such as the breed, geographic region, season and nutrition. A high concentration of total retinol (approx. 2000 $\mu g\ L^{-1}$) was measured in the analyzed samples with both extraction procedures. The absolute datum, obtained after alkaline hydrolysis, was comparable with that calculated adding together the contributions of the found retinoids.

Among the 18 investigated retinoids, only eight forms were detected in cow's milk (see Table 16.4 and Figure 16.1). In agreement with Woollard and Indyk (1989), we found the same proportion of free retinol, confirming that it represents only a small fraction of the vitamin A activity in milk. Comparable values were calculated for retinyl linolenate, retinyl palmitate, and retinyl oleate. Nevertheless, the percentages of retinyl oleate and retinyl palmitate,

Table 16.4   Results of the quantitative analysis of retinoids and carotenoids found in biological raw milk from the *Pontina Plain* (Italy). Amounts (μg) and RAE (μg of RAE), listed in the Table, are referred to for 1 L of bovine milk.

		Biological raw milk			*Pasteurized bovine milk*
		μg	*RAE (μg of RAE)[a]*	*RAE (%)*	*RE (%)[b]*
Vitamin A	Total retinol[c]	1820	1820	98.84	–
Provitamins A	All-*trans*-β-carotene	253	21.10	1.14	–
	All-*trans*-β-cryptoxanthin	6.90	0.290	0.02	–
Non-provitamin	All-*trans*-lutein	10.2	–	–	–
A carotenoids	All-*trans*-zeaxanthin	LOD	–	–	–
Retinoids	Free retinol	37.0	37.00	2.00	2.2
	Retinyl linolenate	260	136.2	7.40	6.2
	Retinyl linoleate	117	61.10	3.32	6.2
	Retinyl myristate	216	124.6	6.77	2.5[d]
	Retinyl oleate	993	516.3	28.0	23
	Retinyl palmitate	1032	563.2	30.6	37
	Retinyl eptadecanoate	273	145.1	7.88	1.9
	Retinyl stearate	155	80.30	4.36	10.8
			1841.4[e]	100	

[a]In 2001, the U.S. Institute of Medicine established a new system, based on RAE (μg of RAE), for defining vitamin A equivalency. The retinol activity equivalency ratios have been set at 12:1 for β-carotene: retinol and 24:1 for other provitamin A carotenoids: retinol. [b]The results of Woollard and Indyk (1989) are reported in this column. Sum of these contributions is 89.8%; the remaining 10.2% was assigned by the authors to retinyl caprilate, retinyl caprate, retinyl pentadecanoate, and retinyl eicosanoate. [c]Result obtained after hot saponification [d]Woollard and Indyk (1989) reported a result referred to retinyl myristate plus retinyl palmitoleate, because the two analytes coeluted on their C18 tandem column system. [e]RAE calculated as sum of provitamin A carotenoid and total retinol contributions.

estimated by our analyses, were similar and they provided for more than half of the total retinol activity in milk (approx. 59%). A significant contribution was provided also by retinyl linolenate, retinyl miristate, and retinyl eptadecanoate (approx. 22%). Different proportions resulted for retinyl linoleate, retinyl miristate, retinyl eptadecanoate, and retinyl stearate; more-over, Woollard and Indyk (1989) found also retinyl caprylate, retinyl caprate, retinyl laurate, retinyl pentadecanoate, and retinyl eicosanoate. These five esters, together with retinaldehyde and retinoic acid, were never detected in the milk samples analyzed during this study.

Comparing the distribution of fatty acids found in retinyl esters and in triglycerides (Belitz *et al* 2004), a good correlation was observed for the long-chain palmitate and oleate residues, whereas high levels of retinyl linolenate and retinyl eptadecanoate were found unexpectedly. A possible explanation

Table 16.5 LC-UV-Vis-MS data for the tentative identification of carotenoids in bovine milk. The LC-DAD-MS hyphenation allows the provisional identification of pigments in bovine milk on the basis of a comparison between the found UV-Vis parameters [absorption maxima, %(III/II), Q-ratio] and those reported in the literature (Britton *et al* 2008). The SRM transitions, chosen for the target carotenoids (all-*trans*-lutein, all-*trans*-zeaxanthin, all-*trans*-β-carotene, and all-*trans*-β-cryptoxanthin), also supported the identification of isomeric compounds (*i.e.* all-*trans*-α-carotene, all-*trans*-α-cryptoxanthin, *etc.*).

Peak no.	Retention time[a] (min) LC-DAD	LC-MS	Compound	SRM transitions (m/z)	Observed λ (nm)	Found %(III/II)	Q-ratio
1	7.71	–	Unknown 1	–	404, 424	–	–
2	9.01	9.27	All-*trans*-lutein	551.4/135.2 551.4/175.0	424, 448, 478	60	–
3	9.44	9.70	All-*trans*-zeaxanthin	569.4/477.2 551.4/135.2	431, 455, 482	25	–
4	10.88	11.14	Zeinoxanthin	553.5/119.1	422, 448, 479	58	55
5	11.34	11.56	All-*trans*-α-cryptoxanthin	553.5/135.1 553.5/119.1	421, 447, 478	60	–
6	11.54	11.80	*cis*-α-Cryptoxanthin	553.5/135.1 553.5/119.1	422, 448, 476	55	–
7	11.95	-	Unknown 2	-	433, 454, 480	23	–
8	12.15	12.41	All-*trans*-β-cryptoxanthin	553.5/135.1 553.5/119.1	430, 454, 480	23	–
9	13.09	–	Unknown 3	–	393, 415, 441	86	–
	13.78	14.04	3-Hydroxy-β-zeacarotene	*[b]	410, 431, 457	33	–
11	14.28	14.54	*cis*-β-Carotene	537.5/119.1 537.5/177.2	346, 430, 451, 480	21	
12	15.39	15.65	13-*cis*-β-Carotene	537.5/119.1 537.5/177.2	345, 426, 448, 471	5	2.6
13	15.75	16.01	All-*trans*-α-carotene	537.5/119.1 537.5/177.2	427, 448, 474	60	–
14	16.13	16.39	All-*trans*-β-carotene	537.5/119.1 537.5/177.2	433, 455, 482	23	–
15	16.69	16.95	β-Zeacarotene	*[b]	410, 430, 449	60	–
16	17.63	17.88	Unknown 4	537.5/119.1 537.5/177.2	Low intensity	–	–
17	18.30	18.56	γ-Carotene	537.5/119.1 537.5/177.2	439, 465, 494	38	–

[a]Analyte retention times recorded on the LC-DAD chromatogram were early of about 0.26 min compared to the corresponding ones, observed on the LC-MS chromatogram. [b]Notwithstanding the slight mass difference, both 3-hydroxy-β-carotene and β-zeacarotene responded to β-cryptoxanthin and β-carotene SRM transitions respectively. In fact, QqQ works in unit mass resolution mode (0.7 ± 0.1 u) and the carotenoids are able to generate several ionic species such as $[M\text{-}H]^+$, $[M]^{+\bullet}$, $[M\text{+}H]^+$.

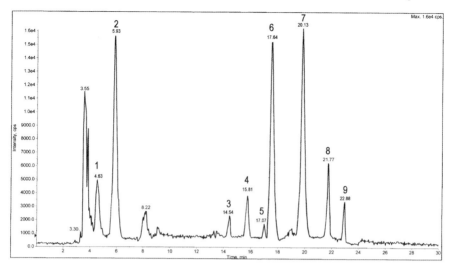

Figure 16.1 LC-MS chromatogram of retinoids occurring in a raw bovine milk sample after direct liquid extraction. The chromatogram, acquired on the basis of conditions described in Section 16.2.3, allowed the identification of the following retinoids according to the criteria reported in Table 16.2 (Section 16.2.6): 1. free retinol; 2. retinyl acetate (IS); 3. retinyl linolenate; 4. retinyl linoleate; 5. retinyl miristate; 6. retinyl oleate; 7. retinyl palmitate; 8. retinyl eptadecanoate; 9. retinyl stearate.

might be deduced considering the fatty acid composition of the membrane lipids of plants and, therefore, of forages. Such composition is relatively simple, as only a few fatty acids account for more than 90% of the total. 56% is constituted by linolenic acid, while the remaining 40% is due to palmitic acid, oleic acid, and linoleic acid (Boufaïed *et al* 2003; Clapham *et al* 2005). Notwithstanding that most linolenic acid is biohydrogenated by rumen microflora to vaccenic acid and, afterwards, to rumenic acid within the mammary gland, a certain amount of linolenic acid might be involved in esterification of retinol. The occurrence of retinyl linolenate confers to bovine milk a further nutritional quality because omega-3 is part of a class of essential unsaturated fatty acids, indispensable for the healthy functioning of the human organism.

The concentration of all-*trans*-β-carotene, the most important provitamin A, was analogous to the expected one [~ 100–200 μg L^{-1} (Belitz *et al* 2004; Hulshof *et al* 2006)], even if large variations have been reported in the literature (Nozière *et al* 2006a). Non-dietary factors responsible for this variability are due to the different analytical methods used for the determination but, above all, to a sometimes inaccurate or confusing terminology (*i.e.* carotenoids, carotene or β-carotene). The breed of the animal and the nature of the forage have a significant effect on the concentration of β-carotene and xanthophylls in milk; surely, pasture feeding

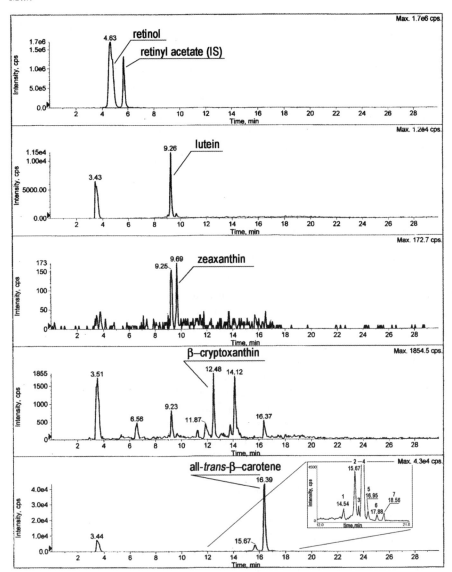

Figure 16.2 LC-MS chromatogram of total retinol and carotenoids occurring in a raw bovine milk sample after hot saponification. The chromatogram, acquired on the basis of conditions described in Section 16.2.3, allowed the identification of retinol and the target carotenoids (lutein, zeaxanthin, β-cryptoxanthin, and all-*trans*-β-carotene) according to the criteria reported in Table 16.1 (Section 16.2.6). Retinyl acetate was used as the IS for retinol. Other carotenoids were identified combining the MS data with those obtained by the DAD (see Table 16.5). On this basis, the labelled peaks in the enlargement were identified as follows: 1. *cis*-β-carotene; 2. 13-*cis*-β-carotene; 3. α-carotene; 4. all-*trans*-β-carotene; 4. β-zeacarotene; 5. γ-carotene.

entails high concentrations of carotenoids with variations related to season (for instance, sunlight exposure decrease their grass concentration as they are UV-sensitive). The content of β-cryptoxanthin, another provitamin A, was approx. 7 μg L^{-1}, whereas lutein, a xanthophyll without provitamin A activity, was the most abundant carotenoid after all-*trans*-β-carotene; zeaxanthin occurred in low concentration only.

Several other pigments were monitored in the analyzed milk samples (see the results of the LC-DAD-SRM screening assay shown in Table 16.5, together with all elements useful for their provisional identification). Their quantitative analysis could not be performed because of the unavailability of the corresponding standards.

The unknown 1 might be a degradation product of chlorophyll *a*, generated in animal rumen, since its UV-Vis spectrum showed an absorption maximum at 422–424 nm, characteristic of the Soret band.

Peak 4 was assigned to zeinoxanthin, a xanthophyll occurring in petals of some flowers, in corn grain and citrus fruits where it is often found together with α- and β-cryptoxanthin. Peaks 5 and 6 were attributed to all-*trans*- and *cis*-α-cryptoxanthin, respectively, but an analogous conclusion could not be drawn for peak 7; in fact, even if unknown 2 was suspected to be a *cis* isomer of β-cryptoxanthin, the low signals from the DAD and MS detectors were not able to support this hypothesis.

Unknown 3 exhibited a characteristic UV-Vis spectrum but its identification was difficult due to the lack of MS data. Peaks 10 and 15 were ascribable to 3-hydroxy-β-zeacarotene and β-zeacarotene respectively, since they had the same UV-Vis spectra (the hydroxyl group does not affect the chromophore absorption), retention times consistent with the expected ones and ion currents shared with the target isomeric carotenoids.

Extracting the ion currents of all-*trans*-β-carotene (16.39 min) from the LC-MS chromatogram, another six peaks were observed at 14.54, 15.65, 16.01, 16.95, 17.88, and 18.56 min (see the inset in Figure 16.2). Five of these compounds were identified by combining UV-Vis and MS data (see Table 16.5) as *cis*-β-carotene, 13-*cis*-β-carotene, all-*trans*-α-carotene, β-zeacarotene, and γ-carotene; due to the too low intensity of the extracted UV-Vis spectrum, peak 16 remained unknown.

The abundances with which the various carotenoids were detected in milk are, for the most part, in agreement with the concentrations measured by other researchers (Chauveau-Duriot *et al* 2010) in forages, where lutein is the prevalent form (230 μg g^{-1}), followed by all-*trans*-β-carotene (60 μg g^{-1}), zeaxanthin (19 μg g^{-1}), neoxanthin (18 μg g^{-1}), 9-*cis*-β-carotene (10 μg g^{-1}), 13-*cis*-β-carotene (8 μg g^{-1}), and violaxanthin (6 μg g^{-1}). Small amounts of α-carotene (Prache *et al* 2003) and β-cryptoxanthin (Chauveau-Duriot *et al* 2010) were also found but their quantification was not reported.

While retinoids occur in milk from different animal species, carotenoids are present in significant concentration only in bovine milk, influencing its colour, sensory properties, and nutritional value. In fact, among ruminants, only cows accumulate high concentrations of carotenoids that are responsible for the yellow coloration of their milk, whereas sheep and goat have a more active rumen and their milk is white, being devoid of these pigments (Nozière *et al* 2006a). Therefore, carotenoids, along with other antioxidants (enzymes, lactoferrin, vitamin C, and vitamin E), contribute to improve the oxidative stability of cow's milk; their action, combined with that of α-tocopherol, as radical scavengers is particularly efficient since they are soluble in the milk fat globule membrane which is considered the major site of auto-oxidation (Lindmark-Månsson and Åkesson 2000).

16.4   Methodological Considerations

The reason for the shortage of information on the retinoid and carotenoid distribution in nature is due to a series of problems related to their analytical determination (the unavailability of standards, their cost, the complexity in developing a simultaneous procedure of extraction, and analysis that minimizes analyte losses).

At the present, vitamin A analysis is usually carried out by means of milk saponification, in order to convert all retinyl esters into free retinol, the only one form to be then quantified. This way of proceeding simplifies the analysis and it is cheap, but it limits the knowledge about the total content of retinol. Other disadvantages concern the minor stability of free retinol towards light and air, and the creation of artefacts produced by thermal isomerization.

In comparison with the previous works, the results obtained by this study are enforced by the greater identification power of the employed analytical technique, as well as by the more enhanced sensitivity and selectivity. In fact, in a complex matrix such as milk, analyte identification was ensured by the retention time and the two SRM transitions which, even if common to all retinyl esters, were characterized by an analyte-dependent relative abundance (see the ion ratio column in Table 16.2). The identification of carotenoids results were still more reliable, being supported by different SRM transitions (Table 16.1) and by characteristic UV-Vis spectra (Table 16.5).

More information is available in the literature about the occurrence of carotenoids in nature, above all in foods of plant origin, since their chromatographic separation is simpler than the retinoid one and the DAD is a potent tool for their identification. The performance of the presented method took advantage of the LC-DAD-MS/MS hyphenation and it allowed a detailed characterization of the carotenoid profile of bovine milk both in qualitative and quantitative terms.

According to the daily intake of vitamin A recommended by the Food and Agriculture Organization/World Health Organization (1988), levels for infants and children aged 1–9 years are in the range 375 to 500 µg, and 500 to 600 µg for older children and adults. In the light of these recommendations, the consumption of bovine milk may supply a significant portion of the daily intake of vitamin A [up to 1684 µg of retinol activity equivalent (RAE) in 1 L of the analyzed Italian milk; see Table 16.4], proving to be an essential food for people of all ages, but especially for infants and children.

Summary Points

- This Chapter focuses on the liquid chromatography-diode array detection-mass spectrometry (LC-DAD-MS/MS) analysis of retinoids and carotenoids in bovine milk.
- The determination of retinoids and carotenoids in food and biological samples is an analytical challenge due to their chemical heterogeneity, protein binding, low endogenous levels, and sensitivity to light, heat, and oxygen.
- Owing to these difficulties, the literature describes only a few methods addressing their analysis in milk.
- The analytical method described here allowed the definition of an accurate profile of these micronutrients in samples of biological raw milk.
- Eight retinoids were identified and quantified in bovine milk: retinol, retinyl linolenate, retinyl linoleate, retinyl palmitate, retinyl oleate, retinyl miristate, retinyl eptadecanoate, and retinyl stearate.
- The vitamin A activity is mainly represented by similar amounts of retinyl oleate and retinyl palmitate (approx. 59%).
- The remaining 22% comprises retinyl linolenate, retinyl miristate, and retinyl eptadecanoate.
- Retinyl linolenate confers a further nutritional quality to cow's milk; it does not occur in sheep and goat's milk.
- Several carotenes (all-*trans*-β-carotene, 13-*cis*-β-carotene, all-*trans*-α-carotene, γ-carotene, and β-zeacarotene) and xanthophylls (all-*trans*-lutein, all-*trans*-zeaxanthin, all-*trans*-β-cryptoxanthin, and all-*trans*-α-cryptoxanthin) were screened.
- All-*trans*-β-carotene and lutein were shown to be the two most abundant carotenoids.
- Carotenoid occurrence in bovine milk may be a tool in feed system traceability.
- According to the daily intake of vitamin A recommended by the FAO/WHO, the consumption of bovine milk may supply a significant portion of the vitamin, along with other micro- and macronutrients, to people of all ages.

Key Facts

Key Features of Liquid Chromatography-Tandem Mass Spectrometry

This list descibes the key facts of the liquid chromatography-tandem mass spectrometry technique, and its advantages in multi-analyte extraction and determination.

- Chromatographic techniques have the great advantage of performing quantitative multi-analyte determination.
- The combination of liquid chromatography with mass spectrometry (LC-MS) allows, in addition to a more definitive identification, the quantitative determination of compounds that could not be fully resolved chromatographically.
- The selectivity of this hyphenated technique reduces problems due to interfering peaks from matrix components, while its sensitivity allows the simplification of the sample pre-treatment step.
- Although the mild ionization conditions of the atmospheric pressure ionization (API) interfaces can usually guarantee the determination of molecular weight, the lack of fragmentation precludes useful structural information.
- A solution to this problem is the use of tandem MS.
- Quadrupole, quadrupole ion trap (IT) and time of flight (TOF) are, in order, the analyzers most used for the LC coupling, either alone or combined to give tandem mass spectrometers as the triple quadrupole (QqQ) and hybrid instruments, such as the quadrupole/time of flight (QqTOF), the quadrupole ion trap/time of flight (QITTOF) and the quadrupole/linear ion trap (QqQLIT).
- QqQ is an excellent MS detector for carrying out quantitative analyses of target compounds in the selected reaction monitoring (SRM) acquisition mode.
- When the main purpose is the detection and the identification of unknown substances in a real matrix, a QqQ instrument provides other MS^2 acquisition modes, such as product ion scan (PIS), neutral loss scan (NLS) and precursor ion scan (PrIS).

Key Features of Saponification

This list describes the key facts of saponification as a sample pre-treatment for the analysis of fat-soluble vitamins and carotenoids in foods, including application conditions and consequences on the analyte stability.

- Fat-soluble vitamins occur in foods within the lipid fraction, composed mainly of triglycerides and partly of sterols, phospholipids and other lipoidal constituents. These substances, having solubility properties analogous to those of the fat-soluble vitamins, complicate their isolation and constitute a source of interference during the following analysis.
- Hot saponification is one of the most used procedures for the extraction of fat-soluble vitamins and carotenoids, especially from foods with a high fat content such as milk.
- It is an effective tool for removing the majority of fat material, hydrolyzing ester linkages of glycerides, phospholipids, and esterified sterols. Moreover,

alkaline hydrolysis frees bound forms of vitamins and xanthophylls (for instance, esterified and protein-bound forms).

- Normally, saponification is carried out with a mixture of ethanol and an aqueous potassium hydroxide solution in the presence of an antioxidant for 30 min at approx. 80 °C. The unsaponifiable fraction (fat-soluble vitamins, carotenoids, sterols, *etc.*) is extracted from alkaline digest by liquid–liquid extraction using a water-immiscible organic solvent.
- Saponification can be used for vitamins A, D, and E; however, it is not suitable for K vitamers which are decomposed quickly in alkaline media at high temperatures.

Definition of Words and Terms

Atmospheric pressure chemical ionization (APCI): This is an ionization technique based on acid–base reactions in the gas phase and suitable for detection of low-polarity compounds such as fat-soluble vitamins and carotenoids.

Chromophore: A chemical group of a molecule capable of selective light absorption.

Flow injection analysis (FIA): This work mode entails the direct injection of an analyte into a mass spectrometer coupled with a liquid chromatograph without using a chromatographic column.

Full width height maximum (FWHM): This is a parameter used for measuring resolution in the case of isolated single peak. A unit resolution corresponds to a peak width at half height of 0.7 ± 0.1 u.

Hypochromic effect: A decrease in absorbance.

Hypsochromic shift: This is a displacement of absorption maximum (λ_{max}) to shorter wavelenghts (higher frequencies).

Limit of detection (LOD): LOD is the lowest amount/concentration of a substance which can be distinguished from noise within a stated confidence limit. The LOD is usually estimated from the mean of a blank plus three times the corresponding standard deviation.

Limit of quantitation (LOQ): LOQ is the lowest amount/concentration of a substance which can be quantified with a suitable precision. The LOQ is usually estimated either as three times the established LOD value or as the mean of a blank plus ten times the corresponding standard deviation.

Selected reaction monitoring (SRM): This is a conventional scan mode of a QqQ for carrying out quantitative analysis when the mass spectrometer is coupled with a liquid chromatography system: Q1 selects the precursor ion (usually the pseudo-molecular ion), while Q3 selects the product ion generated in the collision cell. Another name for this scan mode is multi reaction monitoring (MRM).

Standard addition method: The method of standard addition is an analytical procedure used to determine the unknown concentration of an analyte in a real sample when blanks are unavailable. The sample is split into several aliquots of the same volume. With the exception of the first aliquot, all of the others are spiked with increasing volumes of a standard solution. Finally, the volume of all of the aliquots is adjusted to a final value with a selected solvent. The unknown concentration in the first unspiked aliquot can be therefore extrapolated.

Triple quadrupole: This is a tandem mass spectrometer in space incorporating three quadrupoles: Q1 and Q3 work as ion analyzers, while Q2 is located in the collision cell where it acts as focusing device.

List of Abbreviations

APCI	atmospheric pressure chemical ionization
BHT	butylated hydroxytoluene
C18	octadecyl silica
DAD	diode array detector
HPLC	high-performance liquid chromatography
IS	internal standard
KOH	potassium hydroxide
LC	liquid chromatography
LOD	limit of detection
LOQ	limit of quantitation
MS	mass spectrometry
MS/MS or MS^2	tandem mass spectrometry
QqQ	triple quadrupole
RAE	retinol activity equivalent
SRM	selected reaction monitoring
u	atomic mass unit
UV	ultraviolet
UV-Vis	ultraviolet-visible
λ_{cis} or λ_B	wavelength of a cis-peak in the UV region
λ_{max} or λ_{II}	wavelength of the middle absorption peak of a carotenoid in the UV-Vis spectrum

References

Ball, G. F. M. (ed.), 2006. Vitamin A: Retinoids and the Provitamin A Carotenoids. In: Vitamins in Foods. Analysis, Bioavailability, and Stability. CRC Press, Taylor & Francis Group, Boca Raton, FL, USA, pp. 39–105.

Bates, C. J. and Heseker, H., 1994. Human bioavailability of vitamins. Nutrition Research Reviews. 7: 93–127.

Belitz, H.-D., Grosch, W. and Schieberle P., 2004. Milk and dairy products. In: Food Chemistry. 3rd revised edn. Springer-Verlag, Berlin, Heidelberg, Germany, pp. 505–550.

Boufaïed, H., Chouinard, P. Y., Tremblay, G. F., Petit, H. V., Michaud, R. and Bélanger, G., 2003. Fatty acids in forages. I. Factors affecting concentrations. Canadian Journal of Animal Science. 83: 501–511.

Britton, G., Liaaen-Jensen, S. and Pfander H., 2008. Carotenoids Handbook. Birkhäuser Verlag, Basel, Switzerland.

Capote, F. P., Jimenez, J. R., Mata Granados, J. M. and Luque de Castro, M. D., 2007. Identification and determination of fat-soluble vitamins and metabolites in human serum by liquid chromatography/triple quadrupole mass spectrometry with multiple reaction monitoring. Rapid Communications in Mass Spectrometry. 21: 1745–1754.

Chauveau-Duriot, B., Doreau, M., Nozière, P. and Graulet, B., 2010. Simultaneous quantification of carotenoids, retinol, and tocopherols in forages, bovine plasma, and milk: validation of a novel UPLC method. Analytical Bioanalytical Chemistry. 397: 777–790.

Clapham, W. M., Foster, J. G., Neel, J. P. and Fedders, J. M., 2005. Fatty acid composition of traditional and novel forages. Journal Agricultural and Food Chemistry. 53: 10068–10073.

Gentili, A. and Caretti, F., 2011. Evaluation of a method based on liquid chromatography–diode array detector–tandem mass spectrometry for a rapid and comprehensive characterization of the fat-soluble vitamin and carotenoid profile of selected plant foods. Journal of Chromatography A. 1218: 684–697.

Gundersen, T. E., Bastani, N. E. and Blomhoff, R., 2007. Quantitative high-throughput determination of endogenous retinoids in human plasma using triple-stage liquid chromatography/tandem mass spectrometry. Rapid Communications in Mass Spectrometry. 21: 1176–1186.

Hulshof, P. J. M., van Roekel-Jansen, T., van de Bovenkamp, P. and West, C. E., 2006. Variation in retinol and carotenoid content of milk and milk products in The Netherlands. Journal of Food Composition and Analysis. 19: 67–75.

Lindmark-Månsson, H. and Åkesson, B., 2000. Antioxidative factors in milk. British Journal of Nutrition. 84: S103–S110.

Nozière, P., Graulet, B., Lucas, A., Martin, B., Grolier, P. and Doreau, M., 2006a. Carotenoids for ruminants: from forages to dairy products. Animal Feed Science and Technology. 131: 418–450.

Nozière, P., Grolier, P., Durand, D., Ferlay, A., Pradel, P. and Martin, B., 2006b. Variations in carotenoids, fat-soluble micronutrients, and color in cows' plasma and milk following changes in forage and feeding level. Journal of Dairy Science. 89: 2634–2648.

Ollilainen, V., Heinonen, M., Linkola, E., Varo, P. and Koivistoinen, P., 1989. Carotenoids and retinoids in Finnish foods: dairy products and eggs. Journal of Dairy Science. 72: 2257–2265.

Prache, S., Priolo, A. and Grolier, P., 2003. Persistence of carotenoid pigments in the blood of concentrate-finished grazing sheep: its significance for the traceability of grass-feeding. Journal of Animal Science. 81: 360–367.

Rodriguez-Amaya D. B. and Kimura M., 2004. HarvestPlus Handbook for Carotenoid Analysis. Harvestplus Technical Monograph 2. Available at: http://www.dfid.gov.uk/R4D/PDF/Outputs/Misc_Crop/tech02.pdf. Accessed 11 May 2011.

Tai, C.-Y. and Chen, B. H., 2000. Analysis and stability of carotenoids in the flowers of daylily (*Hemerocallis disticha*) as affected by various treatments. Journal of Agricultural and Food Chemistry. 48: 5962–5968.

Woollard, D. C. and Indyk, H., 1989. The distribution of retinyl esters in milks and milk products. Journal of Micronutrient Analysis. 5: 35–52.

Food and Agriculture Organization/World Health Organization. (1988). Requirements of Vitamin A, Iron, Folate and Vitamin B12. Report of a Joint FAO/WHO Expert Consultation, FAO Food and Nutrition Series No. 23, Food and Agriculture Organization, Rome. Available at: http://whqlibdoc.who.int/publications/2004/9241546123.pdf. Accessed 11 May 2011.

CHAPTER 17

HPLC-DAD-MS (ESI⁺) Determination of Carotenoids in Fruit

PASQUALE CRUPI[a], VICTOR R. PREEDY[b] AND DONATO ANTONACCI*[a]

[a] CRA - Agricultural Research Council - Research Unit for Viticulture and Enology in Southern Italy, via Casamassima 148 - 70010 Turi (BA), Italy;
[b] Diabetes and Nutritional Sciences, School of Medicine, King's College London, Franklin Wilkins Buildings, 150 Stamford Street, London SE1 9NU, UK
*E-mail: donato.antonacci@entecra.it

17.1 Introduction to Carotenoids

17.1.1 Carotenoids in Foods: Presence and Structural Features

Carotenoids are natural pigments, synthesized in plants, fungi and bacteria. In foods they confer varying yellow to red colorations. Their distinctive characteristic is an extensive conjugated double-bond system, which serves as the light-absorbing chromophore in an inclusive range between 400 and 550 nm. Moreover, the chromophore is responsible for the instability of these compounds. For example, they are easily oxidized when exposed to the air and modified by the presence of mineral acids. Furthermore, light or other chemico-physical agents (*i.e.* temperature and acids) will affect the double

Food and Nutritional Components in Focus No. 1
Vitamin A and Carotenoids: Chemistry, Analysis, Function and Effects
Edited by Victor R Preedy

bonds in the natural *trans* configuration of carotenoids to form several *cis–trans* steroisomers.

Food carotenoids are usually C_{40} tetraterpenoids built from eight C_5 isoprenoid units. The basic linear and symmetrical skeleton of their molecule consists of a central position, with 22 carbon (C) atoms, and two terminal units of 9 C atoms. These terminal units can be acyclic (*i.e.* lycopene) or can be cyclized at one or both ends (*i.e.* α, β-carotene or γ-carotene) (Figure 17.1).

The cyclical terminal units can also be attached to oxygenated functional groups, resulting in a myriad of structures such as alcohols, ketones, epoxides, *etc.*

Traditionally, carotenoids are classified into two main structural groups: (a) carotenes, characterized by being composed of only carbon and hydrogen, and (b) xantophylls, characterized by the presence of different oxygenated functional groups in addition to carbon and hydrogen. Over 100 different carotenoids have been found in fruits and vegetables (Oliver and Palou 2000).

Fruit carotenoids are usually more concentrated in the peel than in the pulp. They generally have bicyclical structures and, according to the kind and the degree of fruit ripening, they can be present as esters with fatty acids. Among the carotenes, β-carotene is the most widespread, either as a minor or as a major constituent (*e.g.*, in apricots, mangos, cherries, carrots and grapes). α-Carotene and γ-carotene are present in lower concentrations, whereas δ-carotene is less frequently encountered. Of the acyclic carotenes, lycopene (the principal pigment of many red-fleshed fruits, such as watermelon and tomato) and ζ-carotene are the most common.

As regards xanthophylls, (all-*trans*)-lutein is the most widespread, whereas lesser amounts of zeaxanthin and epoxyxanthophylls, neoxanthin, violaxanthin, luteoxanthin, lutein-5,6-epoxide can also be present (Figure 17.2).

Of course, the type of carotene or xanthophyll in fruit will depend on the species, but their level is affected by the variety or cultivar, stage of ripening, climate and geographic site of production, agronomic practices and the degree of sunlight exposure. For instance, although fruit ripening is usually

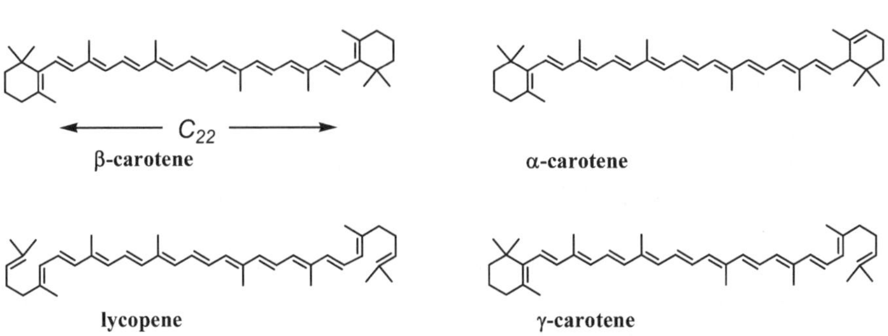

Figure 17.1   Carotenoid structures. The Figure shows the typical polyenic structures of acyclic carotenoids (*i.e.* lycopene), monocyclic carotenoids (*i.e.* γ-carotene) and bicyclic carotenoids (*i.e.* α-carotene and β-carotene).

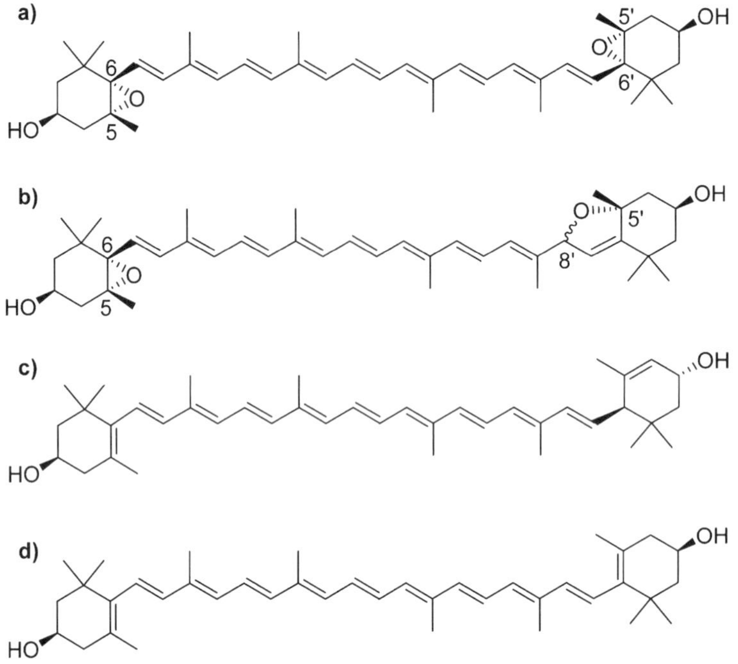

Figure 17.2 Structures of main xanthophylls. The Figure shows the structures of xanthophylls from different classes: (a) 5,6-epoxyxanthophylls (violaxanthin), (b) 5,8-epoxyxanthophylls (luteoxanthin), (c) lutein and (d) zeaxanthin.

accompanied by enhanced carotenogenesis in red pepper or mango (Mercadante *et al* 1998), it is known that in grapes, the level of carotenoids increases up to véraison (the ripening onset) and decreases during fruit maturation (Crupi *et al* 2010a; Razungles *et al* 1988). The degree of irrigation and physicochemical features of the soil also influences the content of carotenoids in grapes (Oliveira *et al* 2006; Steel and Keller 2000).

17.1.2   Biosynthesis of Carotenoids

In plants carotenoids are produced in organelles called plastids. These carotenoids are synthesized through the recently identified 1-deoxy-D-xylulose-5-phosphate (DOXP) pathway. The linear lycopene is produced from DOXP through a series of well-defined biochemical steps. The cyclization of lycopene forms various carotenes having one or two rings. The asymmetric hydroxylation of α-carotene and β-carotene gives rise to lutein and zeaxanthin, respectively (Bramley 2002).

   Zeaxanthin epoxydation forms violaxanthin, which is involved in the "xanthophyll cycle" that is activated only at the end of fruit véraison and allows photosystem protection (Baumes *et al* 2002). The stereoselective

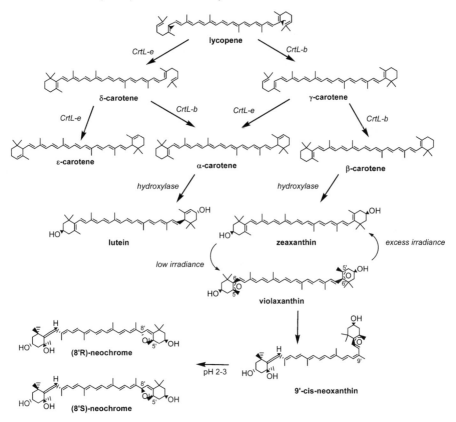

Figure 17.3 Cyclization and hydroxylation reactions of carotenes to form xanthophylls, epoxidation of xanthophylls and isomeric rearrangements of epoxyxanthophylls. The Figure shows the cyclization of lycopene to form monocyclic and bicyclic carotenes and the asymmetric hydroxylation of α-carotene and β-carotene to lutein and zeaxanthin, respectively. Zeaxanthin and its epoxydation product, violaxanthin, are involved in the "xanthophyll cycle". The subsequent violaxanthin transformation to neoxanthin is also shown; under acidic condition this latter 5,6-epoxyxanthophyll can form two neochrome diasteroisomers. Abbreviations: *CrtL-e*, lycopene ε-cyclase; *CrtL-b*, lycopene β-cyclase.

rearrangement of violaxanthin forms (9'-*cis*)-neoxanthin. Under acidic conditions (pH 2–3) a 5,6-epoxyxanthophyll rearrangement can form 5,8-epoxyxanthophyll isomers, such as (8'-R/S)-neochrome (Asai *et al* 2004) (Figure 17.3).

17.1.3 Physiological and Nutritional Properties of Carotenoids

In plants the carotenoids are associated with chlorophyll and absorb light energy at specific wavelengths. They protect the photosynthetic apparatus, dissipating excess light energy, and reduce the reactivity of harmful species such as "singlet" oxygen and the excited (*i.e.* higher energy state) chlorophyll (Frank and Codgell 1993). Carotenoids impart nutritional benefits *via* their function as provitamin A and/or their antioxidant properties (Kotake-Nara *et al* 2001).

17.2   Carotenoid Analysis in Fruit

17.2.1   General Analytic Procedure

The analysis of carotenoids usually includes a series of fundamental steps: sampling and sample preparation, extraction with organic solvents, saponification and washing, evaporation and concentration, chromatographic separation and consequently identification and quantification of specific compounds.

The structural complexity and diversity of the carotenoids with their conformational isomeric forms (Section 17.1.2) implies their analysis is rather complicated. Their characteristic conjugated system of double bonds is the main reason the carotenoids are unstable to light and heat, but also to oxygen and acids. This instability is a problem which should not be underestimated. In their natural environment in plants, carotenoids are relatively well protected, being incorporated in lipoproteins or membranes. However, if they are extracted and transferred into stock solutions, they readily undergo isomerization and degradation. Therefore a number of precautions must be taken during the pre-treatment of samples. These include minimising the exposure of samples to direct light and using antioxidants, such as *tert*-butyl hydroxytoluene (BHT) or *tert*-butyl hydroxyanisole (BHA) in the extraction steps (Oliver and Palou 2000).

A standard procedure for the extraction of carotenoids from fruits does not exist. Rather the extraction of carotenoids generally depends on the food matrix and compound solubility in solvents. For instance, in grape, aprotic organic solvents of middle polarity (*i.e.* acetone or diethyl ether/hexane, 1:1 mixture) are usually preferred to extract both hydrocarbons and oxygenated tetraterpenoids directly from samples which have been previously frozen with liquid nitrogen (-196 °C) and pulverized (Oliveira *et al* 2006). A neutralizing agent (*e.g.*, calcium carbonate or magnesium carbonate dibasic) may be added during the extraction to neutralize acids liberated from the fruit sample itself. This latter step avoids the potential isomerization and rearrangement of 5,6-epoxy- to 5,8-epoxycarotenoids due to the acid medium (Figure 17.3).

It is known that during "fruit ripening" the carotenoids can be partly esterified by fatty acids. In addition, the degree of esterification depends on the number of hydroxyl moieties present in the xanthophylls (Hornero-Mendez and Minguez-Mosquera 2000). Traditionally, after the extraction step, alkaline saponification is carried out in order to hydrolyze the esters to liberate free carotenoids. However, saponification itself can cause the degradation and total or partial loss of carotenoids (Deli *et al* 1996; Minguez-Mosquera and Pérez-

Gálvez 1998). As a consequence, many chromatographic methods for the simultaneous determination of free and esterified carotenoids have been developed (Oliver and Palou 2000).

17.2.2 HPLC Separation of Carotenoids

17.2.2.1 Separation

Because of the potential for automation and ease of use, high-performance liquid chromatography (HPLC) employing various detection systems such as diode array detector (DAD) or mass spectrometry (MS) is currently the method of choice for carotenoid analysis (van Breemen 1997). Both normal- and reversed-phase systems are used, either in isocratic or gradient elution modes. Reversed-phase systems are generally preferred because normal-phase HPLC has several disadvantages, namely, a lower column stability, a poorer reproducibility of retention times and a longer time required for column equilibration (Feltl *et al* 2005).

Depending on the desired information (for example, to determine only provitamin A carotenoids or a complete carotenoid composition including *cis* and *trans* isomers of provitamin A and non-provitamin A carotenoids) a high- or low-resolution chromatographic separation may be carried out. For research purposes, the best separations of various carotenoids have been attained on a C_{30} chemically bonded phase. This stationary phase has the highest separation selectivity and can facilitate the quantification of differing structural and geometrical isomers (Craft 1992). Moreover, the C_{30} stationary phase is capable of resolving geometrical carotenoids in which *cis* bonds are present at the same carbon number but at opposite ends of the molecule (Emenhiser *et al* 1995).

For the mobile phase, ternary mixtures of methanol, *tert*-butyl methyl ether and water containing a small proportion of triethylamine (0.05% v/v) may be used. This protects the carotenoids during chromatographic analysis by minimizing the effects of acidity generated by the free silanol groups present on the silica support, overall allowing higher recoveries from C_{30} columns (Emenhiser *et al* 1995).

17.2.2.2 Detection

The reliability of determining carotenoids in complex mixtures is influenced not only by the separation method, mentioned above, but also by the detection mode. Ultraviolet-visible (UV/Vis) detection is by far the most common because carotenoids absorb strongly in the visible region between 400 and 500 nm (all-*trans* isomers), while *cis*-isomers also exhibit absorption in the near UV region, around 330 nm. Moreover, this detection type is readily compatible with gradient elution. Both DADs and variable wavelength detectors (VWDs), usually operating at 447 nm, are a good compromise, as the absorption maxima of various carotenoids are around this wavelength. DAD- and VWD-based systems have similar detection limits and reproducibility of results. However, DAD can yield more useful spectroscopic information for analyte identification, giving spectra with more substantial data (Crupi *et al* 2010a).

Coupling MS detection to DAD is very useful because it enables the structures of carotenoids to be elucidated on the basis of its molecular mass and fragmentation. Most of the methods applied to carotenoids use the atmospheric pressure ionization (API) mode, even in the absence of protonation sites in carotenoids. Indeed, although ions produced by electrospray ionization (ESI), a typical API technique, are usually preformed in solution by acid–base reactions (*i.e.* $[M+nH]^+$), carotenoid ions are probably formed by a field desorption mechanism (van Breemen 1997). As a result of this unusual ionization process, abundant molecular radical cations, $M^{·+}$, with little fragmentation, are produced. This greatly facilitates molecular weight confirmation and enhances the sensitivity of detection by two orders of magnitude compared to UV/Vis detection (Feltl *et al* 2005).

In particular, because of easy oxidation through loss of an electron, the polyene β-carotene (without hetero-atoms) is normally ionized by a radical process, producing $[M]^{·+}$ prevalently. Conversely, xanthophylls are suitable for both ionization processes, producing molecular and protonated species in ESI-MS (Crupi *et al* 2010b).

17.2.3 DAD and ESI$^+$-MS Identification and Quantification of Carotenoids

The chromatographic behavior and the UV/Vis absorption spectra (namely, the position of absorption maximum, λ_{max}, and the spectral fine structure) provide the first useful information for carotenoid identification. MS analyses, which yield molecular weight and characteristic fragmentation patterns, may then provide final confirmation of individual carotenoid identities when used in conjunction with retention and spectral characteristics (van Breemen 1995). In this sense, it is necessary to take into account the difficulty in identifying carotenoids based only on their UV/Vis spectra, as many carotenoids possess very similar or even identical UV/Vis spectra. The information provided by MS is of great help, as it enables one to differentiate compounds with diverse molecular masses. On the other hand, some carotenoids are characterized by being very similar in their chemical composition, having the same molecular weight and thus making it impossible to correctly identify them (*e.g.*, the isomers) using MS information. In these cases, the information provided by DAD analyses allows their unequivocal identification.

17.2.3.1   Carotenoid Profile in Fruit: Practical Aspects

Here we describe the practical aspects of obtaining a profile of carotenoids in fruits. Separation and identification of carotenoids were carried out using a

HPLC 1100 (Agilent Technologies, USA) equipped with a DAD system and a XCT-trap Plus mass detector (Agilent Technologies, USA) coupled with a pneumatic nebulizer-assisted electrospray liquid chromatography-mass spectrometry (LC-MS) interface. The reversed stationary phase employed was a YMC pack C_{30} (YMC Inc., Wilmington, NC, USA), 5 μm [250 × 3mm internal diameter (i.d.)], with a pre-column C_{30}, 5 μm (20 × 3mm i.d.). The following gradient system was used with H_2O (solvent A), methanol (solvent B) and *tert*-butyl methyl ether (solvent C) in the presence of 0.05% of triethyl amine (TEA) to the three LC mobile phases: 0–2 min, A/B/C, 40:60:0 (%); 5 min, A/B/C, 20:80:0 (%); 10 min, A/B/C, 4:81:15 (%); 60 min, A/B/C, 4:11:85 (%). The flow was maintained at 0.2 mL min^{-1}; the sample injection was 10 μL.

Analyses were stopped after 70 min, and the eluent was brought back to the initial composition within the next 5 min, whereas the equilibration time set to stabilize the column was 10 min. The flow rate and the elution program were controlled by an LC ChemStation 3D software programme (Hewlett-Packard, USA). The detection wavelength for the UV/Vis was set at 447 nm, and spectrophotometric spectra of carotenoids were registered from 250 to 650 nm. The peak width was >0.1 min (2 s) and the slit was 4 nm.

The positive electrospray mode was used for the ionization of molecules with a capillary voltage at -4000 V and a skimmer voltage at 40 V. The nebulizer pressure was 15 psi and the nitrogen flow rate was 5 L min^{-1}. The temperature of the drying gas was 350 °C. In the full scan mode, the monitored mass range was from m/z (mass-to-charge ratio) 100 to 1200. Tandem MS (MS^2) was performed by using helium as the collision gas at a pressure of 4.6 × 10^{-6} mbar. Collision-induced dissociation (CID) spectra were obtained with an isolation width of 4.0 m/z for precursor ions and a fragmentation amplitude of 0.6 V for epoxy-xanthophylls and of 1.0 V for the other carotenoids.

Compound identification was achieved by combining different information as follows: positions of absorption maxima (λ_{max}), the degree of vibrational fine structure (% III/II), the ratio of the absorbance of the *cis* peak to the absorbance of the second absorption band in the visible region, which is known as the Q ratio or D_B/D_{II} (Melendez-Martinez *et al* 2006), the capacity factor values k' and mass spectra. These were compared with those from pure standards and/or interpreted with the help of structural models already hypothesized in the literature.

Quantification of xanthophylls and carotenes was made by using the calibration curves of pure standards, (all-*trans*)-lutein and β-carotene with the regression coefficient (R^2) = 0.9972 and 0.9985, respectively. Quantification was performed as described by Zulueta *et al* (2007). Briefly, the chromatogram was separated into two parts: all of the carotenoids up to k' = 4.78 and including (all-*trans*)-lutein were quantified as such, and the remaining *cis/trans* carotenes isomers were quantified with the β-carotene standard. Finally, the concentrations of individual carotenoids were added together to obtain a total amount.

17.2.3.2   Carotenoid in Fruit: Concentrations and Profiles

Figure 17.4 shows the chromatogram of a non-saponified red grape extract, and typifies the high resolving capacity of a C_{30} column and the need to conduct a multidetector analysis to correctly identify a carotenoid profile in fruit. The observed sequence of chromatographic peaks reflects decreasing polarity of the eluted compounds—more polar xanthophylls are eluted first.

Four isomers of lutein were identified in some typical fruits (apricots, strawberries, peaches and grapes) grown in the Apulia region of Southern Italy, by comparing retention times and UV/Vis spectra with those of reference substances or literature data (Mendes-Pinto *et al* 2004). Moreover, their mass spectrometric behavior with a fragmentation of *m/z* 568.9 to yield dehydrated product ions at *m/z* 550.9 and 532.9 due to the loss of one and two water molecules, respectively (Table 17.1), confirmed these assignments.

The relative intensities of the dehydrated fragment ions differ and thus reflect the structural characteristics of the hydroxylated end groups of isomeric carotenoids, allowing the distinction between lutein and zeaxanthin. Consequently, zeaxanthin showed a base peak at *m/z* 567.9 [M]$^{.+}$ in the MS^1 experiment with the protonated molecular ion at 568.9 [M+H]$^+$ and a dehydrated product ion at *m/z* 550.9 of low abundance. Zeaxanthin was not detected in strawberry samples (Table 17.2), In contrast, lutein, possessing a 3-hydroxy-ε end group (Figure 17.2), showed an [M+H-H$_2$O]$^+$ ion at *m/z* 550.9 as the most abundant signal and another product ion at *m/z* 532.9 ([M+H-

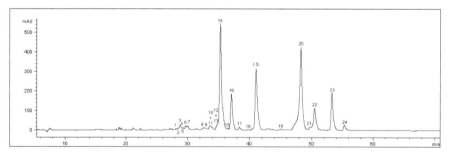

Figure 17.4   HPLC-DAD profile of carotenoids in mature red grape. The Figure shows the chromatographic separation of carotenoids based on the following conditions. Column, YMC pack C30, 5µm, 250 × 3.0 mm i.d.; detection at 447 nm; flow rate, 0.2 mL min^{-1}; ternary gradient elution system of water/methanol/*tert*-butyl methyl ether; injected volume, 10 µL. Peaks: (1) violaxanthin; (2) (8′*R*)-neochrome; (3) (9′-*cis*)-neoxanthin; (4) (8′*S*)-neochrome; (5) lutein-5,6-epoxide; (6) luteoxanthin; (7) unknown; (8) (8′*S*)-auroxanthin; (9) flavoxanthin; (10) chrysantemaxanthin; (11) lutein-like structure; (12) lutein-like structure; (13) (8′*S*)-auroxanthin; (14) (all-*trans*)-lutein; (15) lutein-like structure; (16) zeaxanthin; (17) (9Z) or (9′Z)-lutein; (18) (9Z) or (9′-*cis*)-lutein; (19) pheophytin b-like structure; (20) pheophytin b; (21) pheophytin a-like structure; (22) pheophytin a; (23) β-carotene; (24) (9-*cis*)-β-carotene. I.S., internal standard, β-apo-8′-carotenal.

Table 17.1 HPLC-DAD-MS (ESI⁺) characteristics of carotenoids in fruits. This Table groups the chromatographic (k'), spectroscopic (λ_{max}; %III/II; D_B/D_{II}), and spectrometric ([M]⁺; [M+H]⁺; MS² fragments) characteristics, useful to identify carotenoids isomers.

Compound	k'	λ_{max} (nm)	%(III/II)[a]	D_B/D_{II}[b]	[M+H]⁺ (m/z)	[M]⁺ (m/z)	MS² product ions (m/z)
Violaxanthin[c,d]	3.15	416; 439; 469	86	—	601.5	600.1	583.5, 565.5, 509.5, 491.5, 221.1
(8'R)-Neochrome[e,f]	3.19	400; 422; 450	89	—	601.5	600.1	583.2, 565.3, 509.5, 221.1
(9'-cis)-Neoxanthin[c,d]	3.21	414; 436; 464	86	—	601.5	600.1	583.2, 565.3, 509.5, 221.1
(8'S)-Neochrome[e,f]	3.30	400; 422; 450	89	—	601.5	600.1	583.2, 565.3, 509.5, 221.1
Lutein-5,6-epoxide[c,d]	3.35	416; 439; 468	88	—	585.4	584.2	567.1, 492.3, 244.9
Luteoxanthin[e,f]	3.47	399; 422; 448	95	—	601.5	600.1	583.2, 221.1
(8'R)-Auroxanthin[e,f]	3.72	380; 402; 426	98	—	601.5	600.1	583.5, 565.5, 509.5, 491.5, 221.1
Flavoxanthin[e,f]	3.79	398; 422; 448	112	—	585.4	584.2	567.1, 492.3, 244.9
Chrysanthemaxanthin[e,f]	3.89	398; 422; 448	108	—	585.4	584.2	567.1, 492.3, 244.9
(8'S)-Auroxanthin[e,f]	4.00	380; 402; 426	98	—	601.5	600.1	583.5, 565.5, 509.5, 491.5, 221.1
(all-trans)-Lutein[c,d]	4.05	(422); 445; 472	42	—	568.9	567.9	550.9, 532.9, 476.4, 429.4
Zeaxanthin[c,d]	4.31	(425); 450; 475	22	—	568.9	567.9	550.9, 532.9, 476.4, 429.4
(9-cis) or (9'-cis)-Lutein[g]	4.47	330; (418); 440; 468	52	0.075	568.9	567.9	550.9, 532.9, 476.4, 429.4
β-Cryptoxanthin	4.78	(423); 450; 473	25	—	553.1	552.2	535.2, 460.9
α-Carotene	6.51	(420); 443; 470	40	—	536.9	535.9	444.2, 430.3, 399.3
β-Carotene[c,d]	6.67	(428); 452; 478	25	—	536.9	535.9	444.2, 430.3, 399.3
(9-cis)-β-Carotene[c,d]	6.94	342; (424); 446; 474	17	0.03	536.9	535.9	444.2, 430.3, 399.3

[a]%III/II represents the fine vibrational structure. [b]Q ratio, typical of cis isomers. [c]Identification by comparison with UV-Vis spectrum of the standard compound. [d]Identification by comparison with MS spectrum of the standard compound. [e]Identification by comparison with UV-Vis spectrum of the "parent" standard compound obtained by acidification of the respective 5,6-epoxyxanthophyll. [f]Identification by comparison with MS spectrum of the "parent" standard compound obtained by acidification of the respective 5,6-epoxyxanthophyll. [g]Identification by LC-MS is consistent with the results of Mendes-Pinto et al (2004).

Table 17.2 Carotenoids content of typical fruits [expressed in µg (100g of fresh weight)$^{-1}$] from the Apulia region of Southern Italy. n.d., not detected; tr, trace.

Compound	Red grape $m^a \pm s^b$	White grape $m \pm s$	Apricot $m \pm s$	Peach $m \pm s$	Strawberry $m \pm s$	Cherry $m \pm s$
Violaxanthin[c]	3.0 ± 0.7	2.3 ± 0.5	85 ± 13	105 ± 18	n.d.	n.d.
(8'R)-Neochrome[c]	2.6 ± 0.5	2.9 ± 0.7	44 ± 10	15 ± 4	n.d.	n.d.
(9'cis)-Neoxanthin[c]	4.1 ± 0.9	3.7 ± 0.9	77 ± 11	85 ± 19	n.d.	n.d.
(8'S)-Neochrome[c]	2.6 ± 0.4	tr	25 ± 9	18 ± 8	n.d.	n.d.
Lutein-5,6-epoxide[c]	8.9 ± 0.7	4.4 ± 0.5	n.d.	n.d.	n.d.	n.d.
Luteoxanthin[c]	7.1 ± 1.0	1.3 ± 0.4	n.d.	Tr	n.d.	n.d.
(8'R)-Auroxanthin[c]	1.8 ± 0.4	2.9 ± 0.6	n.d.	Tr	n.d.	n.d.
Flavoxanthin[c]	2.1 ± 0.5	2.2 ± 0.4	n.d.	n.d.	n.d.	n.d.
Chrysanthemaxanthin[c]	2.7 ± 0.7	2.1 ± 0.4	n.d.	n.d.	n.d.	n.d.
(8'S)-Auroxanthin[c]	2.9 ± 0.6	2.3 ± 0.6	n.d.	Tr	n.d.	n.d.
(all-trans)-Lutein	48.1 ± 1.3	55.6 ± 1.8	105 ± 18	98 ± 19	8 ± 0.9	80 ± 20
Zeaxanthin[c]	4.4 ± 0.9	18.7 ± 1.6	88 ± 9	170 ± 20	n.d.	110 ± 18
(9-cis)-Lutein[c]	5.0 ± 0.8	7.7 ± 1.7	51 ± 8	11 ± 2	tr	9.1 ± 1.1
β-Cryptoxanthin[c]	n.d.	n.d.	210 ± 20	150 ± 18	n.d.	n.d.
α-Carotene[d]	n.d.	n.d.	36 ± 5	n.d.	n.d.	130 ± 17
β-Carotene[d]	140 ± 40	110 ± 60	1580 ± 120	160 ± 20	28 ± 7	610 ± 30
(9-cis)-β-Carotene[d]	18 ± 5	18 ± 7	350 ± 30	22 ± 6	tr	n.d.
Total[e]	250 ± 50	230 ± 80	2600 ± 200	830 ± 130	36 ± 8	940 ± 90

[a]Means of three replicates. [b]Standard deviations, $n = 3$. [c]Expressed as lutein equivalents. [d]Expressed as β-carotene equivalents. [e]Sum of identified carotenoids.

$2H_2O]^+$) in the MS^1 experiment (Table 17.1), which was hardly observed for zeaxanthin (Crupi *et al* 2010b).

However, mass spectrometric patterns alone were insufficient to identify the others isomers of lutein because they exhibited a similar MS^1 fragmentation pattern as lutein, and the CID of their $[M]^{\cdot+}$ ions all resulted in dehydrated radical ions at m/z 550.5 $[M-H_2O]^{\cdot+}$ and m/z 476.4 $[M-C_7H_8]^{\cdot+}$, originating probably from in-chain fragmentation by the loss of toluene (Clarke *et al* 1996), and m/z 429.4 which corresponded to loss of the terminal ring (Table 17.1) (Maoka *et al* 2002). Therefore, the fine structure of the UV/Vis spectra was used for the differentiation of the geometrical isomers of lutein. The (all-*trans*)-lutein showed an absorption spectrum, with λ_{max} at 445 and 472 nm with a shoulder at 422 nm and less fine structure (%III/II = 42), that was typical of a carotenoid with ten conjugated double bonds, nine in the polyene chain and one in a β-ring, whilst (9-*cis*)- or (9'-*cis*)-lutein were tentatively identified on the basis of the low intensity of their *cis* peaks (D_B/D_{II} = 0.075 and 0.067 respectively), and the hypsochromic shift of their absorption maxima [~4 nm with respect to the (all-*trans*)-lutein] (Table 17.1).

Also, to distinguish structural and geometrical isomers of carotene, *i.e.* α-carotene (only identified in apricot and cherry samples, Table 17.2), β-carotene and (9-*cis*)-β-carotene, which were characterized by the same MS^1 spectrum with a molecular ion at m/z 535.9 $[M]^{\cdot+}$ and a small fragment at m/z 444.2 $[M-92]^{\cdot+}$ due to the loss of toluene (Table 17.1), their spectroscopic properties were utilized. Indeed, the absorption spectrum of β-carotene showed the λ_{max} at 452 and 478 nm, and a shoulder at 428 nm, whereas the spectrum of (9-*cis*)-β-carotene showed a hypsochromic shift of absorption maxima (~4 nm with respect to the β-carotene) and a low *cis* peak (D_B/D_{II} = 0.03) at 342 nm. Conversely, the absorption spectrum of α-carotene resembled that of lutein, having the same chromophore (Table 17.1).

The 5,6-epoxy-xanthophylls violaxanthin, (9'-*cis*)-neoxanthin and lutein-5,6-epoxide and 5,8-epoxy-xanthophylls diastereoisomers (*R/S*)-neochrome, (*R/S*)-auroxanthin, luteoxanthin, flavoxanthin and chrysanthemaxanthin, particularly present in grapes, were also identified (Table 17.1). It is worth noting that an acid matrix, such as must (pH ~ 3.0–3.5), can promote the epoxide–furanoid rearrangement (Figure 17.3). Indeed, from the spectro-photometric behavior of a model solution of the three 5,6-epoxy-xanthophylls (3 μg mL⁻¹ in ether/hexane, 1:1) added with 500 μL of a tartaric acid solution (15 g L⁻¹), to resemble a "green berries" must condition, it can be shown that after 180 min (that is the time needed for carotenoids to be extracted from grapes), the corresponding 5,8-epoxy-xanthophylls are completely formed (Figure 17.5). Therefore, to avoid interpreting the 5,8-epoxy-xanthophylls as artifact compounds, special care was taken to neutralize the extraction solution of carotenoids from grapes, by using magnesium carbonate basic $[(MgCO_3)*4Mg(OH)_2*5H_2O]$.

The mass spectra of 5,6-epoxy-xanthophylls and corresponding 5,8-epoxy-xanthophylls were characterized by the same pseudo-molecular ions, $[M]^{\cdot+}$ and

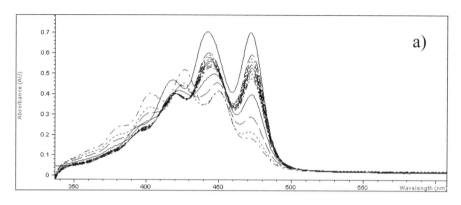

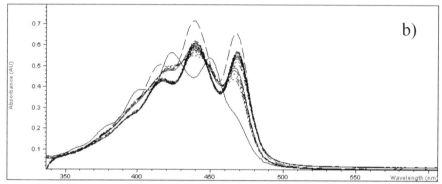

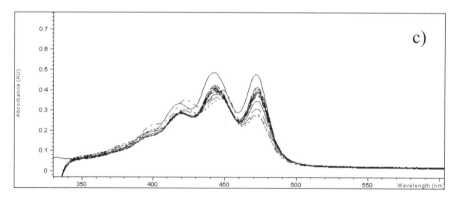

Figure 17.5   UV/Vis spectra of the three 5,6-epoxy-xanthophylls (3 µg mL^{-1} in ether/
hexane, 1:1) added with 500 µL of a tartaric acid solution (15 g L^{-1}).
The Figure shows the acid catalyzed epoxide rearrangement of 5,6-
epoxyxanthophylls to 5,8-epoxyxanthophylls and, in particular, (a)
violaxanthin to give luteoxanthin and subsequently (*R/S*)-auroxanthin;
(b) (9'-*cis*)-neoxanthin to give (*R/S*)-neochrome; (c) lutein-5,6-epoxyde
to give flavoxanthin and chrysanthemaxanthin.

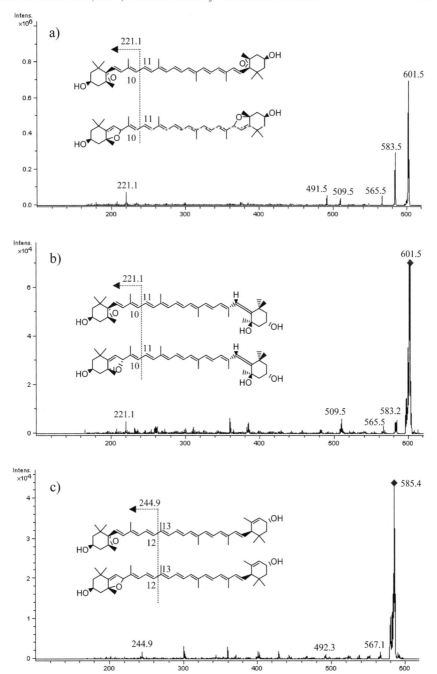

Figure 17.6 CID MS² spectra of [M+H]⁺ of 5,6- and 5,8-epoxyxanthophylls. The Figure shows the similarity of MS fragmentation pattern of (a) violaxanthin and auroxanthin; (b) (9′-*cis*)-neoxanthin and neochrome; (c) lutein-5,6-epoxyde and flavoxanthin, respectively.

[M+H]$^+$, and fragmentation pattern (Table 17.1, Figure 17.6). The regio-isomers, violaxanthin and auroxanthins spectra showed the signals at *m/z* 583.5 [M+H-18]$^+$ (elimination of water) and *m/z* 221.1, resulting from the cleavage of the C10–C11 bond in the polyene chain (Maoka *et al* 2004), from the epoxy end group, as the main product ions (Figure 17.6a). Moreover, CID MS^2 spectra of [M+H]$^+$ ions of the isomers, lutein-5,6-epoxide, and flavoxanthin and chrysanthemaxanthin (Figure 17.6c) showed the major product ions at *m/z* 567.1 (water loss), and *m/z* 492.3 due to the elimination of a neutral molecule of toluene from the polyene chain. In addition, they also exhibited a diagnostic ion at *m/z* 244.9, due to the cleavage of the C12–C13 bond in the polyene chain, indicating the presence of an epoxy substituent in a β-ring with a hydroxyl group (Maoka *et al* 2004). Finally, the presence of a [M+H]$^+$ ion at *m/z* 601.5 and [M]$^{·+}$ molecular ion at *m/z* 600.1 in mass spectra of some compounds (Table 17.1) were indicative of the epoxy-xanthophylls isomers, (8′-*R*)-neochrome, (9′-*cis*)-neoxanthin and (8′-*S*)-neochrome, respectively (Figure 17.6b).

UV/Vis spectra were used to confirm the difference between 5,6- and 5,8-epoxy-xanthophylls regio-isomeric compounds. Because of the presence of one 5,8-furanoid group, the (*R/S*)-neochrome, luteoxanthin, flavoxanthin and chrysanthemaxanthin showed hypsochromic shifts of ∼15–20 nm as compared to (9′-*cis*)-neoxanthin, violaxanthin and lutein-5,6-epoxide, respectively. On the contrary, the presence of two 5,8-furanoid groups in the (*R/S*)-auroxanthin provoked higher hypsochromic shifts of ∼40 nm as compared to violaxanthin (Table 17.1, Figure 17.5).

β-Cryptoxanthin, (essentially the mono-hydroxylated β-carotene) was identified in apricot and peach but not grapes, strawberries or cherries (Table 17.2). Its UV/Vis spectrum was consistent with a carotenoid with 11 conjugated double bonds, nine in the polyene chain and two in β-rings, the λ$_{max}$ being at longer wavelengths (450 and 475 nm with shoulder at 423 nm) and much less definition of the spectral structure (% III/II = 25) compared to those of lutein. The mass spectrum was characterized by a protonated molecular ion [M+H]$^+$ at *m/z* 553.1 and a pseudomolecular ion [M]$^{·+}$ at *m/z* 552.2, and MS^2 fragments at *m/z* 535.2 [M+H-H$_2$O]$^+$ and at *m/z* 460.9 [M+H-C$_7$H$_8$]$^+$, corresponding to the loss of water and toluene, respectively (Azevedo-Meleiro and Rodriguez-Amaya 2004).

Mean values of carotenoid concentrations [expressed in μg (100 g of fresh weight)$^{-1}$] in some typical fruits grown in the Apulia region of Southern Italy (grapes, apricots, peaches, strawberries and cherries) are reported in Table 2.

The analyzed samples showed generally higher (all-*trans*)-lutein and β-carotene contents compared to other compounds. In apricots, peaches and cherries, consistent levels of zeaxanthin and β-cryptoxanthin were also found, in agreement with previous studies (Ben-Amotz and Fishler 1998; Hart and Scott 1995). It is worth noting that in grapes β-carotene concentrations were approx. 2–3-fold higher than lutein, which could be due to the viticultural region (Apulia). This contrasts to data derived from other studies on grapes from other geographical regions in which the xanthophyll levels are higher than the carotenes (Crupi *et al* 2010b).

Lesser amounts of 5,6- and 5,8-xanthophylls were also determined in the fruits, although the concentrations of violaxanthin, neoxanthin and neochrome were 10–50-fold higher in apricots and peaches than in grapes (Table 17.2).

Summary Points

- This Chapter focuses on high-performance liquid chromatography-diode array detection-mass spectrometry with electrospray ionization (HPLC-DAD-MS; ESI⁺) analysis of carotenoids in fruit.
- Carotenoids are pigments synthesized in plastids through the 1-deoxy-D-xylulose-5-phosphate (DOXP) pathway.
- Carotenoids have important nutritional function as provitamin A and antioxidants.
- Simultaneous determination of structural and geometrical isomers of carotenes and xanthophylls is possible using hyphenated analytical techniques (HPLC-DAD-MS).
- However, the isolation of standards from natural sources must be encouraged for accurate carotenoid identification and quantification.
- Several carotenes (α-carotene, β-carotene, 9-*cis*-β-carotene) and xanthophylls ((all-*trans*)-lutein, zeaxanthin, 9-*cis*-lutein, β-cryptoxanthin) were identified and quantified in different fruits including grapes, apricots, peaches, strawberries and cherries.
- 5,6-Epoxyxanthophylls (violaxanthin, 9'-*cis*-neoxanthin, lutein-5,6-epoxide) and 5,8-epoxyxanthophylls [(*R/S*)-neochrome, luteoxanthin, (*R/S*)-auroxanthin, flavoxanthin, chrysanthemaxanthin] were also identified, particularly in grapes.
- Higher amounts of β-carotene and (all-*trans*)-lutein were found, especially in apricots and peaches.

Key Facts

Key Facts of HPLC-DAD

This list describes the basic concepts of liquid chromatographic separation, including the function of the stationary phase and the mobile phase, and the most commonly used HPLC detector.

- High-performance liquid chromatography (HPLC) is a type of chromatographic separation in which the sample is dissolved in a liquid mobile phase which continually fluxes through an immiscibile stationary phase contained in a column.

- The most popular column material is reversed phase, in which the stationary phase is non-polar (or less polar than in the mobile phase) and the analytes are retained until eluted with a sufficiently polar solvent (in the case of an isocratic mobile phase) or solvent mixture (in the case of a mobile phase gradient).
- The analytes are distributed in different proportions between the mobile phase and stationary phase: the molecules mostly retained by the stationary phase stir more slowly with the flow of the mobile phase, conversely those less retained stir more quickly.
- The analytes separate in discrete peaks and can be qualitatively and/or quantitatively analyzed by suitable detectors.
- Diode array detectors (DADs) respond to a specific property of the analyte (*i.e.* UV absorbance) and allow on-line spectra acquisition to be obtained. It is used to measure a wide range of wavelengths (generally from 190 to 950 nm) using a twin-lamp (tungsten-deuterium) design.
- The array consists of 1024 diodes, each of which measures a different narrow-band spectrum. Measuring the variation in light intensity over the entire wavelength range yields an absorption spectrum.

Key Facts of ESI-MS

This list describes the basic concepts of mass spectrometry, focusing on the interface between liquid chromatographic separation of analytes and MS detection.

- Mass spectrometry (MS) is an advanced analytical technique that characterizes complex mixtures of analytes according to the detection and separation of ions generated from them. This allows both qualitative and quantitative profiles to be obtained.
- The *mass spectrum* shows the mass of the molecule and the masses of pieces from it. In the bar-graph from a spectrum, the abscissa indicates the mass (actually the *m/z*, the ratio of mass to the number of charges of the ions employed), and the ordinate indicates the relative intensity.
- A mass spectrometer consists of an ion source, *i.e.* electrospray ionization (ESI), generally interfaced with a separation system (*i.e.* HPLC), that converts the analytes into gas ions. A mass analyzer (*i.e.* quadrupole, ion trap, *etc.*) separates the ions based on their *m/z* ratio, and a detector converts the ionic beam into an electronic signal. All these instrumental parts are inserted in a high vacuum system (10^{-8} Torr).
- In electrospray ionization (ESI), the effluent is directed through a nebulizing needle from the HPLC system into a high-voltage field where charged droplets are formed. The charged droplets are then dried and the analyte ions are desorbed and, finally, transported to the mass analyzer through a series of vacuums and ion-focusing elements.

Definition of Words and Terms

Absorption maximum: This is the wavelength (λ_{max}) which corresponds to the maximum absorbance in the UV/Vis spectrum of the analyte.

Capacity factor (k'): This is an important parameter often used to monitor the migration rate of an analyte through a column.

Carotenes: C_{40} compounds characterized by an extensive conjugated double-bond system, comprised only carbon and hydrogen atoms.

Collision-induced dissociation (CID): The most common method of fragmentation in which the precursor ion is dissociated in a collision chamber containing a high pressure of inert gas (helium), before the mass spectrum is acquired.

Epoxyxanthophylls: A type of xanthophylls having one or two epoxy groups at the 5,6- or 5,8- position.

HPLC-DAD-MS: High-performance liquid chromatography-diode array detection-mass spectrometry with electrospray ionization. A very powerful multi-stage analytical technique, that allows sensitive separation, identification (*via* the HPLC platform) and quantification of analytes (*via* the DAD-MS platforms).

Isomers: Compounds with the same molecular formula, that is the same type and number of atoms, but different structures. They are classified in two main groups: (1) structural isomers, in which the atoms are joined together in different ways, and (2) stereoisomers, in which only the geometrical positioning of atoms around a double bond (*i.e. cis–trans* isomers) or the spatial disposition of atoms (*i.e.* diastero-isomers) differ.

Molecular ion: $[M]^{+}$, the *m/z* signal on the mass spectrum that corresponds to the molecular weight of the analyte.

Molecular radical cation: $[M]^{\cdot+}$, corresponds to a molecular ion with an odd electron that is probably formed by a field desorption mechanism.

Protonated molecular ion: $[M+H]^{+}$, is the hydrogenated molecular ion usually preformed in solution by acid–base reactions.

Xanthophyll cycle: A cycle that allows photosystem protection through the dissipation of excess light excitation energy in plants. The cycle can be reversed. It can adapt to different light conditions.

Xanthophylls: C_{40} compounds characterized by an extensive conjugated double-bond system and by the presence of different oxygenated groups (ethers, ketones, epoxides, *etc.*).

List of Abbreviations

API	atmospheric pressure ionization
CID	collision-induced dissociation
DAD	diode array detector
DOXP	1-deoxy-D-xylulose-5-phosphate
ESI	electrospray ionization

HPLC   high-performance liquid chromatography
HPLC-DAD
-MS-ESI   high-performance liquid chromatography-diode array detec-
                    tion-mass spectrometry with electrospray ionization
i.d.   internal diameter
LC-MS   liquid chromatography-mass spectrometry
$[M]^+$ molecular radical cation
$[M+H]^+$ protonated molecular ion
MS   mass spectrometry
MS^2 tandem mass spectrometry
m/z   mass-to-charge ratio
UV/Vis   ultraviolet-visible
VWD   variable wavelength detector
λ_{max} maximum absorption wavelength

References

Asai, A., Terasaki, M. and Nagao, A., 2004. An epoxide–furanoid rearrange-
    ment of spinach neoxanthin occurs in gastrointestinal tract of mice and *in
    vitro*: formation and cytostatic activity of neochrome stereoisomers.
    Journal of Nutrition. 134: 2237–2243.

Azevedo-Meleiro, C. H. and Rodriguez-Amaya, D. B., 2004. Confirmation of
    the identity of the carotenoids of tropical fruits by HPLC-DAD and
    HPLC-MS. Journal of Food Composition and Analysis. 17: 385–396.

Baumes, R., Wirth, J., Bureau, S., Gunata, Y. and Razungles, A., 2002.
    Biogeneration of C_{13}-norisoprenoid compounds: experiments supportive
    for an apo-carotenoid pathway in grapevines. Analityca Chimica Acta.
    458: 3–14.

Ben-Amotz, A. and Fishler, R., 1998. Analysis of carotenoids with emphasis
    on 9-*cis*-β-carotene in vegetables and fruits commonly consumed in Israel.
    Food Chemistry. 62: 515–520.

Bramley, P. M. 2002. Regulation of carotenoid formation during tomato fruit
    ripening and development. Journal of Experimental Botany. 53: 2107–
    2113.

Clarke, P. A., Barnes, K. A., Startin, J. R., Ibe, F. I. and Shepherd, M. J.,
    1996. High performance liquid chromatography/atmospheric pressure
    chemical ionization-mass spectrometry for the determination of carote-
    noids. Rapid Communications in Mass Spectrometry. 10: 1781–1785.

Craft, N. E., 1992. Carotenoid reversed-phase high-performance liquid
    chromatography methods: reference compendium. Methods in
    Enzymology. 213: 185–205.

Crupi, P., Coletta, A., Milella, R. A., Palmisano, G., Baiano, A., La Notte, E.
    and Antonacci, D., 2010a. Carotenoid and chlorophyll derived compounds
    in some wine grapes grown in Apulian region. Journal of Food Science. 75:
    S191–S198.

Crupi, P., Milella, R. A. and Antonacci, D., 2010b. Simultaneous HPLC-DAD-MS (ESI+) determination of structural and geometrical isomers of carotenoids in mature grapes. Journal of Mass Spectrometry. 45: 971–980.

Deli, J., Matus, Z. and Tóth, G., 1996. Carotenoid composition in the fruits of *Capsicum annuum* Cv. Szentesi Kosszarvú during ripening. Journal of Agricultural and Food Chemistry. 44: 711–716.

Emenhiser, C., Sander, L. C. and Schwartz, S. J., 1995. Capability of a polymeric C30 stationary phase to resolve *cis–trans* carotenoid isomers in reversed phase liquid chromatography. Journal of Chromatography A. 707: 205–216.

Frank, H. and Codgell, R. J., 1993. Photochemestry and function of carotenoids in photosynthesis. In (Young, A. and Britton, G., ed.) Carotenoids in Photosynthesis. Chapman and Hall, London, UK

Feltl, L., Pacáková, V., Štulík, K. and Volka, K., 2005. Reliability of carotenoid analyses: a review. Current Analytical Chemistry 1: 93–102.

Hart, D. J. and Scott, K. J., 1995. Development and evaluation of an HPLC method for the analysis of carotenoids in foods, and the measurement of the carotenoid content of vegetables and fruits commonly consumed in the UK. Food Chemistry. 54: 101–111.

Hornero-Mendez, D. and Minguez-Mosquera, M. I., 2000. Xanthophyll esterification accompanying carotenoid overaccumulation in chromoplast of *Caspicum annum* ripening fruits is a constitutive process and useful for ripeness index. Journal of Agricultural and Food Chemistry. 48: 1617–1622.

Kotake-Nara, E., Kushiro, M., Zhang, H., Sugawara, T., Miyashita, K. and Nagao, A., 2001. Carotenoids affect proliferation of human prostate cancer cells. Journal of Nutrition. 131: 3303–3306.

Maoka, T., Fujiwara, Y., Hashimoto, K. and Akimoto, N., 2002. Rapid identification of carotenoids in a combination of liquid chromatography/ UV-visible absorption spectrometry by photodiode-array detector and atmospheric pressure chemical ionization mass spectrometry (LC/PAD/ APCI-MS). Journal of Oleo Science. 51: 1–9.

Maoka, T., Fujiwara, Y., Hashimoto, K. and Akimoto, N., 2004. Characterization of epoxy carotenoids by fast atom bombardment collision-induced dissociation MS/MS. Lipids. 39: 179–183.

Melendez-Martinez, A. J., Britton, G., Vicario, I. M. and Heredia, F. J., 2006. HPLC analysis of geometrical isomers of lutein epoxide isolated from dandelion (*Taraxacum officinale* F.Weber ex Wiggers). Phytochemistry. 67 : 771–777.

Mendes-Pinto, M. M., Silva Ferreira, A. C., Oliveira, M. B. and De Pinho, P. G., 2004. Evaluation of some carotenoids in grapes by reversed- and normal-phase liquid chromatography: a qualitative analysis. Journal of Agricultural and Food Chemistry. 52: 3182–3188.

Mercadante, A. Z. and Rodriguez-Amaya, D. B., 1998. Effects of ripening, cultivar differences and processing on the carotenoid composition of Mango. Journal of Agricultural and Food Chemistry. 46: 128–130.

Minguez-Mosquera, M. I. and Pérez-Gálvez, A. 1998. Study of lability and kinetics of the main carotenoid pigments of red pepper in the de-esterification reaction. Journal of Agricultural and Food Chemistry. 46: 566–569.

Oliver, J. and Palou, A., 2000. Chromatografic determination of carotenoids in foods. Journal of Chromatography A. 881: 543–555.

Oliveira, C., Barbosa, A., Silva Ferreira, A. C., Guerra, J. and De Pinho, P. G., 2006. Carotenoid profile in grapes related to aromatic compounds in wines from Douro region. Journal of Food Science. 71: S1–S7.

Razungles, A., Bayonove, C. L., Cordonnier, R. E. and Sapis, J. C., 1988. Grape carotenoides: changes during the maturation period and localisation in mature berries. American Journal of Enology and Viticulture. 39: 44–48.

Steel, C. C. and Keller, M., 2000. Influence of UV-B irradiation on the carotenoid content of *Vitis vinifera* tissues. Biochemical Society Transactions. 28: 883–885.

van Breemen, R. B., 1995. Electrospray liquid chromatography–mass spectrometry of carotenoids. Analytical Chemistry. 67: 2004–2009.

van Breemen, R. B., 1997. Liquid chromatography/mass spectrometry of carotenoids. Pure and Applied Chemistry. 69: 2061–2066.

Zulueta, A., Esteve, M. J. and Frigola, A., 2007. Carotenoids and color of fruit juice and milk beverage mixtures. Journal of Food Science. 72: C457–C463.

CHAPTER 18

Thin-layer Chromatographic Analysis of Pro-vitamin A Carotenoids

ALAM ZEB

Department of Biotechnology, University of Malakand, Chakdara, Pakistan
E-mail: Alamzeb01@yahoo.com

18.1 Introduction

The name pro-vitamin A carotenoids is derived from the fact that they produce one or two molecules of vitamin A upon hydrolysis, oxidation or digestion in the body. Vitamin A is present in our daily diet in the form of retinyl ester, retinol, retinal, 3-dehydroretinol, and retinoic acid. It can be obtained from animal liver, milk and milk products, fish, and meat or in the form of carotenoids from plant foods, which can be biologically transformed into vitamin A (pro-vitamin A). The potential toxicity from the overdose of vitamin A is limited using pro-vitamin A, because it is converted into vitamin A only when required by the body. Many factors such as the amount, type, and physical form of the carotenoids, intake of fat, vitamin E, and fiber present in diet, proteins and zinc status and the presence of certain chronic diseases can influence the absorption and utilization of pro-vitamin A and ultimately the bioavailability.

The pro-vitamin A carotenoids include the most commonly available carotenoids such as α- and β-carotene, α- and β-cryptoxanthin, β-zeacarotene, and 8-apo-β-carotenal (Figure 18.1). Of the total of 700 carotenoids, approx. 50 carotenoids have been found to have pro-vitamin A activity (Rodriguez-

Food and Nutritional Components in Focus No. 1
Vitamin A and Carotenoids: Chemistry, Analysis, Function and Effects
Edited by Victor R Preedy
© The Royal Society of Chemistry 2012
Published by the Royal Society of Chemistry, www.rsc.org

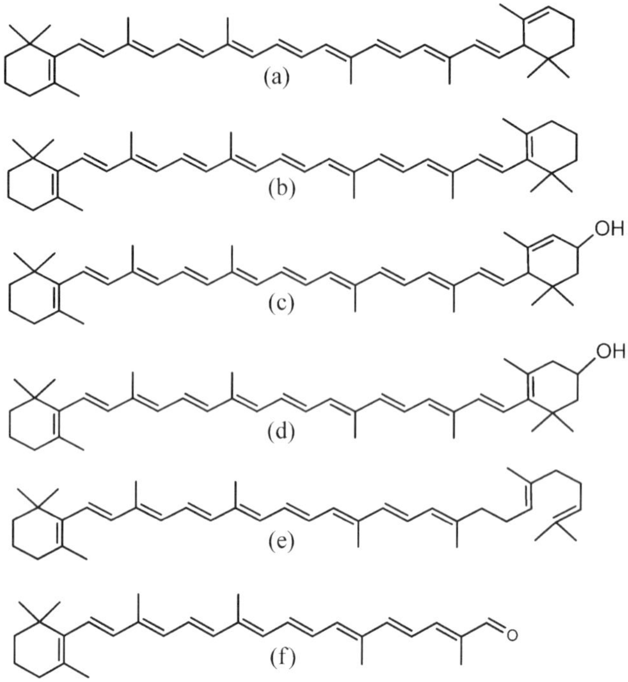

Figure 18.1   Structure of common pro-vitamin A carotenoids: (a) α-carotene, (b) β-carotene, (c) α-cryptoxanthin, (d) β-cryptoxanthin, (e) β-zeacarotene, and (f) 8-apo-β-carotenal.

Amaya 1997). The relative bio-potencies of only a few of these carotenoids have been estimated using *in vivo* studies. Among carotenoids, β-carotene is considered the most important in terms of bioactivity and widespread nature. Chemically, one-half of β-carotene is equivalent to vitamin A, thus it is a potent pro-vitamin A compound to which 100 % activity is assigned. As a general rule, the presence of an unsubstituted β-ring and a minimum of 11 carbon polyene chain are the minimum requirement for pro-vitamin A activity (Britton 1995). β-Carotene and β-cryptoxanthin are the major carotenoids of many fruits such as peach, nectarine, orange-fleshed papaya, persimmon, and tomato. Carrot and some varieties of squash and pumpkin, red palm, and the Brazilian palm fruit buriti (*Mauritia vinifera*) are good sources of α-carotene (Rodriguez-Amaya *et al* 2008). Red palm oil and sea buckthorn seed oils are the richest source of pro-vitamin A in terms of β-carotene (Zeb and Mehmood 2004; Zeb and Murkovic 2010a). In addition to their pro-vitamin A activity, these carotenoids can also perform several other important functions in human beings. Because of their important biological role, pro-vitamin A carotenoids are extracted from their sources and analyzed using different separation techniques. Some of the recent review articles (Marston 2007; Oliver and Palou 2000) describe in detail the uses of all the chromatographic techniques for the analyses of pro-vitamin A carotenoids or related compounds. This chapter, however, focuses on the analysis of pro-vitamin A carotenoids using thin-layer chromatography (TLC) from biological samples.

18.2 TLC Analysis

TLC is widely used separation technique in food and pharmaceuticals, ranging from the study of composition, adulteration, contaminants, degraded and decomposed products of carbohydrates, proteins, lipids, and vitamins. It is an efficient technique in monitoring the progress of chemical tests for the identification purposes as well as for the quantification of carotenoids in biological samples (Evans *et al* 2004; Zeb and Murkovic 2010b). Since TLC is fast, effective and relatively cheap, it is often primarily used for the separation and isolation of individual classes of molecules. We have recently reviewed (Zeb and Murkovic 2010a) the uses of TLC in the analysis of carotenoids from the animal and plant sources. The uses of stationary and mobile phase have been described. TLC has been found to be a more useful technique for the separation and quantification of β-carotene during thermal oxidation of edible oils (Zeb and Murkovic 2010b). TLC involves the uses of two important components, *i.e.* the stationary phase and the mobile phase, which play significant role in the pro-vitamin A carotenoids determination from the biological materials.

18.2.1 Stationary Phases

Most of the earlier TLC methods used a silica stationary phase and a non-polar mobile phase for the separation of carotenoids (Cserhati and Forgacs 2001; Cserhati *et al* 1993). Alumina and diatomaceous earth did not find application in the analysis of such pigments. It has been observed that carotenoids extracted from red pepper can easily be separated on silica gel plate using a mobile phase of petroleum ether/acetone (3:1) as shown in Figure 18.2 (Zeb and Murkovic 2010a). The pro-vitamin A carotenoid β-carotene was found to elute earlier than the non-pro-vitamin A capsanthin. The possible reason may be due to the presence of hydrogen bonding in the case of capsanthin (Figure 18.3). The progress and commercialization of new stationary TLC phases such as octyl-, octadecyl-, cyano-, diol-, and aminopropyl derivatives of silica significantly improved the separation capacity and the mechanism of separation of TLC. Poole (2003) showed some of characteristics of thin layers and eluents systems for various TLC analyses. High-performance TLC (HPTLC) silica gel was one of the efficient stationary phases in the separation of β-carotene (Zeb and Murkovic 2010b).

18.2.1.1 *Normal Phase TLC Analysis of Carotenoids*

As a general rule, the stationary phases are polar and mobile phases are non-polar or of low polarity in normal phase TLC (NP-TLC). The pigments

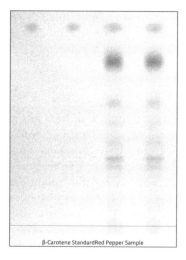

Figure 18.2 A representative HPTLC plate showing the separation of β-carotene from other carotenoids in a red pepper sample. The sample of red pepper extract in acetone was analysed using HPTLC plate and developed in petroleum ether/hexane/acetone (2:1:1, v/v/v). The β-carotene was confirmed using R_f values and spectra. Data are from Zeb and Murkovic (2010a), with permission from Publisher.

composition of the marine algae *Codium fragile* has been quantitatively determined using MgO/CaSO₄ (1:4) TLC plate and 4% (v/v) *n*-propanol in petroleum ether. The extract was found to contain α- and ε-carotenes,

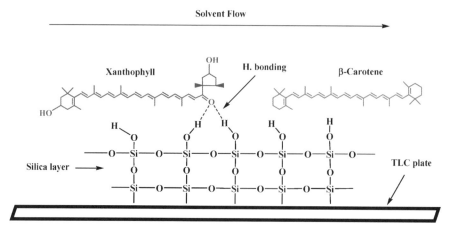

Figure 18.3 Schematic representation of the possible mechanism of separation of pro-vitamin A carotenoid and non-pro-vitamin A carotenoid. Oxygenated carotenoids have strong interactions with the polar layer of the silica gel surface of TLC and thus are separated later than pro-vitamin A carotenoids such as β-carotene. Data are from Zeb and Murkovic (2010a), with permission from Publisher.

siphonoxanthin, siphonein, neoxanthin, and violaxanthin. These carotenoids were found to be present in low concentrations, whereas siphonoxanthin and its ester siphonein account for as much as 60% of the total carotenoids present (Benson and Cobb 1981). Similarly, the carotenoids of different tomato fruits were separated using various adsorbents and solvents systems (Zeb and Murkovic 2010a). It was revealed that the most suitable TLC system for separation and identification was the combination of the MgO/hyflosupercel/cellulose (10:9:1) layer and the solvent system of *n*-hexane/isopropanol/methanol (100:2:0.2) as shown in Table 18.1. In yellow tomato, β-carotene and a small amount of lycopene were identified with different R_f values. In another study a simple and rapid method for the isolation and estimation of carotene compositions of tomatoes, carrots and green vegetables was described (Premachandra, 1985). The data obtained for the individual carotenoids was in close agreement with those obtained after the column chromatographic separation of the same sample. The method was found useful in the separation of vitamin A alcohol from its esters, and retinoic acid from retinoic acid anhydride. Recently, Zeb and Murkovic (2010b) developed a TLC method for the quantification and degradation of β-carotene in edible oils. The method was successfully used for the quantification of other carotenoids and the results were compared with the standard high-performance liquid chromatography (HPLC) data. It has been found that silica gel is a suitable stationary phase for the separation of pro-vitamin A carotenoids.

Table 18.1 Separation of tomato carotenoids using different adsorbents and solvent systems. The stationary phase produces significant effects in the separation of β-carotene from tomato extract. MgO–cellulose has good separation with hexane and acetone. The h R_f values represent the percentage values and are equivalent to R_f values multiplied by 100. HSC, Hyflosupercel. Data are from Zeb and Murkovic (2010a), with permission from Publisher.

TLC System		[*]h R_f value
Adsorbent	Solvent (v/v) n-*Hexane/acetone*	β-*Carotene*
MgO–HSC	80:20	85
MgO–cellulose	80:20	83
MgO–silica	70:30	70
HSC–cellulose	100:0	80
HSC–silica	100:0	60
Cellulose	100:0	83
Cellulose–silica	100:0	62
Silica	100:0	88

18.2.1.2   Reversed Phase TLC Analysis of Carotenoids

In reversed phase TLC (RP-TLC) a non-polar stationary phase and an aqueous moderately polar mobile phase are commonly used. Reversed phase TLC has limited uses in the analysis of carotenoids. Hayashi *et al* (2002; 2003) reported two TLC methods for the analysis of food colourants in tomato, orange, and marigold. The extracted carotenoids were lycopene, β-cryptoxanthin, and lutein, which were analysed by a reversed phase TLC plate (RP-18F$_{254}$S) with developing solvent systems of acetonitrile/acetone/hexane (11:7:2) and acetone/water (9:1). It was found that reversed phase TLC is a useful separation technique for separation of carotenoids in foods.

18.2.2   Mobile Phases

Pro-vitamin A carotenoids are usually extracted from biological samples using acetone, or a mixture of hexane with petroleum ether and ethanol (Jaime *et al* 2005). Mixtures of carotenoids were separated using petroleum ether/acetone (75:25, v/v) and hexane/acetone in ratio of 4:1 (Barua 2001). β-Carotene and lutein have been separated using petroleum/diethyl ether/acetic acid in the ratio of 80:20:1, and petroleum ether/acetonitrile/methanol in the ratio of 1:2:2 have been used. In the majority of the analyses petroleum ether and acetone were found to be used as major organic eluent (Lichtenthaler *et al* 1982; Martin *et al* 2005; Ren and Zhang 2008). The second major solvent was hexane in the TLC analysis of carotenoids from plant sources (Liu *et al* 2004; Ren and Zhang 2008). Similarly, petroleum ether/acetone with a different ratio was used for the determination of carotenoids isomers, β-carotene, oxidized carotenoids, vitamin A, and carotenoids mixtures (Zeb and Murkovic 2010a). Petroleum ether/hexane/acetone (2:1:1, v/v/v) was used for the quantitative determination and degradation of β-carotene in sunflower oil (Zeb and Murkovic 2010b). The inability to completely separate individual isomer of β-carotene was the main limitation of this solvent system. Recently, TLC was used for the investigation of the purity of the sample of carotenoids extracted from gac (*Momordica cochinchinensis*) in comparison with synthetic β-carotene (Cao-Hoang *et al* 2011). The synthetic β-carotene displayed only one band (R_f = 0.73) when submitted to silica gel-based TLC after a 15 min development in petroleum ether. The carotenoid band from the sample was then scraped off and dissolved in hexane for the investigation of ultraviolet (UV) spectral purity.

18.2.3   Applications of Scanning Densitometry

Today, different scanning densitometric instruments and softwares are available in the commercial market. Some software directly control and analyse sample on the TLC plate (*e.g.* CAMAG, Shimadzu), while others can analyse the image (Just TLC, UN-Scan-IT) obtained after plate development through a camera (Zeb and Murkovic 2010a). In comparison with expensive HPLC, these are less expensive, and easier to use and to interpret the data. However, the application of scanning densitometry in carotenoid analysis is rarely reported. An analytical method was established for the separation and identification of food colours, turmeric oleoresin, gardenia yellow, and annatto extract in foods using reversed phase TLC/scanning densitometry. The method involves clean-up of the colours with a C18 cartridge, separation of the colours by reversed phase C18-TLC using acetonitrile/tetrahydrofuran/oxalic acid (7:8:7) as a solvent system, and measurement of visible absorption spectra of the colours using scanning densitometry without isolation of the colours. A total of 89 commercial foods were analysed, and their chromatographic behaviour and spectra were observed. It was found that the spots always gave the same R_f values and spectra similar to the standards with good reproducibility. The method was considered to be useful for the rapid analysis of turmeric oleoresin, gardenia yellow, and annatto extract (including annatto, water-soluble) in foods (Zabkiewicz *et al* 1968), while scanning densitometry was found to be useful method to quantify carotenoids. Similarly, the carotenoid composition of fruit extract from *Rosa canina* was assessed by TLC and HPLC and the results were compared. The extract was separated on silica plates in two steps. The first involve the use of 15% acetone in petroleum ether and second using 100% petroleum ether. The chromatograms were analysed using a Shimadzu CS-9000 dual wavelength flying spot scanner. Both chromatograph analyses revealed β-carotene, lycopene, β-cryptoxanthin, rubixanthin, zeaxanthin, and lutein as major carotenoids (Hodisan *et al* 1997). In another study, TLC was performed on pre-coated silica gel 60 F_{254} aluminium sheets 20 × 20 cm (Merck, Germany) for the determination of the carotenoid profile in grape, musts, and fortified wines. The solvent system employed was acetone/hexane, 3:7 (v/v) and a pre-run in 2.5% (w/v) solution of citric acid in methanol. Qualitative analysis of the TLC plates was performed using an imaging densitometer (model Q5-700, Bio-Rad, USA) (Pinho *et al* 2001). Similarly, chlorophyll and carotenoid contents were determined by HPTLC in Tuo cha by Zhong-xi *et al* (2005). Scanning densitometry was carried out using a CAMAG TLC scanner. The method, however, gives no information about the quantitative analysis of carotenoids. Similarly, Bundit *et al* (2008) presented a TLC densitometric method for carotenoid determination in the serum of fancy carp (*Cyprinus carpio*). The method gives less information about the reproducibility and validation. Scanning densitometry using the CAMAG system was more useful in the study of the degradation of β-carotene in edible oils (Zeb and Murkovic, 2010b). A typical representative densitogram (CAMAG TLC Scanner, winCATS software) of carotenoid profiles of paprika is given by Figure 18.4. It was found that densitometry is a good technique for the analysis of carotenoids.

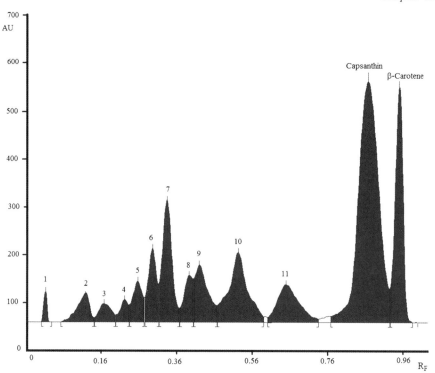

Figure 18.4   Typical densitogram of carotenoid separation from red pepper using a HPTLC plate. The densitogram was obtained using a CAMAG scanner and winCATs software shows that pro-vitamin A carotenoid (β-carotene) moves faster than other carotenoids. Separation and identification of the individual peaks was obtained using absorption spectra at multi-wave lengths. Data are redrawn from Zeb and Murkovic (2010a), with permission from Publisher.

18.3   Advantages of TLC Analysis of Pro-vitamin A Carotenoids

Pro-vitamin A carotenoids are more susceptible to oxidization or degradation if exposed to intense light, heat or kept for a long time, therefore rapid analysis is required as well as the proper storage conditions. The following points make TLC as good tool compared with other chromatographic techniques for carotenoid determination:

- Carotenoids can be analysed quantitatively and qualitatively in a very short time.
- A large number of samples can be measured in a single run, which is not possible in other chromatographic techniques.
- The analyses are usually simple, accurate, and reproducible.

- The introduction of the high performance thin layer plate increases the efficiency of separation.
- It has greater sensitivity for most of the carotenoids.

Summary Points

- Pro-vitamin A carotenoids are the most important organic pigments present in living organisms.
- They are found in common foods and vegetables.
- Petroleum ether, acetone, and hexane are the major mobile phases used for the extraction and analysis of TLC.
- Silica gel-based thin layers have wide applications in carotenoids analysis.
- TLC has excellent potential to be the first choice for the analysis of pro-vitamin A carotenoids in biological samples.
- The coupling of TLC with other techniques such as mass spectroscopy, scanning densitometry, and image analysis provides precise analysis of pro-vitamin A carotenoids.

Key Facts

Key Facts of Stationary Phase

- The thin layer of substances present on the piece of glass, plastic or aluminum, which have the potential to adsorb a mixture of substance and elute them on the basis of different properties.
- The stationary phase is usually made of silica gel, cellulose or aluminum oxide or their derivatives.
- Silica gel-based stationary phases have excellent ability of separation.
- The stationary phase is available in different thickness and size on different kind of plates.
- High-performance thin layer plates are now the most effective in separation of carotenoids.

Key Facts of Hydrogen Bonding

- The bonding formed by the attraction of hydrogen with the other high electronegative elements such as oxygen, nitrogen, and halogen.
- It is one of the main properties found in biological substances and water.
- An increase in hydrogen bonding results an increase in boiling points.
- The length of the hydrogen bonds is dependent on bond strength, temperature, and pressure of the medium.

- It may be symmetric or asymmetric depending on the space between the two identical atoms.

Definitions of Words and Terms

β-Carotene: It is a hydrocarbon tetraterpene commonly present in vegetables and fruits. It has 100% pro-vitamin A activity.

Codium fragile: It is one of algae commonly available in sea beaches. It is also known as Dead man's fingers.

β-Cryptoxanthin: It is a hydroxyl derivatives of β-carotene. It is present in papaya, apples, egg yolk, and butter. It is a pro-vitamin A carotenoid.

Cyprinus carpio (Fancy carp): It is a common fresh water fish found most commonly in Europe and Asia.

Densitometry: When a light is passed through a substance, it absorbs some part of light and the remaining is passed by and collected on a film. The amount of material is quantitatively analysed from the optical density of a light-sensitive substance.

Momordica cochinchinensis: It is also known as gac or sweet gout. It is commonly grown in Southeast Asian countries. It contains more lycopene than tomato and more β-carotene than carrot by mass.

Normal phase thin-layer chromatography (NP-TLC): Normal phase thin-layer chromatography uses a polar stationary phase and a non-polar or relatively less polar solvent(s). It is commonly used for the analysis of non-polar or less polar hydrocarbon compounds.

β-Ring: A β-ring is 1,1,5-trimethyl-5-cyclo hexene. The presence of a β-ring is one of the main parameters for a carotenoid having pro-vitamin activity.

Thin-layer chromatography (TLC): A kind of separation technique (chromatography), which uses a thin layer of adsorbent on the surface of a sheet.

Tuo Cha: It is one of green tea commonly used in Yunnan province of China.

List of Abbreviations

HPLC	high performance liquid chromatography
HPTLC	high-performance thin-layer chromatography
TLC	thin-layer chromatography

Acknowledgements

The author thanks Prof. Dr Haiyan Zhong from the Faculty of Food Science and Technology, Central South University of Forestry and Technology, China, for translating the Chinese literature and CAMAG Switzerland for providing the latest software and guidance.

References

Barua, A. B., 2001. Improved normal-phase and reversed-phase gradient high-performance liquid chromatography procedures for the analysis of retinoids and carotenoids in human serum, plant and animal tissues. Journal of Chromatography A. 936: 71–82.

Benson, E. E. and Cobb, A. H., 1981. The separation, identification and the quantitative determination of photopigments from siphonaceous marine alga *Codium fragile*. New Phytologist. 88: 627–632.

Britton, G. 1995. UV/Visible spectroscopy. In: Britton, G., Liaaen-Jensen, S. and Pfander, H., (ed.) Carotenoids: Spectroscopy, Volume 1B. Birkhäuser Verlag, Basel, Switzerland, pp. 13–62.

Bundit, Y., Jintasataporn, O., Areechon, N. and Tabthipwon, P., 2008. Validated TLC- densitometric analysis for determination of carotenoids in Fancy Carp (*Cyprinus carpio*) serum and the application for pharmaco-kinetic parameter assessment. Songklanakarin Journal of Science and Technology. 30: 693–700.

Cao-Hoang, L., Phan-Thi, H., Osorio-Puentes, F. J. and Wache, Y., 2011. Stability of carotenoid extracts of gac (*Momordica cochinchinensis*) towards co-oxidation. Protective effect of lycopene on β-carotene. Food Research International. 44: 2252–2257.

Cserhati, T. and Forgacs, E., 2001. Liquid chromatographic separation of terpenoid pigments in foods and food products. Journal of Chromatography A. 936: 119–137.

Cserhati, T., Forgacs, E. and Hollo, J., 1993. Separation of color pigments of capsicum-annuum by adsorption and reversed-phase thin-layer chromato-graphy. Journal of Planar Chromatography. 6: 472–475.

Evans, R.T., Fried, B. and Sherma, J., 2004. Effects of diet and larval trematode parasitism on lutein and β-carotene concentrations in planorbid snails as determined by quantitative high performance reversed phase thin layer chromatography. Comparative Biochemistry and Physiology, Part B Biochemistry Molecular Biology. 137: 179–186.

Hayashi, T., Oka, H., Ito, Y., Goto, T., Ozeki, N., Itakura, Y., Matsumoto, H., Otsuji, Y., Akatsuka, H., Miyazawa, T. and Nagase, H., 2002. A reversed-phase TLC/scanning densitometric method for the analysis of tomato, orange, and marigold colors in food. Journal of Liquid Chromatography and Related Technologies. 25: 3151–3165.

Hayashi, T., Oka, H., Ito, Y., Goto, T., Ozeki, N., Itakura, Y., Matsumoto, H., Otsuji, Y., Akatsuka, H., Miyazawa, T. and Nagase, H., 2003. Simultaneous analysis of carotenoid colorings in foods by thin layer chromatography. Journal of Liquid Chromatography and Related Technologies. 26: 819–832.

Hodisan, T., Socaciu, C., Ropan, I. and Neamtu, G., 1997. Carotenoid composition of *Rosa canina* fruits determined by thin-layer chromato-graphy and high-performance liquid chromatography. Journal of Pharmaceutical and Biomedical Analysis. 16: 521–528.

Jaime, L., Mendiola, J. A., Herrero, M., Soler-Rivas, C., Santoyo, S., Senorans, F. J., Cifuentes, A. and Ibanez, E., 2005. Separation and characterization of antioxidants from *Spirulina platensis* microalga combining pressurized liquid extraction, TLC, and HPLC-DAD. Journal of Separation Science. 28: 2111–2119.

Lichtenthaler, H. K., Borner, K. and Liljenberg, C., 1982. Separation of prenylquinones, prenylvitamins and prenols on thinlayer plates impregnated with silver nitrate. Journal of Chromatography. 242: 196–201.

Liu, H. L., Kao, T. H. and Chen, B. H., 2004. Determination of carotenoids in the Chinese medical herb Jiao-Gu-Lan (*Gynostemma pentaphyllum* MAKINO) by liquid chromatography. Chromatographia. 60: 411–417.

Marston, A. 2007. Role of advances in chromatographic techniques in phytochemistry. Phytochemistry. 68: 2785–2797.

Martin, D. L., Fried, B. and Sherma, J., 2005. The absence of β-carotene and the presence of biliverdin in the medicinal leech *Hirudo medicinalis* as determined by TLC. Journal of Planar Chromatography. 18: 400–402.

Oliver, J. and Palou, A., 2000. Chromatographic determination of carotenoids in foods. Journal of Chromatography A. 881: 543–555.

Pinho, P. G., Ferreira, A. C. S., Pinto, M. M., Benitez, J. G. and Hogg, T. A., 2001. Determination of carotenoid profiles in grapes, musts, and fortified wines from Douro varieties of *Vitis vinifera*. Journal of Agriculture and Food Chemistry. 49: 5484–5488.

Poole, C. P., 2003. Thin-layer chromatography: challenges and opportunities. Journal of Chromatography A. 1000: 963–984.

Premachandra, B. R. 1985. A simple TLC method for the determination of pro-vitamin A content of fruits and vegetables. International Journal of Vitamins and Nutrition Research. 55: 139–47.

Ren, D., and Zhang, S., 2008. Separation and identification of the yellow carotenoids in *Potamogeton crispus* L. Food Chemistry. 106: 410–414.

Rodriguez-Amaya, D. B. 1997. Carotenoids and food preparation: the retention of provitamin A carotenoids in prepared, processed, and stored foods. Opportunities for Micronutrient Policy Research Institute (IFPRI) and International Centre for Tropical Agriculture (CIAT). pp. 10–32.

Rodriguez-Amaya, D. B., Kimura, M., Godoy, H. T. and Amaya-Farfan, J., 2008. Updated Brazilian database on food carotenoids: factors affecting carotenoid composition. Journal of Food Composition and Analysis. 21: 445–463.

Zabkiewicz, J. A., Keates, R. A. B. and Brooks, C. J. W., 1968. Leaf extract purification by silver nitrate-silica gel thin-layer chromatography. Biochemical Journal. 109: 929–930.

Zeb, A. and Mehmood, S., 2004. Carotenoids contents from various sources and their potential health applications. Pakistan Journal of Nutrition. 3: 199–204.

Zeb, A. and Murkovic, M., 2010a. Thin-layer chromatographic analysis of carotenoids in plant and animal samples. Journal of Planar Chromatography. 23: 99–103.

Zeb, A. and Murkovic, M., 2010b. High performance thin layer chromato-graphic method for monitoring the thermal degradation of β-carotene in sunflower oil. Journal of Planar Chromatography. 23: 35–39.

Zhong-xi, X., Kun-bo, W. and Bo-hua, J., 2005. Analysis of chlorophylls and carotenoids in Tuo Chaby high performance thin-layer chromatography. Journal of Yunnan Agriculture University. 20: 384–387 (in Chinese).

CHAPTER 19

Extraction of Carotenoids From Plants: A Focus on Carotenoids with Vitamin A Activity

ANITA OBERHOLSTER

Department of Viticulture and Enology, University of California, One Shields Avenue, Davis, CA 95616-5270, USA
E-mail: aoberholster@ucdavis.edu

19.1 Introduction

Most dietary vitamin A is obtained from plants, and there are about 50 naturally occurring carotenoid compounds with vitamin A biological activity. These are the so-called provitamin A carotenoids (pVACs), which are broken down in the body to yield retinol, the active form of vitamin A (Yeum and Russell 2002). Of these, β-carotene, α-carotene, and β-cryptoxanthin are the most important (Figure 19.1).

In general, the carotenoids can be classified into two great groups: carotenes, which are strictly hydrocarbons (*e.g.* β-carotene), and xanthophylls (*e.g.* β-cryptoxanthin), which are derived from the former and contain oxygenated functions. The presence of a large number of conjugated double bonds in the carotenoid molecule makes numerous geometric isomers possible (*Z–E* isomers, also called *cis–trans*). In natural sources, carotenoids occur mainly in the all-*trans* (all-*E*) configuration (Chandler and Schwartz 1987). Isomerization of *trans*-carotenoids to *cis*-isomers is promoted by contact with

Food and Nutritional Components in Focus No. 1
Vitamin A and Carotenoids: Chemistry, Analysis, Function and Effects
Edited by Victor R Preedy
© The Royal Society of Chemistry 2012
Published by the Royal Society of Chemistry, www.rsc.org

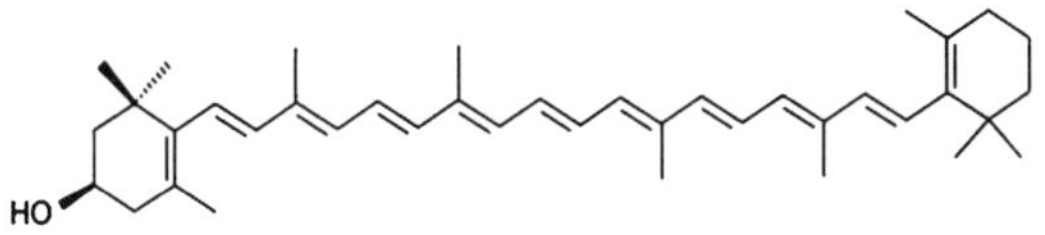

β-cryptoxanthin

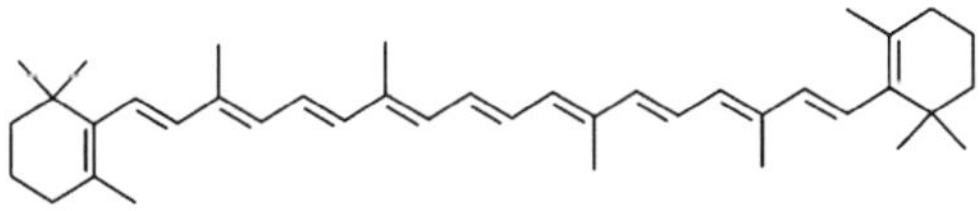

β-carotene

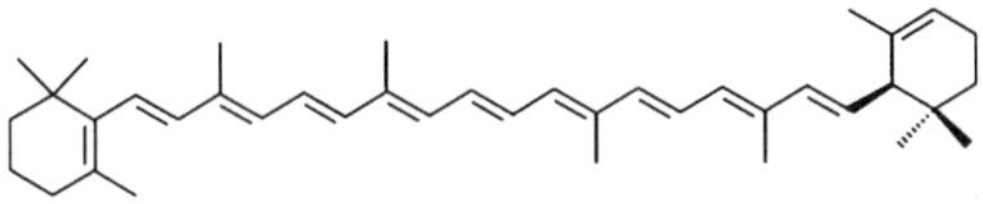

α-carotene

Figure 19.1 Structures of the main provitamin A carotenoids. β-Cryptoxanthin, β-carotene and α-carotene are the main provitamin A carotenoids present in plants.

acids, heat treatment, and exposure to light (Rodriguez-Amaya *et al* 2008; Van den Berg *et al* 2000).

In green plant tissues, including green fruits and vegetables, the chloroplasts contain similar combinations of carotenoid pigments, with β-carotene as the predominant carotene, followed by the xanthophylls, lutein, violaxanthin, and neoxanthin. Small amounts of zeaxanthin, γ-carotene, β-cryptoxanthin, and antheraxanthin are also present. During fruit ripening, the chloroplasts change to chromoplasts, which is associated with the fruit ripening process. This coincides with substantial synthesis of carotenoids during ripening, resulting in a simultaneous change in the carotenoid profile of the fruit (Figure 19.2). The xanthophylls in the chromoplasts are usually esterified with different fatty acids (Mínquez-Mosquera *et al* 2008). The amount and composition of carotenoids in fruits or vegetables depends on what carotenoid biosynthesis genes are present and active. In addition to the carotenoids that accumulate in the ripe fruit, some chloroplast carotenoids may remain from the green pre-ripening stage (Khachik 2009).

19.2 General Properties

The carotenoids are lipophilic substances, and thus generally insoluble in aqueous medium, except in certain cases where highly polar functional groups are present. Carotenoids are chromophores with seven or more double bonds which give these molecules the capacity of absorbing light in the visible range, resulting in the display of colours spanning from yellow to red, *via* a great variety of orange tones. Moreover, the polyene chain makes the carotenoid

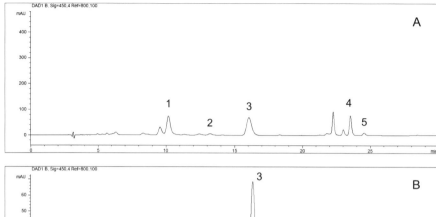

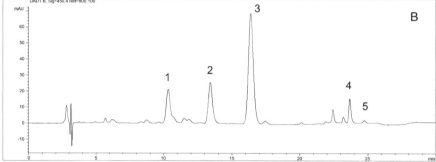

Figure 19.2   Example of the changes in carotenoid profiles with fruit ripening (grapes). Reverse-phase (RP) HPLC profiles of the major carotenoids in Merlot grapes pre-veraison (A) and post-veraison (B): (1) lutein; (2) 5,8-epoxy-β-carotene; (3) β-apo-caroten-8-al (internal standard); (4) β-carotene; (5) 9-*cis*-β-carotene (Kamffer 2009).

molecule extremely susceptible to isomerizing and oxidizing conditions such as light, heat, or acids.

Carotenoids *in situ* in vegetables and fruit are usually more stable than when they are isolated (Britton and Khachik 2009; Gouveia and Empis 2003). Oxidation, either enzymic or non-enzymic, is the main cause of destruction of carotenoids. When fruits and vegetables are cut, chopped, shredded or pulped, it increases exposure to oxygen and remove the physical barriers that normally keep apart the carotenoids and oxidizing enzymes such as lipoxygenase.

Geometrical isomerization, which occurs particularly during heat treatment, increases the proportion of Z isomers and may alter the biological activity, but the total carotenoid content is not greatly changed. Exposure to oxygen during storage, drying, and processing of plant tissue and/or carotenoid extracts leads to the generation of peroxides and oxidizing free radicals and can cause serious losses of carotenoids (Britton and Khachik 2009).

The list below describes the general precautions necessary to minimize carotenoid changes and losses during extraction (Britton and Khachik 2009; Mínquez-Mosquera *et al* 2008).

i.   Oxygen should be excluded as much as possible by replacing air with a vacuum or inert gas.

ii. Antioxidants [0.1% butylated hydroxytoluene (BHT), 5% pyrogallol, ascorbic acid, sodium ascorbate or ascorbyl palmitate] should be added to extracting solutions to protect the carotenoids from oxidation.

iii. *Trans*-to-*cis* isomerization (*E/Z*) is promoted at higher temperatures. Therefore, the use of solvents with low boiling points is recommended, in addition to working at the lowest practical temperature. For rotary evaporation, 40 °C should not be exceeded.

iv. Samples should be stored dry under an inert gas or vacuum and solutions should be stored at temperatures below −20 °C, preferably below −70 °C.

v.   All sunlight should be avoided. Analytical steps should be completed in subdued light or under gold fluorescent light. Solutions should be kept in low actinic glassware whenever possible.

vi. Acid must be avoided and all solvents must be acid-free. Addition of triethylamine (TEA) or *N,N*-di-isopropylethylamine (DIPEA) at 0.001 to 0.1% is useful to neutralize low acid levels occurring in some solvents.

19.3   Preparation of the Sample

The sample to be analyzed should be as fresh as possible and should be damage free to ensure that the pigment fraction has not been modified. If the analysis is not going to be performed immediately, the sample should be stored at −20 °C short-term and long-term at −70 °C (Van den Berg *et al* 2000). Freezing (the more rapid the better) generally preserve the carotenoids but subsequent slow thawing, especially of unblanched products, can be detrimental. It is also advisable to remove oxygen with inert gas flushing or a vacuum before freezing of samples and to use light-protected containers. Blanching of plant products before freezing inactivates lipoxygenase and protects carotenoids from oxidative degradation during freezer storage.

It has been general practice to lyophilize samples with a high water content (Mínquez-Mosquera *et al* 2008). However, several researchers have found degradation of carotenoids occurs during lyophilization of fruits and vegetables (Craft *et al* 1993; Kamffer *et al* 2010; Park 1987). This is probably due to concentration of the acid in the matrix when the water is removed. In contrast, Davey *et al* (2006) found no difference between the use of fresh *versus* lyophilized fruit. Whether or not lyophilization is appropriate depends on the sample matrix. If the sample is already lyophilized or dehydrated, it will also have to be rehydrated for the extraction. The weight of the sample needed for analysis will depend on the carotenoid content. For samples with high carotenoid concentrations, 2–3 g are usually taken, increasing this to 10 g when the water content is high.

19.4   Choice of Solvent and Extraction

The high number of carotenoid pigments in nature (more than 650) and their structural variability, as well as the wide variety of sample matrixes, makes it practically impossible to describe a general methodology for their analysis. Britton *et al* (1995), Feltl *et al* (2005) and Rodríguez-Bernaldo de Quirós and Costa (2006) described and evaluated different procedures used for the extraction of carotenoids from a range of sample matrixes.

In general, the choice of the organic solvent to be used for extraction depends on the nature of the material to be extracted and the solubility properties of the major carotenoids expected to be present. A variety of organic solvents, including acetone, ethanol, methanol, tetrahydrofuran (THF), diethyl ether, ethyl acetate, hexane, and mixtures thereof are used. Acetone is the extraction solvent most commonly utilized (Edelenbos *et al* 2001; Gouveia and Empis 2003; Rodriguez-Amaya *et al* 2008). For food samples that contain a fairly large amount of water, it is desirable to use an organic solvent that is miscible with water to optimize the release of carotenoids from the matrix and prevent the formation of emulsions. Acetone and THF are the recommended solvents for this method. It is advisable in all instances to use an antioxidant in the extraction solvent to prevent the formation of peroxides which could lead to degradation of the carotenoids (Khachik 2009). The extraction of carotenoids must be carried out as quickly as possible, avoiding exposure to light, oxygen, high temperatures, and pro-oxidant metals, in order to minimize autoxidation and *cis–trans* isomerization (Marsili and Callahan 1993; Van den Berg *et al* 2000).

The sample is homogenized with the extracting solvent to facilitate contact between the solvent and the plant material to be extracted. The addition of sodium, magnesium or calcium carbonate [0.1–1 g (g of sample)$^{-1}$] helps to neutralize acids liberated from tissue samples (Hart and Scott 1995; Van den Berg *et al* 2000). The extraction temperature should not be allowed to exceed ambient temperatures. Homogenization and extraction are normally carried out at a low temperature by using ice or liquid nitrogen. However, sometimes higher temperatures are needed to facilitate extraction from the sample matrix (Howe and Tanumihardjo 2006). The supernatant is collected by filtration or centrifugation and the retained residue is extracted until the extract is colourless. The combined filtrate is concentrated by evaporation using temperatures below 40 °C to avoid carotenoid degradation. Depending on the complexity of the sample matrix and the analysis, the carotenoid concentrate is further purified by partitioning between a water-immiscible organic solvent and water to remove water soluble impurities. The most commonly used organic solvents for this are *t*-butyl methyl ether (TBME) and diethyl ether, but dichloromethane, ethyl acetate or hexane/dichloromethane (3:1) can be used. Salt (NaCl) can be added to break emulsions. Approximately 0.1% of an amine such as DIPEA or TEA may be added to the organic solvent to neutralize any traces of acids that may remain in the extract. This is particularly necessary when the organic solvent is dichloromethane because chlorinated solvents may contain traces of HCl. The organic layer is then removed, dried over Na_2SO_4 or $MgSO_4$, and re-dissolved in an appropriate volume of a solvent suitable for further analyses. In Table 19.1 some examples of carotenoid extraction conditions are presented.

Usually an internal standard such as β-apo-8′-carotenal is added during homogenization of the sample. In extraction protocols where heat is needed to facilitate extraction of carotenoids from the sample matrix, it may be necessary to add the internal standard after heat treatment. Howe and Tanumihardjo (2006) found that the internal standard deteriorated faster than the carotenoids in the sample matrix during heat treatment, resulting in an overestimation of the carotenoid content.

Several researchers have developed micro-extraction methods, adapting the amount of sample and volume of extraction solvents used to fit in a 2 mL centrifuge tube (Davey *et al* 2006; Serino *et al* 2009; Taylor *et al* 2006). In some cases this enabled analysis of the extract without a concentration step (Serino *et al* 2009). Micro-extraction methods expedite extraction times and allow simultaneous extraction of interested compounds with often less degradation products. In Figure 19.3 a schematic representation of a possible micro-extraction method is given using the main steps described for carotenoid extraction from plant material.

19.5   Saponification

High-performance liquid chromatography (HPLC) is the main method used for the separation and quantification of carotenoids. Carotenoid extracts are often saponified (alkaline hydrolysis) to simplify their separation by removing substances, such as chlorophylls (when their analysis are not required) and lipids, which could interfere with the chromatographic detection. In addition, saponification hydrolyzes the fatty acid esters of xanthophylls present in many ripe fruits (Khachik 2009). The disadvantage is that it involves extra handling steps including the partitioning of the saponified carotenoid species, drying down, and re-solubilization of extracts. In addition, a loss of total carotenoid content during saponification has been reported (Davey *et al* 2006; Khachik *et al* 1986). Fernandez *et al* (2000) compared the use of alkali saponification and enzymatic hydrolysis to determine the total carotenoid content of Costa Rican crude palm oil. Similarly, Delgado-Vargas and Paredes-López (1997) compared results obtained for marigold flower meal. Both studies found greater concentration of carotenoids using enzymatic hydrolysis. The most sensitive compounds to alkaline treatments are xanthophylls, particularly the epoxycarotenoids (Khachik *et al* 1986).

For most fruits and vegetables, moderate saponification conditions are satisfactory. This involves treatment of the extracts with alcoholic potassium or sodium hydroxide (10%, w/v, in ethanol or methanol) in the dark for 1–3 h or overnight, under an inert atmosphere such as nitrogen or argon (Rodríguez-Bernaldo de Quirós and Costa 2006). For high-fat foods such as maize, olives,

Table 19.1 Examples of carotenoid extraction protocols of plants. BHT, butylated hydroxytoluene; CaCO₃, calcium carbonate; DCM, dichloromethane; MgCO₃, magnesium carbonate; MeOH, methanol; NaCl, sodium chloride; NaSO₄, sodium sulphate; THF, tetrahydrofuran; 2-PrOH, 2-propanol; TBME, *t*-butyl methyl ether.

Sample	Analyte	Extraction conditions	Reference
Arabidopsis thaliana leaves	Neoxanthin, violaxanthin, *cis*-neoxanthin, antheraxanthin, zeaxanthin, β-carotene, 9-*cis*-β-carotene	Extract with MeOH, add 50 mM Tris/HCl (pH 8.0) containing 1 M NaCl, mix. Partition with chloroform (× 2) and centrifuge. Pool chloroform extracts, dry and dissolve in ethyl acetate/MeOH (1:4) containing 0.1% (w/v) BHT for analysis.	Taylor *et al* (2006)
Brussels, beans, broccoli, cabbage, carrots, cauliflower, greens, leeks, lettuce, marrow, parsley, peas, peppers, sweetcorn, spinach, spring onions, tomato, watercress	Lutein, zeaxanthin, β-cryptoxanthin, lycopene, α-carotene, β-carotene, 9-*cis*-β-carotene	Add MgCO₃, homogenize with THF:MeOH (1:1). Filter and re-extract with THF/MeOH (1:1) × 3. Add 5% (w/v) NaCl and partition with petroleum ether containing 0.1% BHT (× 3). Pool, dry and dissolve in DCM. If necessary the extract was saponified (Table 19.2).	Hart and Scott (1995)
Carrots and tomato paste	Lutein, zeaxanthin, β-cryptoxanthin, lycopene, α-carotene, β-carotene	Extract sample with hexane (× 3), centrifuge and wash combined extracts with H₂O. Dry and reconstitute in dichloroethane/MeOH (1:1, v/v).	Mills *et al* (2007)
Grapes	Lutein, zeaxanthin, 5,8-epoxy-β-carotene, β-carotene, 9-*cis*-β-carotene	Extract with diethyl ether/hexane (1:1) × 2, vortex 30 min, centrifuge. Pool organic phase, dry and dissolve in ethyl acetate/MeOH (1:4) for analysis.	Kamffer *et al* (2010)

Table 19.1   (*Continued*)

Sample	Analyte	Extraction conditions	Reference
Mango	β-Carotene, 9-*cis*-β-carotene, 13-*cis*-β-carotene	Homogenize sample with celite, $CaCO_3$ and MeOH. Wash twice with MeOH and filter. Add 10% (w/v) NaCl and partition with acetone/hexane (1:1, v/v). Repeat until aqueous layer is colourless. Remove acetone with H_2O wash (× 2). Add 0.1% BHT and dry over $NaSO_4$. Dissolve in 2-PrOH for analysis.	Pott *et al* (2003)
Peas	Neoxanthin, 9-*cis*-neoxanthin, violaxanthin, lutein epoxide, lutein, β-carotene	Homogenize sample with cold acetone, centrifuge. Repeat extraction 3 times. Pool extracts, filter and analyse.	Edelenbos *et al* (2001)
Tropical fruits (red and yellow tree tomato and naranjilla)	Lutein, zeaxanthin, β-cryptoxanthin, β-carotene	Add $MgCO_3$ and extract sample with ethanol/hexane (4:3, v/v) containing 0.1% BHT. Wash organic phase further with 10% (w/v) NaCl and then H_2O. Dry hexane extract over $NaSO_4$ and dissolve in TBME/MeOH (4:1, v/v) before analysis. Saponify and analyse again (Table 19.2).	Mertz *et al* (2009)

and palm oils, higher temperatures and an additional base [17–80%, w/v, potassium hydroxide (KOH)] may be needed to ensure complete saponification (Fernandez *et al* 2000; Howe and Tanumihardjo 2006; Mínquez-Mosquera and Gandul-Rojas 1994). After saponification the carotenoids are extracted with ethyl ether, diethyl ether, *n*-hexane, petroleum ether or combinations thereof (Rodríguez-Bernaldo de Quirós and Costa 2006). The extracts are then washed with water until all alkali is removed, followed by drying of the organic layer. Examples of different saponification conditions used in carotenoid analysis are presented in Table 19.2.

Saponification should be avoided if it is not necessary for separation of the carotenoids of interest. Several researchers have successfully analysed plant tissues with low fat content by direct solvent extraction (Davey *et al* 2006; Kamffer *et al* 2010; Pott *et al* 2003; Serino *et al* 2009). Enzymatic treatment is also another possibility if improved extraction is needed.

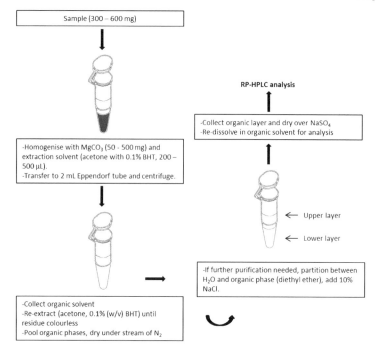

Figure 19.3   Example of a micro-extraction procedure for carotenoids. $MgCO_3$, magnesium carbonate; BHT, butylated hydroxytoluene; N_2, nitrogen, NaCl, sodium chloride; $NaSO_4$, sodium sulphate.

19.6   Supercritical Fluid Extraction

Supercritical fluid extraction (SFE) has been used as an alternative method to traditional liquid extraction for the isolation of carotenoids from food samples, since this technique shows several advantages. It is non-toxic, non-flammable, and environmentally acceptable, and the evaporation step is not required (Marsili and Callahan 1993). SFE is an extraction technique, which exploits the solvent properties of fluids above their critical point. SFE is frequently carried out with carbon dioxide (CO_2) (Turner *et al* 2001). The critical pressure (P_c) and temperature (T_c) for CO_2 are 31.1 °C and 74 bar, respectively. Here, the distinction between the gas and the liquid phases has disappeared, and the resulting supercritical fluid has one uniform density. The density and hence the solvent strength can be changed by varying the temperature and pressure. CO_2 have a low polarity which is not optimal for the extraction of more polar compounds. However, the extraction efficiency of CO_2 can be improved by adding an organic modifier such as methanol (MeOH), ethanol (EtOH), hexane or dichloromethane (Careri *et al* 2001). Marsili and Callahan (1993) evaluated several modifiers and found EtOH to be the best modifier for β-carotene extraction. Samples should be well homogenized and some authors have reported that the removal of water

Table 19.2 Different saponification conditions used in carotenoid analysis of plants. BHT, butylated hydroxytoluene; DCM, dichloromethane; EtOH, ethanol; KOH, potassium hydroxide; MgCO₃, magnesium carbonate; MeOH, methanol; NaCl, sodium chloride; NaSO₄, sodium sulphate; TBME, t-butyl methyl ether; THF, tetrahydrofuran.

Sample	Analyte	Saponification conditions	Reference
Fruits and peppers	Lutein, zeaxanthin, β-cryptoxanthin, lycopene, α-carotene, β-carotene, 9-*cis*-β-carotene	DCM extract mixed with equal volume 10% KOH in MeOH (under N₂ in the dark) for 1 h at room temperature. Carotenoids extracted as described before (Table 19.1).	Hart and Scott (1995)
Green vegetables (broccoli, brussels spouts, cabbage, kale, spinach)	Neoxanthin, violaxanthin, 9-*cis*-neoxanthin, neochrome, lutein epoxide, lutein, β-carotene, 15,15'-*cis*-β-carotene	Homogenize with NaSO₄ and MgCO₃ and extract with THF. Filter and re-extract with THF until colourless. Pool, dry and partition in petroleum ether and H₂O. Wash H₂O layer with petroleum ether until colourless, pool organic phase. Saponify with 30% KOH under nitrogen at room temperature for 3 h. Partition into saturated NaCl solution and petroleum ether. Remove organic layer and wash aqueous layer with ether. Combine organic layers, wash with H₂O and dry over NaSO₄. Dissolve in hexane and analyse.	Khachik *et al* (1986)
Maize	Lutein, zeaxanthin, β-cryptoxanthin, α-carotene, β-carotene	Extract sample with EtOH containing 0.1% BHT in 85 °C waterbath for 5 min. Add 80% (w/v) KOH and extract for additional 15 min. Cool extracts on ice, extract 3 × with hexane. Combine organic layers and wash with H₂O. Extract aqueous layer twice with hexane. Dry combined organic layers. Reconstitute in MeOH/dichloroethane (50:50, v/v) and analyse.	Howe and Tanumihardjo (2006)
Palm oil	Estimated total carotenoids using spectrophotometry	Dissolve palm oil in DCM with 0.04% (v/v) BHT. Mix with 11% (w/v) KOH in MeOH under N₂ for 4 h at room temperature. Extract with petroleum ether × 4. Filter and wash 4 times with 8% (w/v) NaCl solution. Final wash with H₂O. Dry and dissolve in hexane.	Fernandez *et al* (2000)
Tropical fruits (red and yellow tree tomato and naranjilla)	Lutein, zeaxanthin, β-cryptoxanthin, β-carotene	Hexane extract (Table 19.1) was saponified using 10% methanolic KOH, overnight at room temperature, protected from light under N₂. Partition by adding H₂O and wash hexane layer with H₂O until free of alkali. The methanolic KOH was extracted 3 × with DCM. Pool and wash extracts until free of alkali. Dry extracts over NaSO₄ and dissolve in TBME/MeOH (4:1, v/v) before analysis.	Mertz *et al* (2009)

Table 19.3   Examples of SFE conditions used in carotenoid analysis. BHT,
butylated hydroxytoluene; CH₃CN, acetonitrile; EtOH, ethanol;
MeOH, methanol; 2-PrOH, 2-propanol; THF, tetrahydrofuran.

Sample	Analyte	Supercritical fluid	Optimal SFE conditions	Reference
Carrots	α-Carotene, β-carotene	CO_2 with 10% EtOH modifier	Extraction at 50 °C and 304 bar at a flow-rate of 0.5 mL min^{-1} for 60 min. Collect extract in hexane/acetone (9:1, v/v) containing 0.005% (w/v) BHT.	Barth *et al* (1995)
Marjoram	Lutein, β-carotene,	CO_2 without modifier	Optimal extraction at 50 °C and 450 bar at a flow-rate of 2 mL min^{-1} for 22.7 min. Dissolve extract in CH₃CN/MeOH/2-PrOH (39:43:18, v/v/v).	Vági *et al* (2002)
Spirulina Pacifica algae	Zeaxanthin, β-cryptoxanthin, β-carotene	CO_2 with 15% EtOH modifier	Optimal temperature for individual carotenoids/zeaxanthin (80 °C), β-cryptoxanthin (76 °C), β-carotene (60 °C). Extraction at 350 bar at a flow-rate of 2 mL min^{-1} for 60 min. Collect extract in THF containing 1% (w/v) BHT.	Careri *et al* (2001)
Tomato paste waste	β-Carotene, lycopene	CO_2 with 5% EtOH modifier	Extraction at 65 °C and 300 bar at a flow-rate of 4 kg h^{-1} for 120 min. Collect extract in hexane.	Baysal *et al* (2000)

from samples facilitates extraction (Turner *et al* 2001). A range of SFE
conditions have been successfully used to extract carotenoids from plant tissue
by means of a variety of temperature and pressure combinations (40–86 °C,
280–450 bar) (Rodríguez-Bernaldo de Quirós and Costa 2006). In Table 19.3
some examples of SFE extraction conditions for carotenoid analysis are
given.

   SFE and traditional liquid extraction processes were compared for the
extraction of carotenoids from carrots (Barth *et al* 1995) and *Spirulina Pacifica*
algae (Careri *et al* 2001). In both cases SFE obtained similar or higher yields
than those attained in liquid extraction.

   There is no standard procedure for carotenoid extraction from plant tissue
or even for specific sub-categories of plant materials. This is an indication of

the complexities involved in carotenoid extraction, the wide variety of plant matrixes and the great range of carotenoids that can be found in these samples. However, there are a number of options available that can be explored to develop a reliable method for the extraction of the carotenoids of interest.

Summary Points

- The following precautions should be adhered to during preparation of samples:
  - Exclude oxygen, heat and light as far as possible during processing and storage.
  - Avoid dehydration of plant tissue if possible and add calcium or magnesium carbonate during homogenization to neutralize released plant acids.
  - Add internal standard such as β-apo-8′-carotenal as early as possible to compensate for loss of carotenoids during extraction.

- Two methods are mostly used for carotenoid extraction from plant tissue: liquid extraction and SFE.
- An antioxidant should be added to the extraction solvent during liquid extraction.
- Non-chlorinated extracting solvents are recommended, such as acetone, THF, hexane, and diethyl ether.
- Avoid saponification of carotenoid extracts if possible.
- The use of CO_2 as supercritical fluid and EtOH (5–10%) as modifier is recommended when using SFE for carotenoid extraction.
- Optimize extraction during SFE by adjusting the temperature (40–86 °C) and pressure (280–450 bar).

Key Facts

Key Facts Regarding the Importance of Carotenoids in Plants and their Extraction

- Symptoms of vitamin A deficiency include night blindness, loss of appetite, skin lesions, lack of growth, and increased susceptibility to infections.
- Food sources rich in carotenoids include carrots, sweet potato, pumpkin, dark-green leafy vegetables, and various fruits.
- In low-income, impoverished populations, up to 82% of the dietary vitamin A is derived primarily from plant sources as provitamin A carotenoids.
- Heinrich W. F. Wackenroder isolated β-carotene for the first time in 1831 from the roots of carrots and named the substance 'carotin'.

- Carotenoids are sensitive to acid, light, temperature, and oxidation. Carotenoid identification and quantitation may be influenced by processing and storage artefacts due to exposure to these conditions.

Definitions of Words and Terms

Antioxidants: Chemical compounds that protect other compounds against oxidation.

Blanching: The process where food is placed in boiling water for a short period before it is rapidly cooled in iced or cold water to stop the cooking process.

Chlorophyll: The green pigment that is very important for photosynthesis as it captures light energy.

Chloroplasts: The sites in cells where photosynthesis (conversion of carbon dioxide into organic compounds, especially sugars by using light energy) takes place and where chlorophyll and other pigments are stored.

Chromaphore: The structural feature of a molecule responsible for its UV or visible absorption.

Chromoplasts: The plastids (sites in cells) where different pigments are synthesized and stored.

Isomerization: The chemical process by which a compound is transformed into one of its isomers. Isomers have the same chemical composition but differ in their structure or configuration, often resulting in different chemical and physical properties.

Lipophilic: The ability of a compound to dissolve in non-polar solvents such as hexane.

Provitamin A acitivity: A compound that can be broken down to retinol, the form in which vitamin A is absorbed in the body.

Supercritical fluid: This is not a gas or a liquid and is basically an intermediate of the two extremes. By increasing the pressure or temperature of a gas above its critical point (this is where the properties of the gas and liquid phases approach each other), it is possible to give the gas liquid-like densities and solvation strengths.

List of Abbreviations

2-PrOH	iso-propanol
BHT	butylated hydroxytoluene
$CaCO_3$	calcium carbonate
CH_3CN	acetonitrile
CO_2	carbon dioxide
DCM	dichloromethane
DIPEA	*N,N*-di-isopropylethylamine
EtOH	ethanol
HCl	hydrochloric acid

HPLC	high-performance liquid chromatography
IS	internal standard
KOH	potassium hydroxide
MeOH	methanol
MgCO$_3$	magnesium carbonate
N$_2$	nitrogen
NaCl	sodium chloride
NaSO$_4$	sodium sulphate
pVACs	provitamin A carotenoids
RP-HPLC	reverse phase high performance liquid chromatography
SFE	supercritical fluid extraction
TBME	*t*-butyl methyl ether
TEA	triethylamine
THF	tetrahydrofuran

References

Barth, M. M., Zhou, C., Kute, K. M. and Rosenthal, G. A., 1995. Determination of optimum conditions for supercritical fluid extraction of carotenoids from carrot (*Daucus carota L.*) tissue. Journal of Agricultural and Food Chemistry. 43: 2876–2878.

Baysal, T., Ersus, S. and Starmans, D., 2000. Supercritical CO$_2$ extraction of β-carotene and lycopene from tomato paste waste. Journal of Agricultural and Food Chemistry. 48: 5507–5511.

Britton, G. and Khachik, F., 2009. Carotenoids in Food. In: Britton, G., Pfander, H. and Liaaen-Jensen, S. (ed.) Carotenoids, Vol. 5. Birkhäuser Verlag, Basal, Switzerland, pp. 45–66.

Britton, G., Liaaen-Jensen, S. and Pfander, H. (ed.), 1995. In: Carotenoids Vol. 1A: Isolation and Analysis. Birkhaüser Verlag, Basal, Switzerland.

Careri, M., Furlattini, L., Mangia, A., Musc, M., Anklam, E., Theobald, A. and von Holst, C., 2001. Supercritical fluid extraction for liquid chromatographic determination of carotenoids in *Spirulina pacifica* algae: a chemometric approach. Journal of Chromatography A. 912: 61–71.

Chandler, L. A. and Schwartz, S. J., 1987. HPLC Separation of *cis–trans* carotene isomers in fresh and processed fruits and vegetables. Journal of Food Science. 52: 669–672.

Craft, N. E., Wise, S. A. and Soares, J. H., 1993. Individual carotenoid content of SRM-1548 total diet and influence of storage-temperature, lyophilization, and irradiation on dietary carotenoids. Journal of Agricultural and Food Chemistry. 41: 208–213.

Davey, M. W., Keulemans, J. and Swennen, R. L., 2006. Methods for the efficient quantification of fruit provitamin A contents. Journal of Chromatography A. 1136: 176–184.

Delgado-Vargas, F. and Paredes-López, O., 1997. Enzymatic treatment to enhance carotenoid content in dehydrated marigold flower meal. Plant Foods for Human Nutrition. 50: 163–169.

Edelenbos, M., Christensen, L. P. and Grevsen, K., 2001. HPLC determination of chlorophyll and carotenoid pigments in processed green pea cultivars (*Pisum sativum* L.). Journal of Agricultural and Food Chemistry. 49: 4768–4774.

Feltl, L., Pacáková, V., Štulík, K. and Volka, K., 2005. Reliability of carotenoids analyses: a review. Current Analytical Chemistry. 1: 93–102.

Fernandez, R. X. E., Shier, N. W. and Watkins, B. A., 2000. Effect of alkali saponification, enzymatic hydrolysis and storage time on the total carotenoid concentration of Costa Rican crude palm oil. Journal of Food Composition and Analysis. 13: 179–187.

Gouveia, L. and Empis, J., 2003. Relative stabilities of microalgal carotenoids in microalgal extracts, biomass and fish feed: effect of storage conditions. Innovative Food Science and Emerging Technologies. 4: 227–233.

Hart, D. and Scott, K., 1995. Development and evaluation of an HPLC method for the analysis of carotenoids in foods, and the measurement of the carotenoid content of vegetables and fruits commonly consumed in the UK. Food Chemistry. 54: 101–111.

Howe, J. A. and Tanumihardjo, S. A., 2006. evaluation of analytical methods for carotenoid extraction from biofortified maize (*Zea mays* sp.). Journal of Agricultural and Food Chemistry. 54: 7992–7997.

Kamffer, Z. 2009. Effect of some viticultural parameters on the profile of carotenoids and chlorophylls in grape berries of *Vitis vinifera* cv. Merlot. Master of Agricultural Sciences. Department of Viticulture and Oenology, Stellenbosch University. p. 124.

Kamffer, Z., Bindon, K. A. and Oberholster, A., 2010. Optimization of a method for the extraction and quantification of carotenoids and chlorophylls during ripening in grape berries (*Vitis vinifera* cv. Merlot). Journal of Agricultural and Food Chemistry. 58: 6578–6586.

Khachik, F., 2009. Analysis of Carotenoids in Nutritional Studies. In: Britton, G., Pfander, H. and Liaaen-Jensen, S. (ed.) Carotenoids, Vol. 5. Birkhäuser Verlag, Basel, Switzerland, pp. 7–44.

Khachik, F., Beecher, G. R. and Whittaker, N. F., 1986. Separation, identification, and quantification of the major carotenoid and chlorophyll constituents in extracts of several green vegetables by liquid chromatography. Journal of Agricultural and Food Chemistry. 34: 603–616.

Marsili, R. and Callahan, D., 1993. Comparison of a liquid solvent extraction technique and supercritical fluid extraction for the determination of α- and β-carotene in vegetables. Journal of Chromatographic Sciences. 31: 422–428.

Mertz, C., Gancel, A.-L., Gunata, Z., Alter, P., Dhuique-Mayer, C., Vaillant, F., Perez, A. M., Ruales, J. and Brat, P., 2009. Phenolic compounds, carotenoids and antioxidant activity of three tropical fruits. Journal of Food Composition and Analysis. 22: 381–387.

Mills, J. P., Simon, P. W. and Tanumihardjo, S. A., 2007. β-Carotene from red carrot maintains vitamin A status, but lycopene bioavailability is lower relative to tomato paste in Mongolian gerbils. Journal of Nutrition. 137: 1395–1400.

Mínquez-Mosquera, M. I. and Gandul-Rojas, B., 1994. Mechanism and kinetics of carotenoid degradation during the processing of green table olives. Journal of Agricultural and Food Chemistry. 42: 1551–1554.

Mínquez-Mosquera, M. I., Hornero-Méndez, D. and Pérez-Gálvez, A. 2008. Carotenoids and provitamin A in functional Foods. In: Hurst, W. J. (ed.) Methods of Analysis for Functional Foods and Nutraceuticals. CRC Press, Boca Raton, USA, pp. 280–340.

Park, Y., 1987. Effect of freezing, thawing, drying and cooking on carotene retention in carrots, broccoli and spinach. Journal of Food Science. 52: 1022–1025.

Pott, I., Marx, M., Neidhart, S., Muhlbauer, W. and Carle, R., 2003. Quantitative determination of β-carotene stereoisomers in fresh, dried, and solar-dried mangoes (*Mangifera indica* L.). Journal of Agricultural and Food Chemistry. 51: 4527–4531.

Rodriguez-Amaya, D. B., Kimura, M., Godoy, H. T. and Amaya-Farfan, J., 2008. Updated Brazilian database on food carotenoids: factors affecting carotenoid composition. Journal of Food Composition and Analysis. 21: 445–463.

Rodríguez-Bernaldo de Quirós, A. and Costa, H. S., 2006. Analysis of carotenoids in vegetable and plasma samples: A review. Journal of Food Composition and Analysis. 19: 97–111.

Serino, S., Gomez, L., Costagliola, G. and Gautier, H., 2009. HPLC assay of tomato carotenoids: validation of a rapid microextraction technique. Journal of Agricultural and Food Chemistry. 57: 8753–8760.

Taylor, K. L., Brackenridge, A. E., Vivier, M. A. and Oberholster, A., 2006. High-performance liquid chromatography profiling of the major carotenoids in *Arabidopsis thaliana* leaf tissue. Journal of Chromatography A. 1121: 83–91.

Turner, C., King, J. W. and Mathiasson, L., 2001. On-line supercritical fluid extraction/enzymatic hydrolysis of vitamin a esters: a new simplified approach for the determination of vitamins a and e in food. Journal of Agricultural and Food Chemistry. 49: 553–558.

Van den Berg, H., Faulks, R., Granado, H. F., Hirschberg, J., Olmedilla, B., Sandmann, G., Southon, S. and Stahl, W., 2000. The potential for the improvement of carotenoid levels in foods and the likely systemic effects. Journal of the Science of Food and Agriculture. 80: 880–912.

Vági, E., Simándi, B., Daood, H., Deák, A. and Sawinsky, J., 2002. Recovery of pigments for *Origanum majorana* L. by extraction with supercritical carbon dioxide. Journal of Agricultural and Food Chemistry. 50: 2297–2301.

Yeum, K. J. and Russell, R. M., 2002. Carotenoid bioavailability and bioconversion. Annual Review of Nutrition. 22: 483–504.

CHAPTER 20

Quantification of Carotenoids, Retinol, and Tocopherols in Milk and Dairy Products

BEATRICE DURIOT* AND BENOIT GRAULET

INRA, UR1213 Herbivores Research Unit, Team Microbial Digestion and Absorption, Theix, F-63122 Saint Genès Champanelle, France
*E-mail: Beatrice.Duriot@tours.inra.fr

20.1 Introduction

Vitamin A, vitamin E and carotenoids are fat-soluble micronutrients that play an important role in human health. Vitamin A, mainly represented by all-*trans* retinol, is necessary for normal vision, cellular differentiation, immune function, growth and reproduction. Vitamin A deficiency is a major public health problem, especially in developing countries since its prevalence in preschool-age children was 44% in Africa and 50% in South East Asia in 2005. Vitamin A deprivation can cause xerophthalmia, a very serious disease which causes firstly degradation of vision in darkness and ultimately blindness (World Health Organization 2005). Among the fat-soluble vitamins, vitamin A, as free or esterified retinol, is provided only by the consumption of food constituents of animal origin, liver and dairy products; milk and dairy products can be in some countries, such as The Netherlands, according to daily habits and to age, the first source of vitamin A intake (Hulshof *et al* 2006). However, although vegetables do not contain vitamin A, they are a source of carotenoids, some liposoluble pigments, among which the carotenes, for example, all-*trans* β-carotene, possess

Food and Nutritional Components in Focus No. 1
Vitamin A and Carotenoids: Chemistry, Analysis, Function and Effects
Edited by Victor R Preedy

provitamin A activity. So, fruit and vegetables consumption can also provide vitamin A (in the form of their metabolic precursors) and help to protect against vitamin A deficiency. Carotenoids also have other biological properties such as antioxidant activity and a protective effect in the prevention of diseases such as cancer, heart disease or age-related macular degeneration that affects vision. Carotenoids from ruminant diet (mainly from fresh grass) can be partially transferred into milk where they exert also organoleptic effects. Indeed, as pigments, carotenoids influence notably the color in the fat of dairy products (Nozière *et al* 2006a). Finally, they could also be used for traceability of dairy products origin and composition, *i.e.* to differentiate milks from different species: buffalo milk does not have the same carotenoids composition as cow's milk; or to discriminate between animals fed fresh grass *vs.* those receiving preserved forages.

Dairy products also contain vitamin E that is well known for its antioxidant properties and for promoting immunity. Vitamin E, the generic name of tocopherols and tocotrienols, is the most important liposoluble antioxidant for human health, protecting polyunsaturated fatty acids from oxidation in cell membranes and in plasma lipoproteins (Bramley *et al* 2000). Therefore it helps to protect against degenerative diseases in adult. It is also important for the oxidative stability of milk (Barrefors *et al* 1995), especially in the current research trying to increase the level of linolenic acid in milk fat for better nutritional quality of dairy products.

For all these reasons, it appears to be important to quantify these micronutrients to try to improve the nutritional quality of milk and dairy products.

20.2 Usual Concentrations in Milk and Dairy products and Variation Factors

Vitamin E and carotenoids present in milk mainly originate from diet, whereas vitamin A is not present in feedstuffs, thus it mainly originates from pro-vitamin A carotenoids conversion into retinol in the cow's tissues (mainly the intestinal cell wall), and from the balance between retinol storage and mobilization (mainly in the liver). The concentrations and the composition of these components in milk are the consequence of their bioavailability from feedstuffs (content, digestibility of the matrix and absorption efficiency) and their metabolism by the animal tissues (storage, utilization, mobilization and excretion), which are both also dependent on several physiological or environmental factors. Therefore, according to feeding, season, animal species and breed, concentrations and compositions of carotenoids and vitamins A and E in milk can vary (Nozière *et al* 2006a).

The usual concentrations of these components in bovine milk comprise 0.7–12 µg (g of fat)$^{-1}$ (0.03–0.6 µg ml^{-1}) for vitamin A, between 0 and 26 µg (g of fat)$^{-1}$ (0–1 µg ml^{-1}) for vitamin E and between 1.3 and 15 µg (g of fat)$^{-1}$ (0.04–0.5 µg ml^{-1}) for carotenoids (Chauveau-Duriot *et al* 2010; Nozière *et al*

2006a). All-*trans* retinol is the main form of vitamin A present in milk, the *cis* isomer of retinol being sometimes observed in a very low amount in cow's and goat's milk (Nozière *et al* 2006a). α-Tocopherol, the main biological form of vitamin E, is also the most common form of vitamin E observed in milk, but γ-tocopherol is sometimes detected (Chauveau-Duriot *et al* 2010; Romeu-Nadal *et al* 2006). All-*trans* β-carotene represents 75–85% of the total carotenoids present in cow's milk, other carotenoids being lutein, zeaxanthin, β-cryptoxanthin, 13-*cis* β-carotene and 9-*cis* β-carotene (Chauveau-Duriot *et al* 2010; Nozière *et al* 2006a).

Several studies have shown that concentrations of vitamins A and E and carotenoids vary according to the nature of forage fed to ruminant. Milk from cows receiving grass silage is richer in retinol (Martin *et al* 2004), tocopherols (Havemose *et al* 2004) and carotenoids than milk from cows fed corn silage (Havemose *et al* 2004; Martin *et al* 2004). The preservation of forage also has an influence, conserved forages as hay being poorest in carotenoids than grass silage [99 μg (g of dry matter (DM))$^{-1}$ *vs.* 457 μg (g of DM)$^{-1}$, respectively] (Chauveau-Duriot *et al* 2005). These differences are reflected in milk, from 1.5–5.8 μg of β-carotene (g of fat)$^{-1}$ (Nozière *et al* 2006a), since it has been shown that carotenoids concentration in milk are related to their dietary intake, even if the transfer of carotenoids from plasma to milk seems saturated when carotenoids concentration is very high (Figure 20.1) (Calderón *et al* 2007a).

Concentrations of vitamins A and E and carotenoids in milk are also dependent on animal species. Indeed, caprine milk is richer (30%) in retinol and α-tocopherol than bovine milk, and conversely β-carotene is 30% higher in cow's milk than goat's milk which does not contain this component in significant amounts (Lucas *et al* 2008). Similar results were observed with buffalo's milk compared with cow's milk: buffalo's milk contains higher amount of retinol and tocopherol than cow's milk, and does not contain carotenoids (Mihaiu *et al* 2010). One of the most evident (but not the sole) explanations for this between-breeds difference may be that goats and buffalos

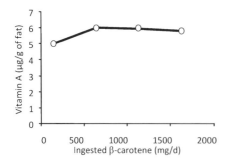 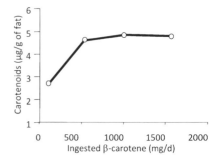

Figure 20.1   Relationship between β-carotene dietary intake and vitamin A and carotenoids concentrations in milk This Figure presents the increasing of concentrations of vitamin A and carotenoids in cow's milk with increasing ingested β-carotene. When β-carotene intake is very high, saturation seems to appear (adapted from Calderón *et al* 2007a).

have a better efficiency for carotenoid conversion into retinol than cows. As mentioned below, this variation according to species can be important for traceability of cheeses, because determination of all-*trans* β-carotene can allow to discriminate cow's milk cheese from buffalo or goat's milk cheese.

In the same species, breeds have also an influence on the concentrations of these components. In cows, β-carotene is higher in Jerseys's milk than in Friesians's milk (Winkelman *et al* 1999), whereas, in goats, retinol content is higher in Parbatsar than in Marwari's milk (Bohra *et al* 2010).

Concentrations of liposoluble micronutrients in milk also vary according to the stage of lactation. Colostrum secreted just after birth is very rich in vitamins A and E, and in carotenoids (Calderón *et al* 2007b). Then, in cow's milk, concentrations in these components strongly decrease in the first 10 days after calving and then do not vary or increase very slowly until the end of lactation (Figure 20.2) (Calderón *et al* 2007b; Duriot *et al* 2010).

Knowing that liposoluble microconstituents of cheese are not dependent on the type of cheese-making technology (Lucas *et al* 2008), the factors of variation cited above also have an influence on the composition and the concentration of carotenoids, vitamins A and E in cheeses and have to be taken into account to improve the nutritional quality of milk but also of dairy products, especially cheeses. Recent results show that concentrations in cheeses, from cows receiving grass from mountain grassland, are on average 4.2 µg (g of fat)$^{-1}$ for vitamin A (only all-*trans* retinol), 2.96 µg (g of fat)$^{-1}$ for vitamin E (96% of α-tocopherol) and 3.2 µg (g of fat)$^{-1}$ for carotenoids. The major carotenoids are lutein (37%) and all-*trans* β-carotene (36%); zeaxanthin, β-cryptoxanthin, 13-*cis* β-carotene and 9-*cis* β-carotene are also present in different proportions (10, 4, 10 and 3% respectively) (Graulet *et al* 2011). These

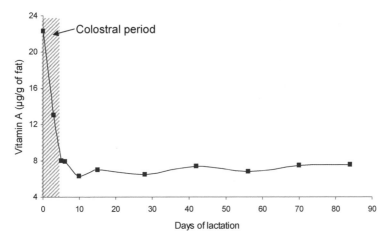

Figure 20.2   Variations in vitamin A concentration in milk during the first 3 months of lactation. Colostrum is very rich in vitamin A. The concentration of vitamin A in milk strongly decreases in the first 10 days after calving and then stays constant (adapted from Duriot *et al* 2010).

results differ from those obtained by Lucas *et al* (2008), especially in the proportion of lutein which is higher in this more recent study; these differences could be explained by the advances in analytical methods.

Considering that classical chromatographic analysis do not allow the efficient separation of lutein and zeaxanthin, of isomers of β-carotene, and of tocopherols, knowing that these molecules have distinct biological activities, analytical advances were necessary to allow a precise quantification of these compounds in milk and in dairy products. The development of a more resolutive, efficient, quick analytical method would be welcome to spare time and money.

20.3   Determination

Simultaneous determination of carotenoids and vitamins A and E using high-performance liquid chromatography (HPLC) have been described by several authors in plasma (Karppi *et al* 2008; Thibeault *et al* 2009) and in forages (Hao *et al* 2005), but only few authors have described the simultaneous determination of these components in milk (Calderón *et al* 2007a; Chauveau-Duriot *et al* 2010). More often, in milk, authors have described the extraction and determination of carotenoids only (Indyk 1987), or of vitamins A and E, or of one of each (Jensen and Nielsen 1996; Meglia *et al* 2006). Determination of these components in dairy products is unusual, but Lucas *et al* (2008) and Blanco *et al* (2000) have recently reported concentration values for these components in cheeses and in butters, respectively.

Here, we will describe the simultaneous determination of vitamins A and E and carotenoids in milk and in dairy products: cheese and butter.

20.3.1   Extraction Procedures

Simultaneous determination needs simultaneous extraction of these components. Procedures of extractions have to be adapted according to biological matrixes. For example, for plasma, that is a relatively simple matrix: a deproteinization, generally performed by adding ethanol, followed by purification with an organic solvent is adequate to extract simultaneously vitamins A and E and carotenoids; petroleum ether can be used for this extraction (Milne and Botnen 1986), but hexane is the most used (Miller and Yang 1985; Thibeault *et al* 2009). For forage samples, however, a longer procedure is used with a first extraction step using acetone to disrupt the vegetable cells followed by a saponification to eliminate chlorophyll esters (Britton 1995; Chauveau-Duriot *et al* 2010).

20.3.1.1   Milk

By comparison to plasma or forages, milk is a complex matrix, because of the high amount of lipids that can interact with lipophilic components. When only vitamins A and E are required to be investigated, simple methods can be applied, such as direct saponification (Meglia *et al* 2006), solvent extraction only (Romeu-Nadal *et al* 2006), or supercritical fluid extraction (Turner and Mathiasson 2000). The method described by Chauveau-Duriot *et al* (2010) to extract carotenoids and vitamins A and E simultaneously was adapted from the extraction method for carotenoids and vitamin A as described by Nozière *et al* (2006b). This method has been improved to obtain higher recoveries for vitamin E also (unpublished data). Indeed, adequate conditions are necessary on the one hand to protect vitamin E and on the other hand to release it because it could be associated to several matrix components such as membranes, lipoproteins and fat droplets (Rupérez *et al* 2001). Different conditions were tested (Table 20.1). Internal standards were added at the beginning of extraction to calculate extraction recoveries. This internal standard should not be present in the sample and should not be destroyed by extraction procedure; echinenone was used for vitamin A and carotenoids recoveries, and δ-tocopherol for vitamin E recovery. Tocopheryl acetate and retinyl acetate are often used for vitamin E and vitamin A recoveries respectively, but it was not possible here because of the saponification step during extraction which would hydrolyze them. Firstly, antioxidants were used to overcome oxidation, such as butylated hydroxytoluene (BHT) or acid ascorbic, have been tried, and they were not necessary as described previously by Rupérez *et al* (2001). Then, the conditions of saponification were tested after a solvent extraction using hexane. Saponification at 60 °C for 1 h with 10% KOH in ethyl alcohol (w/v) was chosen; 80 °C was too strong for carotenoids and vitamin A which were destroyed, and 37 °C does not improve vitamin E extraction. Then, solvents of different natures such as hexane, acetone/chloroforme (30:70), hexane/ethyl acetate (9:1) and proportions in relation to sample were tested. Hexane/ethyl acetate (9:1) in the same sample volume gave the best recovery values. When ethyl acetate is added to hexane, extraction solvent becomes more polar, interacts more with other components of the matrix (Romeu-Nadal *et al* 2006; Rupérez *et al* 2001), and improves the recovery. Increasing the shaking steps (frequency and time) is also very important to free up vitamin E from the lipid matrix (Rupérez *et al* 2001). Vortexing for 1 min followed by 10 min of shaking with a platform shaker gave good recoveries. Finally, the sample and final solvent volumes were adjusted.

The conditions used were described in Chauveau-Duriot *et al* (2010): 2 ml of milk were deproteinized by adding the same volume of ethyl alcohol (containing internal standards). After 1 min of vortexing and 10 min of agitation on a platform shaker, the same volume of *n*-hexane/ethyl acetate (9:1) was added to extract lipophilic components. After another 1 min of vortexing and 10 min of agitation on the platform shaker, the samples were centrifuged at 1000 *g* for 5 min, and the resulting organic phase was collected. This extraction step was repeated once, and the two organic phases were pooled. In the resulting organic phase, 2 ml of ethanol/water (90:10) were added to extract xanthophylls and vitamins E to preserve them from saponification. After 1 min

Table 20.1   Conditions tested to optimize simultaneous extraction of vitamins
A and E and carotenoids from milk samples. This Table describes
all of the conditions tested for the development of present the
extraction procedure for carotenoids, and vitamins A and E from
bovine milk. The tests were performed by a spike with internal
standards in milk samples at the beginning of the procedure, and
quantifying the corresponding compounds at the end of the
procedure (expressed as the percentage of recovery of the spike).
Different kind of tests were performed such as adding antioxidants, saponification (temperature, length and solvent), extraction solvent composition, mechanical action to favor extraction
and final volume of the extract. (Duriot, unpublished data).

		Recoveries (%)	
Conditions tested		*Carotenoids Vitamin A*	*Vitamin E*
Antioxidants	Without	78.59 ± 2.16	42.88 ± 3.12
	BHT	68.58 ± 6.6	44.49 ± 0.65
	Acid ascorbic (200 mg)	73.88 ± 2.21	47.11 ± 1.75
	Acid ascorbic (500 mg)	71.63 ± 4.69	45.40 ± 2.81
Saponification	37 °C, 1 h	73.05 ± 4.45	46.54 ± 4.10
KOH (10%) in pure ethanol	37 °C, 2 h	76.49 ± 0.50	51.02 ± 1.86
	60 °C, 30 min	72.65 ± 2.13	51.52 ± 0.77
	60 °C, 1 h	73.46 ± 0.42	54.24 ± 1.41
	60 °C, 2 h	65.07 ± 3.65	52.77 ± 0.17
	80 °C, 20 min	65.95 ± 0.77	50.31 ± 0.01
	80 °C, 45 min	64.74 ± 1.48	52.91 ± 0.71
	80 °C, 1 h	55.23 ± 2.52	50.84 ± 0.80
KOH (10%) in water/ ethanol (45:55)	80 °C, 20 min	31.55 ± 3.09	34.09 ± 4.47
	80 °C , 45 min	29.04 ± 2.34	32.75 ± 5.02
	80 °C, 1 h	35.36 ± 1.02	35.38 ± 0.51
Extraction solvents	Hexane	73.1 ± 2.3	62.4 ± 5.8
	Acetone/chloroform	30.6 ± 4.8	32.7 ± 3.4
	Hexane/ethyl acetate (9:1)	71.2 ± 5.6	72.8 ± 4.1
Vortex and shaking	30 s	72.28 ± 6.08	36.50 ± 2.97
	1 min	99.3	66
	1 min + 10 min on platform	93.8	80.7
Volumes (sample/ final)	2 ml/300 µl	109	72.05 ± 7.9
	2 ml/200 µl	100	61.93 ± 3.10
	1 ml/200 µl	88	63.89 ± 9.4

of vortexing and centrifugation at 1000 *g* for 5 min at room temperature, the lower ethanolic phase was collected in a new tube. This step was repeated once. The resulting lower phase was pooled with the previous one, and they both were evaporated under nitrogen. The remaining upper hexanic phase was also evaporated under nitrogen, and the dry residue was saponified in 2 ml of a solution of 10% KOH in ethyl alcohol (w/v) for 1 h at 60 °C in a shaking water bath under darkness. The reaction was stopped by adding 2 ml of water on an ice bath. Then, 2 ml of *n*-hexane/ethyl acetate (9:1) were added for carotenes and vitamins purification. This step was repeated twice, and the three successive hexanic phases were pooled with the ethanolic phases evaporated previously. They were evaporated under nitrogen and the final dry residue was dissolved first in 30 µl of tetrahydrofurane followed by 270 µl of acetonitrile/dichloromethane/methanol (75:10:15), transferred into a 2 ml glass screw-top vial, and 10 µl was injected in the UPLC system. Extraction recoveries were up to 70% for carotenoids and vitamins A and E.

Extraction of these components are always carried out in the dark because of their sensitivity to light.

20.3.1.2 Dairy Products

The extraction method used for simultaneous determination of carotenoids, retinol and tocopherols in milk has been adapted to purify these components from cheese (Graulet *et al* 2011) or from butter (Chauveau-Duriot, unpublished data). Different specifications were necessary because of their high amount of lipids.

20.3.1.2.1 Cheeses

Three ml of *n*-hexane/ethyl acetate (9:1) were firstly added to 750 mg of lyophilized and ground cheese to begin the extraction of the lipophilic components. After 1 min of vortexing and 10 min of agitation on a platform shaker, samples were deproteinized by adding 3.5 ml of ethyl alcohol containing internal standards (echinenone and δ-tocopherol), and vortexing for 1 min. Then 3 ml of water were added and after another 1 min of vortexing and 10 min of agitation on the platform shaker, samples were centrifuged at 1000 *g* for 5 min, and the resulting organic phases were collected. The extraction step, with 3 ml of *n*-hexane/ethyl acetate (9:1), was repeated once, and the two organic phases were pooled. On the resulting organic phase, 3 ml of ethanol/water (90:10) were added to extract xanthophylls and vitamins E to preserve them from saponification. After 1 min of vortexing and centrifugation at 1000 *g* for 5 min at room temperature, the lower ethanolic phase was collected in a new tube. This step was repeated once, and the two lower phases were evaporated under nitrogen. The remaining upper hexanic phase was also evaporated under nitrogen, and the dry residue was saponified in 2 ml of a solution of 10% KOH in ethyl alcohol (w/v) for 2 h at 60 °C in a shaking water bath under darkness. The reaction was stopped by adding 2 ml of water on an ice bath. Then, 2 ml of *n*-hexane/ethyl acetate (9:1) were added for the purification of carotenes and vitamins. This step was repeated twice, and the three successive hexanic phases were pooled with the ethanolic phases evaporated previously. They were evaporated under nitrogen and the final dry residue was dissolved first in 30 μl of tetrahydrofurane followed by 270 μl of acetonitrile/dichloromethane/methanol (75:10:15), and transferred into a 2 ml glass screw-top vial and 10 μl was injected in the UPLC system. Extraction recoveries were approx. 90% for carotenoids, retinol and tocopherols.

20.3.1.2.2 Butter

The same extraction was used for butter as for milk, except with the two following modifications. Firstly, the extraction was performed from 200 mg of butter melted at 37 °C in a water bath. Secondly, the length of saponification was increased to 3 h instead of 1 h. Following these analytical conditions, mean extraction recoveries were approx. 75% for carotenoids and vitamins E.

20.3.2   Quantification

20.3.2.1   HPLC and U-HPLC

HPLC has become the reference method to analyze these components (Table 20.2). The simultaneous determination of retinol, tocopherols and carotenoids needs to use specific conditions which allow the separation of carotenoids, because the latter is more difficult than for retinol or tocopherols due to the numerous xanthophylls and geometrical isomers of carotenes, especially in complex matrices such as animal feed, milk or dairy products. In milk, all-*trans* retinol is the only form present of vitamin A, and, concerning tocopherol isomers, the major difficulty is to separate β- and γ-tocopherol, but β-tocopherol is not present in milk. So, the separation of vitamins does not present an important problem. The simultaneous determination of these different components involves also a simultaneous detection. Diode array detector (DAD) is the detector that is used most often (Chauveau-Duriot *et al* 2010, Chávez-Servín *et al* 2006; Karppi *et al* 2008), because it is able to record a specific spectral range and to extract specific wavelengths; carotenoids are generally monitored at 450 nm, retinol at 325 nm, and vitamins E at 292 nm. Moreover, this detector allows identification of peaks using their spectral absorption pattern which is characteristic of one molecule. Multi-wavelength detectors have also been used (Miller and Yang 1985). Some methods using two types of detector have been described, the DAD or ultraviolet/visible (UV/Vis) spectrophotometric detector is coupled with another detector, generally, a fluorescence detector for vitamins E and A detection (Jensen and Nielsen 1996; Meglia *et al* 2006); the detection of tocopherols is generally performed with excitation and emission wavelengths of approx. 296 and 330 nm, respectively,

Table 20.2 Different HPLC or U-HPLC methods described to analyze retinol, tocopherols and carotenoids alone, or simultaneously. The Table presents a bibliographic overview of methods used to quantify simultaneously the carotenoids, vitamins A and E from several matrices. Analytical conditions, column chemistry and detection mode are listed. ELSD, evaporative light scattering detector; NI, not indicated; PDA, photodiode array.

Authors	Matrix	Components analysed	Column	Mobile phase	Detector	Flow (ml min^{-1})	Run time (min)
Rodas-Mendoza et al (2003)	Infant formula	Retinol Vitamins E	RP C18 Spherisorb ODS2, 250 × 4.6 mm, 5 μm	Methanol (100%)	DAD	1	20
Escrivá et al (2002)	Milk	Retinol Vitamins D Vitamins E	RP C18 Spherisorb ODS2, 250 × 4.6 mm, 5 μm	Methanol (94%) Water (6%)	UV	NI	20
Romeu-Nadal et al (2006)	Human milk	Vitamins E (α and γ)	C18 Pinnacle II, 500 × 2.1 mm, 3 μm	Acetonitrile/methanol/dichloromethane (60:38:2)	UV/Vis	0.2	6
Romeu-Nadal et al (2006)	Human milk	Vitamins E (α and γ)	C18 Pinnacle II, 500 × 2.1 mm, 3 μm	Methanol	ELSD	0.2	6
Chávez-Servín et al (2006)	Infant milk formula	Vitamins E (α,γ and δ)	NP Pinnacle II silica, 500 × 2.1 mm, 3 μm	Hexane/ethyl acetate (99.5:0.5)	PDA	0.4	25
Albalá-Hurtado et al (1997)	Infant milk formula	Retinol α-Tocopherol	RP C18 Tracer Spherisorb ODS2, 25 × 4.6 mm, 5 μm	Water/acetonitrile/methanol (4:1:95)	UV	NI	25
Meglia et al (2006)	Cow's milk	Vitamins E	Chiralcel OD-H, 25 × 4.6 mm, 5 μm	Heptane + 2-propanol (0.065ml L^{-1})	Fluorescence	1	NI

Table 20.2 (*Continued*)

Authors	Matrix	Components analysed	Column	Mobile phase	Detector	Flow (ml min^{-1})	Run time (min)
Turner and Mathiasson (2000)	Milk powder	Retinol α-Tocopherol	RP Lichrospher RP-18, 250 × 4 mm, 5 μm	Methanol/water (96:4)	UV	1	30
Hulshof et al (2006)	Cow's milk and dairy products	Retinol Carotenoids	C18 Vydac 218TP53, 250 × 3.2 mm, 5 μm	Methanol/water/THF/triethylamine (87.9:10:2:0.1) from 0.25 to 0.5 min Methanol/water/THF (92.4:7.5:0.1:0) from 0.5 to 20 min	PDA	0.7	20
Hulshof et al (2006)	Cow's milk and dairy products	Retinol	NP BDS Hypersil CN, 150 × 3 mm, 5 μm	Hexane + 0.01% BHT/isopropanol (98.5:1.5)	PDA	0.7	20
Liu et al (1998)	Human milk	Retinol Carotenoids	C18 YMC, 250 × 4.6 mm, 5 μm	95% A: Acetonitrile/THF (85:15) + BHT + TEA (0.05%) 5% B: Methanol/ammonium acetate (50 mM) + TEA (0.05%)	UV	2.5	13
Jensen and Nielsen (1996)	Cow's milk	Retinol	HS-5-Silica, 125 × 4 mm, 5 μm	Heptane + 2-propanol (60 ml L^{-1})	Fluorescence	NI	NI
		Tocopherols	HS-5-Silica, 125 × 4 mm, 5 μm	Heptane + 2-propanol (60 ml L^{-1})	Fluorescence	NI	NI
		β-Carotene	Supelcosil LC-NH2, 250 × 4.6 mm, 5 μm	Heptane + 2-propanol (50 ml L^{-1}) + triethylamine (1 ml L^{-1})	UV/Vis	NI	NI

Table 20.2 (*Continued*)

Authors	Matrix	Components analysed	Column	Mobile phase	Detector	Flow (ml min^{-1})	Run time (min)
Nozière *et al* (2006b) according to Lyan (2001)	Milk	Retinol Carotenoids	2 columns in series: RP C18 Nucleosil, 150 × 4.6 mm, 3 μm RP C18 Vydac TP54, 250 × 4.6 mm, 5 μm	Acetonitrile/dichloromethane/ ammonium acetate (0.05 M) in methanol/water	PDA	2	36
Chauveau-Duriot *et al* (2010)	Milk Plasma Forages	Retinol Carotenoids Vitamins E	Acquity UPLC HSS T3, 150 × 2.1 mm, 1.8 μm	A: Acetonitrile/ dichloromethane/methanol (75:10:15) B: Acetate ammonium (0.05 M) in water	PDA	0.4	46

and with excitation and emission wavelengths of 344 and 472 nm for retinol. Evaporative light scattering detection is sometimes used for vitamin E determination in milk but it is less sensitive than the UV/Vis detector (Romeu-Nadal 2006).

Some normal phase (NP) HPLC methods have been described for separating carotenoids or liposoluble vitamins (Chávez-Servín *et al* 2006; Hao *et al* 2005); this method is interesting for milk because it makes saponification, to remove fat, unnecessary. But, reversed phase (RP) HPLC is more frequently used. Different RP monomeric or polymeric C18 columns are available: Zorbax, Vydac 201 TP, Spheri-5-RP-18 and Spheri-5-ODS (Oliver and Palou 2000). An option chosen by some authors is the combination of two columns in series: for example, Lyan *et al* (2001) used a 150 × 4.6 mm, RP C18, 3 μm, Nucleosil, and a 250 × 4.6 mm, RP C18, 5 μm, VydacTP54, to separate 13 carotenoids present in human plasma in 50 min; in another study, Karppi *et al* (2008) used a pair of 150 × 4.6mm, 4 μm, Synergy Hydro RP 80A allowing the determination of retinol and α-tocopherol, and the separation and determination of six carotenoids in plasma in a 35 min run. Finally, a C30 stationary phase is also described to have a superior resolution for carotenoid isomer separation (Sander *et al* 1994), for that reason, it has become a method of choice for carotenoids analysis, even if time analysis is longer (approx. 50 min).

In the last decade, rapid liquid chromatography such as ultra-high-performance liquid chromatography (UPLC or U-HPLC) has been developed on the basis of columns with 1.7 μm or 1.8 μm particles, so operating at very high pressure. It promised higher sensitivity, faster analysis time and better resolution than HPLC. Chauveau-Duriot *et al* (2010) have published one of the first methods of simultaneous determination of carotenoids, vitamins A and E with an UPLC system in milk, forages, and plasma.

In this study, a 150 × 2.1 mm Acquity UPLC HSS T3, 1.8 μm column was used with a gradient of acetonitrile/dichloromethane/methanol (75:10:15) (A) and acetate ammonium 0.05 M in water (B). The linear gradient consisted of 75:25 (A/B) at initial conditions, 75:25 (A:B) from 0 to 20 min, 100:0 (A/B) from 20 to 21 min, 98:2 (A/B) from 21 to 30 min, 98:2 (A/B) from 30 to 44 min, and finally returned to initial conditions. The flow rate was 0.4 ml min^{-1}, and the run time was 45 min. The column temperature was maintained at 35 °C using a column oven. The apparatus was equipped with a DAD settled for scanning at between 210 and 600 nm, and carotenoids, vitamins A and vitamins E were detected at 450, 325 and 292 nm, respectively.

Moreover, the authors also compared the chromatography conditions designed for the quantification of carotenoids, vitamins A and E with the UPLC system to those of an HPLC system. The method using UPLC offers higher sensitivity (limits of detection and quantification are lower), especially for vitamins A and E, and a better resolution than HPLC. The increase in this latter parameter with UPLC makes it possible to optimize the separation of the compounds present in matrices such as forages. Indeed, carotenoids in these types of matrices are very difficult to separate because of the similarities in their chemical properties (especially polarity). Thus, they require specific column length (150 mm) and analytical conditions (25% of water at start). The UPLC conditions presented in this work gave the best quality results for separation of forage-based carotenoids among those tested, but did not produce a reduction in analysis time (45 min), compared with HPLC. A similar problem exists with HPLC using C30 column. Methods were optimized for good discrimination of geometrical isomers of carotenoids (Figure 20.3) and, so, reducing the time was not the first aim. For plasma, milk or dairy products samples, the relatively few types of carotenoid present make it viable to use a shorter run time analyse using other column chemistry with a smaller length. However, with the perspective of feed traceability through plasma, milk or dairy products analysis, it seems important to analyze milk, plasma and dairy products samples under the same chromatographic conditions as forages. In the same way, if only vitamins A and E had been analyzed, another column and mobile phase could have been used, and time analysis could have been reduced. For example, a simultaneous quantification in plasma of vitamins A

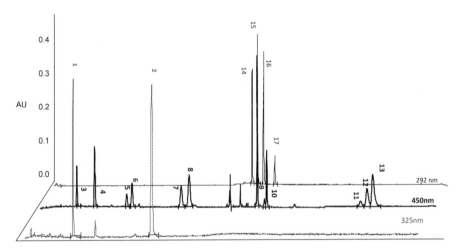

Figure 20.3 Chromatograms of standards of vitamins A and E and carotenoids. Standards of vitamin A monitored at 325 nm: all-*trans* retinol (1) and retinyl acetate (2); of carotenoids monitored at 450 nm: neoxanthin (3), violaxanthin (4), lutein epoxide (5), antheraxanthin (6), zeaxanthin (7), lutein (8), β-cryptoxanthin (9), echinenone (10), 13-*cis* β-carotene (11), 9-*cis* β-carotene (12) and all-*trans* β-carotene (13); and of vitamins E monitored at 292 nm: δ-tocopherol (14), γ-tocopherol (15), α-tocopherol (16) and tocopheryl acetate (17). Chromatographic conditions: column: 150 x 2.1 mm Acquity UPLC HSS T3, 1.8 μm; gradient: linear gradient 75:25 [acetonitrile/dichloromethane/methanol, 75-10-15; acetate ammonium (0.05 M) in water] (A/B) from 0 to 20 min, 100:0 (A/B) from 20 to 21 min, 98:2 (A/B) from 21 to 30 min, 98:2 (A:/B) from 30 to 44 min; flow rate: 0.4 ml/min. AU, absorbance units.

and E only with an UPLC system has been described by Citová *et al* (2007), with a very short time analysis (less than 3 min).

Technological advances (C30 column, UPLC) involve better resolution, therefore more detectable peaks; although some of them are unidentified, because of the lack of standards available. Identification could be made easier by coupling mass spectrometry (MS) or nuclear magnetic resonance (NMR) detectors to HPLC or UPLC.

20.3.2.2 Alternative Methods

In some particular situations, HPLC or UPLC can not be used because of their cost, or because they are not easy to use fieldwork. Therefore, alternative methods can be proposed to determine simultaneously vitamins A and E, and carotenoids.

20.3.2.2.1 Spectrophotometry

Spectrophotometric methods have been described for a long time, or more recently to determine retinol and/or carotene in plasma (Bessey *et al* 1946), in foodstuffs (Rodriguez-Amaya *et al* 1988) and in milk (Calderón 2007c). Calderón (2007c) has adapted the method described by Nozière et al (2006b) to determine carotenoid content in milk by spectrophotometry. The study compared 68 milk samples known to have a large variation in carotenoid content. Extraction had to be modified to be in agreement with detectable concentrations by the spectrophotometer and with sample volume necessary for spectrometric cells. The principal changes were: (1) the assay volume of sample: 6 ml of milk; (2) the volume of ethyl alcohol used for deproteinisation (6 ml); (3) the volume of *n*-hexane (4 ml) added to extract lipophilic components; (4) the disappearance of the step to extract xanthophylls from organic phase and preserve them from saponification; (5) the final volume: the dry residue was dissolved in 1.5 ml of solvent. Carotenoid content was determined by spectrophotometry at 450 nm using molecular coefficients of β-carotene (2 560 $\varepsilon^{1\%}$1 cm), as it is the main component found in milk, and, moreover, xanthophylls have not been protected against saponification. As xanthophylls represent 15% of milk's carotenoids, the results were multiplied by 1.15. The slope of the regression between values obtained with HPLC and spectrophotometry was not significantly different to 1, even if carotenoids content was few underestimated by spectrophotometry (Figure 20.4).

Similarly, this spectrophotometric method could also be optimized to determinate or to estimate simultaneously vitamins A and E, and carotenoids.

20.3.2.2.2 Color Index

A method to determine the color of milk has been described (Calderón *et al* 2007b). No extraction is necessary, so it is a rapid and economical method and

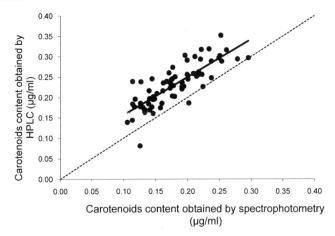

Figure 20.4 Comparison between a spectrophotometric and an HPLC method for carotenoids quantification in cow's milk. Slope of regression between carotenoids contents in cow's milk quantified by spectrophotometry (X) and by HPLC (Y) is $Y = 0.064\,(\pm0.015, P < 0.001) + 0.929\,(\pm0.082, P < 0.001)\,X$; $N = 68$, RSM $= 0.031$, $R^2 = 0.66$. (Calderón 2007c, reproduction with the authorization of the author).

easy to use in fieldwork. The color index (CI) is determined by reflectance, using a spectrocolorimeter. The CI corresponded to the absolute value of the integral of the translated spectrum between 450 and 530 nm (Figure 20.5). There is a relationship between CI and β-carotene in milk which appears linear within the usual target observed concentrations. Therefore this method can give a rapid idea of the carotenoid content, but not composition, in milk.

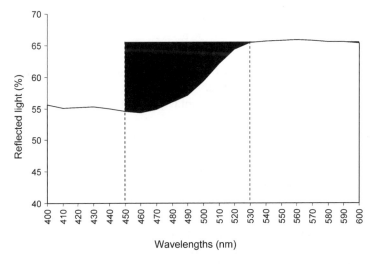

Figure 20.5 Reflectance spectrum of milk. The CI is the black upper area between 450 and 530 nm (10 nm steps). (Calderón *et al*, 2007c, reproduction with the authorization of the author).

The same technique can be used for plasma, with reflectance being replaced by absorbance (Calderón *et al* 2007b).

20.4   Conclusion

Different methods for the determination of vitamins A and E and carotenoids have been described here. Spectroscopic methods are easy to use in fieldwork and not very expensive, but allow the determination or estimation only of the global concentrations of these components. Chromatographic methods are more expensive and mainly used in the research laboratory, but they allow the detailed composition of vitamins A, E and carotenoids to be obtained, and to calculate the concentrations of each known component.

Milk appears to be a good source of vitamins A (as retinol and carotenes). Carotenoids and vitamin A have strong metabolic links, as some carotenoids have provitamin A activity. Together with vitamin E (that are very important liposoluble antioxidants), they protect against diseases linked to oxidative stress, for example. Moreover, in milk and dairy products, they can help to protect polyunsaturated fatty acids from oxidation. Therefore, they have high (direct or indirect) nutritional effects for the consumer and they can also be used for traceability. This is why the simultaneous determination of vitamins A and E, and carotenoids in milk seems to be of particular interest. Moreover, simultaneous determination is time-saving: only one extraction for the three components' families, only one injection and one run analysis on U-HPLC.

It would be interesting to determine the other liposoluble vitamins (D and K) in the same time, but it appears to be difficult because of their low concentrations in milk by contrast with those of vitamins A and E, and carotenoids.

Summary Points

- Vitamin A, vitamin E and carotenoids are fat-soluble micronutrients that have distinct biological activities.
- Vitamin A, vitamin E and carotenoids play an important role in human health.
- Milk and dairy products are a major source of vitamin A intake in human nutrition.
- It appears to be important to identify and quantify these micronutrients to evaluate the nutritional quality of milk and dairy products.
- Milk and dairy products are very complex matrices because of their amount of lipids and proteins that can interact with lipophilic component.
- Simultaneous extraction and quantification of carotenoids and vitamins A and E represents a real time-saving.
- This Chapter focuses on methods describing simultaneous extraction of vitamin A, vitamins E and carotenoids from milk and dairy products (cheese and butter).

- A method of separation and simultaneous quantification of these components, using UPLC, is also presented.

Key Facts

Key Facts of Simultaneous Extraction Vitamins A and E and Carotenoids

- The extraction procedure is specific to each matrix.
- The first step of extraction procedure of liposoluble components is often a step of deproteinization.
- Saponification is necessary to extract vitamin A and carotenoids from milk and dairy products.
- Conditions used for saponification are not the same for milk, cheese and butter.
- Specific conditions are required to release vitamin E that could be associated to several matrix components, such as membranes, and protecting it from saponification.

Key Facts of Simultaneous Determination of Vitamins A and E and Carotenoids

- HPLC is the reference method to analyse these components, and the most often used is RP-HPLC. It allows the determination of the complete chemical composition of these components.
- In the last decade, UPLC (an improved method of liquid chromatography) has been developed promising higher sensitivity, faster analysis time and better resolution than HPLC.
- Forages contain a lot (content and molecular complexity) of carotenoids, especially numerous xanthophylls, and isomers of carotenes that can be difficult to separate.
- Carotenoids, especially xanthophylls and geometrical isomers of carotenes, are very difficult to separate. However, the separation of vitamins is relatively easy because retinol is the only form of vitamin A in milk and dairy product, and a good resolution of isomers of tocopherols can be obtained efficiently with simple experimental procedures.
- In an aim of traceability, it is important to use the same chromatographic method for all matrixes (feedstuffs, milk, dairy products, plasma, *etc.*). Chromatographic conditions have to be developed for carotenoids from the most complex matrix.

Definition of Words and Terms

Carotenoids: Carotenoids are natural pigments present in vegetables. There are two classes of carotenoids (xanthophylls and carotenes). Some of them are precursors of vitamin A, but they have their own biological effect in vision, reproduction or antioxidant activity. In this manner, they have protective effects against diseases such as cancers, heart diseases or age-related macular degeneration of the retina.

Forages: Forages are feedstuffs in ruminant nutrition composed of plants eaten fresh or conserved (hay, straw, silage) as pasture, crop residue or immature cereal crops.

High-performance liquid chromatography (HPLC): HPLC is a liquid chromatography technique used to separate and quantify different components from a sample. The sample is forced by a liquid mobile phase through a stationary phase (column) at high pressure (HPLC). The particle size of the stationary phase most often used is 3 or 5 μm. Separation of the different components occurs according to their polarity.

Saponification: Saponification is a method to hydrolyse triacylglycerols (esters of fatty acid) and release fatty acid salts and glycerol. Sodium hydroxide or potassium hydroxide is often used for alkaline saponification. It is necessary to perfom a saponification to extract liposoluble components, such as fat globules in milk or dairy products, from some matrixes.

Traceability: Traceability is the ability to verify all steps in a production process. In recent years, in animal production, methods for tracing diets of ruminants via milk or meat composition have been developed.

UPLC: UPLC is an evolution of HPLC method but particle size of stationary phase is smaller (less than 2 μm), so it operates at very high pressure, promising higher sensitivity, better resolution and faster analysis time than HPLC.

Vitamin A: Vitamin A is a fat soluble micronutrient that is important in human health, for vision, cellular differentiation, immune function, growth and reproduction. Milk is a very important source of vitamin A (preformed and as pro-vitamin A carotenoids).

Vitamin E: Vitamin E is a group of liposoluble compounds, including tocopherols and tocotrienols, which are mainly present in cellular membranes. Their main action is their antioxidant activity that helps to fight against acute stress but also to protect from degenerative diseases.

List of Abbreviations

BHT	butylated hydroxytoluene
DAD	diode array detector
DM	dry matter
CI	color index
HPLC	high-performance liquid chromatography
NP	normal phase
RP-HPLC	reversed phase HPLC
UPLC/U-HPLC	ultra-high-performance liquid chromatography
UV	ultraviolet
Vis	visible

References

Albalá-Hurtado, S., Novella-Rodríguez, S., Veciana-Nogués, M. T. and Mariné-Font, A., 1997. Determination of vitamins A and E in infant milk formulae by high-performance liquid chromatography. Journal of Chromatography A. 778: 243–246.

Barrefors, P., Granelli, K., Appelqvist, L. A. and Bjoerck, L., 1995. Chemical characterization of raw milk samples with and without oxidative off-flavor. Journal of Dairy Science. 78: 2691–2699.

Bessey, O. A., Lowry, O. H., Brock, M. J. and Lopez, J. A., 1946. The determination of vitamin A and carotene in small quantities of blood serum. Journal of Biological Chemistry. 166: 177–188.

Blanco, D., Fernandez, M. P. and Guttierrez, M. D., 2000. Simultaneous determination of fat-soluble vitamins and provitamins in dairy products by liquid chromatography with a narrow-bore column. Analyst. 125: 427–431.

Bohra, H. C., Khan, H. C., Patel, A. K. and Mathur, B. K., 2010. Milk carotene and retinol levels in cattle and goats. Indian Veterinary Journal. 87: 45–46.

Bramley, P. M., Elmadfa, I., Kafatos, A., Kelly, F. J., Manios, Y., Roxborough, H. E., Schuch, W., Sheehy, P. J. A. and Wagner, K. H., 2000. Review. Vitamin E. Journal of the Science of Food and Agriculture. 80: 913–938.

Britton, G., 1995. Example 1: higher plants. In: Britton, G. Liaaen-Jensen, S. and Pfander H. (ed.), Carotenoids, Vol. 1A. Birkhaüser, Basel, Switzerland, pp. 201–214.

Calderón, F., Chauveau-Duriot, B., Pradel, P., Martin, B., Graulet, B., Doreau, M. and Nozière, P., 2007a. Variations in carotenoids, vitamins A and E, and color in cow's plasma and milk following a shift from hay diet to diets containing increasing levels of carotenoids and vitamin E. Journal of Dairy Science. 90: 5651–5664.

Calderón, F., Chauveau-Duriot, B., Martin, B., Graulet, B., Doreau, M. and Nozière, P., 2007b. Variations in carotenoids, vitamins A and E, and color in cow's plasma and milk during late pregnancy and the first three months of lactation. Journal of Dairy Science. 90: 2335–2346.

Calderón, F., 2007c. Evolution des concentrations en caroténoïdes du plasma et du lait chez la vache laitière : « Effets de l'alimentation et du stade de lactation ». Mémoire de thèse. Université de Rennes - ENSA Rennes. 145 pp.

Chávez-Servín, J. L., Castellote, A. I. and López-Sabater, M. C., 2006. Simultaneous analysis of vitamins A and E in infant milk-based formulae by normal-phase high-performance liquid chromatography-diode array detection using a short narrow-bore column. Journal of Chromatography A. 1122: 138–143.

Chauveau-Duriot, B., Thomas, D., Portelli, J. and Doreau, M., 2005. Carotenoid content in forages: variation during conservation. Rencontres autour des Recherches sur les Ruminants. 12: 117.

Chauveau-Duriot, B., Doreau, M., Noziere, P. and Graulet, B., 2010. Simultaneous quantification of carotenoids, retinol, and tocopherols in forages, bovine plasma, and milk: validation of a novel UPLC method. Analytical and Bioanalytical Chemistry. 397: 777–790.

Citová, I., Havlíková, L., Urbánek, L., Solichová, D., Nováková, L. and Solich, P., 2007. Comparison of a novel ultra-performance liquid chromatographic method for determination of retinol and α-tocopherol in human serum with conventional HPLC using monolithic and particulate columns. Analytical and Bioanalytical Chemistry. 388: 675–681.

Duriot, B., Pradel, P., Nozière, P., Troquier, O., Martin, B., Cirie, C. and Graulet, B., 2010. Carotenoid and vitamin A colour and concentration variations in plasma and cow's milk during lactation. Rencontres autour des Recherches sur les Ruminants. 17: 400.

Escrivá, A., Esteve, M.J., Farré, R. and Frígola, A., 2002. Determination of liposoluble vitamins in cooked meals, milk and milk products by liquid chromatography. Journal of Chromatography A. 947: 313–318.

Graulet, B., Chatelard, C., Lepetit, M., Blay, B., Hulin, S., Duriot, B. and Martin, B., 2011. Carotenoids, vitamins A and E concentrations in Saint Nectaire cheeses according to process and season. Proceedings of the 10th International Meeting on Mountain Cheese, 14–15 September 2011, Dronero, Italy. University of Turin Editors, pp. 51–52.

Havemose, M. S., Weisbjerg, M. R., Bredie, W. L. P. and Nielsen, J. H., 2004. Influence of feeding different types of roughage on the oxidative stability of milk. International Dairy Journal. 14: 563–570.

Hao, Z., Parker, B., Knapp, M. and Yu, L. L., 2005. Simultaneous quantification of α-tocopherol and four major carotenoids in botanical materials by normal phase liquid chromatography-atmospheric pressure chemical ionization-tandem mass spectrometry. Journal of Chromatography A. 1094: 83–90.

Hulshof, P. J. M., van Roekel-Jansen, T., van de Bovenkamp, P. and West, C. E., 2006. Variation in retinol and carotenoid content of milk and milk products in The Netherlands. Journal of Food Composition and Analysis.19: 67–75.

Indyk, H., 1987. The rapid determination of carotenoids in bovine milk using HPLC. Journal of Micronutrient Analysis. 3: 169–183.

Jensen, S. K. and Nielsen, K. N., 1996. Tocopherols, retinol, β-carotene and fatty acids in fat globule membrane and fat globule core in cow's milk. Journal of Dairy Research. 63: 565–574.

Karppi, J., Nurmi, T., Olmedilla-Alonso, B., Granado-Lorencio, F. and Nyyssönen, K., 2008. Simultaneous measurement of retinol, α-tocopherol and six carotenoids in human plasma by using an isocratic reversed-phase HPLC method. Journal of Chromatography B. 867: 226–232.

Liu, Y., Xu, M. J. and Canfield, L. M., 1998. Enzymatic hydrolysis, extraction, and quantification of retinol and major carotenoids in mature human milk. Journal of Nutrition and Biochemistry. 9: 178–183.

Lucas, A., Rock, E., Agabriel, J., Chilliard, Y. and Coulon, J. B., 2008. Relationships between animal species (cow *versus* goat) and some nutritional constituents in raw milk farmhouse cheeses. Small Ruminant Research. 74: 243–248.

Lyan, B., Azaïs-Braesco, V., Cardinault, N., Tyssandier, V., Borel, P., Alexandre-Gouabau, M. C. and Grolier, P., 2001. Simple method for clinical determination of 13 carotenoids in human plasma using an isocratic high-performance liquid chromatographic method. Journal of Chromatography B. 751: 297–303.

Martin, B., Fedele, V., Ferlay, A., Grolier, P., Rock, E., Gruffat, D. and Chilliard, Y., 2004. Effects of grass-based diets on the content of micronutrients and fatty acids in bovine and caprine dairy products. In: Lüscher, A., Jeangros, B., Kessler, W., Huguenin, O., Lobsiger, M., Millar, N. and Suter, D. (ed.) Land Use Systems in Grassland Dominated Regions. Vdf, Zürich, Switzerland, 9: 876–886.

Meglia, G. E., Jensen, S. K., Lauridsen, C. and Persson Waller, K., 2006. α-Tocopherol concentration and stereoisomer composition in plasma and milk from dairy cows fed natural or synthetic vitamin E around calving. Journal of Dairy Science. 73: 227–234.

Mihaiu, M., Pintea, A., Bele, C., Lapusan, A., Mihaiu, R., Dan, S. D., Taulesco, C. and Ciupa, A., 2010. Investigations on the nutritional and functional value of the buffalo milk. Bulletin of University of Agricultural Sciences and Veterinary Medicine Cluj-Napoca. 67: 155–160.

Miller, K. W. and Yang, C. S., 1985. An isocratic high-performance liquid chromatography method for the simultaneous analysis of plasma retinol, α-tocopherol, and various carotenoids. Analytical Biochemistry. 145: 21–26.

Milne, D. B. and Botnen, J., 1986. Retinol, α-tocopherol, lycopene, and α- and β-carotene simultaneously determined in plasma by isocratic liquid chromatography. Clinical Chemistry. 32: 874–876.

Nozière, P., Graulet, B., Lucas, A., Martin, B., Grlier, P. and Doreau, M., 2006a. Carotenoids for ruminants: from forages to dairy products. Animal Feed Science and Technology. 131: 418–450.

Nozière, P., Grolier, P., Durand, D., Ferlay, A., Pradel, P. and Martin, B., 2006b. Variations in carotenoids, fat soluble micronutrients, and color in cows' plasma and milk following changes in forage and feeding level. Journal of Dairy Science. 89: 2634–3648.

Oliver, J. and Palou, A., 2000. Chromatographic determination of carotenoids in foods. Journal of Chromatography A. 881: 543–555.

Rodas Mendoza, B., Morera-Pons, S, Castellote Bargalló, A. I. and López-sabater, M. C., 2003. Rapid determination by reversed-phase high-performance liquid chromatography of vitamins A and E in infant formulas. Journal of Chromatography A. 1018: 197–202.

Rodriguez-Amaya, D. B., Kimura, M., Godoy, H. T. and Arima, H. K., 1988. Assessment of provitamin A determination by open column chromatography/visible absorption spectrophotometry. Journal of Chromatography Science. 26: 624–629.

Romeu-Nadal, M., Morera-Pons, S., Castellote, A. I. and López-Sabater, M. C., 2006. Determination of γ- and α-tocopherols in human milk by a direct high-performance liquid chromatographic method with UV-Vis detection and comparison with evaporative light scattering detection. Journal of Chromatography A. 1114: 132–137.

Rupérez, F. J., Martín, D., Herrera, E. and Barbas, C., 2001. Chromatographic analysis of α-tocopherol and related compounds in various matrices. Journal of Chromatography A. 935: 45–69.

Sander, L. C., Sharpless, K. E., Craft, N. E. and Wise, S. A., 1994. Development of engineered stationary phases for the separation of carotenoids isomers. Analytical Chemistry. 66 : 1667–1674.

Thibeault, D., Su, H., MacNamara, E. and Schipper, H. M., 2009. Isocratic rapid liquid chromatographic method for simultaneous determination of carotenoids, retinol, and tocopherols in human serum. Journal of Chromatography B. 877: 1077–1083.

Turner, C. and Mathiasson, L., 2000. Determination of vitamins A and E in milk powder using supercritical fluid extraction for sample clean-up. Journal of Chromatography A. 874: 275–283.

World Health Organization, 2005. Global Prevalence of Vitamin A Deficiency in Populations at Risk 1995–2005. WHO, Geneva, Switzerland, 55 pp.

Winkelman, A. M., Johnson, D. L. and MacGibbon, A. K. H., 1999. Estimation of heritabilities and correlations associated with milk color traits. Journal of Dairy Science. 82: 215–224.

CHAPTER 21

Simultaneous Ultra-high-performance Liquid Chromatography for the Determination of Vitamin A and Other Fat-soluble Vitamins to Assess Nutritional Status

FERNANDO GRANADO-LORENCIO*,
INMACULADA BLANCO-NAVARRO AND
BELÉN PÉREZ-SACRISTÁN

Unidad de Vitaminas, Bioquimica Clinica, Edificio Laboratorios (Peine 7)
Planta 1ª, Hospital Universitario Puerta de Hierro-Majadahonda, c/Maestro
Joaquin Rodrigo, 2, 28222, Madrid, Spain
*E-mail: fgranado.hpth@salud.madrid.org

21.1 Introduction

Diet constitutes a key modifiable factor in the development and maintenance of health. Malnutrition (both over- and under-nutrition) plays a critical role in morbidity and mortality, and therefore the evaluation of nutritional status is a key point to improve the health of the individuals and populations.

Nutritional status is the balance between the intake of nutrients by an organism and the expenditure of these in the processes of growth, reproduction

Food and Nutritional Components in Focus No. 1
Vitamin A and Carotenoids: Chemistry, Analysis, Function and Effects
Edited by Victor R Preedy

and health maintenance. Because this process is highly complex and quite individualized, nutritional status assessment can be directed at a wide variety of aspects of nutrition. These range from nutrient levels in the body, to the products of their metabolism and to the functional processes they regulate.

Nutritional status can be measured for individuals as well as for populations. While accurate measurement of individual nutritional status is required in clinical practice, population measures are more important in research and can be used to identify populations at risk for nutrition-related health outcomes and to evaluate interventions.

21.2   Fat-soluble Vitamins and Related Compounds

21.2.1   Vitamin A

Vitamin A is a generic term for a large number of related compounds. Retinol, retinal, retinoic acid and related compounds are known as retinoids, while β-carotene and other carotenoids that can be converted by the body into retinol are referred to as provitamin A carotenoids (Figure 21.1).

Vitamin A is an essential nutrient needed for the normal functioning of the visual system, maintenance of cell function for growth, epithelial integrity, red blood cell production, immunity and reproduction. Vitamin A deficiency (VAD) is a major nutritional concern in poor societies. While an inadequate dietary intake of vitamin A or β-carotene probably reveals an important and preventable cause of VAD in a population, it is not an indicator of vitamin A status (WHO, 2009). Similarly, although the distribution of serum retinol concentrations below appropriate cut-offs are considered to reflect inadequate states of vitamin A, a low concentration of retinol in the circulation is not considered a VAD disorder.

Carotenoids constitute one of the major groups of dietary phytochemicals that, apart from their provitamin A activity, display several other biological activities including antioxidant capacity, blue light filtering, modulation of immune function, and regulation of cellular differentiation and apoptosis (Bendich and Olson1989; Stahl and Sies 2005). In human beings, the most widely studied and well understood nutritional role for carotenoids is their provitamin A activity. Vitamin A can be produced within the body from certain carotenoids, notably β-carotene. On the other hand, high or low levels of carotenoids in serum do not constitute a causal factor for any disease but they have been consistently associated with several non-transmissible diseases (*i.e.* cancer and cardiovascular and age-related eye diseases) (Bendich and Olson1989). More recently, individual carotenoids have attracted much attention due to their potential nutritional and clinical relevance, *i.e.*, the effectiveness of lutein in increasing macular pigments and improving visual function; the *in vitro*, animal and human evidence suggesting a unique anabolic effect on bone calcification and osteoporosis prevention by β-cryptoxanthin; or the role of lycopene in prostate cancer, bone resorption and oxidative stress

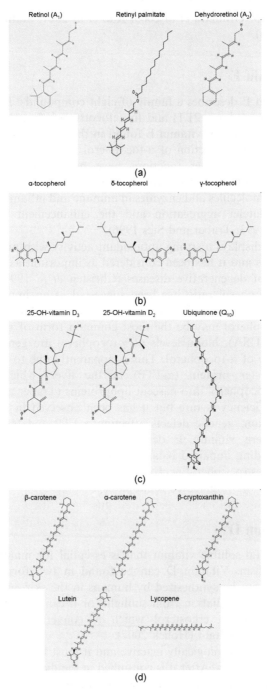

Figure 21.1   Chemical structures of fat-soluble vitamins and related compounds. (A) Vitamin A; (B) vitamin E; (C) vitamin D and coenzyme Q$_{10}$; and (D) carotenoids.

(Fraser *et al* 2005; Granado *et al* 2003; Rao *et al* 2007; Stahl and Sies 2005; Yamaguchi 2006).

21.2.2   Vitamin E

The term vitamin E describes a family of eight compounds: four tocopherols (α-, β-, γ- and δ-) (Figure 21.1) and four tocotrienols (α-, β-, γ- and δ-). α-Tocopherol is the form of vitamin E found in the largest quantities in blood and tissues. The main function of α-tocopherol in humans appears to be as antioxidant, although other functions have been identified including its capacity to inhibit the activity of protein kinase C, the effect on the expression and activities of molecules and enzymes in immune and inflammatory cells, the inhibition of platelet aggregation and the enhancement of vasodilation (Morrisey *et al* 1993; Traber and Sies 1996).

γ-Tocopherol displays a strong antioxidant activity with possible physiological implications and it has been considered as important as α-tocopherol in the prevention of degenerative diseases (Christen *et al* 1997). Also, the α-tocopherol/γ-tocopherol ratio has been suggested as an important discriminator between subjects with disease and controls (Ohrvall *et al* 1996). Although γ-tocopherol may be the most common form of vitamin E in the some diets (*i.e.* USA), blood levels of γ-tocopherol are generally ten times lower than those of α-tocopherol. This is apparently due to the action of α-tocopherol transfer protein (α-TTP) in the liver, which preferentially incorporates α-tocopherol into nascent lipoproteins (Traber and Sies 1996).

Vitamin E deficiency is rare but it has been observed in individuals with severe malnutrition, genetic defects affecting α-TTP and fat malabsorption syndromes. Severe vitamin E deficiency results mainly in neurological symptoms, including impaired balance and co-ordination (ataxia), injury to the sensory nerves (peripheral neuropathy), muscle weakness (myopathy) and damage to the retina (pigmented retinopathy).

21.2.3   Vitamin D

Vitamin D is a fat-soluble vitamin that is essential for maintaining normal calcium metabolism. Vitamin D can be found in two forms. Vitamin D_3 (cholecalciferol) can be synthesized by humans in the skin upon exposure to ultraviolet B (UVB) radiation from sunlight, or it can be obtained from the diet. Plants synthesize ergosterol which is converted into vitamin D_2 (ergocalciferol) by UV light (Holick 2007).

Vitamin D itself is biologically inactive and it must be metabolized into its biologically active forms. After it is consumed in the diet or synthesized in the epidermis of skin, vitamin D enters the circulation and is transported to the liver. In the liver, both forms of vitamin D (D_3 and D_2) are hydroxylated to form 25-hydroxy-vitamin D_3 and D_2, respectively (Figure 21.1). Calcidiol (25-OH-vitamin D) constitutes the major circulating form of vitamin D and is increased after exposure to sunlight or dietary intake of vitamin D, making the serum 25-OH-vitamin D concentration the best indicator of vitamin D nutritional status.

Vitamin D deficiency is an important concern in clinical settings as hypovitaminosis D has been associated with important short- and long-term health effects, including rickets, osteomalacia and the risk of osteoporosis, and common chronic diseases such as diabetes, cardiovascular conditions and cancer (Bischoff-Ferrari *et al* 2006; Holick 2007; WHO 2003). However, at present, there is no consensus among clinical laboratories on the optimum reference intervals for 25-OH-vitamin D to classify patients with moderate to severe vitamin D deficiency (Binkley *et al* 2010; Bischoff-Ferrari *et al* 2006; Kennel *et al* 2010).

21.2.4   Coenzyme Q_{10}

Coenzyme Q_{10} (Q_{10} or ubiquinone) (Figure 21.1) is a fat-soluble compound primarily synthesized by the body and also consumed in the diet. It is a key component of the mitochondrial respiratory chain and, in addition to its role in oxidative phosphorylation, Q_{10} is also present in other subcellular fractions and plasma lipoproteins where it may display antioxidant properties (Litarru and Tiano 2010). Endogenous synthesis and dietary intake appear to provide sufficient coenzyme Q_{10} to prevent deficiency in healthy people.

Although cardiovascular disease and statin-related Q_{10} deficiency is the main field of study, observations regarding its effect in other conditions (*i.e.* Parkinson and Huntingtons disease, Friedreichs ataxia, liver disease or pre-eclampsia) has recently gained much attention. Coenzyme Q_{10} functions as an electron carrier and the chemically reduced form (ubiquinol-10) is a free radical scavenger that prevents peroxidation damage to cell membranes and regenerates α-tocopherol. Deficiencies in Q_{10} have been associated with four major clinical phenotypes: encephalomyopathy, infantile multisystemic disease, cerebellar ataxia and pure myopathy (Quinzii *et al* 2008).

21.3   Assessment of Nutritional Status

The evaluation of nutritional status provides unbiased data for diagnosing malnutrition, identifying individuals and groups at risk, establishing appropriate intervention programs and to measure the effectiveness of nutritional programmes once established (Fidanza 1994). From a broad perspective, nutritional status can be approached by direct and indirect methods. *Indirect methods* include ecological, economical, agricultural or socio-cultural evaluations that may provide relevant information regarding the availability and consumption of certain foods (*i.e.* milk production, food imports and food consumption/per capita). For *direct assessment*, four types of evaluations may be used: anthropometrical measurements, clinical examinations, dietary assessment and biochemical markers (Table 21.1).

Table 21.1   Examples of available approaches to assess nutritional status of fat-soluble vitamins and related compounds in humans. CIC, conjunctival impression cytology; LDL, low-density lipoprotein; PTH: parathyroid hormone; UL-10; ubiquinol-10; UN-10; ubiquinone-10. (Adapted from different sources.)

Micronutrient	*Clinical*	*Biological*	*Dietetic*	*Ecological*
Vitamin A	Xerophthalmia Bitots spots Night blindness	Static: Serum retinol Serum RBP Functional: Visual adaptation RDR/MRDR Isotopic labelling Histological: CIC	Dietary assessment	Food availability Growth rates Diarrheal episodes Conflict areas (*i.e.* war)
Carotenoids	Carotenodermia	Static: Serum carotenoids Functional: Serum metabolites DNA protection LDL oxidation	Dietary assessment	Food availability Food production imports
Vitamin E	Neurological dysfunction	Static: Serum α-tocopherol α-Tocopherol/ cholesterol ratio Functional: Red blood cell hemolysis Pentane exhalation LDL oxidation Inmunocompetence Histological: Intestine biopsy	Dietary assessment	Dietary oils production/ imports
Vitamin D	Bow legs Rickety rosary	Static: Serum 25-OH- vitamin D Functional: PTH Alkaline phosphatase Calcium	Dietary assessment	Latitude Sun exposure
Coenzyme Q_{10}	Myopathy Neuropathy	Static: UL-10 UN-10 UL-10/UN-10 ratio		

Each approach displays advantages and disadvantages when used for individuals and populations, and vary substantially regarding its applicability and costs. *Anthropometrical measurements* are relatively non-invasive methods that assess the size or body composition of an individual [*i.e.* body mass index

Table 21.2 Association between biomarkers in different matrices as indicators of nutritional status over time. (Adapted from different sources.)

	Days	*Weeks*	*Months*	*Years*
Tocopherols	Serum		Adipose tissue	
Retinol	Serum		Liver	
Carotenoids	Serum		Adipose tissue	

(BMI), weight and height]. These are exact and easy to obtain, although they are dependent on physical activity and diet, and reflect long-term energy balance. Similarly, *clinical exams* investigate at the clinical consequences (*i.e.* deficiency or excess) due to nutrient intake inadequacy, which usually appears after a long-term imbalance. Assessment of *dietary intake* involves measuring food consumption, the calculation of nutrient(s) intake by using food composition tables and its comparison with reference or recommended nutrient intakes. These methods, however, are prone to significant bias and errors.

Among the alternatives for assessing the nutritional status of vitamins in humans, the use of *biochemical markers* is considered a reliable approach to be used in groups to identify subjects at risk and to diagnose sub-clinical malnutrition at early stages (Van den Berg 1994). Overall, biochemical methods constitute essential tools to assess nutritional status as they are objective, specific and sensitive, and thus they are free of many errors associated with other approaches (*i.e.* dietary assessment). Nevertheless, many biomarkers only reflect slightly its relationship with diet or nutritional status due to metabolic control or the presence of other factors/nutrients that may affect its concentrations. Additionally, biomarkers measured in different matrices (*i.e.* serum *versus* tissues) may provide information regarding short- or long-term nutritional status (Table 21.2).

In general, biochemical approaches can be divided into *direct (static) methods*, which measure the concentrations of analytes in different biological matrices (*i.e.* serum, tissues), and *indirect (functional or dynamic) tests* that evaluate *in vitro*, *ex vivo* or *in vivo* the response of a related parameter or physiological function (Table 21.1). Thus, on selecting the adequate approach, it is important to bear in mind the context in which the data will be used; clinical settings (*i.e.* to characterize individuals), public health context (*i.e.* to monitor malnutrition and nutritional interventions) and research studies.

21.3.1 Nutritional Status of Vitamin A

VAD is a major nutritional concern, especially in lower income countries, and a major contributor to morbidity and mortality from infections (WHO 2009). On a population level, the main objective of assessing vitamin A status is to determine the magnitude, severity and distribution of VAD (WHO 2009).

Overall, vitamin A status may be evaluated by biological, functional and histological indicators. As shown in Table 21.1, biochemical assessment methods available include retinol in plasma and other biological matrices (*i.e.* breast milk, tears, dried blood spots), the relative dose response (RDR) and the modified relative dose response (MRDR) tests, and the deuterated retinol isotope dilution test (Tanumijhardo 2004). The most frequently used biomarker is serum retinol, even when its concentration is homeostatically regulated and it is not related to the food intake or to the level of vitamin A body stores (*i.e.* liver), except in extreme hypo- or hyper-vitaminosis A. Also, retinol-binding protein (RBP) is a negative acute-phase protein, and thus serum retinol and RBP concentrations will fall during infection. Moreover, the status of other nutrients (*i.e.* iron deficiency) may also negatively affect serum retinol concentrations, and iron deficiency may decrease the mobilization of vitamin A from liver storage (Tanumihardjo 2004).

Two response tests, RDR and MRDR, provide more specific information about the vitamin A status of the individuals and populations from deficiency through marginal status (Olson 2001). Breast milk retinol concentrations have also been proposed as a population measure of vitamin A status. While a unique indicator for lactating women, the status of the mother can usually be predictive of the infant status (Tanumihardjo 2004). Finally, isotope dilution techniques employing vitamin A labelled with deuterium or ^{13}C are used to determine total body stores of vitamin A in populations at risk (Olson 2001).

21.3.2 Clinically Relevant Cut-Off Points for Vitamin A Markers in Serum

Vitamin A status can be classified into five categories: deficient, marginal, satisfactory, excessive and toxic. The deficient and toxic states are characterized by clinical signs, whereas the other three states are not. Serum retinol reflects only very low or very high liver vitamin A stores and could be decreased below normal levels in some situations (*i.e.* total parenteral nutrition and retinoid therapy), although liver vitamin A stores are likely to be adequate. Nevertheless, measuring serum retinol concentrations constitutes a major approach to assess vitamin A status in a population, with values below a cut-off of 0.70 μmol L^{-1} representing VAD, and below 0.35 μmol L^{-1} indicating severe VAD. Although there is not yet an international consensus, a serum retinol concentration below 1.05 μmol L^{-1} has been proposed to reflect low vitamin A status, particularly among pregnant and lactating women (WHO 2009).

On the other hand, traditionally, carotenoids have not received much attention in clinical laboratories, except, possibly, as a tool for differential diagnosis (*i.e.* pseudojaundice) or perhaps in relation to the use of β-carotene in some photosensitive disorders (*i.e.* erythropoietic porphyria). Hypercarotenemia (>300 μg dL^{-1}), with or without carotenodermia, is usually due to increased dietary or supplement intake, but does not specifically cause any clinical disorder, although it may be present in several conditions (*i.e.* diabetes mellitus, anorexia nervosa). Similarly, hypocarotenemia does not constitute a causal factor for any disease, although it may be secondary to malabsorption syndromes or radiotherapy, and it is consistently associated with a high risk for non-transmissible diseases (*i.e.* cancer, cardiovascular and age-related eye diseases) (WHO 2003).

21.4 Nutritional Status of Vitamin A and Other Fat-soluble-related Compounds: a Multi-marker Approach

As pointed out, the information given by one biomarker should be carefully interpreted. This often requires additional data, not directly related to the compound, but necessary to ensure that the considered biomarker is valid. Thus, whenever possible, the combined use of several biomarkers should be favoured.

In recent decades, much research has been performed on fat-soluble vitamins (A, E and D) and related compounds because of their biological activity and their potential role in the prevention and treatment of different diseases. From a viewpoint of public health and disease prevention, the simultaneous assessment of several vitamins, chemical forms and bioactive related compounds (*i.e.* retinol, retinyl esters, carotenoids and coenzyme Q_{10}) is essential to evaluate dietary intake which becomes critical to study the diet–health relationship. In addition, many micronutrients are involved in the homeostatic regulation of overarching processes intimately involved in whole body metabolism, including oxidative–reductive and inflammatory pathways (German and Watkin 2004). Thus, a global approach is required to assess both dietary exposition, nutritional status and the response to an intervention, evaluating other components that may be involved or may modify a response (oxidation, inflammation) or risk (German and Watkin 2004; van Ommen *et al* 2008).

In this context, the application of novel approaches, and time- and cost-reducing methods in clinical laboratories for routine practice is of great interest. Moreover, metabolic profiling, defined as the quantification of metabolites involved in the same metabolic pathway, has become an important tool for determining steady-state concentrations of metabolites and studying the regulation of the corresponding pathways (Aronov *et al* 2008). In this context, the use of analytical methods for the simultaneous determination of several vitamers with different biological activities (*i.e.* 25-OH-vitamin D_2 and D_3) and metabolites (*i.e.* "nutrimetabonomics") become highly relevant for biochemical and nutritional studies (Aronov *et al* 2008; Collino *et al* 2009).

21.4.1 Assessment of Fat-soluble Nutritional Status: Analytical Perspective

The more useful laboratory indices of micronutrient status are usually indicators of long-term dietary intake (Table 21.2). However, as stated, for most nutrients there is no simple relation between intake and indices of status

(Van den Berg *et al* 1994), and vitamin pool size and half-lives may vary considerably for the different vitamins.

In a health context, management of individuals mostly relies on laboratory data and thus, the decision to intervene, the approach to be used and the frequency of monitoring will be affected by the established cut-offs and the health endpoint; that is, to avoid deficiency (a clinical endpoint, *i.e.* xerophthalmia, rickets) or to achieve an optimal status to reduce the risk for chronic diseases (a preventive/public health goal). Thus, selection of the most adequate analytical method relates to the objective of the study, sample accessibility or the size of the group/population to be assessed. At present, direct methods (static) (Table 21.1) are still considered useful and sensitive for routine assessment and the evaluation of the depletion/repletion of body deposits, while for the functional approach, standardization and validation of the methods is required.

Nowadays, different analytical technologies can be used to quantify many of these parameters in clinical settings [*i.e.* high-performance liquid chromatography (HPLC), chemiluminescent immunoassay or radioimmunoassay (RIA)]. However, even the use of commercially available kits (*i.e.* for vitamins A and E) does not cover all the needs such as the determination of related compounds (*i.e.* carotenoids), toxic forms (*i.e.* retinyl esters) or different vitamers (*i.e.* δ- and γ-tocopherols). Moreover, vitamins A and E and 25-OH-vitamin D are often determined by different techniques so that the laboratory has to maintain different lines (equipment, facilities) to cover the demand, increasing the costs considerably (Granado-Lorencio *et al* 2010).

During the last decades, analytical advances have been performed in order to improve the capacity to measure concentrations of fat-soluble vitamins and related compounds in different biological matrices (*i.e.* serum, milk, tissues). Nowadays, chromatographic methods, particularly liquid chromatography, are the methods of choice in many bioanalytical laboratories. Current trends in fast liquid chromatographic separations involve monolith technology, fused core columns, high temperature liquid chromatography and ultra-high-performance liquid chromatography (UHPLC) (Varma *et al* 2010). For the analysis of fat-soluble vitamins and carotenoids, however, high temperature liquid chromatography is not recommended since these analytes are temperature sensitive and on column degradation and isomerisation may occur. Thus, UHPLC has recently become a wide-spread analytical technique in many laboratories which focus on fast and sensitive assays for these compounds.

21.5   Ultra-high-performance Liquid Chromatography (UHPLC)

21.5.1   Basic Concepts

UHPLC refers to a category of analytical separation that retains the practicality and principles of HPLC, while increasing the speed, sensitivity and resolution, and leading, simultaneously, to an improvement in laboratory productivity, efficiency and throughput. This is all enabled by specially designed instruments and sub-2 μm particle packed analytical columns (Novakova and Vickova 2009).

The underlying principles of this technique are governed by the van Deemter equation, an empirical formula that describes the relationship between linear velocity (flow rate) and plate height [height equivalent theoretical plates (HETP) or column efficiency]. According to this formula, as the particle size decreases to less than 2.5 μm, not only is there a significant gain in efficiency, but the efficiency does not diminish at increased flow rates or linear velocities. By using smaller particles, speed and peak capacity (number of peaks resolved per unit time in gradient separations) can be extended to new limits, called ultra-performance liquid chromatography (UPLC) (Swartz 2005; Varma *et al* 2010). The technology takes full advantage of chromatographic principles to run separations using columns packed with smaller particles and/or higher flow rates for increased speed, with superior resolution and sensitivity. As shown in Figure 21.2, the separation of vitamin A (retinol), D (25-OH-D$_3$), tocopherols (γ- and α-tocopherol) and major carotenoids can be accomplished in 3.5 min compared with 15 min using conventional HPLC.

With 1.7 μm particles, half-height peak widths of less than 1 s can be obtained and thus the detector sampling rate must be high enough to capture enough data points across the peak. Conceptually, the sensitivity increase for UPLC detection should be 2–3 times higher than HPLC separations, depending on the detection technique. Mass spectrometry detection is also significantly enhanced by UPLC. Sample introduction is also critical. A fast injection cycle time is needed to fully capitalize on the speed afforded by UPLC and low volume injections with minimal carryover are also required to realize the increased sensitivity benefits (Swartz 2005).

21.5.2 Analysis of Fat-soluble Status in Biological Matrices

The use of UHPLC for assay of biological fluids has proved beneficial in clinical, toxicological or forensic analysis (Citová *et al* 2007) and it is being increasingly used for research purposes. In nutrition, the number of applications is also huge and covers both analytes (*i.e.* biomarkers and metabolites) and matrices to use (*i.e.* reflecting long- *versus* short-term status or minimal invasiveness). In all cases, the methods developed should fulfill chromatographic and analytical quality standards, such as sensitivity, resolution, precision, accuracy, recovery, linearity and limits of detection and quantification, to be applicable in different studies.

On the other hand, sample preparation is usually the rate-limiting step in any bioanalytical assay. Several approaches are used (*i.e.* liquid–liquid, solid-phase or supercritical fluid extraction) and, in many instances, protocols are modified (*i.e.* alkaline hydrolysis) to be applied in different matrices and for distinct purposes. Also, protocols may be further improved by using selective

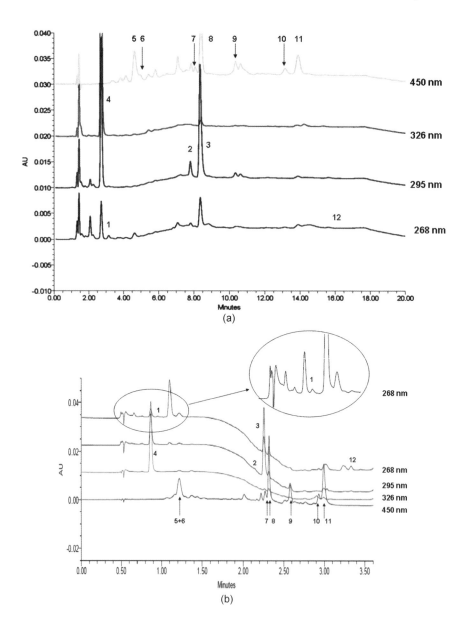

Figure 21.2 Method transference: from HPLC to UHPLC. (A) Chromatograms at different wavelengths of a serum sample obtained by traditional HPLC. (B) Chromatograms of a serum sample performed on a UHPLC system. Peak identification: (1) 25-OH-vitamin D_3; (2) γ-tocopherol; (3) α-tocopherol; (4) retinol; (5) Lutein; (6) zeaxanthin; (7) α-cryptoxanthin; (8) β-cryptoxanthin; (9) lycopene; (10) α-carotene; (11) β-carotene; and (12) coenzyme Q_{10}.

detection such as fluorescence, mass spectrometry and pre- and post-column derivatization. The combination and optimization of all these variables (*i.e.* mass tagging sample-multiplexed assay) allow the development of highly accurate and sensitive methods using minimal amounts of sample (*i.e.* dried blood spots) and high throughput in analytical laboratories.

21.5.2.1 Applications in Clinical Practice

UHPLC provides a new approach that is being increasingly applied in clinical settings to study biochemical markers of vitamin status and metabonomic profiling (Aronov *et al* 2008; Citová *et al* 2007; Ding *et al* 2010; Granado-Lorencio *et al* 2010; Li and Franke 2009; Paliakov *et al* 2009; Stepman *et al* 2011; Want *et al* 2010). Until now, UHPLC methods for routine fat-soluble analysis in human fluids have studied combinations of markers of vitamin A (retinol, β-carotene), vitamin E (α- and γ-tocopherol), vitamin D metabolites [*i.e.* 25-OH-vitamin D_3 and D_2, 24-25-$(OH)_2$-vitamin D] and coenzyme Q_{10} (Aronov *et al* 2008; Citová *et al* 2007; Ding *et al* 2010; Paliakov *et al* 2009; Stepman *et al* 2011). Compared with these UHPLC methods, we used a UPLC system [HSS T3 column (2.1 × 100 mm; 1.8 µm)] with gradient elution and multiple ultraviolet-visible (UV-Vis) detection, with the aim of being applicable for routine clinical evaluations and metabolic studies (Granado-Lorencio *et al* 2010). Moreover, the additional advantages of the technique (*i.e.* low sample volume, speed and reduced costs) made it also appropriate as screening tool for population-based studies.

The developed protocol, for example, allows the simultaneous determination of retinol and different retinyl esters, making it adequate for dietary and bioavailability studies assessing post-prandial response (including conversion of β-carotene into retinol). Additionally, the simultaneous determination of both retinol and ester forms is clinically relevant. In fasting conditions, serum retinyl palmitate (as the major circulating ester form) is assumed to account for <10% of total circulating retinol in serum. Thus, a greater proportion of ester forms in serum is considered to be an indicator of vitamin A toxicity (Figure 21.3), although it may be also found in liver damage (*i.e.* alcoholism) or rare genetic disorders (*i.e.* absence of RBP).

Major carotenoids present in human serum and tissues include lutein, zeaxanthin, α- and β-cryptoxanthin, lycopene, α- and β-carotene, although others are also present in minor quantities (*i.e.* anhydrolutein, neurosporene, phytoene). As mentioned, carotenoids are not routinely measured in clinical practice, although the carotenoid profile in serum may provide relevant information about dietary habits (*i.e.* fruit and vegetables consumption) and absorption disturbances (*i.e.* hypocarotenemia in malabsorption syndromes). Hypercarotenemia (serum total levels >300 µg dL^{-1}) is usually related to the use of carotenoid-containing supplements and diets rich in vegetables and fruits (*i.e.* infants) (Figure 21.4). Nevertheless, although extremely rare, hypercarotenemia may also appear when there is an inability to convert

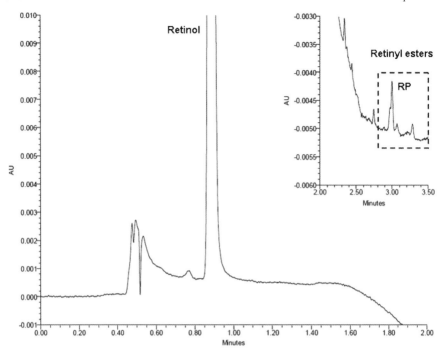

Figure 21.3 Serum chromatogram of a patient with retinyl esters in serum. Under fasting conditions, ester forms of vitamin A should represent less than 10% of total circulating retinol. The chromatographic profile at 326 nm shows the presence of retinyl esters in a patient with high levels of retinol in serum ($>160 \ \mu g \ dL^{-1}$). The presence of significant amounts of retinyl esters may be indicative of vitamin A toxicity. RP, retinyl palmitate.

provitamin A carotenoids into retinol (*i.e.* deficit of β-carotene 15,15′-dioxygenase). In this condition, the simultaneous determination of retinol, retinyl esters and individual carotenoids in the post-prandial state after, for example, β-carotene overload, may contribute to the diagnosis by identifying the underlying causes of the hypercarotenemia (*i.e.* dietary intake *versus* metabolic disturbance).

In clinical settings, the assessment of vitamin D status is increasingly demanded and the treatment of the patients for vitamin D deficiency is becoming routinely used. In many countries, vitamin D_2-containing supplements are the preferred option and thus individual determination of both 25-OH-vitamin D_3 and D_2 is necessary for evaluating treatment adherence and efficacy (Figure 21.5). Regardless of the biopotency of both forms to improve the vitamin D status, the ability to determine both circulating forms may be important to classify subjects according to their status and treat them accordingly.

Similarly, the simultaneous determination of α-, γ- and δ-tocopherol may be relevant to assess not only vitamin E status in serum but also to evaluate

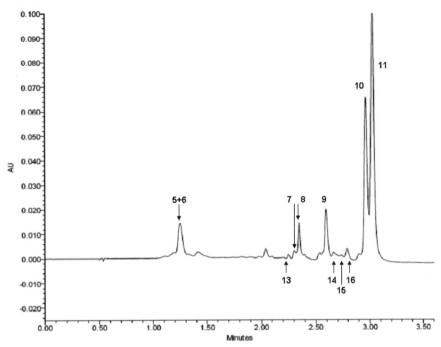

Figure 21.4 UHPLC serum chromatogram of a 2-year old infant with hypocarotenemia. Levels of retinol, vitamin E and vitamin D were within the
reference range (chromatograms not shown). Concentrations of total
carotenoids, α- and β-carotene were indicative of hypercarotenemia. In
addition to the major carotenoids, other minor carotenoids were
tentatively identified by retention time and UV-Vis on-line spectra.
Peak identification as in Figure 21.2; (13) anhydrolutein; (14) neuro-
sperene; (15) γ-carotene; and (16) ξ-carotene. Wavelength: 450 nm.

dietary exposition on a long-term basis, *i.e.* evaluating α-/γ-tocopherol ratio in
adipose tissue. Although α-tocopherol is preferentially incorporated into
nascent very-low-density lipoproteins (VLDLs), which, eventually, supply
vitamin E to the tissues, other vitamin E forms are also transported and
transferred to other lipoproteins and extrahepatic tissues during post-prandial
metabolism (Traber and Sies 1996). Thus, both γ-tocopherol content and the
α-/γ-tocopherol ratio in human plasma and tissues may provide relevant
information regarding dietary exposure and risk of disease (Ohrvall *et al* 1996;
Traber and Sies 1996).

Finally, the simultaneous analysis of fat-soluble vitamins and related
compounds (*i.e.* carotenoids and coenzyme Q_{10}) in the same sample provide a
more wide-spread picture of the nutritional status of an individual. In addition,
the use of a unique serum or tissue sample is less invasive and may provide
specific and more complete information regarding the dietary habits, fat-
soluble vitamin status and antioxidant status of the individual (a multi-marker
approach) (Figure 21.6). This whole picture allows the clinician/researcher to

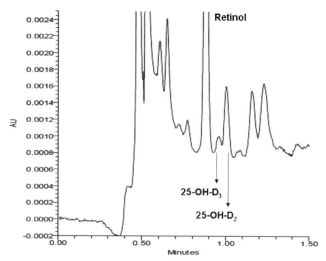

Figure 21.5 Serum chromatogram of a patient following vitamin D_2 treatment. The simultaneous presence of 25-OH-vitamin D_3 and 25-OH-vitamin D_2 can be clearly observed. Contrary to commercially available immunological methods for vitamin D evaluation, the chromatographic resolution allows to quantify separately 25-OH-vitamin D_3 (mostly from endogenous origin) and 25-OH-vitamin D_2 (mostly from supplement intake) so that adherence to the treatment can be assessed. The sum of both vitamers may provide a more accurate picture of the vitamin D nutritional status of the patients.

reliably interpret data of multiple markers in terms of the adequacy of the diet, efficacy of the treatment and risk of disease.

21.5.2.2   *Applications in Research Studies*

Nutritional and metabolic profiling has become an important tool for studying the relationship between diet and health. In this context, the use of analytical methods for the simultaneous determination of a variety of related compounds becomes highly relevant for biochemical and nutritional studies.

Using the developed UHPLC conditions in samples of human serum and animal tissues, we could tentatively identify other components (by chromatographic behaviour and UV-Vis on-line spectra) such as vitamin A_2 (dehydroretinol) and its ester forms, canthaxanthin, γ-carotene, ξ-carotene, neurosporene, phytofluene and phytoene. Considering all the compounds potentially resolved using the UHPLC system, *in vitro* and *in vivo* bioavailability (absorption and conversion), metabolic and balance studies can be performed. For example, using carotenoid profile in human faeces, both qualitative and (semi-)quantitative information regarding the changes during food digestion (*i.e.* degradation, isomerization, oxidation products) of individual vitamers and related compounds (*i.e.* carotenoids) can be obtained (Figure 21.7). This

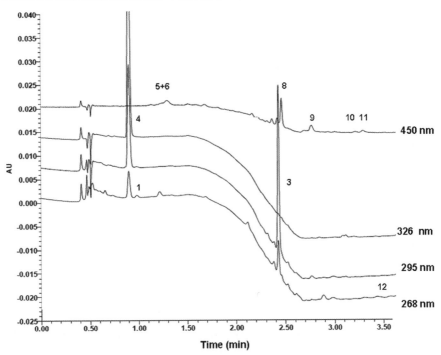

Figure 21.6 Multi-wavelength serum profile of a patient after obesity surgery. While supplement use is routinely prescribed in these patients (vitamins A, E and D), the simultaneous determination of vitamins and carotenoids provides a more complete picture of the nutritional status. Retinol and vitamin D may be within reference ranges, whereas α-tocopherol may be low due to the drop in lipids levels (as carriers of α-tocopherol) usually observed in these patients. Because of the homeostatic control of serum retinol, malabsorption degree and dietary intake restrictions are better reflected by the low serum carotenoid levels, shown at 450 nm. Peak identification as in Figure 21.2.

information is not only relevant from a physisological viewpoint but also for the design and development of more appropriate functional foods by, *i.e.*, changing the food matrix to improve bioavailability or avoid nutrient interactions.

In animals, vitamin E, vitamin A and carotenoids have gained much attention because of their role as antioxidants and inmunostimulants, in growth development and reproduction. However, despite their biological relevance, the nutritional requirements, the physiological processes or the tissue deposition and function are still poorly understood in many species. Using a multi-marker approach, we addressed the presence of several vitamers and related compounds in different tissues of wild lizards. As reported in mammals, the liver seems to store vitamin A (as ester forms) in this species, whereas the detection of both γ- and α-tocopherol suggests their presence in the diet and the lack of selective absorption for both vitamers (Figure 21.8).

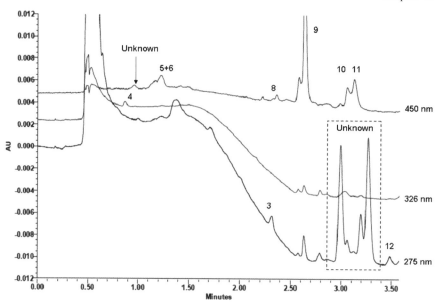

Figure 21.7   Chromatographic profile of human faeces. As part of the human balance studies and malabsorption syndromes, the determination of the multi-marker profile in faeces may provide relevant information regarding dietary intake and metabolic changes of nutrients during gastrointestinal digestion. Additionally, the presence of several unidentified peaks (at 275 nm) may be intestinal microbiota-related compounds and metabolites (referred to as "unknown peaks"). Peak identification as in Figure 21.2.

21.6   Concluding Remarks

Nutritional status is the balance between the intake of nutrients by an organism and the expenditure of these in the processes of growth, reproduction and health maintenance. In addition, within a context of "beyond deficiency", much research has been performed on fat-soluble vitamins (A, E and D) and related compounds (*i.e.* carotenoids, coenzyme Q_{10}) because of their biological activity and possible use for prevention and treatment of different diseases. Thus, the evaluation of nutritional status becomes a key point to improve the health of the individuals and the populations.

The UHPLC method constitutes a high-throughput approach for biomarker profiling, allowing the rapid and low-cost determination of vitamins A, E and D (including vitamers and ester forms) and the major carotenoids in a single run. Following the principles of conventional HPLC, the UHPLC approach provides a rapid tool for the simultaneous and routine evaluation of relevant biomarkers of nutritional status and dietary exposure. Additionally, the higher sensitivity of the system allows the minimization of sample volume and optimization of specimen collection (less invasive techniques), which may be of interest for some patients and settings (*i.e.* children and neonates), in addition

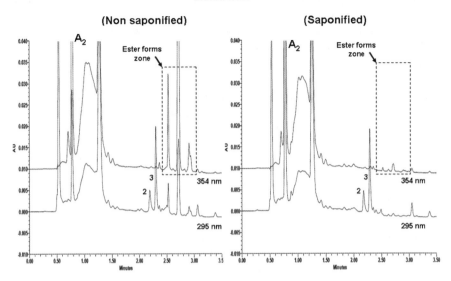

Figure 21.8 Chromatogram of a lizard liver extract (non-saponified and saponified extracts). Ester forms have same absorption spectra but longer retention times than free forms. Therefore, saponification of extracts will result in esters hydrolysis and, consequently, saponified samples show fewer peaks and an increase of free forms. Vitamin A_2 and ester forms of vitamin A_2 were tentatively detected (by chromatographic behaviour and UV-Vis on-line spectra; 354 nm) in the liver of wild lizards. α-Tocopherol is the predominant compound showing significantly higher concentrations than γ-tocopherol in liver. Peak identification as in Figure 21.2.

to substantially reducing costs and wastes without affecting imprecision and accuracy. UHPLC technology has already proved beneficial in clinical, toxicological, forensic and metabonomic analysis. The higher sensitivity of the system and the lower costs make it also applicable and affordable for population-based studies.

Summary Points

- Diet constitutes a key modifiable factor in the development and maintenance of health.
- Nutritional status is the balance between the intake of nutrients by an organism and the expenditure of these.
- The evaluation of nutritional status provides unbiased data for diagnosing malnutrition.
- Nutritional status may be assessed by anthropometrical, clinical, dietary and biochemical methods.
- The most frequently used biomarker to assess vitamin A status is serum retinol.

- Serum 25-hydroxy-vitamin D concentration is the best indicator of vitamin D nutritional status.
- UHPLC utilizes smaller particle size at pressures significantly higher than the conventional HPLC pressures.
- A decrease in particle size increases efficiency and pressure, and reduces the time of analysis.
- Sample preparation is usually the rate-limiting step in bioanalytical assays.
- UHPLC is a rapid tool for the simultaneous and routine evaluation of biomarkers of fat-soluble nutritional status.

Key Facts

Key Facts of Nutritional Status

- Nutritional status is the balance between the intake and the expenditure of nutrients for growth, reproduction and health maintenance.
- The evaluation of nutritional status provides unbiased data to identify individuals and groups at risk.
- The use of biochemical markers is considered a reliable approach to detect sub-clinical malnutrition at early stage.
- Biochemical approaches can be divided into direct (static) and indirect (functional or dynamic) tests.
- The simultaneous assessment of vitamins and related compounds (multi-marker approach) is essential to study diet–health relationships.

Key Facts of Vitamin A Status

- Vitamin A deficiency (VAD) is a major nutritional concern in poor societies.
- Vitamin A status may be evaluated by biological, functional and histological indicators.
- The most frequently used biomarker to assess vitamin A status is serum retinol.
- Vitamin A status can be classified into five categories: deficient, marginal, satisfactory, excessive and toxic.
- Serum retinol concentration below 1.05 μmol L^{-1} has been proposed to reflect low vitamin A status.

Key Facts of Ultra-high-performance Liquid Chromatography (UHPLC)

- UHPLC reduces the analysis time by using columns with packing particle size less than 2 μm.

- The underlying principles of this evolution are governed by the van Deemter equation.
- The key advantages of UHPLC include increased speed of analysis, higher separation efficiency and higher sensitivity.
- UPLC retains the practicality and principles of HPLC, allowing an improvement in productivity and reducing costs.
- UHPLC provides a new approach to study biochemical markers of vitamin status and metabonomic profiling.

Definitions of Words and Terms

Carotenoids: One of the major groups of dietary phytochemicals that, apart from the provitamin A activity, displayed several other biological activities.

Coenzyme Q₁₀ (Q₁₀ or ubiquinone): A fat-soluble compound primarily synthesized by the body, which is a key component of the mitochondrial respiratory chain.

Nutritional status: The balance between the intake of nutrients by an organism and the expenditure of these in the processes of growth, reproduction and health maintenance.

Ultra-high-performance chromatography (UHPLC): A technique that reduces the analysis time by using columns with packing particle size less than 2 μm.

van Deemter equation: The formula that describes the relationship between the linear velocity of mobile phase (flow rate) and height equivalent theoretical plates (HETP or column efficiency).

Vitamin A: A generic term for a large number of related compounds. Retinol, retinal, retinoic acid and related compounds are known as retinoids, whereas β-carotene and other carotenoids that can be converted by the body into retinol are referred to as provitamin A carotenoids.

Vitamin D: A fat-soluble vitamin that is essential for maintaining normal calcium metabolism. Vitamin D can be found in two forms: vitamin D_3, from animal origin, and vitamin D_2, from plant sources.

Vitamin E: The term describes a family of eight antioxidants: four tocopherols (α-, β-, γ- and δ-) (Figure 21.1) and four tocotrienols (α-, β-, γ- and δ-).

List of Abbreviations

HETP	height equivalent theoretical plates
HPLC	high-performance liquid chromatography
MRDR	modified relative dose response
RBP	retinol-binding protein
RDR	relative dose response
α-TTP	α-tocopherol transfer protein
UHPLC	ultra-high-performance liquid chromatography
UPLC	ultra-performance liquid chromatography

UV-Vis   ultraviolet-visible
VAD   vitamin A deficiency

References

Aronov, P. A., Hall, L. M., Dettmer, K., Stephensen, C.B. and Hammock, B. D., 2008. Metabolic profiling of major vitamin D metabolites using Diels–Alder derivatization and ultra-performance liquid chromatography–tandem mass spectrometry. Analytical Bioanalytical Chemistry. 391: 1917–1930.

Bendich, A. and Olson, J.A., 1989. Biological actions of carotenoids. FASEB Journal. 3: 1927–1932.

Binkley, N., Ramamurthy, R. and Krueger, D., 2010. Low vitamin D status: definition, prevalence, consequences, and correction. Endocrinology Metabolism Clinics North America 39: 287–301.

Bischoff-Ferrari, H.A., Giovannucci, E., Willet, W., Dietrich, T. and Dawson-Hughes, B., 2006. Estimation of optimal serum concentrations of 25-hydroxyvitamin D for multiple health outcomes. American Journal Clinical Nutrition. 82: 18–28.

Christen, S., Woodall, A.A., Shigenaga, M.K., Southwell-Keely, P.T., Duncan, M.W. and Ames, B.N., 1997. γ-Tocopherol traps mutagenic electrophiles such as NOx and complements α-tocopherol: physiological implications. Proceedings of the National Academy of Sciences of the USA. 94: 3217–3222.

Citová, I., Havlíková, L., Urbánek, L., Solichová, D., Nováková, L. and Solich, P., 2007. Comparison of a novel ultra-performance liquid chromatographic method for determination of retinol and α-tocopherol in human serum with conventional HPLC using monolithic and particulate columns. Analytical and Bioanalytical Chemistry. 388: 675–681.

Collino, S., Martin, F. P. J., Kochhar, S. and Rezzi, S. M., 2009. Monitoring healthy metabolic trajectories with nutritional metabonomics. Nutrients. 1: 101–110.

Ding, S. J., Schoenmakers, I., Jones, K., Kaulmen, A. and Prenctice, A., 2010. Quantitative determination of vitamin D metabolites in plasma using UHPLC-MS/MS. Analytical and Bioanalytical Chemistry. 398: 779–789.

Fidanza, F., 1994. Nutritional status assessment in perspective. Biblioteca Nutritio et Dieta. 51: 9–18.

Fraser, M., Lee, A. H. and Binns, C. W., 2005. Lycopene and prostate cancer: emerging evidence. Expert Reviews Anticancer Therapy 5: 847–854.

German, J. B. and Watkins, S. M., 2004. Metabolic assessment; a key to nutritional strategies for health. Trends Food Science and Technology. 15: 1549.

Granado, F., Olmedilla, B. and Blanco, I., 2003. Nutritional and clinical relevance of lutein in human health. British Journal of Nutrition. 90: 487–502.

Granado-Lorencio, F., Herrero-Barbudo, C., Blanco-Navarro, I. and Pérez-Sacristán, B., 2010. Suitability of ultra-performance liquid chromatography for the determination of fat-soluble nutritional status (vitamin A, E, D and individual carotenoids). Analytical and Bioanalytical Chemistry. 397: 1389–1393.

Holick, M. F., 2007. Vitamin D deficiency. New England Journal of Medicine. 357: 266–281.

Kennel, K. A., Drake, M. T. and Hurley, D. L., 2010. Vitamin D deficiency in adults: when to test and how to treat. Mayo Clinic Proceedings. 85: 752–757.

Li, X. N. and Franke, A. A., 2009. Fast HPLC-ECD analysis of ascorbic acid, dehydroascorbic acid and uric acid. Journal of Chromatography B. Analytical Technical Biomedical Sciences. 10: 853–856.

Litarru, G. P. and Tiano, L., 2010. Clinical aspects of coenzyme Q10: an update. Nutrition. 26: 250–254.

Morrisey, P. A., Sheehy, P. J. A. and Gaynor, P., 1993. Vitamin E. International Journal of Vitamin and Nutrition Research. 63: 260.

Novakova, L. and Vickova, H., 2009. A review of current trends and advances in modern bio-analytical methods: chromatography and sample preparation. Analytica Chimica Acta. 656: 8–35.

Ohrvall, M., Sundlof, G. and Vessby, B., 1996. γ, but not α, tocopherol levels in serum are reduced in coronary heart disease patients. Journal of Internal Medicine. 239: 111–117.

Olson, J. A., 2001. Vitamin A. In: Handbook of Vitamins Rucker, R. B., Suttie, J. W., McCormick, D. B. and Machlin, L. J. (ed.). Marcel Dekker, NY, USA, pp. 1–50.

Paliakov, E. M., Crow, B. S., Bishop, M. J., Norton, D., George, J. and Bralley, J. A., 2009. Rapid quantitative determination of fat soluble vitamins and coenzyme Q10 in human serum by reversed phase ultra-performance liquid chromatography with UV detection. Journal of Chromatography B. Analytical Technological Biomedical Life Sciences. 1: 89–94.

Quinzii, C. M., López, L. C., Naini, A., Di Mauro, S. and Hirano, M., 2008. Human CoQ10 deficiencies. Biofactors. 32: 113–118.

Rao, L. G., Mackinnon, E. S., Josse, R. G., Murray, T. M., Strauss, A. and Rao, A. V., 2007. Lycopene consumption decreases oxidative stress and bone resorption markers in postmenopausal women. Osteoporosis International. 18: 109–115.

Stahl, W. and Sies, H., 2005. Bioactivity and protective effects of natural carotenoids. Biochimica Biophysica Acta. 1740: 101–107.

Stepman, H. C. M., Vanderroost, A., Stokl, D. and Thienpont, L. M., 2011. Full-scan mass spectral evidence for 3-epi-25-hydroxyvitamin D3 in serum of infants and adults. Clinical Chemistry and Laboratory Medicine. 49: 253–256.

Swartz, M. E., 2005. Ultra Performance Liquid Chromatography (UPLC): An Introduction. Separation Science Redefined 8–14. Available at: www.chromatographyonline.com. Accessed May 2011.

Tanumihardjo, S. A., 2004. Assessing vitamin A status: past, present and future. Journal of Nutrition. 134: 290S–293S.

Traber, M. and Sies, H., 1996. Vitamin E in humans: demand and delivery. Annual Review of Nutrition. 16: 321.

Van den Berg, H., 1994. Functional vitamin status assessment. Biblioteca Nutritio et Dieta. 51: 142–149.

Van Ommen, B., Fairweather-Tait, S., Freidig, A., Kardinaal, A., Scalbert, A. and Wopereis, S., 2008. A network biology model of micronutrient related health. British Journal of Nutrition. 99: S72–S80.

Varma, D., Jansen, S. A. and Ganti, S., 2010. Chromatography with higher pressure, smaller particles and higher temperature: a bioanalytical perspectiva. Bioanalysis. 2: 2019–2034.

Want, E. J., Wilson, I. D., Gika, H., Theodoridis, G., Plumb, R. S., Schockcor, J., Holmes, E. and Nicholson, J. K., 2010. Global metabolic profiling procedures for urine using UPLC-MS. Nature Protocols. 5: 1005–1018.

WHO/FAO Expert Consultation., 2003. Diet, nutrition and the prevention of chronic diseases. WHO Technical Report Series #916. WHO, Geneva, Switzerland.

WHO., 2009. Global prevalence of vitamin A deficiency in populations at risk 1995–2005. WHO Global Database on Vitamin A Deficiency. WHO. Geneva, Switzerland.

Yamaguchi, M., 2006. Regulatory mechanisms of food factors in bone metabolism and prevention of osteoporosis. Yakugaku Zasshi. 26: 1117–1137.

Function and Effects

CHAPTER 22

Distribution and Concentrations of Vitamin A and their Metabolites in Human Tissue

EWA CZECZUGA-SEMENIUK*[a], JANUSZ W. SEMENIUK[b] AND ADRIANNA SEMENIUK[c]

[a] Department of Reproduction and Gynecological Endocrinology, Medical University of Białystok, 24A M. Skłodowskiej-Curie Street, 15-276 Białystok, Poland; [b] Department of Pediatrics, Gastroenterology and Children's Allergology, Medical University of Białystok, Waszyngtona 17 Street, 15-274 Białystok, Poland; [c] Klinikum Vest GmbH, Knappschaftskrankenhaus Recklinghausen, Akademisches Lehrkrankenhaus, der Ruhr-Universität-Bochum, Klinik für Innere Medizin, Kardiologie, Schlafmedizin, Dorstener Strasse 151, 45657 Recklinghausen, Nordrhein- Westfalen, Deutschland
*E-mail: czeczuga@wp.pl

22.1 Introduction

Carotenoids are the most widespread natural pigments. They influence the tint of different parts of the organisms – ranging from slightly yellow to bright red. They accompany plant chlorophylls, contribute to the coloration of flower petals, fruit, and some roots (*e.g.* carrot). They are present in animals, from protozoa to birds' feathers (*e.g.* flamingos or parrots).

Carotenoids maintain various functions in the living organisms. They act as additional aerials in plants and absorb radii not absorbed by the chlorophylls, protect photosynthetic apparatus from ultraviolet light and from lethal

Food and Nutritional Components in Focus No. 1
Vitamin A and Carotenoids: Chemistry, Analysis, Function and Effects
Edited by Victor R Preedy

oxygenation, as well as participate in phototropism and phototaxis. In animals, they are responsible for mating pigmentation, vision and reproduction processes as well as the inhibition of cancerogenesis. In humans, carotenoids are vitamin A precursors. They participate in the vision and reproduction processes, immunological response, and protection against cardiovascular diseases. Their anti-cancerogenous properties are maintained mainly by the potent antioxidant action. They directly or indirectly (as retinoid precursors) influence the metabolism of cells and the proliferation, differentiation, and cancerogenesis.

These pigments are biosynthesized *de novo* only in the cells of bacteria, fungi, and green plants. All heterotrophs (animals and humans) do not synthesize carotenoids, but obtain them from food and transform them into other carotenoids *via* oxidation and reduction.

All carotenoids are divided into two groups:

- Carotenes (hydrocarbons), composed of carbon and hydrogen, *e.g.* α- or β-carotene –$C_{40}H_{56}$
- Xanthophylls (C_{40} –xanthophylls), oxidized carotenes, *e.g.* lutein or zeaxanthin –$C_{40}H_{56}O_2$. They are generally common in fats as free forms, esters with fatty acids, or glycosides with pentoses, or complexes with proteins (Goodwin 1980).

A total of 700 carotenoids have been identified so far, including 50 carotenes. Xanthophylls can be divided into mono- (*e.g.* β- cryptoxanthin), di- (*e.g.* lutein) or polyhydroxy (more than two hydroxy groups) xanthophylls (*e.g.* crustaxanthin), depending on the level of oxidation and the presence of functional groups: hydroxy, metoxy, epoxy, carboxy, ketone and aldehyde. Moreover, there are epoxy xanthophylls (lutein epoxide), aldehyde (rodopinal), monoketone (echinenone), diketone (canthaxanthin) and polyketone (capsorubin), and acids (torularhodin). There are also carotenoids of different chain length called homocarotenoids (*e.g.* sarcinaxanthin), apocarotenoids (β-apo-2'-carotenal), diapocarotenoids (bixin), norcarotenoids (actinioerythrin), and other carotenoids (semi-β-carotenone). Some of them belong to the provitamin A group, *e.g.* α- and β-carotene, β-cryptoxanthin, echinenone, and hydroxyechinenone.

The very first data concerning carotenoid isolation come from the studies of Wackenroder (1831) and Berzelius (1837). The former isolated β-carotene from the carrot root, whereas the latter isolated xanthophylls from yellow autumn leaves. Nevertheless, a broader and more scientific analysis of carotenoids could have been made only after M. Tswiet correctly defined and interpreted the products of pigment separation from leaves using chalk columns in 1903. The whole process was named chromatography (chromatos = pigment and grapho = to write) on the basis of the colorful zones he obtained.

The isolation and analysis of carotenoids not only provide information concerning their existence in particular organisms, but also allow us to observe seasonal changes of carotenoids at different life stages or during neoplastic transformation of cells. Their importance is highlighted in the contemporary literature, with the emphasis on the presence and role of carotenoids and their metabolites—retinoids in the physiological and pathological processes in the human body.

22.2 Vitamin A and Carotenoids in Particular Human Organs

22.2.1 General Reflections

Most analyses concerning the identification of vitamin A and its metabolites in humans refer to the tissues obtained during autopsies and are based on the high-performance liquid chromatography (HPLC) method. Earlier studies were performed with different methods, *e.g.* column chromatography (CC) and/or thin-layer chromatography (TLC). Presently, these methods constitute the preliminary steps before HPLC. HPLC is a sensitive and repeatable method which allows precise separation of carotenoids and quantitative assessment of their concentrations, even in small samples. Reversed-phase (RP) and/or a normal phase (NP) HPLC with ultraviolet/visible (UV/Vis) spectrophotometry are the most commonly used. The extraction and the isolation processes of carotenoids have not changed much for many years. On the other hand, there has been a progress in analysis and identification.

The standard carotenoid identification by UV/Vis spectrophotometry might be presently substituted with new techniques such as mass spectrometry (MS) and nuclear magnetic resonance (NMR) spectroscopy, which allow the assessment of very small samples. The newest method, which allows carotenoid analysis in the living organism without sample collection is non-invasive resonance Raman spectroscopy (RR spectroscopy). It is mainly used in the case of skin and eye tissue samples.

American studies (trifluoroacetic acid procedure and the modified micro-column technique of McLaren), comprising five USA regions, showed that total vitamin A concentrations in liver tissues [in μg of vitamin A (g of wet tissue)$^{-1}$] amounted to 146 ± 151, and the values were comparable in both sexes. High mean values were noted in a group up to 10 years of age and over 50 years of age, with the highest values in a group of 71–80 years of age (207 ± 190). Vitamin A concentrations in other organs are lower (from 0.32 to 1.46) (Raica *et al* 1972). In liver tissue, there was a correlation between vitamin A concentrations and β-carotene concentrations and the total concentrations of carotenoids of the provitamin A group (Schmitz *et al* 1991). Among human tissues, suprarenal glands, liver, and testicles presented with the highest total carotenoid concentrations [investigation of the distribution by CC in μg (g of wet tissue)$^{-1}$]—(20.1 ± 11.9), (8.3 ± 21.3), and (5.0 ± 7.7), respectively. Carotenoid concentrations were also detected in the adipose tissue, pancreas, spleen, kidneys, thyroid gland, lungs, and muscles (Raica *et al* 1972). Further studies (Parker 1989; Schmitz *et al* 1991; Stahl *et al* 1992) also confirm such observations, suggesting the liver, suprarenal glands, and testicles as the organs always accumulating higher carotenoid concentrations than kidneys, ovary or adipose tissue (Kaplan *et al* 1990; Stahl *et al* 1992).

α-Carotene, β-carotene, β-cryptoxanthin, zeaxanthin and lycopene are present in human liver, pancreas, kidneys, suprarenal glands, spleen, heart, testicles, trachea, ovaries, and adipose tissue with various concentrations of zeaxanthin, lycopene, and β-carotene in particular organs (Kaplan *et al* 1990). Also lutein was identified in liver, kidney, and lung tissues. The highest concentrations [in nmol (g of tissue)$^{-1}$] of carotenoids (α-carotene, β-carotene, cryptoxanthin, lutein, and lycopene) were present in liver (21.0 on average) and were 10–30-fold higher than in kidneys (3.1) or lungs (1.9). Carotenes and xanthophylls (β-carotene and lycopene) were the predominant ones, whereas lutein was predominant in some tissues (Schmitz *et al* 1991).

The assessment of isomeric forms of β-carotene and lycopene in testicles, liver and suprarenal glands enabled identification of five geometrical isomers of β-carotene and seven of lycopene. The tissues presented with the mixture of those forms; however, the isomers of β-carotene—9-*cis* form and 13-*cis* form—were dominant. The 15-*cis* form was found at trace levels (Stahl *et al* 1993).

22.2.2   Liver

The studies concerning the identification of vitamin A (retinol), its esters and carotenoids in the liver were also conducted on the material obtained from healthy people. The total vitamin A concentration in such people amounted to 7.8–2860 nmol (g of wet tissue)$^{-1}$, with retinyl palmitate as the dominant vitamin A ester. The total concentration of the identified carotenoids (α- and β-carotene, lycopene, and lutein) amounted to 2.5–67 nmol (g of liver tissue)$^{-1}$ (Tanumihardjo *et al* 1990). The highest mean concentrations [in nmol (g of wet tissue)$^{-1}$] were detected for β-carotene (3.02) and lycopene (1.28) (Stahl *et al* 1992). α-, β- and γ-carotene, β-cryptoxanthin and other xanthophylls and their possible metabolites were also identified in human liver (Khachik *et al* 1998).

22.2.3   Adipose Tissue

Taking into consideration the total amount of adipose tissue in human body and the fat-solubility of carotenoids, adipose tissue constitutes a great reservoir of carotenoid pigments in humans. Easy access to adipose tissue *via* its aspiration with a 1.5 mm needle from the gluteal region and abdomen (umbilical region), thigh, triceps muscle, and little risk of complications have contributed to the fact that such material is fully accessible and useful in the identification of carotenoids.

Studies concerning carotenoid identification in abdominal adipose tissue in humans revealed the predominance of carotenes and xanthophylls in the examined material, which accounted for 44% of the total amount of the identified pigments. Mean values of β-carotene and lycopene concentrations amounted to 0.62 and 0.58 µg (g of tissue)$^{-1}$, respectively, although there were significant individual differences (up to 40-fold) (Parker 1988).

α-Carotene; β-carotene (*trans* + *cis*), all-*trans*, 13-*cis*, 9-*cis*, *cis*; β-cryptoxanthin; lycopene (*trans* + *cis*), all-*trans*, 15-*cis*, 13-*cis*, 9-*cis*, 5-*cis*, *cis*; lutein + zeaxanthin; lutein; zeaxanthin were identified in adipose tissue. Mean values of the concentrations of the particular carotenoids in abdominal adipose tissue were higher than in the gluteal adipose tissue and the lowest in thigh adipose tissue (total carotenoids in pmol mg^{-1}: 5938.7 ± 678.5, 4426.7 ± 400.1, 3507.0 ± 387.9, respectively) (Chung *et al* 2009).

Not only the location of adipose tissue but also other factors, especially sex, general amount of adipose tissue, the type of diet, age, and the diseases which destabilize the number of carotenoids influence the concentration of carotenoids. The EURAMIC study (European study of Antioxidants, Myocardial Infarction and Cancer of the breast) conducted in eight European countries and Israel (Virtanen *et al* 1996) showed the capacity of carotenoids in male study participants to account for 50–76% of the carotenoids in women. Predictors of the concentrations of particular carotenoids in men's adipose tissue were: waist circumference for α-carotene; age, waist circumference, and alcohol use for β-carotene; age, body mass index (BMI), and waist circumference for lycopene level; whereas in women predictors were: waist circumference for α-carotene, BMI for β-carotene, and alcohol consumption for lycopene level.

Zhang *et al* (1997) demonstrated that the concentrations of lutein/zeaxanthin and β-carotene were lower in samples of breast cancer adipose tissue in comparison to benign disease samples. There were no differences in carotenoids and retinol levels in four anatomical locations of the breast (upper medial, upper lateral, lower medial, and lower lateral quadrant), or in their concentrations [in µg (g of tissue)$^{-1}$ and µg (g of fat)$^{-1}$] between samples from anatomical locations and those adjacent to the tumor (Rautalahti *et al* 1990).

Lutein, zeaxanthin, cryptoxanthin, α-carotene, and isomeric forms of β-carotene (*trans-*, 9-*cis* and 13-*cis* forms) and lycopene (*trans*) were identified in breast adipose tissue of women with a benign breast tumor or breast cancer. Statistically significant differences were found for 9-*cis* β-carotene. Its concentration in adipose tissue in benign lesions was significantly lower than in cancers (0.37 ± 0.03/0.47 ± 0.04 µmol kg^{-1}). The dominating carotenoids were: cryptoxanthin (35.9%), β-carotene (26.1%), and lutein (19.3%) (Yeum *et al* 1998). Other studies showed the accumulation of 13 carotenoids in the adipose tissue surrounding benign and malignant breast lesions, with concentrations being significantly higher in comparison with neoplastic lesions. Among them, α-carotene, β-carotene, and β-cryptoxanthin were detected. There was a 100% presence of β-carotene, 95% of β-cryptoxanthin, and 50% of α-carotene. The β-carotene content ranged from 3.26 to 7.94% in samples. Epoxy carotenoids (in %) were the dominant ones, contrary to the pigments from the provitamin A group (Czeczuga-Semeniuk *et al* 2003). Parker (1993) described the extraction and analysis of carotenoids in adipose tissue in 44-year old healthy woman. He demonstrated a unique similarity in the composition and concentration of carotenoids between adipose tissue of thigh and breast.

22.2.4 Skin

In the facial skin of healthy women (from facial surgery) lutein, lycopene, α- and β-carotene, retinol, and retinyl palmitate were identified with the commonly used HPLC method (saponification in KOH/methanol). Concerning the fact that such tissues may cause difficulties during homogenization, a non-saponification procedure was developed with collagenase solution. This new method allowed additional identification of β-cryptoxanthin, zeaxanthin, and *cis*-β-carotene in the material examined. The same carotenoids, retinol and retinyl palmitate, were also identified in cervical, ovarian, abdominal sarcoma, and kidney tumor disease (Peng *et al* 1993). The skin of smokers (from the upper thin area) contains lower mean of α- and β-carotene, and *cis*-β-carotene concentrations [in ng (g of wet tissue)$^{-1}$]. Similar results were obtained in buccal mucosa cells (Peng *et al* 1995).

HPLC techniques and spectrometry are important and useful techniques for detecting carotenoids. However, they require large tissue samples. In the case of healthy skin this is practically impossible. Carotenoids can be identified with a non-invasive optical method—RR spectroscopy. The optical signals identify the molecular structure of carotenoids, excluding other pigments, *e.g.* melanin.

With HPLC and a laser spectroscopic technique, Hata *et al* (2000) demonstrated that human abdominal skin accumulates 18 carotenoids and metabolites, including *e.g.* α-carotene, β-carotene, γ-carotene, and β-cryptoxanthin (by RP-HPLC). The results obtained with RR spectroscopy correlate with HPLC analysis. In RR spectroscopy carotenoid concentrations show differences depending on the anatomical location of the measurement. The highest mean concentrations were detected for palm, followed by forehead and inner arm. The lowest concentrations were obtained in volar arm and dorsal hand in comparison with the palm region (values statistically significant). Also pre-cancerous lesions (actinic keratosis), neoplastic lesions of the skin (basal cell carcinoma), and perilesional skin had lower carotenoid concentrations as compared to healthy skin (Hata *et al* 2000). Similar results were obtained in patients with psoriasis (Lima and Kimball 2010).

22.2.5 Eye

Except for vitamin A, whose insufficiency or lack is responsible for impaired photosensitivity of the retina receptors—stamens (the so-called night-blindness), lutein and zeaxanthin are carotenoids which are necessary for the proper physiological function of the human eye.

Studies with human donor eyes and HPLC method have demonstrated that lutein and zeaxanthin and their metabolites are present in almost every examined structure of the eye, except vitreous, cornea, and sclera. Retina presents with the highest carotenoid concentrations and their metabolites (Khachik *et al* 2002b). Except for lutein and zeaxanthin in human ciliary body there are β-cryptoxanthin, γ-carotene, α- and β-carotene, and in retinal pigment epithelium/choroid (RPE/choroid): α- and β-carotene (Bernstein *et al* 2001). In healthy lenses, there were interpersonal differences of the retinol and retinyl ester concentrations, and also between the two lenses of the same person. The comparison of mean concentrations of retinol and retinyl esters [in ng (g of wet weight)$^{-1}$] found in healthy human lenses and lenses with cataract in two ethnic groups (American normal and cataract, and Indian cataract) showed that the retinol concentrations were statistically significantly higher in the Indian group in comparison with the other groups. There were no statistically significant differences between the American groups. β-Carotene was not detected in the tissues examined. Interpersonal differences of retinol and retinyl ester concentrations were found in healthy lenses and also between two lenses of the same person (Yeum *et al* 1995).

22.2.6 Female Reproductive System

Tissues of the female reproductive system present lower concentrations of β-carotene than other organs (Palan *et al* 1989). Pigment concentrations in the ovary and endometrial tissues are higher than in the cervix, myometrium, and fallopian tube. Few literature data concerning neoplastic tissues suggest that mean β-carotene concentrations in the tissues of carcinomas of the ovary and vulva were higher than in others (Palan *et al* 1994).

22.2.6.1 Cervix Uteri

In normal and precancerous cervical tissue (similarly to ovarian tissue), lutein, zeaxanthin, β-cryptoxanthin, lycopene, α-carotene, β-carotene, and *cis*-β-carotene were identified, with the highest concentrations of lycopene, β-carotene, and isomeric *cis*-β-carotene form. High concentrations of retinol and retinyl palmitate were also obtained (Peng *et al* 1993). The highest mean concentrations (in nmol g^{-1}) of β-cryptoxanthin (0.120), β-carotene (0.091) and α-carotene (0.041) were identified in cervical dysplasia [cervical intraepithelial neoplasia, moderate grade (CIN II)] tissues. α- and β-carotene concentrations correlated with the concentrations observed in plasma samples (Gamboa-Pinto *et al* 1998).

A complex carotenoid pigment profile in ovary and uterus tissues was presented by Czeczuga-Semeniuk and Wołczyński (2005; 2008a). The carotenoid pigments were isolated using CC, TLC and HPLC. Sixteen carotenoids belonging to all four chemical groups were identified (Straub 1987), including hydrocarbon carotenoids—carotenes (α-carotene, β-carotene), hydroxy carotenoids (β-cryptoxanthin), and ketocarotenoids (echinenone, hydroxyechinenone). The studies included the assessment of the total content of major carotenoid and the percentage of carotenoids belonging to the provitamin A group. The studies concerned normal and pathological tissues. This was a qualitative and quantitative analysis of carotenoid pigments correlating with histopathological classification of lesions in the uterus and ovary.

22.2.6.2   Uterus

A total of 13 carotenoids were identified in the tissues of endometrium and myometrium as well as in tumors of the uterine corpus. Among other carotenoids, β-carotene, and β-cryptoxanthin were detected in 100% of the study samples. Hydroxyechinenone was present only in the tissues with histopathological diagnosis of endometrium folliculare and also in the group of endothelial tumors and other endothelial lesions, most often in the group with endometroid carcinoma. Only one study sample of nontypical hyperplasia complex contained α-carotene.

The normal endometrium contained the highest carotenoid concentration during the follicular phase [9.942 ± 1.561 μg (g of tissue)$^{-1}$; in comparison to the secretory phase 4.177 ± 1.160 μg (g of tissue)$^{-1}$]. On the other hand, carotenoids of the provitamin A group constituted the highest percentage of pigments in the mucosa of the uterus in the luteal phase. The average concentrations of carotenoids in both normal and pathological tissues were comparable [6.229 and 6.920 μg (g of tissue)$^{-1}$] (Czeczuga-Semeniuk and Wołczyński 2008a).

In normal myometrium, β-carotene concentrations are over 2-fold higher than in the leiomyoas group (Palan *et al* 1989; 1994). Czeczuga-Semeniuk and Wołczyński (2008a) reported that both groups presented with high percentage of carotenoid concentrations of the provitamin A group (14.0 and 12.6%, respectively). The highest percentage of carotenoids of the provitamin A group was in mesenchymal tumors (11.69%). Similarly to the group of unchanged tissues, the dominant carotenoids were: lutein epoxide (37.14%), mutatoxanthin (28.57%), β-cryptoxanthin (7.14%), and violaxanthin (5.71%) (Czeczuga-Semeniuk and Wołczyński 2008a).

22.2.6.3   Ovary

In normal and pathological ovarian tissues 14 carotenoids were identified, including provitamin A carotenoids: β-carotene, β-cryptoxanthin, echinenone, and hydroxyechinenone (9.8–26.5%). The carotenoid composition in these tissues did not differ between one another. Irrespective of histological classification, β-carotene and β-cryptoxanthin were identified in all the tissues examined. The total carotenoid concentration was relatively low [mean 1.753 μg (g of tissue)$^{-1}$]. In the group of benign mucinous tumors, the mean carotenoid concentration was higher (1.042), whereas these values in the group of malignant mucinous tumors (1.703) did not differ from the values obtained in the group with the unchanged ovarian tissue (1.756). Epoxy carotenoids were the predominant ones (Czeczuga-Semeniuk and Wołczyński 2005).

22.2.7 Mammary Gland

The mammary gland constitutes a great reservoir of carotenoids, which are naturally transferred to the fed child. Ten carotenoids and five metabolites were identified in this tissue in physiological conditions, *e.g.*: α-carotene, β-carotene, ζ-carotene, and α- and β-cryptoxanthin (Khachik *et al* 1998).

The assessment of the total carotenoid concentrations, the percentage of the dominant carotenoids, and the percentage of β-carotene was performed in the tissues of fibroadenoma and ductal infiltrating carcinoma of the breast. The following carotenoids were identified: α- and β-carotene, β-cryptoxanthin, lutein, 3'-epilutein, zeaxanthin, canthaxanthin, astaxanthin, lutein epoxide, antheraxanthin, neoxanthin, violaxanthin, and mutatoxanthin.

β-Carotene was present in 100% of cases and α-carotene in 50% of the analyzed material regardless of its origin. However, the percentage of β-carotene in neoplastic tissues (benign and malignant) was low, amounting to 2.43–4.33%. β-Carotene was not the dominant carotenoid either. The mean total carotenoid concentrations in the cancer tissues were slightly lower in comparison with the concentrations obtained in the benign cases [mean concentrations: 20.433 and 22.889 μg (g of tissue)$^{-1}$, respectively]. In the study material, epoxy carotenoids were the dominant ones (Czeczuga-Semeniuk *et al* 2003).

All analyzed tissues (normal and pathological) obtained from the female reproductive organs (breast, uterus, and ovary) contained β-carotene, 99% of the tissues samples had β-cryptoxanthin, whereas α-carotene was identified in the breast tissues samples.

Provitamin A carotenoids (β-carotene and β-cryptoxanthin) were most often identified. No differences in the prevalence of these carotenoids in the malignant lesions in comparison with the other tissues were found. The dominant carotenoids were: lutein epoxide, mutatoxanthin and violaxanthin (Czeczuga-Semeniuk and Wołczyński 2008b).

22.2.8 Human Milk

HPLC technique (in healthy lactating women) allowed the identification of 13-*cis*-β-carotene, α-carotene, all-*trans*-β-carotene, 9-*cis*/ζ-carotene, and others in human milk. Definitely, all-*trans*-β-carotene presented with the highest mean concentrations (0.916 ± 0.255 μmol L^{-1}). The authors reported an increase in all-*trans*-β-carotene and 9-*cis*-β-carotene in milk after oral supplementation with these carotenoids, which was similar to the increase in serum and buccal mucosa (Johnson *et al* 1997). The complete characteristic of carotenoids and their metabolites in milk was performed by Khachik *et al* (1997). Among them, all-*trans* carotenoids (α-, β-, and γ-carotene) and *cis* carotenoids (β-cryptoxanthin) were detected. Schweigert *et al* (2000) used two methods of extraction (I – saponification and II – ethanol extraction prior to saponification) before separation and quantification by RP-HPLC was used in order to determine the carotenoids in human milk. There were no changes observed in the concentration of the detected provitamin A carotenoids.

22.2.9   Prostate Gland

The tissues obtained from the region of prostate cancer and healthy tissues, in which lycopene, *all*-trans β-carotene, 9-*cis* β-carotene, α-carotene, lutein, *cis* + *trans* α-cryptoxanthin, zeaxanthin, and β-cryptoxanthin were identified with the HPLC method, constituted the analyzed material. The highest all-*trans* β-carotene concentration was in 28% of prostate tissues. Mean concentrations of this carotenoid were higher in the prostate cancer tissues. Mean provitamin A concentration in the prostate cancer tissues amounted to 1.52 nmol g^{-1} (Clinton *et al* 1996). Healthy tissues also contain γ-carotene (Khachik *et al* 2002a).

22.2.10   Sigmoid

To determine tissue concentration of carotenoids, the material for the analysis was taken at the level of rectum and sigmoid colon. Material was obtained during biopsy of the polypous tissue during polypectomy. Using the HPLC technique, the following provitamin A carotenoids were identified in all the tissues (normal and pathological): β-cryptoxanthin, α-carotene, and β-carotene. In the group of tissues obtained from healthy people carotenoid concentrations were higher in rectum than in colon. In the polypus adenomatosus tissues, β-cryptoxanthin concentrations were lower in comparison with the material obtained from rectum and colon of the same person. In people with colonic cancer, cryptoxanthin, and α- and β-carotene concentrations were lower in cancer tissues than in the healthy group. Supplementation with oral β-carotene led to an increase in β-carotene in all the study tissues and in α-carotene in the tissues of patients with a diagnosed cancer (Pappalardo *et al* 1997).

Summary Points

- This Chapter describes the assay of vitamin A and carotenoids belonging to the provitamin A group in human tissues.
- Carotenoids are produced by bacteria, fungi, and green plants. They are delivered to the human body in the form of fruit and vegetables. Carotenoids of the provitamin A group are the precursors of vitamin A in humans.

- The assay of vitamin A and carotenoids is a key element of the analysis of various processes they participate in, which may contribute to the better knowledge of some phenomena or diseases.
- In people, carotenoids are present in many tissues and organs. Their highest concentrations have been identified in suprarenal glands, liver, and testicles. The human liver accumulates not only carotenoids and their metabolites but also vitamin A.
- Carotenoids of all chemical groups are present in human tissues, most often α-carotene, β-carotene, β-cryptoxanthin, zeaxanthin, lycopene, lutein, and other carotenoids, metabolites and isomeric forms of β-carotene and lycopene.
- Adipose tissue is a large carotenoid reservoir in the human body. The predominant forms are carotenes and xanthophylls, with their highest concentrations in the abdominal adipose tissue. Adipose tissue surrounding pathological lesions of the mammary gland accumulates approx. 3.26–7.94% of β-carotene.
- The proper function of the eye structures is connected with the presence of lutein and zeaxanthin (predominant carotenoids) as well as β-cryptoxanthin, γ-carotene, α- and β-carotene in the RPE/choroid, iris and ciliary body, and α- and β-carotene in the RPE/choroid.
- There are 18 carotenoids (including five metabolites) in the human skin. However, in the skin of smokers the mean concentrations of lutein, cryptoxanthin, α- and β-carotene, and *cis*-β-carotene are lower. Carotenoid concentrations in the skin measured with RR spectroscopy present differences dependent on the anatomical localization of the measurement. Pre-neoplastic and neoplastic lesions of the skin have lower carotenoid concentrations.
- All the examined tissues of the female and male reproductive system had carotenoids of the provitamin A group, which were the most often identified pigments. However, there were no differences in their prevalence in cancer tissue as compared to the rest of the tissues. Most normal and pathological tissues of the prostate gland had lycopene and β-carotene, whose mean concentrations as well as total mean concentrations were higher in cancer tissues.
- The concentrations of carotenes were lower in the rectum than in the colon of healthy subjects. The levels of cryptoxanthin, α- and β-carotene were lower in colonic cancer patients than in healthy people.

Key Facts

Key Facts of Provitamin A Carotenoids and Vitamin A History

- In 1831, H. Wackenroder isolated β-carotene from the carrot root (*Daucus carota*) and named it "carotin" = carrot. In 1837, J. Berzelius extracted yellow pigment from the autumn leaves "xanthophyllous" and provided the initial information concerning biosynthesis of carotenoids.

- R. Willstätter described the empirical name of carotene, $C_{40}H_{56}$, and xanthophyll, $C_{40}H_{56}O_2$, in 1907.
- In 1911, M. Tswiet, the inventor of the column chromatography, coined the name of carotenoids.
- In 1912, K. Funk used the term "vitamins" for the first time.
- In 1913, E. McCollum described the growth factor – "factor A", which was soluble in lipids and did not undergo saponification. It was the first identification of vitamin A, long before description of its chemical structure.
- In 1919, H. Steenbock suggested that there was a relation between β-carotene and vitamin A.
- In 1928, P. Karrer described the concept of the provitamin, the substance that may be converted within the body into a vitamin.
- B. von Euler described the relationship between carotenoids and vitamin A in 1929. He noticed that a crystal form of carotene had the same properties as vitamin A.
- T. Moor proved, in the experiments conducted on rats in 1930, that carotene was metabolized into vitamin A and stored in the rat's liver.
- The complete chemical structure of retinol (vitamin A) was described by Karrer *et al* in 1931, 100 years after Wackenroder's discovery.

Definitions of Words and Terms

Carotenoids: The group of natural pigments of yellow orange, red, and red-purple coloring existing in the plants and animals. People do not synthesize these pigments. Carotenoids are divided into carotenes (hydrocarbons), *e.g.* α- and β-carotene, and xanthophylls, oxidized carotene, *e.g.* lutein.

Vitamin A precursor: Carotenoids of the provitamin A group (*e.g.* α-carotene, β-carotene, β-cryptoxanthi, echinenone, and hydroxyechinenone). Their transformation in humans results in the vitamin A synthesis.

Extraction: The separation of carotenoids from the biological material with organic reagents.

Chromatography: The separation of pigments in the extract with various absorbents. Different types of chromatography include: column (CC), thin-layer (TLC), paper, ion-exchange, high-performance (HPLC).

Carotenoid identification: The method based on the determination of the maximum absorption in the visible or infrared light which is characteristic for particular carotenoids – UV/Vis spectrophotometry. Identification might be also performed with MS, NMR, RR spectrophotometry.

List of Abbreviations

BMI	body mass index
CC	column chromatography
HPLC	high-performance liquid chromatography
MS	mass spectrometry

NMR   nuclear magnetic resonance
NP   normal phase
RP   reversed phase
RPE/choroid   retinal pigment epithelium/choroid
RR spectroscopy resonance Raman spectroscopy
TLC   thin-layer chromatography
UV/Vis   ultraviolet/visible

References

Bernstein, P. S., Khachik, F., Carvalho, L. S., Muir, G. J., Zhao, D. Y. and Katz, N. B., 2001. Identification and quantitation of carotenoids and their metabolites in the tissues of the human eye. Experimental Eye Research. 72: 215–223.

Clinton, S. K., Emenhiser, C., Schwartz, S. J., Bostwick, D. G., Williams, A. W., Moore, B. J. and Erdman, Jr, J. W., 1996. *cis-trans* lycopene isomers, carotenoids, and retinol in the human prostate. Cancer Epidemiology, Biomarkers and Prevention. 5: 823–833.

Chung, H. Y., Ferreira, A. L., Epstein, S., Paiva, S. A., Castaneda-Sceppa, C. and Johnson, E. J., 2009. Site-specific concentrations of carotenoids in adipose tissue: relations with dietary and serum carotenoid concentrations in healthy adults. The American Journal of Clinical Nutrition. 90: 533–539.

Czeczuga-Semeniuk, E., Wołczyński, S. and Markiewicz, W., 2003. Preliminary identification of carotenoids in malignant and benign neoplasms of the breast and surrounding fatty tissue. Neoplasma. 50: 280–286.

Czeczuga-Semeniuk, E. and Wołczyński S., 2005. Identification of carotenoids in ovarian tissue in women. Oncology Reports. 14: 1385–1392.

Czeczuga-Semeniuk, E. and Wołczyński S., 2008a. Dietary carotenoids in normal and pathological tissues of corpus uteri. Folia Histochemica et Cytobiologica. 46: 251–258.

Czeczuga-Semeniuk, E. and Wołczyński, S., 2008b. Does variability in carotenoid composition and concentration in tissues of the breast and reproductive tract in women depend on type of lesion? Advances in Medical Sciences. 53: 270–277.

Gamboa-Pinto, A. J., Rock, C. L., Ferruzzi, M. G., Schowinsky, A. B. and Schwartz S. J., 1998. Cervical tissue and plasma concentrations of α-carotene and β-carotene in women are correlated. Journal of Nutrition. 128: 1933–1936.

Goodwin, T.W., 1980. The Biochemistry of the Carotenoids: Vol.1, Plants. Chapman and Hall, London, UK, and New York, USA.

Hata, T. R., Scholz, T. A., Ermakov, I. V., McClane, R. W., Khachik, F., Gellermann, W. and Pershing, L. K., 2000. Non-invasive raman spectroscopic detection of carotenoids in human skin. The Journal of Investigative Dermatology. 115: 441–448.

Johnson, E. J., Qin, J., Krinsky N. I. and Russell, R. M., 1997. β-Carotene isomers in human serum, breast milk and buccal mucosa cells after continuous oral doses of all-*trans* and 9-*cis* β-carotene. Journal of Nutrition. 127: 1993–1999.

Kaplan, L. A., Lau, I. M. and Stein, E. A., 1990. Carotenoid composition, concentrations, and relationships in various human organs. Clinical Physiology and Biochemistry. 8: 1–10.

Khachik, F., Askin, F. B. and Lai, K., 1998. Distribution, bioavailability, and metabolism of carotenoids in humans. In: Bidlack, W. R., Omaye, S. T., Meskin, M. S. and Jahner, D., (ed.) Phytochemicals, A New Paradigm. Technomic Publishing, Lancaster, PA, USA, Chapter 5, pp. 77–96.

Khachik, F., Carvalho, L., Bernstein, P. S., Muir, G. J., Zhao, D. Y. and Katz, N. B., 2002a. Chemistry, distribution, and metabolism of tomato carotenoids and their impact on human health. Experimental Biology and Medicine (Maywood). 227: 845–851.

Khachik, F., de Moura, F. F., Zhao, D. Y., Aebischer, C. P. and Bernstein, P. S., 2002b. Transformations of selected carotenoids in plasma, liver, and ocular tissues of humans and in nonprimate animal models. Investigative Ophthalmology and Visual Science. 43: 3383–3392.

Khachik, F., Spangler, C. J., Smith, Jr, J. C., Canfield, L. M., Pfander, H. and Steck, A., 1997. Identification, quantification, and relative concentrations of carotenoids, and their metabolites in human milk and serum. Analytical Chemistry. 69: 1873–1881.

Lima, X. and Kimball A., 2010. Skin carotenoid levels in adult patients with psoriasis. Journal of the European Academy of Dermatology and Venereology. 25: 945–949.

Palan, P. R., Mikhail, M. and Romney, S. L., 1989. Decreased β-carotene tissue levels in uterine leiomyomas and cancers of reproductive and nonreproductive organs. American Journal of Obstetrics and Gynecology. 161: 1649–1652.

Palan, P. R., Goldberg, G. L., Basu, J., Runowicz, C. D. and Romney, S. L., 1994. Lipid-soluble antioxidants: β-carotene and α-tocopherol levels in breast and gynecologic cancers. Gynecologic Oncology. 55: 72–77.

Pappalardo, G., Maiani, G., Mobarhan, S., Guadalaxara, A., Azzini, E., Raguzzini, A., Salucci, M., Serafini, M., Trifero, M., Illomei, G. and Ferro-Luzzi, A., 1997. Plasma (carotenoids, retinol, α-tocopherol) and tissue (carotenoids) levels after supplementation with β-carotene in subjects with precancerous and cancerous lesions of sigmoid colon. European Journal of Clinical Nutrition. 51: 661–666.

Parker, R.S., 1988. Carotenoid and tocopherol composition of human adipose tissue. The American Journal of Clinical Nutrition. 47: 33–36.

Parker, R.S., 1989. Carotenoids in human blood and tissues. Journal of Nutrition. 119: 101–104.

Parker, R.S., 1993. Analysis of carotenoids in human plasma and tissues. Methods in Enzymology. 214: 86–93.

Peng, Y. M., Peng, Y. S. and Lin, Y., 1993. A nonsaponification method for the determination of carotenoids, retinoids, and tocopherols in solid human tissues. Cancer Epidemiology, Biomarkers and Prevention. 2: 139–144.

Peng, Y. M., Peng, Y. S., Lin, Y., Moon, T., Roe, D. J. and Ritenbaugh, C., 1995. Concentrations and plasma-tissue-diet relationships of carotenoids, retinoids, and tocopherols in humans. Nutrition in Cancer. 23: 233–246.

Raica, Jr, N., Scott, J., Lowry, L. and Sauberlich, H. E., 1972. Vitamin A concentration in human tissues collected from five areas in the United States. The American Journal of Clinical Nutrition. 25: 291–296.

Rautalahti, M., Albanes, D., Hyvönen, L., Piironen, V. and Heinonen, M., 1990. Effect of sampling site on retinol, carotenoid, tocopherol, and tocotrienol concentration of adipose tissue of human breast with cancer. Annals of Nutrition and Metabolism. 34: 37–41.

Schmitz, H. H., Poor, C. L., Wellman, R. B., Erdman, Jr, J. W., 1991. Concentrations of selected carotenoids and vitamin A in human liver, kidney and lung tissue. Journal of Nutrition. 121: 1613–1621.

Schweigert, F. J., Hurtienne, A. and Bathe, K., 2000. Improved extraction procedure for carotenoids from human milk. International Journal for Vitamin and Nutrition Research. 70: 79–83.

Stahl, W., Schwarz, W., Sundquist, A.R. and Sies, H. 1992. *Cis–trans* isomers of lycopene and β-carotene in human serum and tissues. Archives of Biochemistry Biophysics. 294: 173–177.

Stahl, W., Sundquist, A. R., Hanusch, M., Schwarz, W. and Sies, H., 1993. Separation of β-carotene and lycopene geometrical isomers in biological samples. Clinical Chemistry. 39: 810–814.

Straub, O., 1987. Key to Carotenoids. Birkhäuser Verlag, Basel, Switzerland, and Boston, MA, USA.

Tanumihardjo, S. A., Furr, H. C., Amedée-Manesme, O. and Olson, J. A., 1990. Retinyl ester (vitamin A ester) and carotenoid composition in human liver. International Journal for Vitamin and Nutrition Research. 60: 307–313.

Virtanen, S. M., van't Veer, P., Kok, F., Kardinaal, A. F. and Aro, A., 1996. Predictors of adipose tissue carotenoid and retinol levels in nine countries. The EURAMIC Study. American Journal of Epidemiology. 144: 968–979.

Yeum, K. J., Ahn, S. H., Rupp de Paiva, S. A., Lee-Kim, Y. C., Krinsky, N. I. and Russell, R. M., 1998. Correlation between carotenoid concentrations in serum and normal breast adipose tissue of women with benign breast tumor or breast cancer. Journal of Nutrition. 128: 1920–1926.

Yeum, K. J., Taylor, A., Tang, G. and Russell, R. M., 1995. Measurement of carotenoids, retinoids, and tocopherols in human lenses. Investigative Ophthalmology and Visual Science. 36: 2756–2761.

Zhang, S., Tang, G., Russell R. M., Mayzel, K. A., Stampfer, M. J., Willett, W. C. and Hunter D. J. 1997. Measurement of retinoids and carotenoids in breast adipose tissue and a comparison of concentrations in breast cancer cases and control subjects. The American Journal of Clinical Nutrition. 66: 626–632.

CHAPTER 23

Vitamin A Deficiency: An Overview

TERESA BARBER*[a], GUILLERMO ESTEBAN-PRETEL[b], MARIA PILAR MARÍN[b] AND JOAQUÍN TIMONEDA[a]

[a] Departamento de Bioquímica y Biología Molecular, Facultad de Farmacia, Universidad de Valencia, Avda V. Andrés Estellés s/n, 46100-Burjassot, Spain; [b] Sección de Biología y Patología Celular, Centro de Investigación Hospital "La Fe", Avda Campanar, 21, 46009-Valencia, Spain
*E-mail: teresa.barber@uv.es

23.1 Introduction

Vitamin A is a term that strictly refers to the alcoholic form all-*trans*-retinol, although it includes its biologically active derivatives, also known as retinoids, in the broad sense (Figure 23.1), these being its oxidation metabolites, retinaldehyde and retinoic acid (RA), which can exist as all-*trans* and several *cis*-isomers. Vitamin A is involved in essential physiological functions, including vision, immunity, cell differentiation, embryological development and growth, and it also acts as a physiological antioxidant. There are a number of sources of dietary vitamin A. Preformed vitamin A (retinol esters) is abundant in some animal-derived foods, whereas provitamin A (some carotenoids) abounds in fruit and vegetables (Figure 23.2).

Vitamin A deficiency (VAD) is an important nutritional world problem, and has important implications for the global health policy. In addition to protein malnutrition, this is the most common nutritional disorder in the world. VAD has a wide range of undesirable clinical effects which appear when liver reserves are exhausted. The most specific clinical effect of inadequate vitamin A intake is

Food and Nutritional Components in Focus No. 1
Vitamin A and Carotenoids: Chemistry, Analysis, Function and Effects
Edited by Victor R Preedy
© The Royal Society of Chemistry 2012
Published by the Royal Society of Chemistry, www.rsc.org

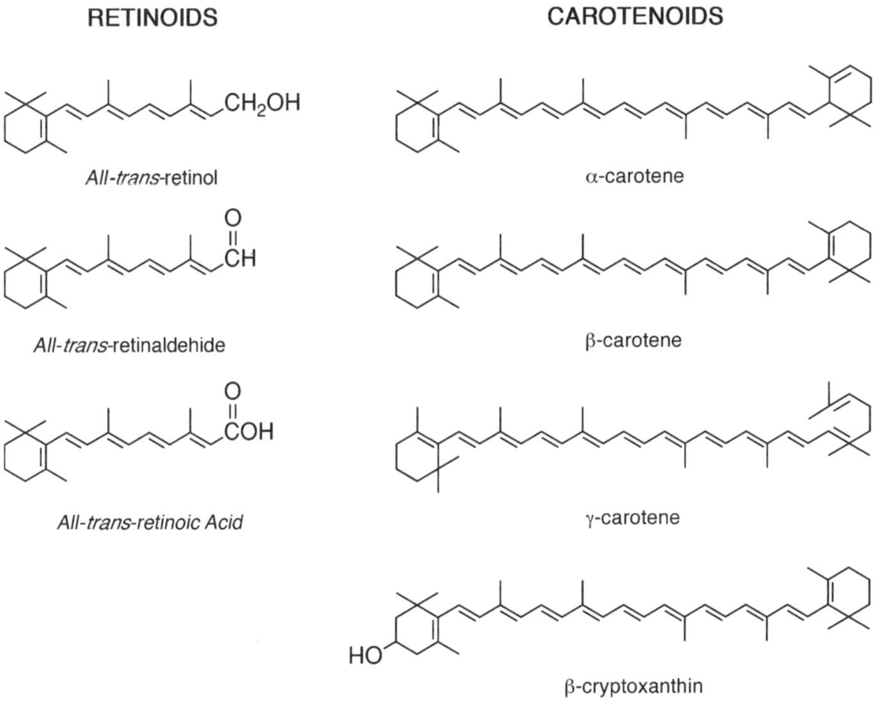

Figure 23.1 Chemical structures of the main natural retinoids and carotenoids. Structures of all-*trans*-retinol, all-*trans*-retinaldehyde, all-*trans*-RA, α-carotene, β-carotene, γ-carotene and β-cryptoxanthin.

xerophthalmia accompanied by nyctalopia. In addition, VAD has been associated with dry skin and hyperkeratosis, increased susceptibility to severe infection and disturbances in cell differentiation, organ development, growth and reproduction (Underwood 1984). Control of VAD involves dose supplementation, vitamin A fortification of commonly consumed food, control of VAD precipitating infections and dietary improvement of natural sources of vitamin A.

This Chapter summarises the information available on VAD at both the molecular and clinical levels, and focuses on its cause, effects and damage of specific tissues.

23.2 Requirements

Requirements for vitamin A are actually expressed in retinol activity equivalents (RAE). One µg of RAE is equal to 1 µg of all-*trans*-retinol and, based on its estimated efficacy of absorption and conversion into vitamin A, it is equal to 12 µg of β-carotene, and to 24 µg of α-carotene or β-cryptoxanthin. However, the efficiency of β-carotene to be converted into retinoids is probably inferior than previously thought. Vitamin A levels could also be

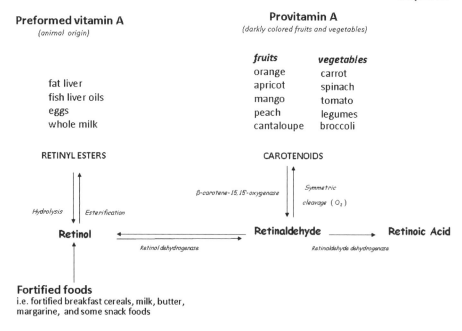

Preformed vitamin A
(animal origin)

Provitamin A
(darkly colored fruits and vegetables)

fruits	*vegetables*
orange	carrot
apricot	spinach
mango	tomato
peach	legumes
cantaloupe	broccoli

fat liver
fish liver oils
eggs
whole milk

RETINYL ESTERS   CAROTENOIDS

β-carotene-15,15'-oxygenase   Symmetric
cleavage (O_2)

Hydrolysis   Esterification

Retinol ⟷ **Retinaldehyde** ⟶ **Retinoic Acid**

Retinol dehydrogenase   Retinaldehyde dehydrogenase

Fortified foods
i.e. fortified breakfast cereals, milk, butter,
margarine, and some snack foods

Figure 23.2   Food sources of vitamin A. Vitamin A is present in foods either as preformed vitamin A, mainly retinyl esters, or provitamin A carotenoids, mainly β-carotene. Some foods are fortified with retinol. Retinol and retinaldehyde may be oxidised to RA.

stated as international units (IU), mainly on food and supplement labels (1 IU is the equivalent of 0.3 μg of retinol or to 0.3 μg of RAE). The recommended dietary allowances (RDAs) for children, men and women are 300–600, 900 and 700 μg of RAE/day, respectively. During pregnancy, the RDA is 750 μg of RAE/day and is 1300 μg of RAE/day during lactation.

A number of factors can influence vitamin A requirement, including the presence and severity of infection and parasites, intestinal disease, iron and zinc status, dietary fat, protein energy malnutrition, alcohol intake and available sources of preformed vitamin A and provitamin A carotenoids.

23.3   Intertissular Transport and Metabolic Transformations of Vitamin A

After absorption by enterocytes, vitamin A is transported mainly in chylomicrons as retinyl ester to the liver, where it is stored in hepatic stellate cells. When needed, these vitamin reserves are mobilised and transported as retinol by the plasma retinol-binding protein (RBP) to other cells (see Figure 23.3 for more details). Within cells, retinol is oxidised reversibly to retinal by the enzymes of the alcohol dehydrogenase and short-chain dehydrogenase/reductase families. RA is formed by the irreversible oxidation of retinal catalysed by retinaldehyde dehydrogenase isoenzymes (Figure 23.4). Several cellular retinol-binding proteins (CRBPI–CRBPIV) participate in these processes, probably as substrate providers to enzymes. Genetic variants of these enzymes or retinoid-binding proteins may lead to inter-individual

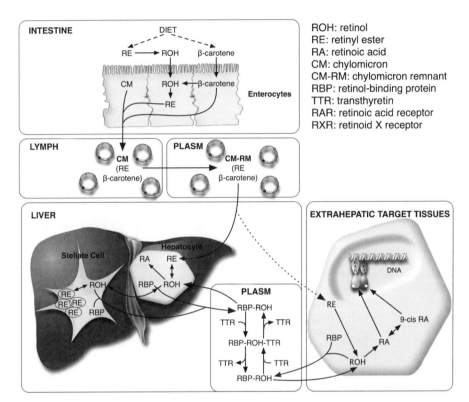

Figure 23.3   Intertissular transport of vitamin A. Vitamin A enters the enterocyte as retinol by a carrier-mediated process and also, like carotenoids, by passive diffusion. Inside the enterocyte, carotenoids are converted into retinol which is esterified (RE) and incorporated, together with a fraction of the absorbed carotenoids, into chylomicrons. The chylomicron remnants formed in plasma by the action of lipoprotein lipase are taken up mainly by hepatocytes in a receptor-mediated process. RE are hydrolysed back to retinol, which is secreted into plasma associated with RBP, oxidised to RA or transferred to stellate cells for storage mainly as esters of palmitic acid (RE). When needed, the vitamin A reserve is mobilised and secreted as a retinol–RBP complex to plasma. This complex associates in plasma with transthyretin (1:1), reducing the glomerular filtration of retinol. From this complex, retinol is delivered to extrahepatic cells by a receptor-mediated process. Inside cells, retinol is oxidised to retinal which plays a role in vision, and to RA which regulates gene transcription by binding to nuclear retinoid receptors (RAR and RXR) that act as transcription factors.

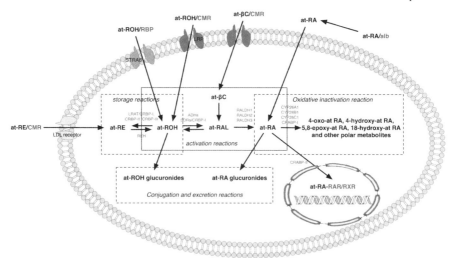

Figure 23.4   Non-visual cell metabolism of vitamin A. All-*trans*-retinyl ester (at-RE)
enters the cells, mainly hepatocytes, as part of chylomicron remnants
(CMR) *via* the low-density lipoprotein (LDL) receptor or via LDL-
related protein (LRP). STRA6 has been suggested to be a receptor for
RBP in many tissues. All-*trans*-retinol–RBP (at-ROH/RBP) binds to the
STRA6 protein located in the plasma membrane, facilitating retinol
uptake. RA is present in plasma at nanomolar concentrations bound to
albumin (at-RA/alb); this may be a minor route for RA incorporation
into cells. Inside the cells, at-RE, at-ROH and at-RAL (all-*trans*-retinal)
are reversibly interconverted by the action of ester hydrolases and
dehydrogenases. at-RAL is irreversibly oxidised by RALDHs to at-RA,
which is further oxidised for excretion by mono-oxygenases of the CYP
family. There are cellular binding proteins for retinol and retinal
(CRBP) and RA (CRABP) which facilitate their metabolism and action.
CRABP-I and CRABP-II seem to be involved in RA degradation and
RA transcriptional activity, respectively.

differences in the sensitivity to VAD. Cells are also able to metabolise all-*trans*-
retinol and all-*trans*-RA to more polar 4-hydroxy- and 4-oxo-metabolites.
These reactions appear to be catalysed by members of the CYP26 family of
cytochrome P450 mono-oxygenases. The metabolites from CYP degradation
of retinol and RA are conjugated in the liver mainly with glucuronic acid, and
are excreted in bile and urine (Blomhoff and Blomhoff 2006; D'Ambrosio *et al*
2011; Ross and Zolfaghari 2011).

23.4   Vitamin A Deficiency

The physiological plasma concentration of vitamin A is 1–2 μM and is under
tight homeostatic control. Consequently, only minor changes take place in
plasma retinol or RBP concentrations over a wide range of adequate liver

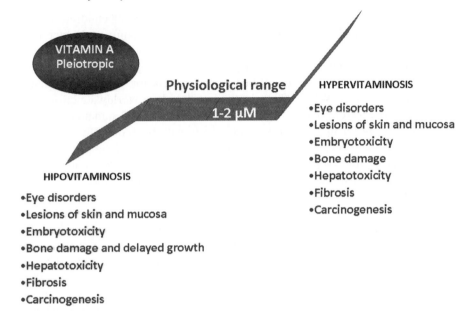

Figure 23.5 Importance of maintaining plasma retinol concentrations within the physiological range. Concentrations of plasma vitamin A, either above or below the physiological range, lead to similar adverse effects.

vitamin A reserves (Underwood 1984). When these reserves fall below a critical concentration, thought to be approximately 20 µg g^{-1} of liver [National Academy of Sciences (NAS) 2006], plasma retinol concentration declines and depends on liver concentration. According to the World Health Organization (WHO), a plasma retinol concentration below 0.70 µM is indicative of VAD. This plasma retinol level is accompanied by tissue concentrations in this vitamin which are low enough to result in adverse health effects. Serum retinol concentrations lower than 0.35 µM are indicative of severe deficiency and are associated with marked increases in the risk of clinical manifestations. Even asymptomatic subclinical forms increase morbidity and mortality from a variety of infections, and, particularly, the incidence and morbidity of respiratory tract diseases (Sommer 2008; Underwood 2004).

It is worth mentioning at this stage that, as with antioxidant capacity, vitamin A concentrations both below and above the physiological range cause adverse effects, which are paradoxically similar in both situations (Figure 23.5).

23.4.1 Epidemiology and Incidence

Epidemiological studies indicate that VAD is currently, along with protein malnutrition, the most serious and common nutritional disorder worldwide. In developing countries, deficiency affects a high percentage of children and is one of the most usual causes of infant mortality in these countries. In fact, it is estimated that 250 million preschool-aged children have biochemical VAD and that 5 million are clinically affected by VAD. In these countries, along with preschool children, other groups of people at high risk for VAD are pregnant and lactating women (WHO 2009). In the last decade, a worldwide effort has been made to control VAD and related diseases, but VAD remains a major public health problem, the impact of which will be moderate to severe depending on the prevalence of deficiency among the population. The latest estimates reveal that based on the prevalence of night blindness and plasma retinol concentrations below 0.70 μM, 45 and 122 countries, respectively, have VAD of public health significance (WHO 2009) (Figure 23.6).

According to the WHO reports, vitamin A intake is low in Eastern countries where rice, which lacks this vitamin, is the major part of diet, is medium in Latin America and Africa, and is higher in Europe and North America in comparative terms. It is important to note that over 20% of the population in the developed world does not reach two-thirds of the recommended intake, and has plasma and liver concentrations of vitamin A lower than those accepted as normal. This situation can be aggravated by the increasingly common tendency to reduce fat intake and to engage in uncontrolled weight-loss diets (Bendich and Langseth 1989). In this context, higher prevalence appears in pregnant women of low social classes, and in adults or children affected by human immunodeficiency virus. Other risk factors for VAD

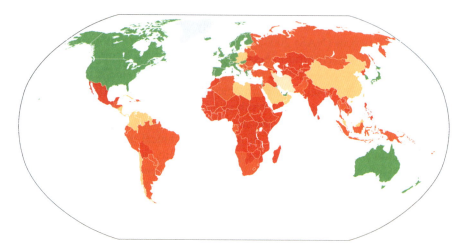

Figure 23.6   Distribution of biochemical VAD in preschool-age children (< 5 years) based on survey data and regression-based estimates [based on WHO (2009)]. Category of the public health significance of the biochemical VAD (prevalence of serum retinol < 0.7 μM) across countries. ☐ None (<2%); ☐ Mild (≥2–<10%); ☐ Moderate (≥10–<20%); ☐ Severe (≥20%); ☐ Gross domestic product ≥US$ 15000 (assumed to be free of VAD of public health significance), no data.

include stress, diseases which affect the intestine's ability to absorb fat, infections, infestations and alcohol abuse (Lieber 2000; WHO 2009).

23.4.2 Vitamin A Deficiency and Tissue Damage

Retinoids are primary players in vision, organ development and function, normal differentiation of epithelia and growth. Xerophtalmia, with night blindness, is one of the first manifestations of VAD and is practically pathognomonic. The involvement of fat-soluble vitamins in maintaining the epithelial morphology of several organs was described a long time ago in experimental animals (Wolbach and Howe 1925). Weanling rats fed on a vitamin A-deficient diet exhibited mucous epithelium alterations, such as squamous metaplasia, cornea ulcerations, sterility and compromised immune function. This situation could be reverted by re-feeding vitamin A-deficient animals with retinol, thus confirming the involvement of vitamin A in the maintenance of organ morphology and cell differentiation. Participation of natural retinoids in embryology and organogenesis was recognised several years later by analysing morphological malformations in the offspring of pregnant rats with VAD. These malformations ranged from organ agenesis to the presence of rudimentary or hypoplastic organs depending on the severity of VAD (Wilson *et al* 1953). Addition of retinol to the diet of females during specific times of pregnancy prevents the appearance of abnormalities, indicating that vitamin A is also critical for normal organ development. The vitamin A-deficient embryo can also be partially rescued by administration of RA, and treatment of normal embryos with an anti-RA monoclonal antibody induces the VAD phenotype. Therefore part, or most, of the actions of vitamin A in mammalian organogenesis are mediated by RA.

Many organs and systems are affected by VAD during their development or in the postnatal state. However, we will consider herein some organs that are not covered in other sections of this book.

23.4.2.1 Eye

Signs of VAD in eyes were among the first diet-related deficiencies to be documented and have been grouped under the term xerophthalmia. Night blindness is the first symptom of VAD, and results when the vitamin A pool in the eye becomes depleted and the concentration in rod cells lowers, and is accompanied by degenerative changes of the retina and dryness of the conjunctiva, producing a greyish pigmentation (Bitot's spots). Night blindness has been known since antiquity; Hippocrates recognised and treated the condition with animal liver in the 4th century BC (Wolf 1996). Ancient authors used the term nyctalopia to mainly define the symptom of impaired dark adaptation. Between night blindness and loss of vision in corneal scarring, there are different degrees of intensity, which are early reversible by appropriate treatment (WHO 2009). During VAD, the amount of mucin-containing goblet cells lowers, and epithelial cells become enlarged and distorted (Tanumihardjo 2011). Afterwards, ulceration and necrosis of the cornea occurs, followed by keratomalacia and blindness.

The recovery of the visual function in vitamin A-deficient patients following vitamin A supplementation has been clearly demonstrated (see Section 23.4.3). The control of blindness in children is considered a high priority within the World Health Organization's VISION 2020 - The Right to Sight Programme (Gilbert and Foster 2001; http://www.vision2020.org/main/ccm; WHO 2010), and one of its goals is to reduce the global prevalence of childhood blindness from 0.75 per 1000 to 0.4 per 1000 children by the year 2020. The control of blindness in children is closely linked to child survival. The number of blind children in the world is estimated to be 1.4 million, and most (approx. 75%) live in the poorest regions of Africa and Asia. In the poorest countries, the principal causes of blindness in children are corneal scarring due to VAD, measles infection and ophthalmia neonatorum. Cataract, retinal diseases and congenital abnormalities affecting the whole eye are important causes of blindness worldwide, while corneal scarring is the single most important cause of avoidable blindness, followed by cataract and perinatal conditions (particularly retinopathy of prematurity) (Gilbert and Foster 2001).

23.4.2.2 Lung

Vitamin A is important in regulating early lung development and alveolar formation. VAD during lung organogenesis leads to bilateral lung hypoplasia, failure of septation of trachea and oesophagus, left lung agenesis or presence of rudimentary lungs.

There is evidence that VAD in mothers during pregnancy could have lasting adverse effects on healthy lungs in their offspring (Checkley *et al* 2010). Similarly in weanling rats fed a vitamin A-deficient diet, alveolar septation significantly lowers, and lungs show areas with emphysematous features (Baybutt and Molteni 2007). Structural and morphological alterations are accompanied by mechanical and functional derangements. An inverse relationship between plasma retinol concentrations and the degree of airway obstruction, assessed by FEV1, has been established in humans (Moravia *et al* 1990).

Many genes are known to respond to RA signalling. Extracellular matrix proteins, especially those of the basement membrane, are involved in tissue mechanics and also in cell proliferation, migration and differentiation, fundamental processes in foetal development, tissue maintenance and injury repair. Consequently, these proteins are potential mediators in the effects of VAD and have received special attention. In murine lung, VAD results in quantitative changes in elastin and collagens, not only during foetal development, but also in postnatal life, which are associated with, and considered to be involved in, the defects of alveolarisation and lung function impairment produced by vitamin deficiency (Baybutt and Molteni 2007). VAD during the growing period also results in emphysemic lungs. The basement membrane thickens (Figure 23.7A, B) and its component macromolecules, such as collagen IV and laminin, are also modified not only quantitatively, but also qualitatively, in VAD rat lungs. Total collagen IV increases and the subunit composition of both collagen IV and laminin is modified (Esteban-Pretel *et al* 2010a, Esteban-Pretel *et al* in press). When considering that these basement membrane macromolecules mediate their cell effects through membrane receptors, namely integrins, and that these receptors can bind differentially to distinct basement membrane macromolecule subunits, their qualitative and/or quantitative modifications may alter cell behaviour and lung function (Miner 2008). Other fundamental components for lung function are

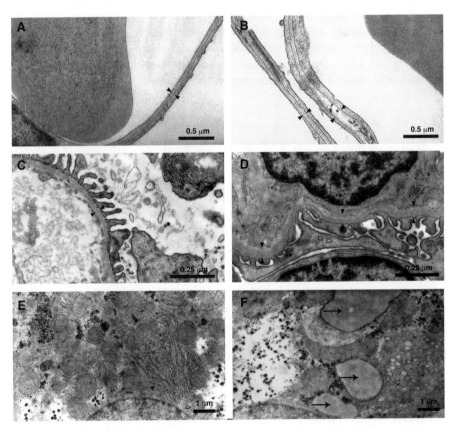

Figure 23.7 Electron micrographs of alveolar basement membranes from lung (A and B) and kidney (C and D), and hepatocytes (E and F) of control (A, C and E) and VAD rats (B, D and F). The micrographs show the thickening and disorganisation of the alveolar (B) and glomerular (D) basement membranes. Inside the VAD alveolar basement membrane, the presence of collagen fibrils can be observed. Lipid vesicles are seen in VAD hepatocytes (F), but not in the control ones (E). Arrows indicate the borders of basement membranes (A–D) and lipid vesicles (F).

the surfactants which line the alveoli and prevent their collapse during respiration. In rat, partial VAD, compatible with maintenance of gestation and absence of major abnormalities in foetuses, leads to decreased surfactant in neonates. The lower capacity of type II pneumocytes to synthesise surfactant also appears when VAD occurs during the growing period.

Some pathological features of chronic lung disease, such as transformation of the mucociliary epithelium into squamous cells which occurs in bronchopulmonary dysplasia, the thinning and destruction of the alveolar septal wall which appears in emphysema, or the alterations in lung extracellular matrix which are present in asthma, are similar to those observed in VAD in experimental animals. These facts, together with findings that RA could partially prevent, or even reverse, experimentally induced emphysema in the rat, that RA could induce *post hoc* septation in rat and mouse (Massaro and Massaro 2000), and that lower concentrations of plasma retinol were detected in preterm infants who developed bronchopulmonary dysplasia (Shenai *et al* 1985), have provided a rationale for the therapeutic use of retinoids and vitamin A supplementation programmes.

23.4.2.3   Kidney

As mentioned earlier, severe VAD during gestation results in death of foetuses or gross kidney malformations. Both retinol and RA are required for normal kidney development in experimental animals. In the VAD pregnant rat, administration of RA is able to maintain apparently normal growth and development of foetuses during the first half of gestation, but most of them are resorbed during the second half. The few foetuses delivered are either stillborn or die soon after birth, and usually present renal anomalies such as horseshoe kidney or hydronephrosis (Takahashi *et al* 1975). The role played by retinol or its metabolites in nephrogenesis is not satisfactorily known. Several transcription factors, growth factors and cell receptors, which are known to be involved in the cross-talk between the uretheric epithelium and metanephric mesenchyme during nephrogenesis, are regulated by RA. Some of them are: Lim-1, a homeodomain transcription factor; c-ret, a receptor tyrosine kinase; midkine, a heparin-binding growth/differentiation factor; and several members of the Hox family of transcription factors. Moreover, there is potential interplay between RA and the major signalling pathways involved in renal development, such as the Wnt/β-catenin, the bone morphogenetic protein (BMP)/activin-like receptor kinases, fibroblast growth factor/fibroblast growth factor receptor 1-2 (FGF/FGFR1-2), glial cell line-derived neurotrophic factor (GDNF)/ret and sonic hedgehog signalling (Reidy 2009). However, the cross-interactions of these signalling molecules and transcription factors between them with RA are complex, and they need to be fully clarified by means of future research.

Even mild VAD in gestating mothers results in smaller kidneys and a lower number of nephrons in the offspring. This is particularly important in the light of several studies conducted in animals which show that a low nephron number leads to hypertension later in life. The association between a low nephron number and hypertension has also been observed in humans. Moreover, a wide ranging nephron number, considered normal, has been described in adult humans (0.3–1.3 million nephrons/kidney). This variation could be due to perturbations in maternal nutrition, such as insufficient vitamin A intake, or to defects in its metabolism. In developed countries where maternal VAD during pregnancy is rare, the variability in nephron numbers could result from inter-individual variations in the efficacy of retinol-metabolising enzymes. This possibility deserves future studies with a view to clarifying the role of the common variants of retinol-metabolising enzymes in nephrogenesis and hypertension.

VAD during the postnatal growing period in the rat results in morphological and molecular alterations of the renal extracellular matrix. These include: thickening of basement membranes, particularly the tubular basement membrane, deposition of collagens I and IV, and modification of the chain composition in collagen IV (Marín *et al* 2005) (Figures 23.7C and 23.7D). These alterations may predispose or lead to kidney malfunction and disease in the long term. However, this possibility needs to be confirmed.

23.4.2.4 Liver

The liver, a major storage site for vitamin A, is responsible for its metabolism by regulating retinoids homeostasis. Hepatic vitamin A concentration can vary markedly depending on dietary intake. In healthy individuals, approximately 90% of vitamin A in the body is stored in the liver, and is concentrated in lipid droplets of hepatic stellate cells, as retinyl esters. The existence of stellate cells in other organs, such as lungs and kidneys, suggests that they may also be adapted to store vitamin A as retinyl esters.

VAD is associated with the activation of hepatic stellate cells into myofibroblast-like cells, enhanced cell proliferation and increased synthesis of extracellular matrix components, which is related with fibrogenic activation, deterioration of the hepatic parenchyma and other adverse effects. Besides deficient dietary vitamin A intake, drugs that induce cytochromes P450 in liver microsomes, and administration of ethanol and other xenobiotics including carcinogens, also provoke a depletion of hepatic vitamin A (Lieber 2000).

Under strictly controlled conditions, chronic VAD has been found to induce hepatic macrovesicular lipid accumulation and to increase the levels of plasma alanine aminotransferase (ALT) (Figures 23.7E and 23.7F). Plasma elevations of ALT have also been described in several hepatic steatosis models, including those induced by an altered retinoid status. The fact that VAD provokes a histologically apparent fatty liver, with hepatocytes revealing enlarged lipid droplets, a rise in plasma adiponectin, a peroxisome proliferator-activated receptor γ-regulated adipocytokine, suggests that VAD induces adipogenesis (Esteban-Pretel *et al* 2010b). At the molecular level, it has been demonstrated that RA signal is indispensable for normal liver architecture development and that functional loss of RA leads to hepatic steatosis, cellular dysplasia and cancer (Yanagitani *et al* 2004). Chronic VAD also increases hepatic amino acid catabolism and induces an accelerated protein breakdown. These results emphasise the importance of maintaining adequate vitamin A consumption not only to prevent liver damage, but to also modulate the energy balance control system. These results also provide an explanation for the role of vitamin A in protein turnover, development and growth (Esteban-Pretel *et al* 2010b).

By considering possible tissue-damage mechanisms through VAD, one attractive and widely investigated possibility lies in its antioxidant activity, especially at low oxygen tensions. VAD produces hepatic oxidative stress, which may play an important role in the initiation and/or progression of hepatic lesion. Moreover, we have shown that VAD causes oxidative damage to liver mitochondria in the rat that can be reverted by vitamin A supplementation (Barber *et al* 2000). However, we observed no alterations in the mitochondrial structure under electron microscopy (Estornell *et al* 2000), DNA fragmentation or cytochrome *c* release from mitochondria in VAD rats (Borrás *et al* 2003). The changes shown in our studies may, thus, represent a pre-apoptotic stage before morphological alterations and DNA fragmentation occur. Interestingly, recent evidence suggests that liver fibrosis, and even cirrhosis, can be reversed; however, the processes involved are presently unclear.

23.4.3   Treatment and Prevention of Vitamin A Deficiency

Vitamin A supplementation has reduced child mortality worldwide and has been one of the most important advances in public health in the last century (Griffiths 2000). In many countries, half of all deaths in children below the age of 5 years may be related to malnutrition. The WHO's goal is the worldwide elimination of VAD and its tragic consequences, including blindness, disease and premature death (WHO 2009).

In a population with chronic VAD, preformed vitamin A, but not β-carotene, proved efficient in ameliorating abnormal dark adaptation in pregnant and lactating women, and reduced death rates among the infants born to mothers with night blindness. Specifically, dark adaptation measured by the pupillary threshold in Nepali women with night blindness improved when liver, fortified rice, amaranth leaves, carrots, or retinyl palmitate were consumed for 6 weeks (Tanumihardjo 2011).

Normalisation of the immune response by supplementation is also an important strategy to combat infant mortality. Vitamin A supplementation in infants and children reduces mortality by approximately 60% in cases of measles, and reduces total mortality by around 23% in children aged 6 months or more according to most studies (Checkley *et al* 2010; WHO 2011). Supplementation successfully complements the two strategies that increase global child health: vaccination and treatment of disease. Routine administration of vitamin A strengthens host defences against infection, which can favour child survival in undernourished populations. Randomised trials in Indonesia, India and Bangladesh have proved that oral supplementation with 50 000 IU of vitamin A in oil shortly after birth lowers death rates during the first year of life by 64%, 23% and 16%, respectively. The control of blindness in children is considered a high priority within the WHO's VISION 2020 -The Right to Sight Programme (see Section 23.4.2.1).

Preliminary studies with provitamin A in adults support β-carotene administration in cancer patients. Increased susceptibility to chemical carcinogens and increased incidence of cancer has been observed in animals with VAD and in individuals with a low vitamin A intake. Epidemiological studies suggest that high consumption of fruit and vegetables is associated with a reduced risk of chronic diseases, including cancer and cardiovascular disease. RA (isotretinoin) is currently valued in more than 30 clinical trials in the treatment of cancer, kidney diseases, emphysema and other lung diseases and, among others, in improving the lipid profile. This justifies the importance of retinoids as a therapeutic tool (http://clinicaltrials.gov/ct2/results?term= retinoic+acid). However, the Beta-Carotene and Retinol Efficacy Trial (Omenn *et al* 1996)—a multicentre, randomised, double-blind, placebo-controlled primary prevention trial—showed an adverse effect of provitamin A and/or vitamin A administration on the risk of death from lung cancer, cardiovascular disease or for any cause in smokers and workers exposed to asbestos. In this context, it is interesting to note that the p53 tumour-suppressor protein changed its expression in the lung at different vitamin A levels (Borrás *et al* 2003). Therefore, the potential benefit of retinoids in anti-tumour therapy needs further studies.

23.5 Perspective and Future Directions

Interest in vitamin A at both the molecular and clinical levels continues, with potentially important implications for global health policy. Recently, the WHO (2011) has reported several approaches to reduce VAD, which include promotion of home gardening, health and nutrition education, fortification of commonly consumed foods, food supplementation programmes and supplementation for at-risk populations with high-dose vitamin A in a capsule or syrup form. The last decade has seen global improvement in VAD control by intensifying the distribution of supplements, fortification of foods, and through horticulture and education programmes. The benefits of supplementation in combating VAD in women and children are well documented (Sommer 1998; Underwood 2004). Linking vitamin A supplementation to routine immunisation programmes is a good quality strategy to increase coverage, and this policy is being adopted by many countries. Industrialised countries already enjoy several vitamin A-fortified foods, including breakfast cereals, milk, butter, margarine, some snack foods and infant formula. Furthermore, certain food products, such as sugar, are being fortified with vitamin A in some developing countries, thus improving vitamin A status in the global population. However, supplementation programmes with a periodic mass distribution have been difficult to sustain because of high distribution costs. Moreover, it has been suggested that obtaining vitamin A from provitamin A sources is ultimately safer than fortification with preformed vitamin A when considering the potential risk of hypervitaminosis A.

Several successful efforts have been made to improve provitamin A content in major crops, such as wheat, rice and potato, which are poor in β-carotene. It appears that "golden" rice, a new type of rice which accumulates a large amount of β-carotene (35 μg of β-carotene per g of rice), has the potential to be a low-cost, broad-coverage intervention, and is also effectively transformed into vitamin A in humans (Tang *et al* 2009). Moreover, transgenic "golden" potato tubers overexpressing three bacterial genes for β-carotene synthesis accumulate the largest amount of β-carotene in the four above-mentioned crops and have clearly a great potential to alleviate the VAD nutritional problem (Diretto *et al* 2010). Thus, genetically modified food may be a cost-effective staple food for combating VAD. Other provitamin A-containing foods should be developed to supply vitamin A-deficient populations with different food habits.

23.6 Conclusions

This Chapter has focused on documenting the importance of maintaining tissue levels of vitamin A within the physiological range. Lack of vitamin A results in a plethora of clinical VAD manifestations which can be reverted by adequate treatment with retinoids. Moreover, there is sufficient experimental evidence to consider retinoids as a hopeful promise in the prevention and therapy of several diseases. Current data reflecting the incidence of VAD on the population, especially in developing countries, is extremely alarming (WHO 2011). Moving from science to a global health policy requires an active management role, and needs a clearly defined series of steps which ensure that the conclusions are valid, and that the scientists and politicians are engaged in this process. Finally, it has been well documented (Blasbalg *et al* 2011; Sommer 1998; Underwood 2004) that the accomplishment of any efforts to translate research into global health interventions depends on continuing research to overcome unexpected obstacles or to simplify implementation of the knowledge.

Summary Points

- This Chapter focuses on the effects of vitamin A deficiency (VAD).
- VAD is a common and serious nutritional world disorder.
- The vitamin A recommended dietary allowance (RDA) for children, men and women is 300–600, 900 and 700 μg of RAE/day, respectively. The RDA increases in pregnancy and lactation.

- The most specific clinical effect and one of the first manifestations of VAD is xerophtalmia with nyctalopia. If untreated, they progress to keratomalacia and blindness.
- In rodents, VAD during embryogenesis leads to foetal death or morphological malformations which range from organ agenesis to rudimentary or hypoplastic organs, depending on the severity of deficiency.
- There is evidence indicating that VAD in mothers during pregnancy could have lasting adverse effects on the lung function of their offspring. In rats, VAD leads to emphysematous lungs, which can be reversed with retinoic acid.
- In rats, mild VAD in mothers during pregnancy results in smaller kidneys, and to a lower number of nephrons in the offspring which may lead to hypertension later in life.
- VAD is associated with activation of hepatic stellate cells, steatosis, deterioration of the hepatic parenchyma associated with fibrogenic activation and other adverse effects.
- Experimental evidence supports the potential use of retinoids as therapeutic agents in the prevention and therapy of several human diseases.

Key Facts

Key Facts of Vitamin A Deficiency (according to the WHO)

- Vitamin A deficiency (VAD) is a public health problem in more than half of all countries, and especially affects pregnant women and young children in developing countries.
- 19 million pregnant women are estimated to have biochemical VAD.
- 250 million preschool children are estimated to have VAD.
- VAD is the leading cause of preventable blindness in children, and increases the risk of disease and death from severe infections.
- Globally, xerophtalmia affects 9.8 million pregnant women and 5 million preschool children.
- An estimated 250 000 to 500 000 VAD children become blind every year, half of whom die within 12 months of losing their sight.
- In communities where VAD exists, improving the vitamin status can, on average, reduce young child mortality by 23% and measles mortality by 50%.

Definitions of Words and Terms

Asthma: From the Greek ασθμα, shortness of breath. A respiratory disease characterised by variable airway obstruction and associated with chronic inflammation in the airways, especially bronchi. One usual histopathological feature is the thickening of the basement membrane in the small airways due to deposition of collagen.

Basement membrane: It is a sheet-like, specialised part of the extracellular matrix which underlies the epithelia and endothelia and separates them from the adjacent connective tissue. It provides a support for cell binding and transmits instructive information to the bound cells. Its main components are collagen IV and laminins.

Bronchopulmonary dysplasia: A chronic lung disease commonly associated with premature birth, particularly among those of very low birth weight receiving mechanical ventilation and oxygen therapy.

Emphysema: A chronic and destructive lung disease. It is characterised by a progressive degradation of the extracellular matrix and the elastic fibres of alveoli and airways, leading to alveolar destruction and over-inflation, reduced gas interchange surface and collapse of the small airways.

Estimated average requirement (EAR): the average daily amount of energy of any nutrient considered sufficient to satisfy the requirements of 50% of healthy individuals in a particular group.

Hepatic stellate cells: A star-shaped cell, previously known as Ito cell and sinusoidal fat-storing cell, which is interspersed with hepatocytes in the liver. These cells comprise up to one-third of the non-parenchymal cells in the liver. They store retinol and regulate retinol homeostasis. When activated, they synthesise the proteins of the extracellular matrix and are potential contributors to liver fibrosis. Recent evidence associates them with other liver diseases such as alcoholic liver steatosis.

Horseshoe kidney: A congenital anomaly in which the two kidneys are fused, usually at the lower pole, giving the appearance of a horseshoe.

Hydronephrosis: A condition characterised by swelling of the kidney ducts due to impairment in the urine flow. It may cause kidney damage, urinary infection and pain.

Keratomalacia: From the Greek κηρας, horn and μαλακια, softness. It refers to a full-melting of the cornea which rapidly progresses to loss of the eye.

Nyctalopia: From the Greek νυξ, night, αλαος, blind and ωψ, eye (night blindness). It refers to the inability to see under low levels of illumination.

Provitamin A: Carotenoids which can be metabolically transformed into vitamin A.

Recommended dietary allowance (RDA): This is the average daily intake of energy or of selected nutrients recommended by the Food and Nutrition Board of the Academy of Sciences/National Research Council to meet the needs of nearly all healthy individuals in a particular group. Nowadays, RDA is integrated into a broader set of dietary recommendations, the dietary reference intake (DRI), developed in collaboration by the United States of America and Canada.

Retinol activity equivalent (RAE): This expresses the amount of a substance that has a biological activity equivalent to 1 μg of retinol.

Ureteric epithelium and metanephric mesenchyme: Embryonal tissues of the intermediate mesoderm in the metanephros which, by interactions and reciprocal inductions, will differentiate and form the kidneys.

Xerophtalmia: Signs in the eye of vitamin A deficiency were among the first diet-related deficiencies documented and have been grouped under the term xerophthalmia, from the Greek ξερos, dry, and οφθαλμos, eye. It refers to the dryness of the eye conjunctiva which is accompanied by its thickening and loss of transparency.

List of Abbreviations

ADH	alcohol dehydrogenase
ALT	alanine aminotransferase
BMP	bone morphogenetic protein
CYP	cytochrome P450 monooxygenase
CRABP	cellular retinoic acid-binding protein
CRBP	cellular retinol-binding protein
EAR	estimated average requirement
FAO	Food and Agricultural Organization
FEV1	Forced expiratory volume in one second
FGF	fibroblast growth factor
FGFR	FGF receptor
GDNF	glial cell line-derived neurotrophic factor
IU	international units
RA	retinoic acid
RAE	retinol activity equivalents
RALDH	retinaldehyde dehydrogenase
RAR	retinoic acid receptor
RE	retinyl esters
RDA	recommended dietary allowance
RBP	retinol-binding protein
RXR	retinoid X receptors
STRA6	stimulated by retinoic acid 6
VAD	vitamin A deficiency
WHO	World Health Organization

References

Barber, T., Borrás, E., Torres, L., García, C., Cabezuelo, F., Lloret, A., Pallardó, F. V. and Viña, J. R., 2000. Vitamin A deficiency causes oxidative damage to liver mitochondria in rats. Free Radical Biology and Medicine. 29: 1–7.

Baybutt, R. C. and Molteni, A., 2007. Vitamin A and emphysema. Vitamins and Hormones. 75: 385–401.

Blasbalg, T. L., Wispelwey, B. and Deckelbaum, R. J., 2011. Econutrition and utilization of food-based approaches for nutritional health. Food and Nutrition Bulletin. 32: S4–S13.

Bendich, A. and Langseth L., 1989. Safety of vitamin A. American Journal of
  Clinical Nutrition. 49: 358–371.

Blomhoff, R. and Blomhoff, H. K., 2006. Overview of retinoid metabolism
  and function. Journal of Neurobiology. 66: 606–630.

Borrás, E., Zaragozá, R., Morante, M., García, C., Gimeno, A., López-Rodas,
  G., Barber, T., Miralles, V. J., Viña, J. R. and Torres, L., 2003. *In vivo*
  studies of altered expression patterns of p53 and proliferative control genes
  in chronic vitamin A deficiency and hypervitaminosis. European Journal of
  Biochemistry 270: 1493–1501.

Checkley, W., West, K., Wise, R., Baldwin, M., Wu, L., LeClerq, S., Christian,
  P., Katz, J., Tielsch, J., Khatry, S. and Sommer A., 2010. Maternal vitamin
  A supplementation and lung function in offspring. The New England
  Journal of Medicine. 362: 1784–1794.

D'Ambrosio, D. N., Clugston, R. D. and Blaner, W. S., 2011. Vitamin A
  metabolism: an update. Nutrients. 3: 63–103.

Diretto, G., Al-Babili, S., Tavazza, R., Scossa, F., Papacchioli, V., Migliore,
  M., Beyer, P. and Giuliano, G., 2010. Transcriptional-metabolic networks
  in β-carotene-enriched potato tubers: the long and winding road to the
  golden phenotype. Plant Physiolgy. 154: 899–912.

Esteban-Pretel, G., Marín, M. P., Renau-Piqueras, J., Barber, T. and
  Timoneda, J., 2010a. Vitamin A deficiency alters rat lung alveolar
  basement membrane. Reversibility by retinoic acid. Journal of
  Nutritional Biochemistry. 21: 227–236.

Esteban-Pretel, G., Marín, M. P., Cabezuelo, F., Moreno, V., Renau-Piqueras,
  J., Timoneda, J., and Barber, T., 2010b. Vitamin A deficiency increases
  protein catabolism and induces urea cycle enzymes in rats. Journal of
  Nutrition. 140: 792–779.

Esteban-Pretel, G., Marín, MP., Renau-Piqueras, J., Sado, Y., Barber, T.,
  Timoneda J. Vitamin A deficiency disturbs collagen IV and laminin
  composition and decreases matrix metalloproteinase concentrations in rat
  lung. Partial reversibility by retinoic acid. Journal of Nutricional
  Biochemistry, in press.

Estornell, E., Tormo, J. R., Marín, M. P., Renau-Piqueras, J., Timoneda, J.
  and Barber T., 2000. Effects of vitamin A deficiency on mitochondrial
  function in rat liver and heart. British Journal of Nutrition. 84: 927–934.

Gilbert, C. and Foster, A., 2001. Childhood blindness in the context of
  VISION 2020 - The Right to Sight. Bulletin of the World Health
  Organization. 79: 227–232.

Griffiths, J.K., 2000. The vitamin A paradox. The Journal of Pediatrics. 2000
  137: 604–607.

Lieber, C.S., 2000. Alcohol: its metabolism and interaction with nutrients.
  Annual Review of Nutrition. 20: 395–430.

Marín, M. P., Esteban-Pretel G., Alonso, R., Sado, Y., Barber, T., Renau-
  Piqueras, J. and Timoneda, J., 2005. Vitamin A deficiency alters the

structure and collagen IV composition of rat renal basement membranes. The Journal of Nutrition. 135: 695–701.

Massaro, G. C. and Massaro, D., 2000. Retinoic acid treatment partially rescues failed septation in rats and mice. American Journal of Physiology. Lung Cellular and Molecular Physiology. 278: L955–L960.

Miner, J. H., 2008. Laminins and their roles in mammals. Microscopy Research and Technique. 71: 349–356.

Moravia, A., Menkes, M. J., Comstock, G. W. and Tockman, M. S., 1990. Serum retinol and airway obstruction. American Journal of Epidemiology. 132: 77–82.

National Academy of Sciences (NAS), 2006. The Essential Guide to Nutrient Requirements. Food and Nutrition Board. National Academic Press. Available at: http://www.nap.edu/catalog.php?record_id=11537#toc. Accessed May 20th, 2011.

Omenn, G. S., Goodman, G. E., Thornquist, M. D., Balmes, J., Cullen, M. R., Glass, A., Keogh, J. P. and Meyskens, F. L., Valanis, B., Williams, J. H., Barnhart, S. and Hammar, S., 1996. Effects of a combination of β-carotene and vitamin A on lung cancer and cardiovascular disease. New England Journal of Medicine. 334: 1150–1155.

Reidy, K. J., 2009. Cell and molecular biology of kidney development. Seminars in Nephrology. 29: 321–337.

Ross, A. C. and Zolfaghari, R., 2011. Cytochrome P450s in the regulation of cellular retinoic acid metabolism. Annual Review of Nutrition. 31: 65–87.

Shenai, J. P., Chytil, F. and Stahlman, M. T., 1985. Liver vitamin A reserves of very low birth weight (VLBW) neonates. Pediatric Research. 19: 892–893.

Sommer, A., 1998. Moving from science to public health programs: lessons from vitamin A. American Journal of Clinical Nutrition. 68:513S–516S.

Sommer A., 2008. Vitamin A deficiency and clinical disease: an historical overview. Journal of Nutrition. 138: 1835–1839.

Takahashi, Y. I., Smith, J. E., Myron, W. and Goodman D. S., 1975.Vitamin A deficiency and fetal growth and development in the rat. The Journal of Nutrition. 105: 1299–1310.

Tang, G., Qin, J., Dolnikowski, G. G., Russell, R. M. and Grusak, M. A., 2009. Golden rice is an effective source of vitamin A. American Journal of Clinical Nutrition. 89: 1776–1183.

Tanumihardjo, S. A., 2011. Vitamin A: biomarkers of nutrition for development. American Journal of Clinical Nutrition. 94: 658S–665S.

Underwood, B.A., 1984. Vitamin A in animal and human nutrition. In Sporn, M. B., Roberts, A. and Goodman, D.S. (ed.) The Retinoids, Vol. 1. Academic Press, New York, USA.

Underwood, B. A., 2004. Vitamin A deficiency disorders: international efforts to control a preventable "pox". Journal of Nutrition. 134: 231S–236S.

Wilson, J. G., Roth C. B. and Warkany J., 1953. An analysis of the syndrome of malformations induced by maternal vitamin A deficiency. Effects of restoration of vitamin A at various times during gestation. The American Journal of Anatomy. 92: 189–217.

Wolbach, S. B. and Howe, P. R., 1925. Tissue changes following deprivation of fat soluble A vitamin. The Journal of Experimental Medicine. 42: 753–777.

Wolf, G., 1996. FASEB J. A history of vitamin A and retinoids. 10:1102–1107.

World Health Organization (WHO), 2009. Global prevalence of vitamin A deficiency in populations at risk 1995–2005. WHO Global Database on Vitamin A Deficiency. World Health Organization, Geneva, Switzerland.

World Health Organization (WHO), 2010. Action plan for the prevention of avoidable blindness and visual impairment, 2009–2013. Available at: http://www.who.int/blindness/ACTION_PLAN_WHA62-1-English.pdf. Accessed May 25th, 2011.

World Health Organization (WHO), 2011. Micronutrients deficiencies. Available at: http://www.who.int/nutrition/topics/vad/en/. Accessed May 30th, 2011.

Yanagitani, A., Yamada, S., Yasui, S., Shimomura, T., Murai, R., Murawaki, Y., Hashiguchi, K., Kanbe, T., Saeki, T., Ichiba, M., Tanabe, Y., Yoshida, Y., Morino, S., Kurimasa, A., Usuda, N., Yamazaki, H., Kunisada, T., Ito, H., Murawaki, Y. and Shiota, G., 2004. Retinoic acid receptor a dominant negative form causes steatohepatitis and liver tumors in transgenic mice. Hepatology. 40: 366–375.

CHAPTER 24

Retinoic Acid Receptors and their Modulators: Structural and Functional Insights

ALBANE LE MAIRE[a], WILLIAM BOURGUET[a], HINRICH GRONEMEYER[b] AND ANGEL R. DE LERA*[c]

[a] INSERM U554 and CNRS UMR5048, Centre de Biochimie Structurale, Universités Montpellier 1 and 2, 34090 Montpellier, France; [b] Department of Cancer Biology, Institut de Génétique et de Biologie Moléculaire et Cellulaire (IGBMC), BP 10142, 67404 Illkirch Cedex, C. U. de Strasbourg, France; [c] Departamento de Química Orgánica, Facultad de Química, Universidade de Vigo, 36310 Vigo, Spain
*E-mail: qolera@uvigo.es

24.1 Introduction

The genetics and molecular biology of the six retinoic acid receptors [subtypes: retinoic acid receptors (RARs) α, β and γ, and retinoid X receptors (RXRs) α, β and γ], members of the nuclear receptor (NR) superfamily, demonstrated that these ligand-dependent transcription factors are master regulators of a plethora of physiological processes, such as embryo development and organ homeostasis (Germain *et al* 2006a; 2006b; Laudet and Gronemeyer 2002). Deregulated retinoic acid signaling can lead to severe pathologies, such as acute promyelocytic leukemia (APL), which originates from the somatic formation of oncogenic RARα fusion proteins. Apart from APL, there is promise for the therapeutic use of retinoids and rexinoids (Altucci and Gronemeyer 2001; Altucci *et al* 2007; Shankaranarayanan *et al* 2009).

Food and Nutritional Components in Focus No. 1
Vitamin A and Carotenoids: Chemistry, Analysis, Function and Effects
Edited by Victor R Preedy
© The Royal Society of Chemistry 2012
Published by the Royal Society of Chemistry, www.rsc.org

At the molecular level RARs and RXRs form heterodimers, which are the actual "workhorses" that respond to RAR ligands. Subtype and RAR–RXR-selective, and function-specific retinoids and rexinoids have been developed (de Lera *et al* 2007). Molecular biological studies and the crystal structures of RAR and RXR ligand-binding domain (heterodimers) in the presence and absence of these various ligands have provided a detailed view of the allosteric effects that are at the basis of ligand action and account for the ability of NRs in general to communicate with other intracellular co-regulatory factors that mediate their action (Bourguet *et al* 2000; Germain *et al* 2002; 2009; Gronemeyer *et al* 2004; le Maire *et al* 2010). RXRs also heterodimerize with other NRs, some of which are "permissive", as they become transcriptionally active in the sole presence of an RXR-selective ligand ("rexinoid"). RAR–RXR heterodimers are "non-permissive" and do not respond to rexinoids alone. However, rexinoid agonists superactivate transcription induced by RAR–RXR in the presence of RAR agonists.

As for all members of the NR superfamily, RAR–RXR heterodimers act as ligand-regulated *trans*-acting transcription factors that bind to *cis*-acting DNA regulatory elements in the promoter regions of target genes. Conceptually, ligand binding does nothing else than modulating the communication functions of the receptor with its intracellular environment, which mainly entails receptor–protein and receptor–DNA or receptor–chromatin interactions (Gronemeyer and Bourguet 2009). Essentially, receptors act as interpreters of the information encoded in the chemical structure of a NR ligand, which they read in the context of cellular identity and cell physiological status, and convert it into dynamic temporally controlled sequences of receptor–protein and receptor–DNA interactions, thereby resulting in the temporally controlled regulation of cognate gene networks. To process input and output information NRs are composed of a modular structure with several domains and associated functions, such as the DNA-binding (DBD) and ligand-binding (LBD) domains. The DBD specifies direct receptor binding to cognate target genes and interprets the DNA-binding site for functional relevance. In heterodimers the DBDs of both subunits contribute to the definition of a cognate RAR–RXR-binding site. The LBD serves as dual input–output information processor, as ligand binding induces allosteric changes of receptor surfaces that represent docking sites for subunits of transcription and/or epigenetic machineries, or enzyme complexes (output).

24.2 Retinoid and Rexinoid Receptor Ligand-Binding Domains: Structure–Function Relationships

24.2.1 Structural Basis of RXR Action and Modulation by Ligands

The structure of the RXR LBD in its ligand-free form (apo) is a single protein domain organized in a primarily helical scaffold of 12 α-helices arranged in three layers and a short (S1–S2) β-turn (Figure 24.1A). Binding of an agonist, *i.e.*, 9-*cis*-retinoic acid (**4.2**) (Figure 24.4) induces a *trans*-conformation that entraps the ligand in the ligand-binding pocket (LBP). In apo-RXR, helix 11 (H11) is almost perpendicular to H10 and is the main contributor to the hydrophobic occupation of the ligand-free binding pocket, whereas the C-terminal H12 protrudes out of the core of the domain. Agonist binding triggers major structural changes affecting helices H3, H6, H11 (which rotates 180° around its own axis) and the "*trans*-activation helix" H12 which seals the LBP. In this holo-RXR conformation (Figure 24.1A), H3, H4 and H12 define a hydrophobic binding groove that specifically recognizes a conserved structural LXXLL motif (X stands for any amino acid) present in most coactivators (CoAs). The NR interaction domain (NRID) of CoAs generally contains several copies of this motif, which adopts a two-turn α-helix. In the RXR LBP, the bound ligands are stabilized through extensive van der Waals contacts and by a network of ionic and hydrogen bonds involving their carboxylate moiety, a conserved arginine in H5 and water molecules (Figure 24.1B). The LBP of RXR is highly restrictive to flexible and elongated ligands. Those that bind must adapt to the available volume by twisting around single bonds as shown in the crystal structures of complexes of RXR with various ligands (de Lera *et al* 2007).

Other types of ligands bind retinoid receptors with high affinity, but fail to stabilize their active conformation and are classified as partial agonists or antagonists depending on their ability to prevent CoA recruitment. In RXR, the effect of agonists, partial agonists and antagonists on the conformation and dynamics of helix H12 has been addressed by nuclear magnetic resonance (NMR) and fluorescence anisotropy techniques (Lu *et al* 2009; Nahoum *et al* 2007). In contrast to agonists, antagonists and partial agonists fail to shift the equilibrium of multiple RXRα H12 conformations observed in the unliganded state to a more compact form. Antagonists such as UVI3003 (**7.28**) (see Figure 24.7) generally feature a bulky side chain that sterically prevents H12 from adopting the active conformation, thus disrupting the interaction surface with CoAs (Nahoum *et al* 2007). Partial agonists also lower the interaction strength between holo-H12 and the LBD surface, which renders the activation function 2 (AF2) helix more dynamic as compared to the agonist-bound situation. However, in contrast with antagonists, partial agonists (*i.e.*, **7.13a**–**7.13c**, see Figure 24.7) exert moderate constraints on H12 so that the presence of CoAs helps stabilization of H12 in the active position (Nahoum *et al* 2007; Pérez-Santín *et al* 2009). This unique feature allows partial agonists to "sense" intracellular coregulator levels and act as context-dependent cell-selective modulators with agonist or antagonist properties. A key signature of all agonist-bound RXRs is the conformation of L436 in H11 (human RXRα numbering; Figure 24.1B). This residue plays a pivotal role in inducing the sharp turn on the LBP volume to accommodate the twisted ligand and in stabilizing the agonist-bound conformation of H12. Partial agonists (**7.13a**–

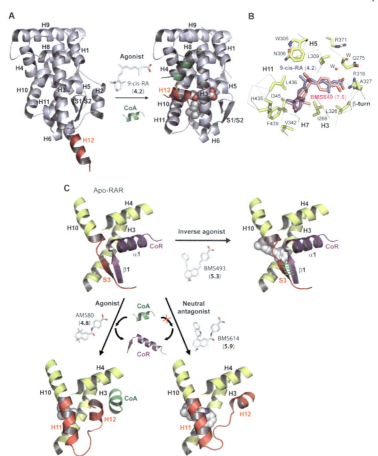

Figure 24.1   Structural basis of RXRs and RARs action. (A) In the unliganded
RXR, H12 (red helix) protrudes out of the core of the LBD. Upon
agonist (9-*cis*-retinoic acid in grey spheres) binding, H12 relocalizes into
the so-called active position, allowing the formation of a surface
specifically recognized by short LXXLL motifs contained in CoA
(green helix). Lys284 (H3) and Glu453 (H12), which generate a charge
clamp interacting with the LXXLL helical motif of CoA, are
highlighted as yellow sticks. (**B**) Detailed view of the RXRα LBP in
which two agonists [9-*cis*-retinoic acid (**4.2**) in blue and BMS649 (**7.5**) in
pink] are superimposed. Some RXR residues (yellow sticks) as well as
some water molecules (w) involved in the interactions with ligands are
drawn and labeled. (**C**) In the unliganded RAR, the β-strand S3
interacts with β1-strand of the CoR (violet). This antiparallel β-sheet
interface is required for the constitutive interaction of RAR with CoRs.
Addition of an inverse agonist [BMS493 (**5.3**)] reinforces this interaction
(green dashed lines). In contrast, agonist binding [AM580 (**4.8**)] induces
the S3 to H11 secondary structure switch, CoR dissociation and CoA
binding. Binding of a neutral antagonist provokes a displacement of
H12 towards the coregulator binding groove and prevents CoA and
CoR recruitment.

7.13c) alter the orientation of L436, which in turn destabilizes the holo-position of H12 and decreases its interaction with CoAs.

24.2.2 Structural Basis of RAR Action and Modulation by Ligands

The structure of RARγ LBD bound to the natural agonist all-*trans*-retinoic acid **4.1** (see Figure 24.4) is reminiscent of that of holo-RXR (Figure 24.1A). However, a particular feature of RAR resides in the strong interaction of its unliganded form with the transcriptional corepressors (CoRs) SMRTs (silencing mediator for retinoid and thyroid hormones receptors) and NCoR (nuclear receptor corepressor). Similar to the NRID of CoAs, the C-terminal region of SMRT and NCoR contains two or three of the so-called "CoR-NR boxes" (CoRNR1-CoRNR3) characterized by the core sequence (I/L)XX(V/I)I (where X denotes any residue), which presents a central extended helical conformation. This sequence, reminiscent of the LXXLL motif found in CoAs, mediates the interaction with the NRs recognition surface formed by residues from H3 and H4 and residues flanking the core sequence strengthen the interaction and determine NR specificity. The molecular basis of the repression function of RAR has been revealed recently through a combination of structural, biochemical and cell-based assays (le Maire *et al* 2010). Whereas H12 is primarily involved in the interaction with CoAs, the discovery of a specific interface between RAR and a fragment (CoRNR1) of NCoR revealed that a secondary-structure transition affecting H11 plays a master role in CoR association and release. The constitutive interaction of RAR with CoRs involves: (i) the formation of an antiparallel β-sheet between β-strand S3 of the receptor and β-strand β1 of the CoRs, and (ii) the binding of the four-turn α-helix α1 of the CoRs to the coregulator groove of RAR (Figure 24.1C). Agonist binding induces the S3 to H11 secondary-structure switch because of a stabilization of the H11 conformation *via* interactions of the ligand with residues V395 and L398, as observed in the crystal structure of RAR in complex with the synthetic agonist AM580 (**4.8**) (see Figure 24.4) (le Maire *et al* 2010). This structural transition was shown to account for the release of the CoRs, the subsequent folding back of H12 and the recruitment of CoAs (Figure 24.1C).

Two types of RAR antagonists referred to as neutral antagonists and inverse agonists, both of which abrogate the transcriptional activity of RAR, have been characterized. While neutral antagonists prevent interaction with any coregulator (CoAs and CoRs), inverse agonists favor CoR recruitment, thus precluding CoA binding (Germain *et al* 2009). In the crystal structure of RAR in complex with the synthetic antagonist BMS614 (**5.9**) (see Figure 24.5), the quinolinyl group of the ligand points towards H12, and the bulky extension protruding between H3 and H11 prevents the active conformation of H12, which instead docks into the coregulator groove (Figure 24.1C). Thus, neutral antagonists block the AF2 of the receptor by preventing the binding of native retinoic acids and disrupting the interaction surface with coregulators. In the

structure of RAR in complex with the inverse agonist BMS493 (**5.3**) (see
Figure 24.5) (le Maire *et al* 2010), the RAR–CoR interaction is strengthened
by stabilization of the β-sheet S3–β1 interface (Figure 24.1C).

24.3 RXR and RAR Heterodimers: Structure and Synergy

Numerous NRs bind to DNA as heterodimers with RXRs, which may be
receptors for endogenous 9-*cis*-retinoic acid (**4.2**), but evidence for alternative
natural RXR ligands has been provided. In contrast to homodimerization,
heterodimerization allows, in principle, fine-tuning of NR action by using
combinatorial sets of ligands and thus provides interesting pharmacological
opportunities. However, although RAR agonists can autonomously activate
transcription through RAR–RXR heterodimers, RXR is unable to respond to
RXR-selective agonists in the absence of an RAR ligand. Consequently, RXR-
selective ligands on their own, although bound to the receptor, could not
trigger RAR–RXR heterodimer-mediated retinoic acid-induced events in
various cell systems (Germain *et al* 2002). Similarly, RXR cannot autonomously respond to its ligands in the corresponding thyroid hormone receptor
(TR) and vitamin D receptor (VDR) heterodimers, unless those heterodimeric
partners are bound by a ligand. The biological significance of this RXR
"subordination" or "silencing" is presumably to avoid confusion between
retinoic acid, thyroid hormone and vitamin D_3 signaling pathways.

24.3.1 Structural Basis of RAR–RXR LBD Heterodimers

Crystal structures containing the RXR LBD in complex with the RAR LBD
bound to various ligands (Bourguet *et al* 2000; Pogenberg *et al* 2005; Sato *et al*
2010) show that the dimeric arrangements of these heterodimers are closely
related, with residues from helices H7, H9 and H10, as well as loops L8–L9
and L9–L10 of each protomer forming an interface (Figure 24.2A) comprising
a network of complementary hydrophobic and charged residues and further
stabilized by neutralized basic and acidic surfaces. In heterodimers the two
subunits are related by a symmetry axis that deviates by 10% from the C2 axis.
Thus, the two protomers do not contribute equally to the heterodimerization
interface and mutants can be generated that separate RXR homodimerization
and heterodimerization. Finally, these studies also revealed that the overall
structures of heterodimeric RXR and partner receptors do not differ
significantly from that of their monomeric forms.

24.3.2 Subordination and Synergy in RAR–RXR Heterodimeric Interactions

In vitro, RAR and RXR ligands are able to individually bind to their
corresponding receptors and, as such, activate RAR–RXR heterodimers.

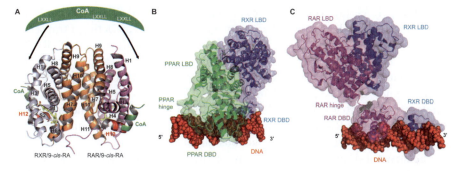

Figure 24.2   Structures of RXR heterodimers. (**A**) Crystal structure of the RARβ–RXRα LBD heterodimer bound to 9-*cis*-retinoic acid and a LXXLL-containing peptide. The dimerization interface comprising helices H7, H9, H10 and H11 as well as loops L8–L9 and L9–L10 of each monomer is colored in orange. The rest of the RXR LBD and the RAR LBD are colored in blue and magenta, respectively. H12 of each monomer is highlighted in red. A schematic representation of a CoA molecule interacting with the heterodimer through several LXXLL motifs is drawn in green. (**B**) Crystal structure of the intact PPARγ–RXRα bound to its DNA response element. The RXRα LBD and DBD domains that do not interact together are colored in light blue. On the contrary, the PPARγ LBD (green) is closely positioned to the two DBDs. The DNA is colored in red. (**C**) Solution structure of RAR (magenta)–RXR (light blue) bound to DNA (red). Contrary to Figure 24.2(B), the heterodimer shows an elongated conformation with separate DBD and LBD domains.

However, it appears that in the usual cellular environment and in the absence of RAR agonist, rexinoids are unable to do so (Germain *et al* 2002; Pogenberg *et al* 2005). This so-called "RXR subordination" phenomenon implies that RXR ligands can activate "permissive" heterodimers on their own [for example, peroxisome proliferator-activated receptor (PPAR)–RXR] but generally not "non-permissive" heterodimers (for example, RAR–RXR), which require the presence of ligands of the partner receptors for activation. The strength of the overall association of RAR–RXR heterodimers with CoAs is dictated by the combinatorial action of RAR and RXR ligands, the simultaneous presence of the two receptor agonists being required for highest binding affinity of CoAs. Subordination of RXR in RAR–RXR heterodimers is due to the inability of rexinoids to induce CoR dissociation from heterodimers (Germain *et al* 2002) and thus to recruit CoAs, because of the inaccessibility of the mutually exclusive binding site of the two coregulator types. CoRs interact with apo-RAR–RXR heterodimers through formation of a specific interface with the RAR subunit as described above. The conformational change from the β-strand S3 to the α-helix H11 upon RAR agonist binding switches the RAR–RXR heterodimers from an "off" to an "on" state, which can be further activated by addition of rexinoids (le Maire *et al* 2010).

Notably, in certain cells and for complexes with the RARβ subtype, RXR subordination can be weak and consequently rexinoids can activate transcription through apo-RAR–RXR. Apart from the specific case of RARβ heterodimers, the only way for RXR to modulate transactivation in response to its own ligand in RAR–RXR heterodimers is thus through synergy with RAR ligands due to increased interaction efficiency of two NR boxes in a single CoA molecule with both holo-RAR and holo-RXR of the heterodimer. Interestingly, also RAR antagonists can synergize with rexinoid agonists and can activate transcription of endogenous target genes.

The structural and functional study of the RAR-RXR heterodimer in the presence of 9-*cis* retinoic acid (**4.2**) and a CoA NRID containing three LXXLL motives (Pogenberg *et al* 2005) indicated that the synergy between receptor agonists would result from the enlargement of the contact area between the heterodimer and the CoA through formation of one interacting surface on each heterodimer subunit (Figure 24.2A). Thus, two LXXLL motifs would mediate the optimal assembly of one CoA on to heterodimeric receptors. In contrast to inverse agonists, which reinforce CoR interaction with RAR and fully prevent any activation of the RXR–RAR heterodimer by rexinoids, neutral antagonists [*e.g.* BMS614 (**5.9**); Figure 24.1C] prevent binding of both CoAs and CoRs to RAR by re-localizing H12 in the coregulator groove and may allow some activation by rexinoids (Germain *et al* 2002).

24.3.3   Structural Description of Full-length RXR Heterodimers

The X-ray structure of the entire PPARγ–RXRα heterodimer bound to the PPAR response element (PPRE) (Chandra *et al* 2008) shows that the hinge region linking the DBD and the LBD of PPARγ interacts with the DNA minor groove, thus dictating heterodimer polarity (Figure 24.2B). In addition, the relative arrangement of receptor domains generates a rather compact conformation and positions the PPARγ LBD so that it contacts both receptor DBDs to stabilize their PPRE binding.

Solution structures determined essentially by small angle X-ray scattering (SAXS) of several heterodimers (including RARα–RXRα, and VDR–RXRα) bound to their cognate response elements (Rochel *et al* 2011) show different spatial domains organization as compared with the previously described crystal structure. In particular, all heterodimer structures show an elongated conformation with separate DBD and LBD domains and thus lack interdomain contacts. Moreover, they emphasize the essential role of the extended hinge regions that connect DBDs and LBDs in establishing and maintaining the integrity of the functional structures. The apparent discrepancy between solution and crystal structures in the case of PPARγ–RXRα heterodimer illustrates the dynamic character of these molecules which can adopt different architectures to adapt the diversity of response elements, ligands or coregulators.

24.4 Selective Retinoid and Rexinoid Receptor Modulators

Only a selection of the large number of existing RAR- and RXR-selective (including the subtype selectivity and the function specificity) retinoids and rexinoids will be presented here in order to illustrate the structure- and function-based design of ligands. The readers should consult more comprehensive recent reviews (Bourguet *et al*, 2010; Dawson and Zhang 2002; Dawson, 2004; de Lera *et al* 2007; Kagechika 2002; Kagechika and Shudo 2005).

24.4.1 RXR *vs*. RAR Selectivity as a Function of the LBP Architectures

The architectures of the LBPs of RAR and RXR are sufficiently different to allow the preferential recognition and binding of topologically distinct families of ret(x)inoids. RAR LBP exhibits a linear I-shape (Figure 24.1C) and binds elongated ligands with rather linear skeletons, either polyenes or synthetic analogues in which some double bonds have been replaced by (hetero)aromatic rings (the so-called arotinoids). In contrast, RXR features an L-shaped pocket (Figure 24.1B), which restricts the binding of ligands to those with twisted polyene skeletons and other scaffolds that deviate from co-planarity by conformational/configurational restrictions.

24.4.2 Modulators of RAR–RXR Heterodimers Acting at the RAR Site

Figure 24.3 summarizes the building blocks that allow to construct the skeleton of the synthetic retinoids by the attachment of a central region or linker to both a hydrophobic and a polar (a carboxylic acid or a bioisostere) termini. The combination of these elements cover the structural space of the vast majority of retinoids reported in the literature (Figures 24.4 and 24.5). The functional groups on the central unit are primarily responsible for the acquisition of subtype selectivity.

RARα (H3-S232; H5-I270; H11-V395) and RARβ (H3-A225; H5-I262; H11-V388) differ by only one amino acid in H3 of the LBP, whereas RARβ and RARγ (H3-A234; H5-M272; H11-A397) differ by two located in H5 and H11. The amino acid divergencies of the LBPs have been exploited to generate synthetic retinoids with RAR subtype selectivity (Figure 24.4). For example, for the thoroughly studied stilbene-based retinoids [derivatives and analogues of TTNPB (**4.3**)], which are conformationally flexible at the linker region and can replace the olefin with functional groups (amide, hydroxyamide, chalcone oxime, hydroxyketone, *etc.*) capable of hydrogen bonding,

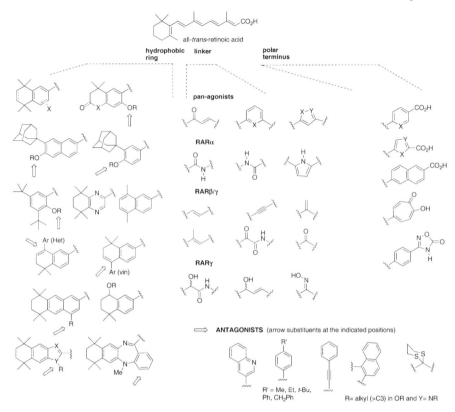

Figure 24.3   The combination of hydrophobic, linker and polar motifs allows to build-up the chemical space covered by these modulators. Subtype selectivity as a result of the functional groups on the linker is explained. Antagonist position (arrows in the hydrophobic scaffolds) and antagonist substituents are also shown.

the following guidelines have been proposed (for other scaffolds, these may vary):

(a)   RARα-selective ligands can establish hydrogen bonds with the subtype-specific S232.
(b)   RARγ-selective ligands can form hydrogen bonds to subtype-specific M272.
(c)   RARβ has a larger more hydrophobic LBP cavity than its paralogues and can accommodate bulkier ligands (Álvarez *et al* 2007; Germain *et al* 2004).

Isotype selectivity has been reported for ligand AC-261066 (**4.17**), the first RARβ2-selective agonist discovered by ultra-high-throughput screening (uHTS), although no structural rationale has been provided for this unique profile.

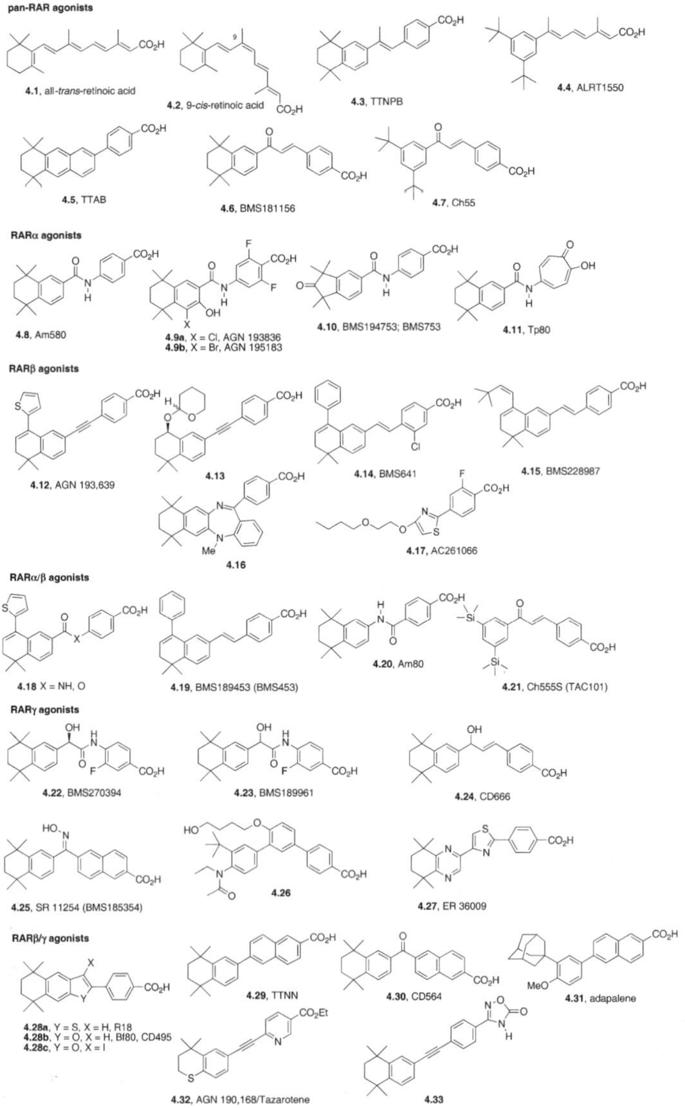

Figure 24.4 Pan-RAR agonists, RAR-subtype selective agonists. Pan-RAR agonists are characterized by conformational flexibility and the lack of subtype-selective functionalities on the linkers that can function as H-bond donors. The hydrogen bond to S232 is the most important structural factor favoring RARα selectivity. RARβ-subtype selectivity is promoted by the addition of bulky groups to the hydrophobic moieties that fill the extra space in the cavity caused by the orientation of the I263 residue away from the ligand, which enlarges the LBP between H5 and H10. RARγ selectivity is optimal for ligands with a longer connector (three-atom linker; containing a hydrogen bond donor to M272) between aryl rings in benzoic acids (or a one atom connector in naphthoic acids), and a greater steric bulk.

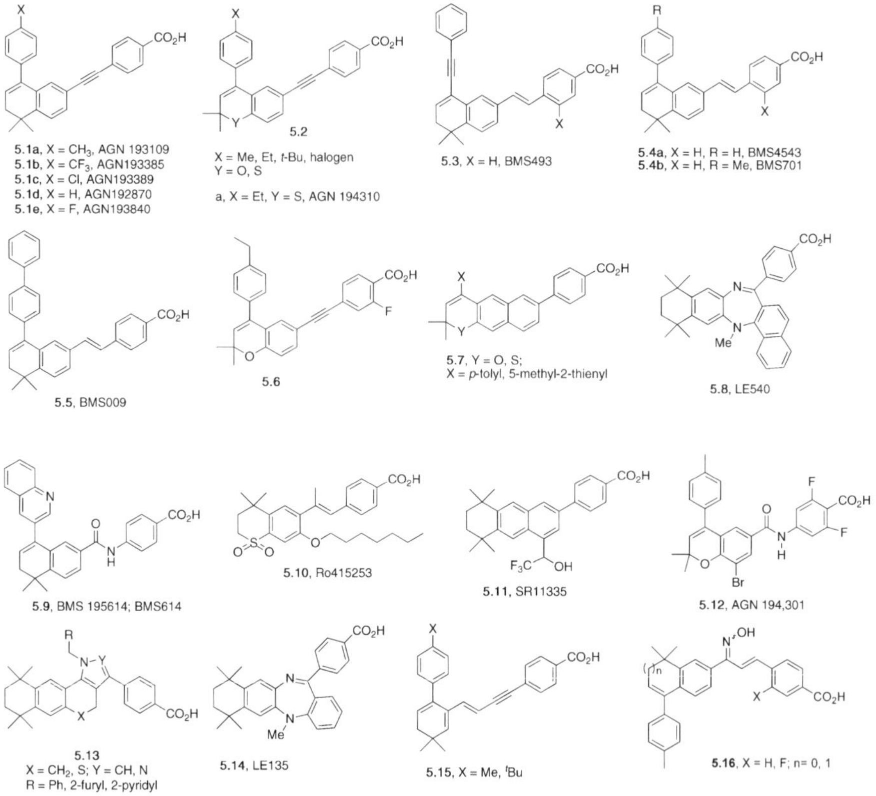

Figure 24.5 Pan and subtype-selective RAR inverse agonists and antagonists. Inverse agonists with diarylacetylene (**5.1**, **5.2**) and stilbene retinoid scaffolds (**5.3**, **5.4**) have *p*-substituents at the C8″-phenyl ring of larger size than those of pan-antagonists.

The transition from agonists to antagonists can be achieved by incorporation of long and/or bulky substituents (aryl, heteroaryl and alkyl groups of size greater than three carbons) attached or fused to carbo/heterocyclic rings that interfere with H12 repositioning and prevent the recruitment of CoAs (Figure 24.3).

Antagonists and inverse agonists (Germain *et al* 2009) display slightly different binding modes in the RAR LBD that have an impact on coregulator recruitment (le Maire *et al* 2010). Figure 24.5 depicts structures of pan-antagonists and inverse agonists (**5.1**–**5.8**) with connectors (double or triple bonds or their aromatic counterparts) that lack hetero-atoms are, therefore, are in general subtype-unselective. Other analogues **5.9**–**5.16** show some subtype selectivity (Álvarez *et al* 2009; Germain *et al* 2009) following the same principles explained for agonists, and there are in general clear structural coincidences between the cores of ligands that display these contrasting biological activities.

24.4.3 Modulators of RAR–RXR Heterodimers Acting at the RXR Site

Figure 24.6 summarizes the chemical diversity in rexinoid structures. Using the same dissection of the native skeleton, rexinoids are composed of three structural units. Compared to RAR ligands, rexinoids display a greater variety of linkers, although they all share the presence of (hetero)aryl rings replacing some of the olefins of native retinoids.

Most of the ligands incorporate structural features that induce the skeletal distortions required to bind the L-shaped LBP of RXR. The most common of those are: (a) aryl rings connected by a saturated or a Csp^2 carbon; (b) trienoic and dienoic acids, their heterocyclic analogues and 9-*cis*-locked retinoids; (c) dibenzodiazepines; (d) diarylamines and diarylsulfides; and (e) biphenyl rings with *ortho* substituents (de Lera *et al* 2007).

Although all residues of helices H3, H5, H7 and H11, and the β-turn, which constitute the LBP of the RXR, are highly conserved in all three subtypes (α, β and γ), agonists that display some RXR subtype-selectivity, derived from NEt-3IP (**7.18b**) and PA024 (**7.17a**), (Figure 24.7) have been recently reported.

Rexinoids that exhibit heterodimer selectivity are called selective RXR synergists as they allosterically increase the potencies of the partner ligands

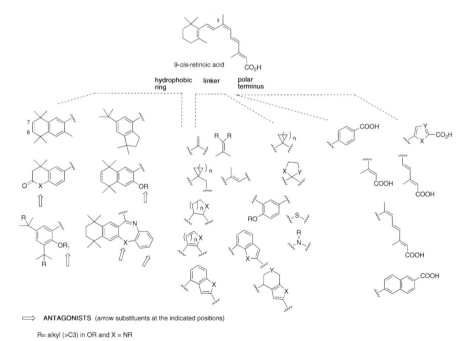

Figure 24.6   The structural diversity of rexinoids. The combination of hydrophobic, linker and polar motifs allow to build-up the scaffolds that cover the chemical space of these modulators. Antagonist position (arrow) and antagonist substitution are also shown.

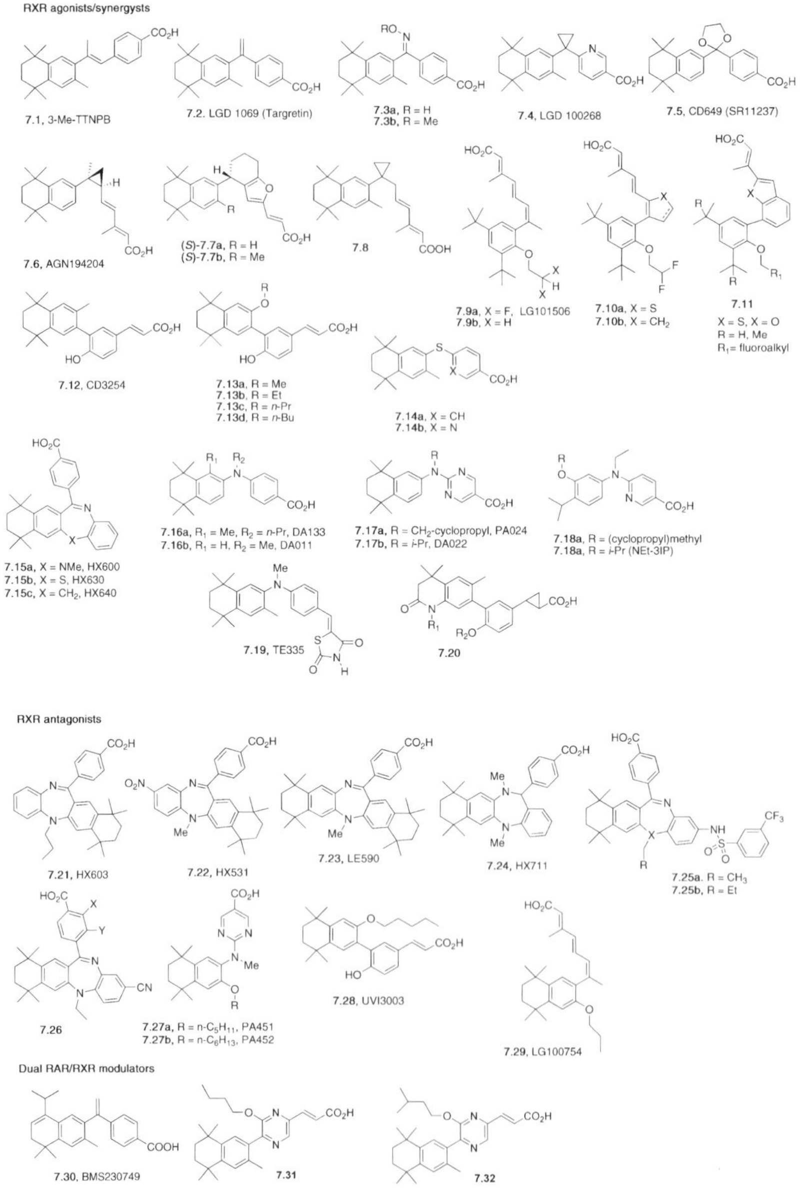

RXR agonists/synergysts

7.1, 3-Me-TTNPB

7.2, LGD 1069 (Targretin)

7.3a, R = H
7.3b, R = Me

7.4, LGD 100268

7.5, CD649 (SR11237)

7.6, AGN194204

(S)-7.7a, R = H
(S)-7.7b, R = Me

7.8

7.9a, X = F, LG101506
7.9b, X = H

7.10a, X = S
7.10b, X = CH₂

7.11
X = S, X = O
R = H, Me
R₁= fluoroalkyl

7.12, CD3254

7.13a, R = Me
7.13b, R = Et
7.13c, R = n-Pr
7.13d, R = n-Bu

7.14a, X = CH
7.14b, X = N

7.15a, X = NMe, HX600
7.15b, X = S, HX630
7.15c, X = CH₂, HX640

7.16a, R₁ = Me, R₂ = n-Pr, DA133
7.16b, R₁ = H, R₂ = Me, DA011

7.17a, R = CH₂-cyclopropyl, PA024
7.17b, R = i-Pr, DA022

7.18a, R = (cyclopropyl)methyl
7.18a, R = i-Pr (NEt-3IP)

7.19, TE335

7.20

RXR antagonists

7.21, HX603

7.22, HX531

7.23, LE590

7.24, HX711

7.25a. R = CH₃
7.25b. R = Et

7.26

7.27a, R = n-C₅H₁₁, PA451
7.27b, R = n-C₆H₁₃, PA452

7.28, UVI3003

7.29, LG100754

Dual RAR/RXR modulators

7.30, BMS230749

7.31

7.32

Figure 24.7 RXR modulators. RXR-selective agonists with Csp³- and Csp²-biaryl connectors, unsaturated side chains, amines, sulfides and heterocyclic-locked analogues. Also, included are skeletons with a central (tetrahydro) benzo-fused heterocyclic scaffold (incorporating furan, thiophene or isoquinoline rings), dienoic acids with hydrophobic rings (pentamethylnaphthalene, meta-di-*tert*-butylbenzene of *tert*-butyl-dimethylindene), trienoic acid scaffolds, and derivatives with benzo-furan, benzothiophene, thiophene or cyclopentenyl rings. Dibenzodiazepines are in general RXR antagonists and synergists.

(Kagechika 2002; Kagechika and Shudo 2005). This important class of rexinoids might act as agonist or antagonist of heterodimers with different partners, and the modulation is sometimes achieved through subtle chemical engineering of their structures. Different antagonistic scaffolds should result in defined actions on coregulator interactions and RXR function (Sato *et al* 2010). Z-Oxime (**7.3a**) and the 2,2-difluoroethyloxyphenyltrienoic acid derivative (**7.9a**) (LG101506) are potent PPARγ–RXR heterodimer activators, although the latter is a RXR homodimer partial agonist and also activates PPARα–RXR heterodimers. Rexinoids with polar dihydro-[1*H*]-quinolin-2-ones or dihydro-[1*H*]-benzoxazin-2-ones cores, such as **7.20**, selectively activate the liver X receptor (LXR)–RXR heterodimers. The dibenzodiazepine HX600 (**7.15a**), at low concentrations, is a synergist of the RAR–RXR heterodimers, and at high concentrations antagonizes RARs. Diarylamines **7.16–7.18** are also RXR agonists that show highly potent retinoid synergistic activity and **7.18** shows selectivity for activation of PPARγ–RXRα heterodimers.

RXR antagonists also feature long or bulky substituents at the central linker that can displace residue T305, disrupting the hydrophobic interactions that stabilize the agonist conformation of H12. The agonist to partial agonist to antagonist transition within a family of rexinoids is a function of the length or bulk of the substituent. For example, incorporation of *O*-alkyl substituents in *o,o'*-disubstituted biaryl cinnamic acids [represented by the selective RXR agonist CD3254 (**7.12**)] led to analogues **7.13a–7.13d** which exhibited partial RXR agonist–antagonist actions (Pérez-Santín *et al* 2009.

24.4.4   Modulation at Both Sites of the RAR–RXR Heterodimer

Compounds that bind both receptors as agonists are rare, given the structural disparity of their LBP. 9-*cis*-Retinoic acid (**4.2**) binds both units as a result of its conformational flexibility and its adaptation to the different binding pockets of RAR and RXR. The dibenzodiazepines HX711 (**7.24**) and HX531 (**7.22**) are antagonists of both RARs and RXRs.

Less common is binding with contrasting activities, such as the agonist of one receptor and the antagonist of the other. This rare occurrence has been demonstrated with ligand **7.30**, which is a RXR agonist and RAR antagonist (Shankaranarayanan *et al* 2009). The family of pyrazine acrylic acid arotinoids are RARα,β-selective inverse agonists (**7.31** is RARβ-selective) and **7.32** is a pan-RAR–RXR inverse agonist (García *et al* 2009).

24.5   Summary and Perspectives

At present our knowledge about the mechanistic principles of retinoid signaling is well advanced concerning the initial events following ligand binding and its ultimate role in the animal. The initial events include allosteric effects on the receptor structure that is the basis for interaction of the receptor heterodimer with co-regulatory and other less well-defined factors

(Gronemeyer *et al* 2004). Mouse genetics has provided information about the implication of retinoic acid and its cognate receptors during development and organogenesis, and in the adult. However, the systems biology of retinoic acid function, *i.e.* the decryption of the temporal networks that are regulated by the activated RAR–RXR heterodimer in a given cell, is still lacking. Also the mechanistic basis of the cross-talk with other signaling pathways, such as those inducing activator protein 1 (AP1) activity, is poorly understood.

Retinoids and rexinoids are among those drugs that are discussed as potential candidates for cancer chemoprevention in patients at risk. The functional RXR–RAR heterodimers are "non-permissive" and do not respond to rexinoids alone. However, rexinoid agonists superactivate transcription induced by RAR–RXR in the presence of RAR agonists (Germain *et al* 2002). Presently, we recognize structural [*e.g.*, heterodimer-selective rexinoids (de Lera *et al* 2007)] and functional [*e.g.*, rexinoid–kinase or rexinoid–growth factor cross-talk) (Altucci *et al* 2005; Shankaranarayanan *et al* 2009)] paradigms, which have turned RXR into an attractive drug target. Moreover, in a randomized placebo-controlled study the loss of RARβ expression, which is considered to be a biomarker of pre-neoplasia, could be significantly reversed in former smokers by treatment with 9-*cis*-retinoic acid but not with 13-*cis*-retinoic acid. The recent finding that *all-trans*-retinoic acid can induce a cancer cell-selective apoptosis program and the observation that the tumor suppressor interferon regulatory factor-1 (IRF-1) mediates the action of retinoic acid on the death ligand TRAIL [tumor necrosis factor (TNF)-related apoptosis-inducing ligand] further underscores the potential of retinoids for chemotherapy and chemoprevention of cancer (Altucci *et al* 2007; Gronemeyer *et al* 2004). In this respect the observation that the combination of TRAIL and vitamin A-acetate targets pre-malignant tumor cells for apoptosis, resulting in inhibition of tumor growth and prolonged survival of adenomatous polyposis coli^{min} (APC^{min}) mice is highly encouraging (Zhang *et al* 2010).

Summary Points

- Retinoid and rexinoid receptors form "non-permissive" RAR–RXR heterodimers, in which RXR is subordinated.
- Ligands modulate the ability of these receptors to interact with a complex network of other intracellular proteins and DNA/chromatin complexes.
- Chemical engineering of subsets of retinoid chemotypes has provided a panel of modulators that display the entire range of ligand modulatory functions, including subtype- and isotype-selectivities.
- Ligand function can be correlated with the structure within a retinoid scaffold. RAR agonists and antagonists/inverse agonists induce different helix H12 positioning and dynamics that impact on the structure of the activation function AF-2 and in coregulator recruitment/dissociation.

- The divergent expression levels of CoAs and CoRs in cells might allow the discovery of selective retinoid modulators [a type of selective nuclear receptor modulators (SNuRMs)] with tissue-selective activities.

Key Facts of Retinoid Receptor Modulation

- RAR–RXR heterodimers act as ligand regulated *trans*-acting transcription factor that bind to *cis*-acting DNA regulatory elements in the promoter regions of target genes.
- Receptors interpret the information encoded in the chemical structure of a nuclear receptor ligand in the context of cellular identity and cell physiological status, and convert it into dynamic temporally controlled sequences of receptor–protein and receptor–DNA interactions, thereby resulting in the temporally controlled regulation of cognate gene networks.
- RAR and RXR are composed of a modular structure with several domains and associated functions, in particular DBD and LBD domains.
- The LBD is a single protein domain organized in a primarily helical scaffold of 12 α-helices arranged in three layers and a short (S1–S2) β-turn.
- Retinoid agonists and antagonists/inverse agonists induce different helix H12 positioning and dynamics that impact on the structure of the activation function AF-2. Agonist and antagonists binding recruit or dissociate CoAs, respectively, whereas inverse agonist stabilize CoRs.
- Advances in chemical design supported by crystal structure determination of liganded RAR and RXR have provided (subtype-selective) potent retinoid receptor modulators.

Definitions of Words and Terms

Agonist: An endogenous substance or a drug that can interact with a receptor and initiate a physiological or a pharmacological response characteristic of that receptor.

Analog: An analog is a drug whose structure is related to that of another drug but whose chemical and biological properties may be quite different.

Antagonist: A drug or a compound that opposes the physiological effect of another. At the receptor level, it is a chemical entity that opposes the receptor-associated response normally induced by another bioactive agent.

Coactivators (CoAs): Proteins that cooperate with nuclear hormone receptors to activate transcription.

Corepressors (CoRs): Proteins that cooperate with nuclear hormone receptors to repress transcription.

Inverse (ant)agonist: A ligand that stabilizes an (in)active conformation of a receptor, thereby decreasing signaling below basal levels.

Partial agonist: An agonist that is unable to induce maximal activation of a receptor population, regardless of the amount of drug applied.

Selective nuclear receptor modulators (SNuRMs): Ligands that selectively modulate different receptor subtypes and/or act in a cell-selective manner.

TRAIL/Apo2L: The tumor necrosis factor (TNF)-related apoptosis-inducing ligand is an important regulators of the extrinsic pathway of apoptosis.

Transactivation: Activation of transcription by the binding of a transcription factor to a DNA-regulatory sequence.

List of Abbreviations

AF	activation function
APL	acute promyelocytic leukemia
CoA	coactivator
CoR	corepressor
DBD	DNA-binding domain
LBD	ligand-binding domain
LBP	ligand-binding pocket
NCoR	nuclear receptor corepressor
NRID	nuclear receptor interaction domain
NR	nuclear receptor
PPAR	peroxisome proliferator-activated receptor
PPRE	PPAR response element
RAR	retinoic acid receptor
RXR	retinoid X receptor
SMRT	silencing mediator for retinoid and thyroid hormones receptor
TRAIL	tumour necrosis factor (TNF)-related apoptosis inducing ligand
VDR	vitamin D receptor

Acknowledgements

Work from our laboratories was partially supported by funds from the the MICINN (SAF-2010-17395-FEDER, AdL), the Xunta de Galicia (INBIOMED; Project 08CSA052383PR from DXI+D+i, AdL), the Association for International Cancer Research (HG), the Ligue National Contre le Cancer (HG; laboratoire labelisé), the French National Research Agency (WB; ANR-07-PCVI-0001-01), and the European Community contracts QLK3-CT2002-02029 "Anticancer Retinoids" (AdL, HG), LSHM-CT-2005-018652 "Crescendo" (HG) and LSHC-CT-2005-518417 "Epitron" (AdL, HG).

References

Altucci, L. and Gronemeyer, H., 2001. The Promise of Retinoids to Fight Against Cancer. Nature Reviews Cancer. 1: 181–193.

Altucci, L., Leibowitz, M. D., Ogilvie, K. M., de Lera, A. R. and Gronemeyer, H., 2007. RAR and RXR modulation in cancer and metabolic disease. Nature Reviews Drug Discovery. 6: 793–810.

Álvarez, S., Germain, P., Alvarez, R., Rodríguez-Barrios, F., Gronemeyer, H. and de Lera, A. R., 2007. Structure, function and modulation of RARβ, a tumour supressor. International Journal of Biochemistry and Cell Biology. 39: 1406–1415.

Álvarez, S., Khanwalkar, H., Álvarez, R., Erb, C., Martínez, C., Rodríguez-Barrios, F., Germain, P., Gronemeyer, H. and de Lera, A. R., 2009. C3 halogen and C8″ substituents on stilbene arotinoids modulate retinoic acid receptor subtype function. Chemistry Medicinal Chemistry. 4: 1630–1640.

Bourguet, W., Vivat, V., Wurtz, J.-M., Chambon, P., Gronemeyer, H. and Moras, D., 2000. Crystal structure of a heterodimeric complex of RAR and RXR ligand-binding domains. Molecular Cell. 5: 289–298.

Bourguet, W., de Lera, A. R. and Gronemeyer, H., 2010. Inverse agonists and antagonists of retinoid receptors. In: Conn, P. M. (ed.) Methods in Enzymology, Vol. 485. Academic Press, pp. 161–195.

Chandra, V., Huang, P., Hamuro, Y., Raghuram, S., Wang, Y., Burris, T. P. and Rastinejad, F., 2008. Structure of the intact PPARγ–RXRα nuclear receptor complex on DNA. Nature. 456: 350–356.

Dawson, M. I., 2004. Synthetic retinoids and their nuclear receptors. Current Medicinal Chemistry – Anti-Cancer Agents. 4: 199–230.

Dawson, M. I. and Zhang, X.-k., 2002. Discovery and design of retinoic acid receptor and retinoid X receptor class- and subtype-selective synthetic analogues of all-*trans*-retinoic acid and 9-*cis*-retinoic acid. Current Medicinal Chemistry. 9: 623–637.

de Lera, A. R., Bourguet, W., Altucci, L. and Gronemeyer, H., 2007. Design of selective nuclear receptor modulators: RAR and RXR as a case study. Nature Reviews Drug Discovery. 6: 811–820.

García, J., Khanwalkar, H., Pereira, R., Erb, C., Voegel, J. J., Collette, P., Mauvais, P., Bourguet, W., Gronemeyer, H. and de Lera, A. R., 2009. Pyrazine arotinoids with inverse agonist activities on the retinoid and rexinoid receptors. Chemistry BioChemistry. 10: 1252–1259.

Germain, P., Iyer, J., Zechel, C. and Gronemeyer, H., 2002. Co-regulator recruitment and the mechanism of retinoic acid receptor synergy. Nature. 415: 187–192.

Germain, P., Kammerer, S., Pérez, E., Peluso-Iltis, C., Tortolani, D., Zusi, F. C., Starrett, J. E., Lapointe, P., Daris, J.-P., Marinier, A., de Lera, A. R., Rochel, N. and Gronemeyer, H., 2004. Rational design of RAR selective ligands revealed by RARβ crystal structure. EMBO Reports. 5: 877–882.

Germain, P., Chambon, P., Eichele, G., Evans, R. M., Lazar, M. A., Leid, M., de Lera, A. R., Lotan, R., Mangelsdorf, D. J. and Gronemeyer, H., 2006a.

The pharmacology and classification of the nuclear receptor superfamily. Retinoic acid receptors (RARs). Pharmacological Reviews. 58: 712–725.

Germain, P., Chambon, P., Eichele, G., Evans, R. M., Lazar, M. A., Leid, M., de Lera, A. R., Lotan, R., Mangelsdorf, D. J. and Gronemeyer, H., 2006b. The pharmacology and classification of the nuclear receptor superfamily. Retinoid X receptors (RXRs). Pharmacological Reviews. 58: 760–772.

Germain, P., Gaudon, C., Pogenberg, V., Sanglier, S., Potier, N., Van Dorsselaer, A., Royer, C. A., Lazar, M. A., Bourguet, W. and Gronemeyer, H., 2009. Differential action on coregulator interaction defines inverse retinoid agonists and neutral antagonists. Chemistry and Biology. 16: 479–489.

Gronemeyer, H. and Bourguet, W., 2009. Allosteric effects govern nuclear receptor action: DNA appears as a player. Science Signaling. 2: pe34.

Gronemeyer, H., Gustafsson, J.-A. and Laudet, V., 2004. Principles for modulation of the nuclear receptor superfamily. Nature Reviews Drug Discovery. 3: 950–964.

Laudet, V. and Gronemeyer, H., 2002. The Nuclear Receptor Facts Book. Academic Press, San Diego, USA, 462 pp.

Kagechika, H., 2002. Novel synthetic retinoids and separation of the pleiotropic retinoidal activities. Current Medicinal Chemistry. 9: 591–680.

Kagechika, H., Shudo, K., 2005. Synthetic retinoids: recent developments concerning structure and clinical utility. Journal of Medicinal Chemistry. 48: 5875–5883.

le Maire, A., Teyssier, C., Erb, C., Grimaldi, M., Alvarez, S., de Lera, A. R., Balaguer, P., Gronemeyer, H., Royer, C. A., Germain, P. and Bourguet, W., 2010. A unique secondary-structure switch controls constitutive gene repression by retinoic acid receptor. Nature Structural and Molecular Biology. 17: 801–807.

Lu, J., Dawson, M. I., Hu, Q. Y., Xia, Z., Dambacher, J. D., Ye, M., Zhang, X.-K. and Li, E., 2009. The effect of antagonists on the conformational exchange of the retinoid X receptor α ligand-binding domain. Magnetic Resonance in Chemistry. 47: 1071–1080.

Nahoum, V., Pérez, E., Germain, P., Rodríguez-Barrios, F., Manzo, F., Kammerer, S., Lemaire, G., Hirsch, O., Royer, C. A., Gronemeyer, H., de Lera, A. R. and Bourguet, W., 2007. Modulators of the structural dynamics of RXR to reveal receptor function. Proceedings of the National Academy of Sciences U.S.A. 104: 17323–17328.

Pérez-Santín, E., Germain, P., Quillard, F., Khanwalkar, H., Rodríguez-Barrios, F., Gronemeyer, H., de Lera, A. R. and Bourguet, W., 2009. Modulating retinoid X receptor with a series of (*E*)-3-[4-hydroxy-3-(3-alkoxy-5,5,8,8-tetramethyl-5,6,7,8-tetrahydronaphthalen-2-yl)phenyl]acrylic acids and their 4-alkoxy isomers. Journal of Medicinal Chemistry. 52: 3150–3158.

Pogenberg, V., Guichou, J.-F., Vivat-Hannah, V., Kammerer, S., Pérez, E., Germain, P., de Lera, A. R., Gronemeyer, H., Royer, C. A. and Bourguet,

W., 2005. Characterization of the interaction between RAR/RXR heterodimers and transcriptional coactivators through structural and fluorescence anisotropy studies. Journal of Biological Chemistry. 280: 1625–1633.

Rochel, N., Ciesielski, F., Godet, J., Moman, E., Roessle, M., Peluso-Iltis, C., Moulin, M., Haertlein, M., Callow, P., Mély, Y., Svergun, D. I. and Moras, D., 2011. Common architecture of nuclear receptor heterodimers on DNA direct repeat elements with different spacings. Nature Structural and Molecular Biology. 18: 564–570.

Sato, Y., Ramalanjaona, N., Huet, T., Potier, N., Osz, J., Antony, P., Peluso-Iltis, C., Poussin-Courmontagne, P., Ennifar, E., Mély, Y., Dejaegere, A., Moras, D. and Rochel, N., 2010. The "Phantom Effect" of the rexinoid LG100754: structural and functional insights. PLoS ONE. 5: e15119.

Shankaranarayanan, P., Rossin, A., Khanwalkar, H., Alvarez, S., Alvarez, R., Jacobson, A., Nebbioso, A., de Lera, A. R., Altucci, L. and Gronemeyer, H., 2009. Growth factor-antagonized rexinoid apoptosis involves permissive PPARγ/RXR heterodimers to activate the intrinsic death pathway by NO. Cancer Cell. 16: 220–231.

Zhang, L., Ren, X., Alt, E., Bai, X., Huang, S., Xu, Z., Lynch, P. M., Moyer, M. P., Wen, X. F. and Wu, X., 2010. Chemoprevention of colorectal cancer by targeting APC-deficient cells for apoptosis. Nature. 464: 1058–1061.

CHAPTER 25
Retinoic Acid in Development

DON CAMERON[a], TRACIE PENNIMPEDE[b] AND
MARTIN PETKOVICH*[a]

[a] Division of Cancer Biology and Genetics, Queen's University Cancer
Research Institute, 10 Stuart St, Boterell Hall, Rm 354, Kingston, Ontario,
Canada; [b] Max Planck Institute for Molecular Genetics, Developmental
Genetics Department, Ihnestrasse 73, 14195 Berlin, Germany
*E-mail: Martin.Petkovich@queensu.ca

25.1 Introduction

Evidence indicates that vitamin A and its active derivatives (retinoids) are critical throughout mammalian development. Retinoids exert biological effects on cell differentiation, proliferation, and morphogenesis by binding to their cognate receptors, the retinoic acid receptors (RARs) and retinoid X receptors (RXRs), resulting in the activation of retinoid-responsive genes. Localized synthesis and catabolism of retinoic acid (RA) by specific retinaldehyde dehydrogenases (Raldhs) and cytochromes P450, family member 26 (CYP26s) respectively, control tissue exposure to RA throughout development.

25.2 RA Metabolism and Signalling

RA synthesis results from the irreversible conversion of retinal to RA, mediated by the RALDH enzymes: RALDH1, -2, and -3 (Figure 25.1). When RA is present in a cell it initiates gene transcription by acting as a ligand for the RARs in heterodimeric complex with one of the RXRs. These RAR–RXR transcription factor complexes bind DNA response elements known as retinoic

Food and Nutritional Components in Focus No. 1
Vitamin A and Carotenoids: Chemistry, Analysis, Function and Effects
Edited by Victor R Preedy
Published by the Royal Society of Chemistry, www.rsc.org

acid response elements (RAREs) located within the promoters of target genes. Within tissues, RA availability is limited by cytochrome P450-mediated metabolism. Catabolism of RA is carried out specifically by the CYP26 family of enzymes (CYP26A1, -B1, and -C1) which catalyze hydroxylation at the C4 and C18 positions of the β-ionone ring of RA, resulting in conversion into more polar inactive metabolites such as 4-oxo-RA, 4-OH-RA, and 18-OH-RA (Chithalen *et al* 2002; White *et al* 2000) to facilitate excretion (Figure 25.1).

25.3 RA in Development

Maternal vitamin A deficiency (VAD) in rodents results in high rates of fetal resorption and death, along with characteristic defects mainly involving the eyes, heart, and genito-urinary tract, along with various other organs. The teratogenic consequences of maternal vitamin A treatment are dependent on both dose and stage of treatment, and resulting congenital malformations include exencephaly, cleft palate, spina bifida, ocular malformations, and limb defects, among others.

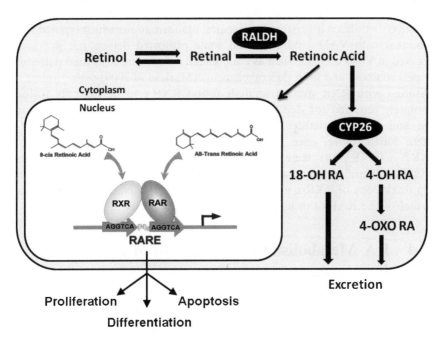

Figure 25.1 Overview of RA signalling and metabolism. The generation of RA involves the conversion of retinol into retinal, and subsequent conversion of retinal into RA. RA is produced in cells expressing Raldh enzymes. RA then signals by binding to nuclear RARs which activates transcription of RA responsive genes (shown for simplicity in the same cell, although RA generally signals through a paracrine mechanism). Cells expressing CYP26 enzymes inactivate RA through oxidation to polar metabolites.

25.3.1   Role of the Retinoid Receptors

Within each of the two receptor subfamilies (RAR and RXR) there are three isotypes (α, β, and γ), each of which is encoded by separate genes and with multiple isoforms, resulting from alternative splicing and the differential use of two promoters [reviewed in Chambon (1996)]. The full family of retinoid receptors represents fourteen different isoforms: RARα1 and RARα2; RARβ1–RARβ4; RARγ1 and RARγ2; RXRα1 and RXRα2; RXRβ1 and RXRβ2; RXRγ1 and RXRγ2.

RAR transcripts are widely distributed throughout the embryo during early morphogenesis [reviewed in Dolle (2009)]. RARα shows nearly ubiquitous expression at these stages, whereas the RARβ and RARγ expression domains are significantly more restricted. RXRα and RXRβ are also widely expressed, whereas RXRγ is restricted to developing muscle and regions of the central nervous system. Mouse knockouts of particular RAR isotypes, or one isoform of a particular isotype, have been thoroughly studied [reviewed in Mark *et al* (2009)]. For the most part, the ablation of a single receptor isotype did not recapitulate the spectrum of defects observed in VAD. Animals which were compound-null for several of the receptor isoforms or isotypes, however, collectively exhibited a variety of congenital malformations which typified those associated with VAD syndrome, and some presented defects not previously observed in VAD animals, such as craniofacial defects and abnormal patterning of spinal column and limb skeletal elements (Mark *et al* 2009).

Studies with RXR mutant animals found RXRα to be the only isoform absolutely required for development. RXR$\alpha^{-/-}$ embryos died during gestation, and displayed defects in development of the eye and heart (Mark *et al* 2009). Furthermore, mice having a single allele of RXRα (*i.e.* RXR$\alpha^{+/-}$/ RXR$\beta^{-/-}$/RXR$\gamma^{-/-}$) were growth deficient but viable, suggesting that virtually all developmental and postnatal roles of the RXRs can be maintained by a single copy of RXRα, implying that RXRα is the main heterodimerization partner of the RARs throughout development (Mark *et al* 2009).

25.4   RA Metabolism in the Control of Embryonic Development

Many studies have suggested that RA can act as a morphogen, providing positional information to cells. Striking examples are the duplicating effect of exogenous RA on developing limbs and the posteriorizing effect on the segmented hindbrain. As the actions of RA became better understood in these contexts, it was apparent that restricting RA levels within developing tissues was essential for proper embryogenesis. This is accomplished through the precise regulation of RA synthesis and catabolism *via* the RALDH and CYP26 enzymes, respectively.

The localized production of RA during development is performed by RALDH2, and to a lesser extent RALDH1 and RALDH3. The predominant role of RALDH2 in synthesizing RA during embryonic development has been established using a knock-out mouse model, which exhibits severe VAD-associated developmental defects and dies at mid-gestation (Niederreither *et al* 1999). Moreover, *Raldh2*$^{-/-}$ embryos fail to activate a highly-sensitive retinoid-responsive *lacZ* reporter transgene (RARE-*lacZ*) which, along with their abnormal phenotype, can be largely rescued following transient maternal treatment with RA. When RA is synthesized in tissues where RALDH is expressed it activates gene transcription unless it is effectively cleared by CYP26 activity. Thus, CYP26-mediated RA catabolism governs tissue sensitivity to RA and acts in concert with RALDH enzymes to establish strictly controlled domains of RA distribution within the embryo.

The expression patterns of *Cyp26a1*, *Cyp26b1*, and *Cyp26c1* during organogenesis are distinct and mainly confined to tissues that are particularly sensitive to RA. *Cyp26a1* is strongly expressed in the developing tail bud from embryonic day 8.5 (E8.5) to around E11.5 (Fujii *et al* 1997; MacLean *et al* 2001); *Cyp26b1* is found in the limb buds throughout the proximal–distal (Pr-Di) axis outgrowth (E9.5–E12.5) (MacLean *et al* 2001; Yashiro *et al* 2004); and each of the *Cyp26* isoforms is differentially expressed in the rhombomeres (segments) of the developing hindbrain, from around E8–E10.5 (Abu-Abed *et al* 2001; MacLean *et al* 2001; Tahayato *et al* 2003). These expression domains define areas where ectopic signalling by endogenous levels of RA can lead to congenital malformations.

25.4.1 CYP26 Blocks RA Exposure to the Early Embryo

Recent studies have shown that CYP26s form a necessary protective barrier to RA in the early embryo, even before the onset of *Raldh* expression. Mouse embryos lacking all three CYP26 enzymes (*Cyp26a1b1c1*$^{-/-}$) displayed a variably expressive phenotype, with the most severely affected mutants exhibiting a complete duplication of the body axis. Pre-gastrulation stage *Cyp26a1b1c1*$^{-/-}$ embryos showed RA signalling activity throughout the epiblast where it is normally absent, as well as *Nodal* expression along the entire anterior–posterior (A-P) axis of the epiblast instead of being restricted to the posterior portion (Uehara *et al* 2009). Consistent with this, expression of *Cyp26s* was found in extra-embryonic tissues in the pre-gastrulation embryo, suggesting that RA metabolism is required to protect the embryo proper from exposure to maternally derived RA. These results illustrate that, collectively, the CYP26s are required to restrict the activity of RA in the early embryo to allow proper patterning and morphogenesis.

25.4.2 CYP26 is Required for Early Head Development

Cyp26a1 is expressed in the anterior epiblast and neural plate, and then specifically in prospective rhombomere 2 (Fujii *et al* 1997). *Cyp26a1* null embryos display an incompletely penetrant exencephaly phenotype and mild posteriorization of the hindbrain, indicating that CYP26A1 also protects regions of the early rostral neural tube from the posteriorizing effects of RA (Abu-Abed *et al* 2001; Sakai *et al* 2001). Moreover, this effect is exaggerated when embryos lacking CYP26A1 are treated with a sub-teratogenic dose of maternally administered RA (Ribes *et al* 2007).

Cyp26c1 is also detected in the early anterior mesenchyme and neural plate, although slightly later than that of *Cyp26a1*, and subsequently in rhombomeres 2 and 4 of the hindbrain (Tahayato *et al* 2003; Uehara *et al* 2007). Although mice lacking *Cyp26c1* develop normally, $Cyp26a1^{-/-}/Cyp26c1^{-/-}$ compound mutants die during gestation and show reductions in head formation, indicating that these two enzymes cooperate to limit RA signalling in the early anterior embryo to ensure proper head development (Uehara *et al* 2007). Moreover, these embryos fail to form migratory cranial neural crest cells (CNCCs), which arise from premigratory CNCCs specified in the developing hindbrain and contribute to many structures of the head (Uehara *et al* 2007). Interestingly, *Cyp26b1* in the most widely expressed of the three CYP26s in the hindbrain, and yet, although $Cyp26b1^{-/-}$ mutants develop severe craniofacial malformations, they do not show any significant defects in hindbrain patterning or specification of migratory CNCCs (MacLean *et al* 2009). These observations indicate that the early expression of *Cyp26a1* and *Cyp26c1* appear to be sufficient to allow for proper head formation and specification of CNCCs.

25.4.3 CYP26A1 and Caudal Development

Cyp26a1 is strongly expressed in the caudal embryo during axis extension where it plays a critical role in preventing the caudal stem zone from premature exposure to RA. Embryos lacking *Cyp26a1* die late in gestation and develop severe defects of posterior structures characteristic of RA teratogenesis, with variable expressivity (Abu-Abed *et al* 2001; Sakai *et al* 2001) (Figure 25.2B). These include spina bifida, abdominoschisis, defective urogenital formation and in most severe cases sirenomelia and complete regression of caudal structures. Expression of the RARE-*lacZ* reporter indicates that in the absence of CYP26A1 function RA activity is detected in the caudal portion of the embryo where it is normally absent (Niederreither *et al* 2002a; Sakai *et al* 2001). Reducing the amount of embryonic RA by *Raldh2* heterozygosity in the *Cyp26a1* mutant background ($Raldh2^{+/-}/Cyp26a1^{-/-}$) restores formation of caudal structures, and allows some mutants to reach adulthood (Niederreither *et al* 2002a) (Figure 25.2C). Not only did this study identify that loss-of-function of the main embryonic RA synthetic enzyme acted as a dominant suppressor of the *Cyp26a1*-null embryonic lethal phenotype, but it demonstrated conclusively that the observed defects upon loss of CYP26A1 function are due to increased levels of RA and not the lack of an active RA metabolite generated by cytochrome P450-mediated oxidation.

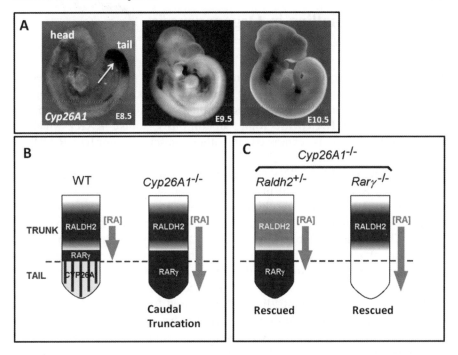

Figure 25.2 CYP26A1, RA and caudal development. (A) Expression of *Cyp26a1* shown by whole mount *in situ* hybridization at E8.5, E9.5, and E10.5. *Cyp26a1* is strongly expressed in the caudal region of the embryo (arrow). (B) Schematic diagram illustrating the disruption of caudal development in the absence of CYP26A1. In wild-type (WT) embryos, *Cyp26a1* expressed caudally prevents RA synthesized by RALDH2 in the trunk region from activating RARγ in the tail bud. In mutant embryos lacking CYP26A1, RA inappropriately enters tail bud tissue where it signals through RARγ, leading to caudal truncation. (C) Caudal defects in *Cyp26a1*$^{-/-}$ embryos can be rescued by decreasing the levels of RA signalling either by loss of one allele of *Raldh2* or by genetic removal of the receptor RARγ.

Embryos treated with exogenous RA during early axis extension (E8.5) also exhibit truncation of caudal structures, and exhibit altered expression of RA-regulated genes in the caudal embryo including *Wnt3a*, *Cdx4*, and *Brachyury* (*T*) and, accordingly, loss of *Cyp26a1* results in alteration in the expression of these tail bud genes (Abu-Abed *et al* 2001). *Rarg*$^{-/-}$ mutant mice are resistant to the caudal defects associated with ectopic RA signalling, indicating that this receptor is responsible for transducing the RA signal in the tail bud. Consistent with this, the *Cyp26a1*$^{-/-}$ phenotype is also rescued by further disruption of *Rarg* (*Cyp26a1*$^{-/-}$/*Rarg*$^{-/-}$), along with restoration of tail bud gene expression (Abu-Abed *et al* 2003) (Figure 25.2C).

25.5 RA Signalling in Organogenesis

25.5.1 Heart

The embryonic heart is the first organ to form which permits study, even in mutants displaying early embryonic lethal phenotypes. The expression of RXRs and RARs has been examined within the context of the developing heart; RXRα and RXRβ are found ubiquitously, unlike RXRγ which is absent, although it is expressed in other muscle primordia. RARβ1/3 transcripts are detected at E11.5 in the heart conotruncal mesenchyme, whereas RARα1 and RARβ2/4 were moderately expressed in the myocardium at E13.5 and RARγ is present in the endocardial cushion tissue.

The embryonic lethality resulting from ablation of RXRα was suspected to be the result of heart abnormalities, although some embryos survived to E16.5. These mice exhibited a variable "thin myocardium" phenotype which was characterized by abnormal thinness of the compact layer, the ventricular septum, and atrium walls (Mark *et al* 2009). Similar phenotypes were observed in RXRα$^{+/-}$/RARγ$^{-/-}$, RXRα$^{-/-}$/RARγ$^{-/-}$, and RXRα$^{-/-}$/RARα$^{-/-}$ compound mutants, but with increased frequency of defects in the ventricular septum, and additional defects of the aorticopulmonary septum and aortic arches. In the RAR compound mutants, persistent truncus arteriosus (PTA), which is the failure of aortic sac division by the aorticopulmonary septum, was frequently observed in RARαγ, RARα/RARβ2 (αβ2 for simplicity), αβ, and α1β animals, and sometimes seen in RARα$^{-/-}$β2$^{+/-}$ and α1β2 fetuses, as well as in RXRα/RARβ and RXRα/RARγ null embryos (Mark *et al* 2009).

The production of RA and the activation of receptors occur in the same regions of the heart, corresponding to regions that do not express *Cyp26a1*. *Cyp26a1* is present in the early embryo (E7.5–E8.0) within the cardiac crescent, heart outflow tract, atria, and sinus venosus (MacLean *et al* 2001); specific areas, wherein, activation of RARE-*lacZ* is not observed (Moss *et al* 1998). *Cyp26a1* and *Cyp26b1* are also expressed at E14.5 within the atrioventricular valve and outflow tract endocardial cells, respectively (Abu-Abed *et al* 2002), but heart defects associated with loss of these enzymes have not yet been reported. Loss of *Raldh2*, however, results in the heart forming as a single dilated cavity with no particular left–right organization, indicating a requirement for RA synthesis in proper heart morphogenesis. An early role for RA signalling in defining populations of cells that give rise to the heart has been observed, where RA signalling acts to restrict the cardiac progenitor field (Keegan *et al* 2005; Ryckebusch *et al* 2008; Sirbu *et al* 2008). Moreover, *Raldh2*$^{-/-}$ embryos show defective patterning of early heart progenitors due to an expansion of fibroblast growth factor 8 gene (*Fgf8*) expression and Fgf target genes (Ryckebusch *et al* 2008; Sirbu *et al* 2008).

25.5.2 Eye

RA has also been shown to be an indispensable molecule with respect to eye development, as conceptuses from VAD dams exhibit various eye defects. Each of the receptor isotypes has been shown to be present in the prenatal eye and throughout its development. RARβ, RARγ, and RXRα, in particular, are abundantly expressed within the periocular mesenchyme (PM).

Ocular defects seen in $RXR\alpha^{-/-}$ mice included abnormal thickening of the corneal stroma, and coloboma of the optic nerve (Mark *et al* 2009). Similar malformations were seen in $RXR\alpha^{+/-}/RAR\gamma^{-/-}$ fetuses, in addition to shortening of the ventral retina. These defects were often less severe than in the $RXR\alpha^{-/-}$ single mutant, although neither $RXR\alpha^{+/-}$ nor $RAR\gamma^{-/-}$ alone exhibited any ocular defects. The severity of these defects was increased in $RXR\alpha^{-/-}/RAR\gamma^{+/-}$ and especially $RXR\alpha^{-/-}/RAR\gamma^{-/-}$ mutants, with a thick layer of mesenchyme burying the eye, bilateral eversion of the retina, and defects of the iris (coloboma and absence of the ventral iris). $RXR\alpha^{-/-}/RAR\alpha^{-/-}$ mutants exhibited retinal eversion, retinal hypoplasia, and coloboma of the iris. RAR compound-knockout mice presented with a number of ocular defects at E18.5 (Mark *et al* 2009). The RARαγ mutants displayed the widest range and most conspicuous defects including coloboma of the retina and optic nerve, unfused eyelids, and lack of anterior chamber, all of which were completely penetrant. Mutants for RARβ2γ, on the other hand, exhibited all of the aforementioned defects, with the exception of unfused eyelids and corneal–lenticular stalk. At E18.5 RARβγ mutants showed abnormal retrolenticular membrane, along with various mesenchymal defects of the stroma, sclera, and anterior chamber.

Distribution of RA within the developing eye (especially the retina) represents a striking example of the coordinated expression of RALDH and CYP26 enzymes. *Raldh1* and *3* are the main sources of RA in the eye, with *Raldh1* being expressed in the dorsal retina and *Raldh3* ventrally (Li *et al* 2000), although *Raldh2* has been shown to be expressed in the PM between E14.5–E15.5 (Niederreither *et al* 1997). In contrast, *Cyp26a1* is expressed within the equatorial region of the retina, where none of the *Raldh* transcripts are observed, and is followed by the expression of *Cyp26c1* (Sakai *et al* 2004). Single knockouts of *Raldh1* (Fan *et al* 2003) or *Raldh3* (Dupe *et al* 2003) have failed to show significant ocular defects, but *Raldh1/3* compound mutants display severe ocular malformations (Matt *et al* 2005). More importantly, these defects were similar to those found following specific ablation of RARβ/γ in the neural crest cell (NCC)-derived PM (Matt *et al* 2005). These data, and the fact that most of the ocular defects observed in retinoid receptor-null fetuses appear to correspond to defects in neural crest-derived structures, suggests that NCC are responsible for mediating retinoid signals within the developing eye.

25.5.3 Limb

A role for RA in limb morphogenesis has been recognized since the earliest studies on RA teratogenesis, where administration of exogenous RA to pregnant dams could induce severe limb deformities in developing embryos. Early patterning of the limb involves two signalling centres [reviewed in Niswander (2003)]. The zone of polarizing activity (ZPA) located at the posterior base of the limb bud controls A-P patterning through secretion of Sonic hedgehog (Shh); and the apical ectodermal ridge (AER) which is a strip of cells along the distal edge of the limb bud that controls limb Pr-Di outgrowth through Fgf signalling. Early experiments using the chick limb bud model revealed that ectopic administration of RA to the anterior portion of the early limb bud could induce mirror image pattern duplications of the A-P axis, identical to that seen upon grafting cells from the ZPA into the anterior limb bud (Tickle *et al* 1982). Although the physiological role of RA signalling in limb patterning and morphogenesis is still not clear, changes in the distribution of RA or its ability to signal in the developing limb by genetic or pharmacological means have dramatic consequences. In early stages of limb development (E10.5), both RARα and RARγ transcripts are distributed throughout the limb bud, whereas RARβ transcripts are present only in the flanking trunk mesoderm (Dolle *et al* 1989). As the limb continues to grow, RARγ is also expressed in central pre-cartilaginous condensations, and RARβ is detectable in the interdigital mesenchyme. Although RARα- and RARγ-null animals have normal limbs, compound RARαγ mutants show several limb malformations (Mark *et al* 2009). In these mutants, the forelimbs displayed a variety of defects such as malformation of the scapula, agenesis of the radius, digit 1 and central carpal bones, and syndactyly. The hindlimbs of these mutants were slightly less affected, showing only malformations of the tibia. Despite the specific expression of RARβ in the interdigital mesenchyme, mice lacking this receptor do not exhibit any limb defects. $RAR\beta^{+/-}/RAR\gamma^{-/-}$ and $RAR\beta^{-/-}/RAR\gamma^{-/-}$ compound mutants, however, display webbing (Mark *et al* 2009).

The exposure of the limb to RA must be finely controlled for proper morphogenesis to occur. Although there is no detectable *Raldh* expression in the developing limb itself at early patterning/outgrowth stages, the high expression of *Raldh2* in the trunk somites during limb bud initiation and outgrowth acts as the presumptive source of RA. Consistent with this is the observation that $Raldh2^{-/-}$ embryos lack limb buds (Niederreither *et al* 1999). Although these *Raldh2*-null mutants die between E9.5 and E10.5, corresponding to the initial period of limb bud development, administration of RA through the maternal food supply can rescue development in a stage- and dose-dependent manner, such that effects of loss of *Raldh2* later in limb development can be examined. Rescued embryos display markedly reduced forelimb growth while the hindlimbs appear unaffected, possibly due to expression of *Raldh3* in the mesonephric region, located near the base of the hindlimb bud (Niederreither *et al* 2002b). *Fgf8*, which is normally evenly distributed along the AER, was up-regulated in the anterior portion of the AER in RA-treated $Raldh2^{-/-}$ embryos, and in extreme cases two distal outgrowths were evident. Shh was also reduced and sometimes abnormally expressed in the anterior portion of the limb bud. More recent studies examining limb bud growth in *Raldh2* and *Raldh3* mutants rescued with a short pulse of RA have suggested that, in fact, RA is dispensable for limb bud patterning and outgrowth (Zhao *et al* 2009), arguing that RA is not required at all for hindlimb development, and that it simply provides a suitably competent environment for the formation of forelimbs through its antagonism of Fgf8 signalling in the developing trunk.

Although the role of endogenous RA in limb bud patterning and outgrowth remains unclear and somewhat controversial, it is clear that local inactivation of RA in the limb bud is absolutely required for proper outgrowth and patterning. Strong expression of *Cyp26b1* is detected in the limb bud during stages of Pr-Di outgrowth which overlap with the retinoid-sensitive period (E9.5–E12.5), whereas neither *Cyp26a1* nor *Cyp26c1* are expressed at appreciable levels in the limb (MacLean *et al* 2001; Tahayato *et al* 2003). As the limb bud emerges, *Cyp26b1* expression is detected throughout the entire bud, and becomes progressively more distalized (MacLean *et al* 2001; Yashiro *et al* 2004) (Figure 25.3A). Mice lacking functional CYP26B1 display severely truncated limbs (phocomelia) and only 2 or 3 digits per limb in both fore and hindlimbs, defects resembling those resulting from excess RA (Figure 3B) (Yashiro *et al* 2004). Recent work has also shown that deletion of *Rarg* in a $Cyp26b1^{-/-}$ background is able to ameliorate phocomelia and digit defects in these mutants (Figure 3B) (Pennimpede *et al* 2010). Since RA production during normal early limb outgrowth is confined to the somites and lateral plate mesoderm, limb defects following loss of CYP26B1 are dependent on diffusion of RA into the limb bud. Although RA in these mutants does not reach the distal-most tip of the limb bud, its expansion into the medial limb bud is sufficient to disrupt the expression of genes essential for normal patterning of the stylopod (humerus/femur), zeugopod (radius and ulna/tibia and fibula), and autopod (wrist and hand/ankle and foot) (Pennimpede *et al* 2010; Yashiro *et al* 2004).

Removal of *Cyp26b1* allows endogenous RA to flood the proximal limb bud during all stages of outgrowth and results in a severe reduction in Pr-Di outgrowth. Since, phocomelia can be consistently produced by RA treatment at E11, a time when the limb segments are already well-defined and chondrogenic precursors have been established, RA treatment must also effect either the survival or differentiation of these cells. Recent work utilizing a conditional knockout of *Cyp26b1* specifically in limb buds has indicated that increased RA caused a reduction in chondroblast differentiation, suggesting that RA signalling interferes with limb mesenchymal cell fate (Dranse *et al* 2011).

It has been suggested that RA acts as a 'proximalizing factor' in the limb, which along with the 'distalizing factor' (AER Fgfs), compose the two signals

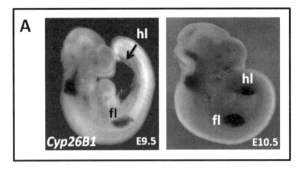

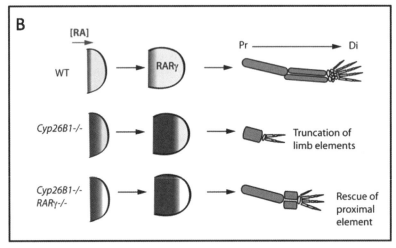

Figure 25.3 CYP26B1 and limb bud outgrowth. (A) Expression of *Cyp26b1* detected
by whole mount *in situ* hybridization at E9.5 and E10.5. *Cyp26b1*
transcripts are strongly expressed in the developing limb buds. (**B**)
Normally, CYP26B1 prevents RA from entering the distal part of the
bud where RARγ is expressed. In *Cyp26b1*$^{-/-}$ mutants, RA can enter
the limb bud, leading to truncation of limb elements. In compound
Cyp26B1$^{-/-}$*;RARγ*$^{-/-}$ mutants, RA can still enter the limb bud, but is
unable to signal through RARγ, which allows development of the
proximal-most element, the stylopod, and partially rescues more distal
elements, the zeugopod, and autopod. fl: forelimb, hl: hindlimb, Pr:
proximal, Di: distal.

that cooperate to pattern the limb Pr-Di axis. This model has been supported
by studies indicating that pharmacological application of RA can up-regulate
the expression of genes within the proximal limb bud and repress the
expression of genes in the distal aspect of the limb (Mercader *et al* 2000). It is
also consistent with the limb truncations observed upon genetic deletion of
AER Fgfs (Mariani *et al* 2008) or CYP26B1. Recent studies in chicken have
supported this notion that limb outgrowth involves two signals—a distal Fgf
signal from the AER, as well as a proximal one; RA, or if not RA, then a

different signal yet to be identified that can be effectively mimicked by RA (Cooper *et al* 2011; Rosello-Diez *et al* 2011).

Taken together, the recent molecular and cell fate analyses on RA teratogenesis in the limb, either following RA treatment or loss of *Cyp26b1* (and *Cyp26b1/Rarg*), have indicated that there are multiple effects of excess RA on limb development.

25.6 RA Signalling and Germ Cells

Mammalian germ cells develop along a sexually dimorphic pathway, according to whether they reside in a testis or an ovary. Part of this sexual dimorphism involves differential timing of the induction of meiosis. In females, meiosis begins during embryogenesis at around E13.5, but ovary cells then arrest in prophase of meiosis I at birth. In males, germ cells enter mitotic arrest during development, and initiate meiosis after birth. This decision faced by germ cells to enter meiosis or arrest in mitosis during development involves the reception of signals from the local environment, and evidence supports a critical role for RA in this decision. In the embryonic ovary, the RA target gene *Stra8* (stimulated by RA gene 8) is expressed in pre-meiotic cells, whereas its expression is absent from the embryonic testis, despite expression of the RA synthetic enzyme *Raldh2* in the mesenephroi of both sexes (Bowles *et al* 2006; Koubova *et al* 2006). *Stra8* is expressed later in males after birth in pre-meiotic germ cells, and cells in both males and females lacking *Stra8* gene function fail to complete meiosis (Anderson *et al* 2008; Baltus *et al* 2006; Mark *et al* 2008). These observations indicate that RA signalling plays an important role in the differential entry into meiosis of germ cells between males and females, but that it is apparently delayed in the male. Indeed, at the same time that *Stra8* exhibits female-specific germ cell expression in the embryo, *Cyp26b1* is expressed in the developing testes but is absent from the developing ovaries, and expression of *Stra8* and *Scp3* (synaptonemal complex protein 3; a marker of meiotic progression) are up-regulated in *Cyp26b1*-null testes at E13.5 (Bowles *et al* 2006; Koubova *et al* 2006). These observations suggested that RA metabolism by CYP26B1 prevents exposure of testicular germ cells in the embryo to inappropriate RA signalling and STRA8 activity, and thus premature entry into meiosis.

Testes of newborn $Cyp26b1^{-/-}$ pups were found to be smaller, and show a striking lack of germ cells (MacLean *et al* 2007). At E13.5, E14.5 and E16.5, many meiotic and apoptotic germ cells were detected in *Cyp26b1* mutant testes, while somatic cells (Sertoli cells and Leydig cells) were apparently unaffected. These results indicate that, not only does CYP26B1 function as a meiosis-inhibiting factor in embryonic testis development, it is also required for germ cell survival, presumably by blocking premature exposure to RA. Indeed, in the absence of CYP26B1, increased RA levels are present in the testes as early as E12.5 (MacLean *et al* 2007).

Taken together, these studies showed that RA signalling is involved in the decision of germ cells to enter meiosis, and thus develop along a male- or female-specific pathway during gonad development, and CYP26B1 is responsible for delaying meiotic entry in male germ cells. However, expression of *Cyp26b1* persists in the developing testis after this decision has been made (Koubova *et al* 2006), suggesting that it plays a further protective role in male germ cell development. This possibility was tested by removing CYP26B1 function at later stages by conditional ablation in specific deletion in Sertoli cells at E15. Removing *Cyp26b1* at this stage resulted in an incompletely penetrant testis degeneration phenotype, as 3-month-old mutant testes contained many seminiferous tubules devoid of germ cells (Li *et al* 2009). Moreover, mutant testes were found to express *Stra8* and contain meiotic germ cells at E16.5. Interestingly, these results indicate that male germ cells are still able to enter meiosis (*i.e.* develop along a female pathway) at E15.5 after sex-specific developmental programs have been initiated, and that CYP26B1 activity is required to maintain these cells in an undifferentiated state throughout development.

25.7   Concluding Remarks

It is clear both from nutritional and genetic studies that RA plays a critical role at many stages in the development of an embryo. From early events such as axial patterning to later stages of organ morphogenesis, normal development depends on the appropriate levels and distribution of RA in both a spatial and temporal specific manner.

Summary Points

- Retinoic acid receptors (RARα, -β, and -γ) are required for many aspects of embryonic development.
- Loss of function of these receptors collectively result in malformations of the developing embryo that resemble severe vitamin A deficiency.
- Retinoic acid (RA) signalling is regulated in the embryo through the control of its distribution by localized synthesis and catabolism.
- Loss of the RA synthetic enzymes results in developmental defects consistent with severe vitamin A deficiency.
- Loss of the RA catabolic enzymes results in defects consistent with embryonic exposure to excess RA.

Key Facts

Key Facts of Embryonic Development and Retinoic Acid

- The proper formation of an embryo is a complex process which involves the co-ordinated control of cellular processes such as growth, proliferation, and death. Cells must also become aware of where in the embryo they reside, and what cell type to become.

- These processes are regulated by many different signalling molecules including secreted proteins, as well as RA.
- Unlike protein-based signals, RA is a small fat-soluble molecule formed enzymatically in the embryo from maternally provided vitamin A (retinol).
- Thus, the distribution of the RA signal in the embryos is achieved primarily through localized expression of synthetic and catabolic enzymes.

Key Facts of Retinoic Acid Signalling and Metabolism

- RA binds to nuclear RARs to modulate gene expression.
- RA is synthesized in the embryo by the retinaldehyde dehydrogenase (RALDH) enzymes.
- RA activity is limited by tissue-specific expression of the RA catabolic enzymes, the CYP26s (cytochrome P450, member 26).

Definitions of Words and Terms

Cytochrome P450: A superfamily of heme-containing enzymes that catalyze the oxidation of a diverse set of substrates.

CYP26: A family of cytochrome P450 enzymes that specifically oxidize RA.

Differentiation: The process during development by which cells become more and more specialized. Cellular differentiation is regulated by extracellular signals such as secreted proteins, and RA.

Morphogen: A secreted signalling molecule that can act on cells by producing distinct cellular responses depending upon its concentration.

Morphogenesis: The process by which an embryo, or embryonic structure forms its shape.

Retinaldehyde dehydrogenase (RALDH): The enzymes responsible for the production RA from retinal.

Retinoic acid receptor (RAR): A member of the family of nuclear receptors that act as ligand dependent transcription factors, which is specifically activated by RA.

Retinoid X receptor (RXR): Another member of the nuclear receptor superfamily, and heterodimerization partner of many other nuclear receptors, including RARs.

Teratogen: A substance that causes birth defects.

List of Abbreviations

AER	apical ectodermal ridge
A-P	anterior–posterior (axis)

CNCC   cranial neural crest cell
CYP26   Cytochrome P450, family member 26
E   embryonic day
Fgf   fibroblast growth factor
NCC   neural crest cell
PM   periocular mesenchyme
Pr-Di   proximal–distal (axis)
RA   retinoic acid
RALDH   retinaldehyde dehydrogenase
RAR   retinoic acid receptor
RARE   retinoic acid response element
RXR   retinoid X receptor
Shh   Sonic hedgehog
Stra8   stimulated by retinoic acid gene 8
VAD   vitamin A deficiency
ZPA   zone of polarizing activity

References

Abu-Abed, S., Dolle, P., Metzger, D., Beckett, B., Chambon, P. and Petkovich, M., 2001. The retinoic acid-metabolizing enzyme, CYP26A1, is essential for normal hindbrain patterning, vertebral identity, and development of posterior structures. Genes and Development. 15: 226–240.

Abu-Abed, S., MacLean, G., Fraulob, V., Chambon, P., Petkovich, M. and Dolle, P., 2002. Differential expression of the retinoic acid-metabolizing enzymes CYP26A1 and CYP26B1 during murine organogenesis. Mechanisms of Development. 110: 173–177.

Abu-Abed, S., Dolle, P., Metzger, D., Wood, C., MacLean, G., Chambon, P. and Petkovich, M., 2003. Developing with lethal RA levels: genetic ablation of Rarg can restore the viability of mice lacking Cyp26a1. Development. 130: 1449–1459.

Anderson, E. L., Baltus, A. E., Roepers-Gajadien, H. L., Hassold, T. J., de Rooij, D. G., van Pelt, A. M. and Page, D. C., 2008. Stra8 and its inducer, retinoic acid, regulate meiotic initiation in both spermatogenesis and oogenesis in mice. Proceedings of the National Academy of Sciences of the United States of America. 105: 14976–14980.

Baltus, A. E., Menke, D. B., Hu, Y. C., Goodheart, M. L., Carpenter, A. E., de Rooij, D. G. and Page, D. C., 2006. In germ cells of mouse embryonic ovaries, the decision to enter meiosis precedes premeiotic DNA replication. Nature Genetics. 38: 1430–1434.

Bowles, J., Knight, D., Smith, C., Wilhelm, D., Richman, J., Mamiya, S., Yashiro, K., Chawengsaksophak, K., Wilson, M. J., Rossant, J., Hamada, H. and Koopman, P., 2006. Retinoid signaling determines germ cell fate in mice. Science. 312: 596–600.

Chambon, P., 1996. A decade of molecular biology of retinoic acid receptors. FASEB Journal. 10: 940–954.

Chithalen, J. V., Luu, L., Petkovich, M. and Jones, G., 2002. HPLC-MS/MS analysis of the products generated from all-trans-retinoic acid using recombinant human CYP26A. Journal of Lipid Research. 43: 1133–1142.

Cooper, K. L., Hu, J. K., ten Berge, D., Fernandez-Teran, M., Ros, M. A. and Tabin, C. J., 2011. Initiation of proximal-distal patterning in the vertebrate limb by signals and growth. Science. 332: 1083–1086.

Dolle, P., 2009. Developmental expression of retinoic acid receptors (RARs). Nuclear Receptor Signalling. 7: e006.

Dolle, P., Ruberte, E., Kastner, P., Petkovich, M., Stoner, C. M., Gudas, L. J. and Chambon, P., 1989. Differential expression of genes encoding α, β and γ retinoic acid receptors and CRABP in the developing limbs of the mouse. Nature. 342: 702–705.

Dranse, H. J., Sampaio A. V., Petkovich, M. and Underhill T. M., 2011. Genetic deletion of *Cyp26b1* negatively impacts limb skeletogenesis by inhibiting chondrogenesis. J Cell Sci. 2011 Aug 15; 124(Pt 16): 2723–2734.

Dupe, V., Matt, N., Garnier, J. M., Chambon, P., Mark, M. and Ghyselinck, N. B., 2003. A newborn lethal defect due to inactivation of retinaldehyde dehydrogenase type 3 is prevented by maternal retinoic acid treatment. Proceedings of the National Academy of Sciences of the United States of America. 100: 14036–14041.

Fan, X., Molotkov, A., Manabe, S., Donmoyer, C. M., Deltour, L., Foglio, M. H., Cuenca, A. E., Blaner, W. S., Lipton, S. A. and Duester, G., 2003. Targeted disruption of Aldh1a1 (Raldh1) provides evidence for a complex mechanism of retinoic acid synthesis in the developing retina. Molecular and Cellular Biology. 23: 4637–4648.

Fujii, H., Sato, T., Kaneko, S., Gotoh, O., Fujii-Kuriyama, Y., Osawa, K., Kato, S. and Hamada, H., 1997. Metabolic inactivation of retinoic acid by a novel P450 differentially expressed in developing mouse embryos. EMBO Journal. 16: 4163–4173.

Keegan, B. R., Feldman, J. L., Begemann, G., Ingham, P. W. and Yelon, D., 2005. Retinoic acid signaling restricts the cardiac progenitor pool. Science. 307: 247–249.

Koubova, J., Menke, D. B., Zhou, Q., Capel, B., Griswold, M. D. and Page, D. C., 2006. Retinoic acid regulates sex-specific timing of meiotic initiation in mice. Proceedings of the National Academy of Sciences of the United States of America. 103: 2474–2479.

Li, H., Wagner, E., McCaffery, P., Smith, D., Andreadis, A. and Drager, U. C., 2000. A retinoic acid synthesizing enzyme in ventral retina and telencephalon of the embryonic mouse. Mechanisms of Development. 95: 283–289.

Li, H., MacLean, G., Cameron, D., Clagett-Dame, M. and Petkovich, M., 2009. *Cyp26b1* expression in murine Sertoli cells is required to maintain male germ cells in an undifferentiated state during embryogenesis. PLoS One. 4: e7501.

MacLean, G., Abu-Abed, S., Dolle, P., Tahayato, A., Chambon, P. and Petkovich, M., 2001. Cloning of a novel retinoic-acid metabolizing cytochrome P450, Cyp26B1, and comparative expression analysis with Cyp26A1 during early murine development. Mechanisms of Development. 107: 195–201.

MacLean, G., Li, H., Metzger, D., Chambon, P. and Petkovich, M., 2007. Apoptotic extinction of germ cells in testes of Cyp26b1 knockout mice. Endocrinology. 2007 Oct; 148(10): 4560–4567. Epub 2007 Jun 21.

MacLean, G., Dolle, P. and Petkovich, M., 2009. Genetic disruption of CYP26B1 severely affects development of neural crest derived head structures, but does not compromise hindbrain patterning. Developmental Dynamics. 238: 732–745.

Mariani, F. V., Ahn, C. P. and Martin, G. R., 2008. Genetic evidence that FGFs have an instructive role in limb proximal–distal patterning. Nature. 453: 401–405.

Mark, M., Jacobs, H., Oulad-Abdelghani, M., Dennefeld, C., Feret, B., Vernet, N., Codreanu, C. A., Chambon, P. and Ghyselinck, N. B., 2008. STRA8-deficient spermatocytes initiate, but fail to complete, meiosis and undergo premature chromosome condensation. Journal of Cell Science. 121: 3233–3242.

Mark, M., Ghyselinck, N. B. and Chambon, P., 2009. Function of retinoic acid receptors during embryonic development. Nuclear Receptors and Signaling. 7: e002.

Matt, N., Dupe, V., Garnier, J. M., Dennefeld, C., Chambon, P., Mark, M. and Ghyselinck, N. B., 2005. Retinoic acid-dependent eye morphogenesis is orchestrated by neural crest cells. Development. 132: 4789–4800.

Mercader, N., Leonardo, E., Piedra, M. E., Martinez, A. C., Ros, M. A. and Torres, M., 2000. Opposing RA and FGF signals control proximodistal vertebrate limb development through regulation of *Meis* genes. Development. 127: 3961–3970.

Moss, J. B., Xavier-Neto, J., Shapiro, M. D., Nayeem, S. M., McCaffery, P., Drager, U. C. and Rosenthal, N., 1998. Dynamic patterns of retinoic acid synthesis and response in the developing mammalian heart. Developmental Biology. 199: 55–71.

Niederreither, K., McCaffery, P., Drager, U. C., Chambon, P. and Dolle, P., 1997. Restricted expression and retinoic acid-induced downregulation of the retinaldehyde dehydrogenase type 2 (*RALDH-2*) gene during mouse development. Mechanisms of Development. 62: 67–78.

Niederreither, K., Subbarayan, V., Dolle, P. and Chambon, P., 1999. Embryonic retinoic acid synthesis is essential for early mouse post-implantation development. Nature Genetics. 21: 444–448.

Niederreither, K., Abu-Abed, S., Schuhbaur, B., Petkovich, M., Chambon, P. and Dolle, P., 2002a. Genetic evidence that oxidative derivatives of retinoic acid are not involved in retinoid signaling during mouse development. Nature Genetics. 31: 84–88.

Niederreither, K., Vermot, J., Schuhbaur, B., Chambon, P. and Dolle, P., 2002b. Embryonic retinoic acid synthesis is required for forelimb growth and anteroposterior patterning in the mouse. Development. 129: 3563–3574.

Niswander, L., 2003. Pattern formation: old models out on a limb. Nature Reviews in Genetics. 4: 133–143.

Pennimpede, T., Cameron, D. A., MacLean, G. A. and Petkovich, M., 2010. Analysis of Cyp26b1/Rarg compound-null mice reveals two genetically separable effects of retinoic acid on limb outgrowth. Developmental Biology. 339: 179–186.

Ribes, V., Fraulob, V., Petkovich, M. and Dolle, P., 2007. The oxidizing enzyme CYP26a1 tightly regulates the availability of retinoic acid in the gastrulating mouse embryo to ensure proper head development and vasculogenesis. Developmental Dynamics. 236: 644–653.

Rosello-Diez, A., Ros, M. A. and Torres, M., 2011. Diffusible signals, not autonomous mechanisms, determine the main proximodistal limb subdivision. Science. 332: 1086–1088.

Ryckebusch, L., Wang, Z., Bertrand, N., Lin, S. C., Chi, X., Schwartz, R., Zaffran, S. and Niederreither, K., 2008. Retinoic acid deficiency alters second heart field formation. Proceedings of the National Academy of Sciences of the United States of America. 105: 2913–2918.

Sakai, Y., Meno, C., Fujii, H., Nishino, J., Shiratori, H., Saijoh, Y., Rossant, J. and Hamada, H., 2001. The retinoic acid-inactivating enzyme CYP26 is essential for establishing an uneven distribution of retinoic acid along the anterio–posterior axis within the mouse embryo. Genes and Development 15: 213–225.

Sakai, Y., Luo, T., McCaffery, P., Hamada, H. and Drager, U. C., 2004. CYP26A1 and CYP26C1 cooperate in degrading retinoic acid within the equatorial retina during later eye development. Developmental Biology. 276: 143–157.

Sirbu, I. O., Zhao, X., and Duester, G., 2008. Retinoic acid controls heart anteroposterior patterning by down-regulating Isl1 through the Fgf8 pathway. Developmental Dynamics. 237: 1627–1635.

Tahayato, A., Dolle, P. and Petkovich, M., 2003. Cyp26C1 encodes a novel retinoic acid-metabolizing enzyme expressed in the hindbrain, inner ear, first branchial arch and tooth buds during murine development. Gene Expression Patterns. 3: 449–454.

Tickle, C., Alberts, B., Wolpert, L. and Lee, J., 1982. Local application of retinoic acid to the limb bond mimics the action of the polarizing region. Nature. 296: 564–566.

Uehara, M., Yashiro, K., Mamiya, S., Nishino, J., Chambon, P., Dolle, P. and Sakai, Y., 2007. CYP26A1 and CYP26C1 cooperatively regulate anterior–posterior patterning of the developing brain and the production of migratory cranial neural crest cells in the mouse. Developmental Biology. 302: 399–411.

Uehara, M., Yashiro, K., Takaoka, K., Yamamoto, M. and Hamada, H., 2009. Removal of maternal retinoic acid by embryonic CYP26 is required for correct Nodal expression during early embryonic patterning. Genes and Development. 23: 1689–1698.

White, J. A., Ramshaw, H., Taimi, M., Stangle, W., Zhang, A., Everingham, S., Creighton, S., Tam, S. P., Jones, G. and Petkovich, M., 2000. Identification of the human cytochrome P450, P450RAI-2, which is predominantly expressed in the adult cerebellum and is responsible for all-trans-retinoic acid metabolism. Proceedings of the National Academy of Sciences of the United States of America. 97: 6403–6408.

Yashiro, K., Zhao, X., Uehara, M., Yamashita, K., Nishijima, M., Nishino, J., Saijoh, Y., Sakai, Y. and Hamada, H., 2004. Regulation of retinoic acid distribution is required for proximodistal patterning and outgrowth of the developing mouse limb. Developmental Cell 6: 411–422.

Zhao, X., Sirbu, I. O., Mic, F. A., Molotkova, N., Molotkov, A., Kumar, S. and Duester, G., 2009. Retinoic acid promotes limb induction through effects on body axis extension but is unnecessary for limb patterning. Current Biology. 19: 1050–1057.

CHAPTER 26

Retinol/Vitamin A Signaling and Self-renewal of Embryonic Stem Cells

JASPAL S. KHILLAN*, HIMANSHU BHATIA AND LIGUO CHEN

Department of Microbiology and Molecular Genetics, 3501 Fifth Ave, University of Pittsburgh PA, USA
*E-mail: Khillan@pitt.edu

26.1 Introduction

Retinol, the alcohol form of vitamin A, is a key dietary component. Mammalian species cannot synthesize retinol/vitamin A and it is absorbed from animal food or from plant sources as β-carotene. Retinol is stored in the liver as retinoyl esters, primarily as retinyl palmitate (Blaner and Olsen 1994) from where it is mobilized into blood plasma by retinyl ester hydrolases (REHs) (Harrison 2005). Retinol is believed to play important role in embryonic development, reproduction, cell differentiation and immune function *via* its potent metabolite retinoic acid. However, a direct function of retinol has never been explored. This Chapter reviews the function of retinol in the self-renewal of embryonic stem (ES) cells.

Pluripotent ES cells derived from blastocysts (Evans and Kaufman 1981; Martin 1981; Thomson *et al* 1998) have unlimited capacity for self-renewal and

Food and Nutritional Components in Focus No. 1
Vitamin A and Carotenoids: Chemistry, Analysis, Function and Effects
Edited by Victor R Preedy

have the potential to form cells representing all the primary germ layers such as endoderm, mesoderm and ectoderm, which can give rise to more than 200 lineages of the adult organism (Thomson *et al* 1998). These cells, therefore, are believed to be an unlimited source of cells for clinical applications for regenerative medicine and to study early mammalian development. ES cells maintain pluripotency *via* complex interactions of extrinsic signaling by leukemia inhibitory factor (LIF) that activates the Jak/Stat3 signaling pathway (Matsuda *et al* 1999; Niwa *et al* 1998; Raz *et al* 1999), the bone morphogenic protein (BMP) pathway (Ogawa *et al* 2007; Ying *et al* 2003), the mitogen-activated protein kinase/extracellular signal-regulated kinase (MAPK/ERK) pathway (Burdon *et al* 2002), the Wnt/β-catenin signaling pathway (Hao *et al* 2006; Ogawa *et al* 2006) and intrinsic transcription factors such as Nanog, Oct4 and Sox2 (Boiani and Scholer 2005).

ES cells from mouse and human embryos differentiate spontaneously and lose pluripotency in prolonged cultures, therefore these cells require highly specialized culture conditions. To prevent spontaneous differentiation, ES cells are co-cultured with mitotically inactive mouse embryonic fibroblasts (MEFs) that support their self-renewal (Robertson 1997). In addition, the medium must be supplemented with LIF in the case of mouse ES cells and basic fibroblast growth factor (bFGF) for human ES cells (Robertson 1987; Thomson *et al* 1998). Recently, however, retinol has been found to maintain the self-renewal of ES cells by activating the endogenous machinery of the cell by overexpressing the key transcription factors Nanog and Oct4 which maintain their pluripotency.

26.2   Regulation of Pluripotency of ES Cells by Retinol

The past two decades of research has shown that retinol is associated with the cell differentiation *via* its metabolite retinoic acid (Gudas 1994). In ES cells, removal of LIF increases the metabolism of retinol into 4-hydroxyretinol and 4-oxoretinol with concomitant differentiation of ES cells without forming retinoic acid (Lane *et al* 1999). New studies, however, have provided evidence that retinol can maintain the pluripotency of ES cells (Chen *et al* 2007; Chen and Khillan 2008).

Mouse ES cells are generally cultured on the MEF feeders in a medium supplemented with LIF to maintain their pluripotency. ES cells stain positive for alkaline phosphatase which is marker for undifferentiated cells. The culture of ES cells for 12 days showed that the untreated cells do not stain for alkaline phosphatase, indicating the differentiation, whereas ES cells treated with 0.125–0.25 μM retinol exhibit strong staining of ES colonies for alkaline phosphatase, indicating that retinol prevents the differentiation of ES cells (Figure 26.1, right-hand panel).

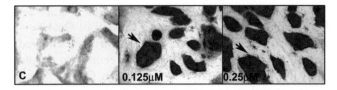

Figure 26.1 Effect of retinol on the pluripotency of ES cells. Mouse ES cells were cultured over MEF feeders for 12 days in the presence of different concentrations of retinol. The cells were then analyzed by staining for AP following the procedures described by the manufacturers. The untreated cells (left-hand panel) exhibited flat cells that were negative for AP staining, whereas the cells treated with 0.125–0.25 μM retinol maintained well-formed colonies that showed strong AP staining, indicating that retinol prevents the differentiation of ES cells.

26.3 Retinol Up-regulates the Expression of Nanog and Oct4

The transcription factors Nanog, Oct4 and Sox2 play critical roles in the maintenance of pluripotency in both human and mouse ES cells (Boiani and Scholer 2005). Nanog specifically is required for the maintenance and self-renewal of ES cells and its overexpression suppresses the differentiation in mouse (Chambers *et al* 2003; Mitsui *et al* 2003) and in human ES cells (Darr *et al* 2006). On the other hand, precise levels of Oct4 must be sustained for the maintenance of pluripotency (Niwa *et al* 2000).

Western blot analysis of ES cells treated with 0.25 μM retinol for 5 days for Nanog and Oct4 showed 3–5-fold overexpression of Nanog (Figure 26.2,

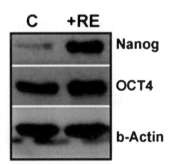

Figure 26.2 Western blot analysis of ES cells treated with retinol. ES cells were cultured for 5 days in ES medium without (C) and with 0.25 μM retinol (+RE). The cells were harvested and total protein was analyzed by Western blot analysis using antibodies specific for Nanog and Oct4. Retinol treatment elevated the expression of Nanog by approx. 5-fold (right-hand lane) as compared with untreated cells (left-hand lane). Retinol also caused the increase in expression of Oct4. β-Actin (b-Actin) was used as a control for the amount of protein loaded for the Western blot.

right-hand lane) compared to normal untreated cells (Figure 26.2, left-hand lane). Similarly, retinol also increases the expression of Oct4 by approx. 3–4-fold (Figure 26.2, right-hand lane) after 5 days of treatment, indicating that prevention of differentiation of ES cells by retinol is mediated by the overexpression of the key regulatory transcription factors. The cells treated with retinoic acid, which is responsible for cell differentiation, did not show any expression of Nanog and Oct4 (data not shown).

Further analysis has revealed that retinol-treated ES cells do not express Nestin, Brachyury and GATA4 (data not shown), which are the markers for differentiated cells representing ectoderm, mesoderm and endoderm, respectively, further supporting the pluripotency of retinol-treated ES cells.

26.4   Self-renewal of ES Cells by Retinol is Independent of Retinoic acid

Retinol circulating in the blood plasma binds with a 21 kDa retinol-binding protein (RBP) and thyroxine-binding protein transthyretin (TTR) to form a complex for delivery to the cytoplasm of the target cell *via* the cell surface receptor Stra6 (Kawaguchi *et al* 2007). In the cytoplasm, retinol binds to the 15 kDa cellular retinol-binding protein (CRBP) and is converted into retinaldehyde by retinol dehydrogenase 1 [alcohol dehydrogenase (Adh1), Class I] and Adh7 (Class IV Adh, previously called Adh4). Retinaldehyde is then irreversibly oxidized to biologically active all-*trans* retinoic acid by retinaldehyde dehydrogenase (RALDH2) which binds to specific cellular retinoic acid-binding proteins (CRABPs) and is shuttled to the nucleus. Retinoic acid then binds to the heterodimer receptors, retinoic acid receptor (RAR) and retinoid X receptor (RXR), in the nucleus. This complex then binds to the promoter elements of retinoic acid-responsive genes (RAREs) to activate the expression of several genes to induce differentiation (Chambon 1996; Balmer and Blomhoff 2002; Lefebvre *et al* 2005).

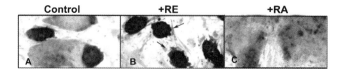

Figure 26.3   Effect of retinoic acid on ES cells. ES cells were treated separately with retinol (RE; 0.25 µM) and retinoic acid (RA; 1.0 µM). After 4 days of culture, the cells were stained for AP. The cells treated with retinoic acid (right-hand panel) showed cells that were flat with a complete absence of staining for AP, whereas the cells treated with retinol (middle panel) maintained the typical morphology of undifferentiated colonies that showed strong positive staining for AP. The untreated cells (left-hand panel) also exhibited flat cells with negative staining for AP. Some undifferentiated colonies were also observed in the control cultures, which also differentiated after prolonged culture.

To investigate if the retinol-mediated self-renewal of ES cells is dependent upon retinoic acid, the treatment of ES cells with 0.25 μM retinol and 1.0 μM retinoic acid separately showed strongly positive alkaline phosphatase (AP)-stained colonies in retinol-treated cells (Figure 26.3, middle panel, +RE). The cells treated with retinoic acid, on the other hand, exhibit complete flat morphology and fail to show any positive staining for AP (Figure 26.3, right panel, +RA). The addition of retinol to the cells treated with retinoic acid did not prevent the differentiation, suggesting that retinol may not be converted into retinoic acid in the ES cells.

26.5   ES Cells Lack Enzymes that Metabolize Retinol into Retinoic Acid

As described above, retinol is metabolized into retinoic acid by Adh1, Adh4 and RALDH2 enzymes. Reverse transcription (RT) PCR analysis of ES cells treated with retinol for Adh1, Adh4, RALDH2, STRA6 and CRBP revealed the absence of these enzymes and proteins, indicating the impairment of retinol metabolism in stem cells.

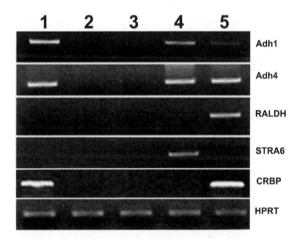

Figure 26.4   The absence of retinol metabolizing enzymes in ES cells. ES cells treated with retinol and retinoic acid, were harvested to isolate total RNA. The RNA was subjected to RT-PCR analysis using specific primers. Liver cells and mouse fibroblast cells were used as controls. Hypoxanthine-guanine phosphoribosyl transferase (HPRT) was used as control for the PCR assay. Lane 1, liver cells; lane 2, ES cells; lLane 3, ES cells treated with retinol; lane 4, ES cells treated with retinoic acid; lane 5 mouse fibroblasts. Normal and retinol-treated ES cells did not express the retinol metabolizing enzymes Adh1, Adh4 and RALDH, as well as STRA6 and CRBP (lanes 2 and 3), whereas retinoic acid treated cells showed expression of Adh1, Adh4 and STRA6 (lane 4). The absence of these enzymes and receptors indicate that ES cells cannot metabolize retinol into retinoic acid.

Normal and retinol-treated ES cells do not express Adh1, ADH4 and RALDH2 enzymes (Figure 26.4), indicating that these cells do not have the capacity to metabolize retinol into retinoic acid (Chen and Khillan 2010). In addition, ES cells do not express the RBP receptor STRA6, which is required for the transport of retinol into cytoplasm. Apart from the absence of retinol metabolizing enzymes, ES cells do not express CRBP as well, indicating the absence of the retinol metabolizing machinery compared with non-stem cells such as adult liver cells and mouse fibroblasts cells (Figure 26.4).

Collectively, these studies demonstrate: (1) retinol cannot be transported to the cytoplasm and (2) the retinol cannot be metabolized to retinoic acid. The possibility of production of trace amounts of retinoic acid, however, cannot be ruled out.

The lack of capacity to metabolize retinol into retinoic acid was further proven by adding retinoic acid into the medium, which resulted in the complete differentiation of ES cells as observed by loss of AP staining (Figure 26.3, right-hand panel). The addition of retinol and retinoic acid together, on the other hand, cannot prevent differentiation, proving that the limiting step in this case is the conversion of retinol into retinoic acid. The cells treated with retinoic acid, on the other hand, express Adh1, Adh4 and STRA6 (Figure 26.4, lane 4), indicating the differentiation.

26.6   Mechanism of Retinol Function in Self-renewal of ES Cells

26.6.1   Retinol Function is Mediated *via* Activation of the PI3K/Akt Signaling Pathway

The phosphoinositide 3-kinase (PI3 kinase) signaling pathway is a key pathway for cell proliferation (Cantley 2002). The cellular response to growth factors generates the second messenger phosphoinositide-3,4,5-trisphosphate (PIP3) by the activation of PI3 kinase. A serine/threonine kinase, Akt/protein kinase B (PKB), which is a downstream effector of PI3 kinase, then binds to PIP3 and is translocated to the inner membrane. Akt/PKB is activated by phosphorylation by a serine/threonine kinase 3-phosphoinisitide-dependent kinase 1 (PDK1) *via* phosphorylation at Thr-308 (Cantley 2002). Akt/PKB is also phosphorylated at Ser-473 by the mammalian target of rapamycin complex 2 (mTORC2) (Huang and Manning 2009), which is required for its full activation, and results in the stimulation of translational machinery of the cell *via* phosphorylation of a key translational regulator p70 ribosomal protein S6 kinase (p70S6K) (Fingar and Blenis 2004).

The treatment of ES cells with LY294002, a reversible inhibitor of PI3 kinase, prevents proliferation of ES cells in the presence of retinol. As shown in Figure 26.5, treatment with LY294002 alone causes complete differentiation of ES cells (panel LY), whereas retinol-treated cells formed AP-positive colonies (panel RE). On the other hand, the cells treated with RE and LY294002 also

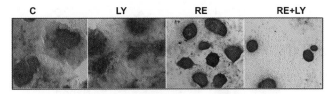

Figure 26.5 Effect of inhibitors on the retinol-mediated self-renewal of ES cells. ES cells were treated with 10 μM LY294002, a reversible inhibitor of PI3 kinase, for 4 days in the presence and absence of 0.25 μM retinol followed by staining for AP. The LY2842002 caused complete differentiation of ES cells (panel LY) similar to control cells (panel C). The cells treated with retinol showed strong positive staining of ES colonies (panel RE). There were very few colonies in the cultures treated with retinol and LY294002, suggesting that LY294002 prevented the proliferation of retinol-treated ES cells (panel RE+LY).

form AP-positive colonies, but their number is dramatically reduced (Figure 26.5, panel RE+LY), indicating that retinol signaling is mediated *via* PI3 kinase.

To prove it further, ES cells were analyzed for the phosphorylated forms of Akt/PKB. Figure 26.6 shows that retinol caused an almost 3-fold elevation in Nanog within 6 h. A dramatic increase in the phosphorylation of Akt at Thr-308 and Ser-473 is observed, proving that the retinol up-regulates Nanog by activating the PI3 kinase/Akt signaling pathway. Figure 26.7 summarizes the involvement of PI3 kinase signaling in the self-renewal of ES cells. The inhibitors such as LY294002 (the inhibitor of PI3 kinase) SH-6 (the inhibitor of Akt) and rapamycin [the inhibitor of the mammalian target of rapamycin (mTOR)] prevented the elevation of Nanog by retinol (Figure 26.7, compare +RE lane with the other lanes). The level of Nanog in the cells treated with the inhibitors is lower than the control cells (Figure 26.7, control lane).

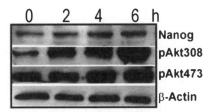

Figure 26.6 Activation of the PI3 K/Akt signaling pathway by retinol. ES cells treated with 0.25 uM retinol were harvested after 2, 4 and 6 h followed by Western blot analysis using antibodies against Nanog and phosphorylated Akt/PKB. Retinol enhanced the expression of Nanog by approx. 3-fold within 6 h. The increase in Nanog was also accompanied by a dramatic increase in the phosphorylation of Akt at Thr-308 (pAkt308) and Ser-473 (pAkt473), proving that retinol regulates the Nanog expression *via* the PI3 kinase signaling pathway.

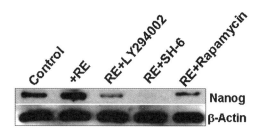

Figure 26.7   Effect of inhibitors of the PI3 kinase signaling pathway on retinol-treated ES cells. ES cells were cultured with 10 μM LY294002, an inhibitor of PI3 K; 10 μM SH-6, an inhibitor of Akt; and 10 μM rapamycin, an inhibitor of mTOR, for 24 h in the presence of 0.25 μM retinol. The total protein isolated from the cell was analyzed by Western blot analysis using antibodies against Nanog. The retinol alone caused a 3–5-fold increase in Nanog, whereas all the inhibitors prevented the up-regulation of Nanog.

26.6.2   Retinol Activates PI3 kinase Signaling Pathway *via* IGF1 Receptor

Growth factors and cytokines activate PI3 kinase *via* many different receptor tyrosine kinases (Cantley 2002). ES cells express several receptor tyrosine kinases, including the insulin-like growth factor 1 (IGF1) receptor and the epidermal growth factor receptor (EGFR). The cells treated with AG1478, a specific inhibitor of EGFR do not show any effect on the level of Nanog (data not shown), suggesting that EGFR is not be involved in the retinol-mediated signaling. Treatment of cells with picropodohyllin (PPP), a specific inhibitor of the IGF1 receptor, blocked the up-regulation of Nanog (Figure 26.8), indicating that retinol engages the IGF1 receptor to activate the PI3 kinase signaling pathway, which is further confirmed by the depletion of phosphorylation of Akt/PKB at Thr-308 and Ser-473.

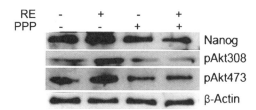

Figure 26.8   The effect of inhibitors of IGF1 receptor on the retinol-mediated up-regulation of Nanog. ES cells were treated with 0.25 μM retinol (RE) in the presence of 10 μM PPP for 24 h followed by Western blot analysis of the total protein levels. As expected, retinol increased the expression of Nanog and phosphorylation of Akt at Thr-308 (pAkt308) and Ser-473 (pAkt473). The addition of PPP, on the other hand, abolished the increase in Nanog and Aktô phosphorylation.

Summary Points

- This Chapter provides evidence on a new function of retinol in stem cell biology.
- Retinol, the alcohol form of vitamin A, plays important role in cell differentiation *via* its potent metabolite retinoic acid. Because of this property, retinol and its derivatives have been used as chemopreventive for many cancers.
- In ES cells, and perhaps in all stem cells, retinol has a direct role in their self-renewal.
- The function of retinol in the self-renewal of ES cells is independent of retinoic acid as these cells do not contain enzymes that metabolize retinol into retinoic acid.
- Retinol maintains the self-renewal of ES cells by elevating the expression of key transcription factors Nanog and Oct4.
- Nanog and Oct4 are necessary for the maintenance of pluripotency of ES cells.
- The elevated expression of these transcription factors is mediated by activating the PI3 kinase signaling pathway, a key signaling pathway in cell proliferation.
- ES cells maintain pluripotency and retain the potential to form all types of cells of ectoderm, mesoderm and endoderm lineages.

Key Facts

Key Facts of Retinol/Vitamin A

- Retinol, the alcohol form of vitamin A, is an important dietary component that plays critical role in the development, cell differentiation, reproduction and in the immune system *via* retinoic acid.
- Retinol is transported into the cytoplasm of the target cell by specific cell surface receptors.
- The target cells express specific enzymes that convert retinol into retinoic acid
- Retinoic acid then executes its function by activating over 300 genes that cause cell differentiation.
- Removal of LIF increases the metabolism of retinol into 4-hydroxy retinol and 4-oxoretinol in ES cells without conversion into retinoic acid that results in the differentiation of ES cells.
- The direct function of retinol is not well understood.

Key Facts of Embryonic Stem (ES) Cells

- ES cells represent pluripotent stem cells that are established from the inner cell mass of the blastocyst embryo.

- ES cells have complete potential to form all types of cells in the body and can contribute to the generation of all the organs.
- ES cells are considered to be an unlimited source of cells to cure degenerative diseases of the cardiovascular, pancreatic and neuronal systems.
- These cells can differentiate spontaneously and lose their pluripotency, therefore require highly specialized culture conditions such as growth factors and embryonic fibroblasts as feeder cells.
- ES cells maintain pluripotency by expressing certain key regulatory factors such as Nanog, Oct4 and Sox2.

Key Facts of Retinol Function in ES Cells

- The past two decades of research has shown that retinol plays a critical role in development, reproduction and cell differentiation.
- The information in this Chapter challenges the old dogma and reviews the evidence on the direct retinoic acid-independent function of retinol.
- This Chapter describes how retinol prevents the differentiation of ES cells by elevating the expression of the key pluripotent cell-specific transcription factors Nanog and Oct4.
- Undifferentiated and pluripotent ES cells with the capability for self-renewal can be maintained for prolonged periods while maintaining the complete potential to differentiate into all types of cells.

Definitions of Words and Terms

Alkaline Phosphatase (AP): This enzyme is a specific marker for undifferentiated stem cells.

Blastocyst: A blastocyst represents the pre-implantation embryo which contains two types of cells: the inner cell mass that forms the actual embryo and the trophectoderm that contributes to the formation of placenta and extra-embryonic membranes.

Embryonic stem (ES) cells: ES cells are pluripotent stem cells derived from the inner cell mass of the blastocyst and have unlimited potential for self-renewal and capacity to form all types of cells in the body.

Mouse embryonic fibroblasts (MEFs): These cells are derived from 14–15-day-old mouse embryos. The cells are used as feeder cells for culturing mouse and human ES cells after they have been mitotically inactivated either by irradiation or by mitomycin C.

Nanog: Nanog is a homeodomain-containing transcription factor that maintains the pluripotency of mouse and human ES cells.

Oct4: Oct4 is a Pou-domain containing transcription factor which is required for the maintenance of pluripotency of ES cells.

Pluripotent stem cells: Cells that have ability to differentiate into more than one type of cells.

Retinoic acid: This is the most potent metabolite of retinol that causes cell differentiation by activating multiple genes.

Retinol: This is the alcohol form of vitamin A which is generally acquired from plant and animal sources. Humans cannot synthesize retinol.

List of Abbreviations

Adh	alcohol dehydrogenase
AP	alkaline phosphatase
CRBP	cellular retinol-binding protein
EGFR	epidermal growth factor receptor
ES cell	embryonic stem cell
IGF1	insulin-like growth factor 1
LIF	leukemia inhibitory factor
MEF	mouse embryonic fibroblast
mTOR	mammalian target of rapamycin
PI3 kinase	phosphoinositide 3-kinase
PIP3	phosphoinositide-3,4,5-trisphosphate
PKB	protein kinase B
PPP	picropodohyllin
RALDH2	retinaldehyde dehydrogenase 2
RBP	retinol-binding protein
RT-PCR	reverse transcription PCR

References

Balmer, J. E. and Blomhoff, R., 2002. Gene expression regulation by retinoic acid. Journal of Lipid Research. 43: 1773–1808.

Blaner, W. S. and Olson, J. A., 1994. Retinol and retinoic acid metabolism. In Sporn, M.B., Roberts, A.B. and Goodman, D.S. (ed.) Retinoids: Biology, Chemistry and Medicine. Raven Press, New York, NY, USA, pp. 229–255.

Boiani, M. and Scholer, H. R., 2005. Regulatory networks in embryo-derived pluripotent stem cells. Nature Reviews in Molecular and Cellular Biology. 6: 872–881.

Burdon, T., Smith, A. and Savatier, P., 2002. Signaling, cell cycle and pluripotency in embryonic stem cells. Trends in Cell Biology. 12: 432–438.

Cantley, L. C. The phosphoinositide 3-kinase pathway, 2002. Science. 31; 296: 1655–1657.

Chambers, I., Colby, D., Robertson M., Nichols, J., Lee, S. S., Tweedle, S. and Smith, A., 2003. Functional expression cloning of Nanog, a pluripotency sustaining factor in embryonic stem cells. Cell. 113: 643–655.

Chambon, P., 1996. A decade of molecular biology of retinoic acid receptors. FASEB Journal. 10: 940–954.

Chen, L. and Khillan, J. S., 2008. Promotion of feeder independent self-renewal of embryonic stem cells by retinol (vitamin A). Stem Cells. 26: 1858–1864.

Chen, L., and Khillan, J. S., 2010. A novel signaling by vitamin A/retinol promotes self renewal of mouse embryonic stem cells by activating PI3K/Akt signaling pathway via insulin-like growth factor-1 receptor. Stem Cells. 28: 57–63.

Chen, L., Yang, M., Dawes, J. and Khillan, J. S., 2007. Suppression of ES cell differentiation by retinol (vitamin A) via the over expression of Nanog. Differentiation. 75: 682–693.

Darr, H., Mayshar, Y. and Benvenisty, N., 2006. Over-expression of NANOG in human ES cells enables feeder-free growth while inducing primitive ectoderm features. Development. 133: 1193–1201.

Evans, M. J. and Kaufman, M. H., 1981. Establishment in culture of pluripotential cells from mouse embryos. Nature. 292: 154–156.

Fingar, D. C. and Blenis, J. T., 2004. Target of rapamycin (TOR): an integrator of nutrient and growth factor signals and coordinator of cell growth and cell cycle progression. Oncogene. 23: 3151–3171.

Gudas, L. J., 1994. Retinoids and vertebrate development. Journal of Biological Chemistry. 269: 15399–15402.

Hao, J., Li, T. G., Qi, X., Zhao, D. F. and Zhao, G. Q., 2006. WNT/β-catenin pathway up-regulates Stat3 and converges on LIF to prevent differentiation of mouse embryonic stem cells. Developmental Biology. 290: 81–91.

Harrison, E. H., 2005. Mechanisms of digestion and absorption of dietary vitamin A. Annual Reviews in Nutrition. 25: 87–103.

Huang, J. and Manning, B. D., 2009. A complex interplay between Akt, TSC2 and the two mTOR complexes. Biochemical Society Transactions. 37: 217–222.

Kawaguchi, R., Yu, J., Honda, J., Hu, J., Whitelegge, J., Ping, P., Wiita, P., Bok, D. and Sun, H., 2007. Membrane receptor for retinol binding protein mediates cellular uptake of vitamin A. Science. 315: 820–825.

Lefebvre, P., Martin, P. J., Flajollet, S., Dedieu, S., Billaut, X. and Lefebvre, B., 2005. Transcriptional activities of retinoic acid receptors. Vitamins and Hormones. 70: 199–264.

Lane M. A., Chen A. C., Roman, S. D, Derguini, F. and Gudas, L. J., 1999. Removal of LIF (leukemia inhibitory factor) results in increased vitamin A (retinol) metabolism to 4-oxoretinol in embryonic stem cells. Proceedings of the National Academy of Sciences of the United States of America. 96: 13524–13529.

Martin, G., 1981. Isolation of a pluripotent cell line from early mouse embryos cultured in medium conditioned by teratocarcinoma stem cells. Proceedings of the National Academy of Sciences of the United States of America. 78: 7634–7638.

Matsuda, T., Nakamura, T., Nakao, K. Arai, T., Katsuki, M., Heike, T. and Yokota, T., 1999. STAT3 activation is sufficient to maintain an undifferentiated state of mouse embryonic stem cells. EMBO Journal. 18: 4261–4269.

Mitsui, K., Tokuzawa, Y., Itoh, H., Segawa, K., Murakami, M., Takahashi, K., Maruyama, M., Maeda, M. and Yamanaka, S., 2003. The homeoprotein Nanog is required for maintenance of pluripotency in mouse epiblast and ES cells. Cell. 113: 631–642.

Niwa, H., Burdon, T., Chambers, I. and Smith, A., 1998. Self renewal of pluripotent ES cells is mediated via activation of STAT3. Genes and Development. 12: 2048–2060.

Ogawa, K., Nishinakamura, R., Iwamats, Y., Shimosato, D. and Niwa, H., 2006. Synergistic action of Wnt and LIF in maintaining pluripotency of mouse ES cells. Biochemical and Biophysical Research Communications. 343: 159–166.

Ogawa, K., Saito, A., Matsui, H., Suzuki, H., Ohtsuka, S., Shimosato, D., Morishita, Y., Watabe, T., Niwa, H., and Miyazono, K., 2007. Activin-Nodal signaling is involved in propagation of mouse embryonic stem cells. Journal of Cell Science. 120: 55–65.

Raz, R., Lee, C. K., Cannizzaro, L. A., d'Eustachio, P. and Levy, D. E., 1999. Essential role of STAT3 for embryonic stem cell pluripotency. Proceedings of the National Academy of Sciences of the United States of America. 96: 2846–2851.

Robertson, E. J., 1987. Embryo derived cell lines. In Teratocarcinoma and Embryonic Stem Cells: A Practical Approach. IRL Press, Oxford, UK, pp. 71–112.

Thomson, J. A., Itskovitz-Eldor, J., Shapiro S. S., Waknitz M. A., Swiergiel J. J., Marshall V. S. and Jones, J. M., 1998. Embryonic stem cell lines derived from human blastocysts. Science. 282: 1145–1147.

Ying, Q. L., Nichols, J., Chambers I. and Smith, A., 2003. BMP induction of Id proteins suppresses differentiation and sustains embryonic stem cell self-renewal in collaboration with STAT3. Cell. 115: 281–292.

CHAPTER 27
Retinoic Acids and their Biological Functions

JOSEPH L. NAPOLI

Metabolic Biology Graduate Group, Nutritional Sciences and Toxicology, University of California, 119 Morgan Hall, MC#3104, Berkeley, CA 94720, USA
E-mail: jna@berkeley.edu

27.1 Introduction

At least four geometric isomers of retinoic acid (RA) occur in tissues: all-*trans*-RA (atRA); 9-*cis*-RA (9cRA); 13-*cis*-RA (13cRA); 9,13-di-*cis*-RA (9,13dcRA) (Kane *et al* 2005). atRA and 9cRA both activate the nuclear receptors, RA receptors α, β and γ (RARα, -β and -γ). atRA also activates the peroxisome proliferator-activated receptor β/δ (PPARβ/δ), whereas 9cRA also activates *in vitro* the retinoid X receptors α, β and γ (RXRα, -β and -γ) (Noy 2007). In contrast, 13cRA and 9,13dcRA, do not bind nuclear receptors with high affinity, and, unlike atRA and 9cRA, do not seem to have direct biological activity, but may act as precursors to atRA or 9cRA. In addition to functioning *via* nuclear receptors, increasing evidence suggests that atRA and 9cRA also function via non-genomic mechanisms (Chen *et al* 2008; Sidell *et al* 2010). Much has been written about the biological functions of atRA; this Chapter will focus on 9cRA.

Food and Nutritional Components in Focus No. 1
Vitamin A and Carotenoids: Chemistry, Analysis, Function and Effects
Edited by Victor R Preedy

27.2 Bioanalytical Analysis of RA Isomers

The four geometric isomers of RA are difficult to resolve, even by high-resolution techniques, such as high-performance liquid chromatography (HPLC). Since the introduction of HPLC into retinoid research, many systems allow the co-migration of various isomers. 9cRA, for example, can co-migrate with atRA or 9,13-dcRA, and 9,13-dcRA can co-migrate with 13cRA or 9cRA, depending on conditions. This can engender costly misinterpretations, because each isomer differs in biological actions. Fortunately, HPLC systems have been devised that resolve these isomers, and have been coupled with

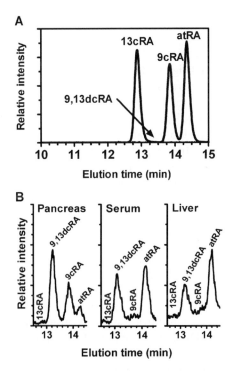

Figure 27.1 Resolution of geometric RA isomers by HPLC. (A) In the example shown, a gradient of 0.1% formic acid in water to 0.1% formic acid in acetonitrile was used with a reverse phase amide column to achieve baseline resolution of RA isomers. Detection and quantification was by MS/MS. MS/MS currently provides the most sensitive method of RA detection. Typically, quantitative MS/MS relies on a triple quadrupole (Q) instrument. Q1 selects the parent ion. Q2 serves as a locus of collisional fragmentation of the parent ion. Q3 selects a product ion representative of the analyte for quantification. MS/MS increases sensitivity by decreasing background as much as 100–1000-fold over a single mass analyzer. LC contributes to specificity and MS/MS improves specificity by assessing both the parent ion and a characteristic product ion. (B) LC-MS/MS chromatograms of RA isomers in mouse pancreas, serum, and liver.

sensitive detectors, mostly based on tandem mass spectrometry (MS/MS) (Kane and Napoli 2010). These assays allow quantification in a variety of biological matrices of as little as 0.05 pmol g^{-1} tissue. Figure 27.1(A) shows an example of resolution by HPLC with detection by MS/MS.

27.3   9cRA as an Endogenous Pancreas Autacoid

Through high affinity (k_d = low nM) binding to both RARs and RXRs, 9cRA has a spectrum of biological actions. Yet, the occurrence of 9cRA as an endogenous retinoid has been questioned. The following excerpts from reviews frame the issue: "A major premise for [the] biological role [of 9cRA] has ... largely been neglected by the research field: unequivocal identification of 9-*cis*-retinoic acid as an endogenous compound has not yet been possible" (Blomhoff and Blomhoff 2006); "...multiple questions about...the existence of an endogenous [RXR] ligand have still to be answered" (Germain *et al* 2006); "The finding of permissive heterodimers raises the question as to whether 9-*cis* RA could actually be a physiological ligand for RXRs" (Mark *et al* 2006); "...in later work 9-*cis*-retinoic acid could not be detected...it would not be unreasonable to propose that 9-*cis*-retinoic acid is not the endogenous ligand for RXR" (Wolf 2006). The data in Table 27.1 illustrate the reason for this concern (Kane *et al* 2008). Although atRA has been quantified in multiple tissues, sensitive assays have not detected 9cRA in the same tissues. In addition, genetic evidence has shown that 9cRA does not serve as a universal RXR activator (Calleja *et al* 2006). The widespread occurrence of 9,13dcRA, however, poses intriguing questions: does 9,13dcRA originate from 9cRA or 13cRA (or both)?; and if from the former, is formation of 9,13dcRA a mechanism to prevent 9cRA from activating RXR, or does 9,13dcRA serve as a reservoir to produce 9cRA "on demand"? These questions have not been

Table 27.1   RA isomers concentrations in mouse serum and tissues (Kane *et al* 2008). Data are from 2- to 4-month-old male SV129 mice fed a diet with 4 IU vitamin A g^{-1}. Values are means $\pm$ S.E.M. (number of mice) expressed as pmol (g of tissue)$^{-1}$ or (ml of serum)$^{-1}$ and were rounded to the nearest whole number. ND, not detected (<0.05 pmol g^{-1}).

Locus	atRA	9cRA	9,13dcRA
Serum	3 ± 0.3 (21)	ND	2 ± 0.3 (15)
Liver	38 ± 3 (18)	ND	23 ± 1 (5)
Kidney	15 ± 2 (30)	ND	12 ± 2 (20)
Adipose (epididymal)	14 ± 2 (18)	ND	5 ± 0.6 (29)
Muscle	2 ± 0.2 (15)	ND	ND
Spleen	7 ± 0.6 (14)	ND	3 ± 0.4 (9)
Testis	9 ± 1 (14)	ND	4 ± 0.4 (5)
Brain	17 ± 4 (19)	ND	18 ± 3 (21)

answered, but recent work has identified 9cRA as an endogenous pancreas autacoid.

A tissue scan relying on liquid chromatography (LC)/MS/MS to establish reference values for atRA, and to set the stage for investigating the contribution of atRA to pancreas function, detected 9cRA as an endogenous retinoid in pancreas in concentrations similar to those of atRA and 9,13-dcRA (Figure 27.1B). Consistent with previous work, 9cRA was not detected in serum or liver (Kane *et al* 2010).

27.4 Pancreas 9cRA Varies Inversely with Blood Glucose

Because 9cRA occurred only in pancreas (of the tissues screened), it made sense to determine the effects of fasting *vs.* feeding on its concentrations. Fed mice had 9cRA levels 36% lower than fasted mice, but showed no differences in atRA or 9,13dcRA (Figure 27.2). A direct test of the effect of glucose on pancreas 9cRA, *i.e.* an intraperitoneal glucose tolerance test (GTT), revealed that glucose itself was sufficient to cause the decrease (Figure 27.3A). Within 15 min of dosing, a bolus of glucose caused a >80% decrease in 9cRA, but had no effect on atRA. Pancreas 9cRA related inversely to serum insulin during the GTT (Figure 27.3B).

27.5 Pancreas Islet β-cells Biosynthesize 9cRA

The impact of blood glucose on pancreas 9cRA and the relationship between 9cRA and serum insulin suggested that pancreas islet β-cells might serve as a source of 9cRA. Retinoids are very susceptible to degradation by light, nucleotides and non-neutral pH conditions. In addition, islets comprise only 1–2% of the pancreas, and RA concentrations are in the nM range. Sensitive and

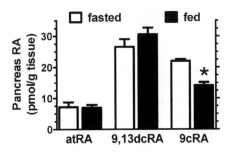

Figure 27.2 Effects of energy status on pancreas RA isomers. The increase in blood glucose after transition from the fasted to the fed state led to a 36% decrease in 9cRA, but no change in atRA or 9,13-dcRA. Means ± S.E.M. of 3 experiments, each with 7–10 mice per group: *$P < 0.003$ relative to fasted value.

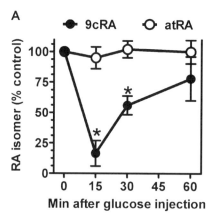

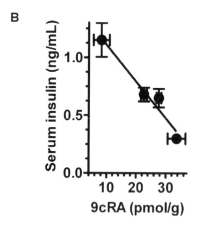

Figure 27.3   Pancreas RA responses to a glucose challenge. (A) Challenging fasted
mice with glucose [intraperitoneal (i.p.), 2 g kg^{-1}] decreased 9cRA
>80% within 15 min, coinciding with the rise in blood glucose (data not
shown), but had no effect on atRA. (B) During the glucose challenge,
9cRA related inversely to serum insulin. Means ± S.E.M. of 2–5
experiments, each with 5–10 mice per group: *P < 0.05 relative to 0 min
value.

isotope-specific antibodies for RA are not available. Measurements would
need to be made by LC-MS to determine whether 9cRA occurs in the
endocrine pancreas (islets). In view of these limitations, to query the
contribution of β-cells to the pancreas 9cRA pool, mice were injected with
streptozotocin, an agent that causes β-cells to undergo apoptosis. Both β-cells,
evaluated by insulin content of islets, and 9cRA decreased linearly with time
after dose (Figure 27.4A). However, the slopes were not equivalent. Islet β-
cells decreased by ~95% and 9cRA deceased by ~70% at 96 h after the dose,
indicating that β-cells contribute ~75–80% of the 9cRA pancreas pool. To
confirm this insight, 9cRA content was quantified in animal models of

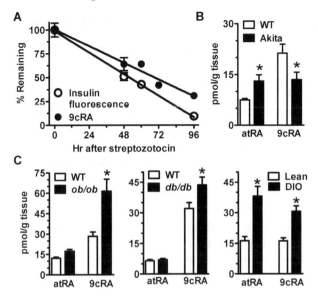

Figure 27.4   Islet β-cell sizes and numbers affect pancreas 9cRA concentrations. (A) Effect of streptozotocin on β-cell numbers and pancreas 9cRA with time after dose. (B) Decreased 9cRA *vs.* increased atRA in pancreas of Akita mice: $*P < 0.05$ *vs.* WT (8 mice per group). (C) RA isomers in pancreas of mice with β-cell hypertrophy: *ob/ob* mice ($*P < 0.008$ *vs.* WT; 8 mice per group); *db/db* mice ($*P = 0.033$ *vs.* WT; 8 mice per group); mice with DIO ($*P < 0.001$ *vs.* lean; 8–10 mice per group). Data represent means ± S.E.M.

increased or decreased numbers or sizes of β-cells. The Ins2Akita mouse (Akita) has a point mutation in the insulin 2 gene, such that the heterozygote has approximately half the number of β-cells as wild-type (WT), reflected by a 46% reduction in pancreas insulin content (Yoshioka *et al* 1997). Pancreata from fed heterozygous Akita mice had 40% lower 9cRA than WT, indicating β-cells as a major source of 9cRA and consistent with the results of the streptozotocin experiment (Figure 27.4B). Interestingly, atRA increased in the Akita mouse. Mice which lack leptin (*ob/ob*) undergo β-cell hypertrophy that enlarges β-cells by ~2-fold. *ob/ob* mice had ~2-fold more 9cRA than WT (Figure 27.4C). Mice which lack the leptin receptor (*db/db*), also experience β-cell hypertrophy, which abates with time. Three-month-old *db/db* mice showed a significant increase in 9cRA, consistent with the somewhat enlarged β-cells at this age. Finally, feeding a high-fat diet to instill diet-induced obesity (DIO) also enlarges β-cells and resulted in increased 9cRA. In this case, atRA increased as well. These data indicate that the pancreas content of 9cRA, but not atRA, depends on the numbers and sizes of β-cells. atRA content seems governed by influences other than by only β-cell numbers and sizes.

27.6   Increased 9cRA in the *Rbp1*-null Mouse

Rbp1 encodes cellular retinol-binding protein, type 1 (Crbp1), a member of the fatty acid binding-protein gene family with widespread tissue expression, which sequesters retinol (vitamin A) with high affinity (k_d values in the low nM range) (Newcomer *et al* 1998). Crbp1 serves as a chaperone that facilitates cellular retinol uptake, modulates retinol storage as retinyl esters, and affects retinoid homeostasis by regulating enzyme activity and retinol access to enzymes that biosynthesize RA (Napoli 2011).

    Studies with the $Rbp1^{-/-}$ mouse revealed that the amount of dietary retinol and the presence of Crbp1 affect glucose tolerance (Kane *et al* 2011). Feeding mice the amount of retinol in stock lab chow (30 units g^{-1} until the last few years), which is much more than needed (~ 2.6 units g^{-1}), *i.e.* a diet copious in vitamin A (CVA), decreased glucose tolerance in both WT and $Rbp1^{-/-}$ mice, relative to mice fed a vitamin A-deficient diet (VAD) (Figure 27.5). It is important to note that mouse pups bred from dams fed a CVA do not become vitamin A-deficient when fed a VAD when weaned. Therefore, the VAD fed

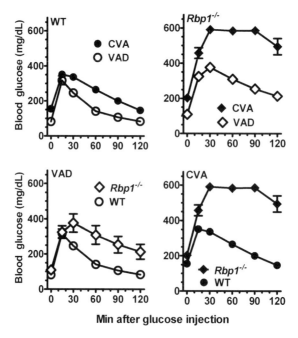

Figure 27.5   Effects of *Rbp1* and dietary vitamin A content on glucose homeostasis. Blood glucose after a 2 g kg^{-1}, i.p., glucose challenge in WT or $Rbp1^{-/-}$ mice fed a diet with CVA or a VAD for 3 months. The CVA had 30 international units (IU) of vitamin A (g of diet)$^{-1}$—the amount in chow diets until recently. A mouse weaned from a dam fed a CVA and then weaned on to a VAD does not become vitamin A-deficient in 3 months. Two-way ANOVA, $P < 0.0001$, $Rbp1^{-/-}$ *vs.* WT mice and CVA *vs.* VAD (7 mice per group).

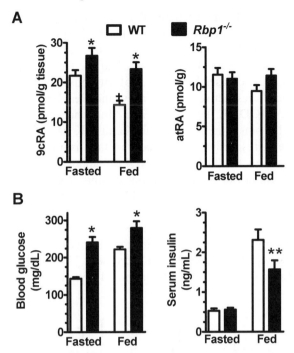

Figure 27.6 *Rbp1* ablation increases pancreas 9cRA and changes glucose homeostasis. (A) Pancreas RA isomers in fed or 12 h-fasted mice. Values are means of 2–3 experiments; *$P < 0.006$ *vs.* WT; $^{+}P < 0.0006$ *vs.* fasted value (8–10 mice per group). Fasted and fed $Rbp1^{-/-}$ 9cRA values did not differ significantly. (B) Blood glucose and serum insulin values of fasted and fed mice. Values are means of 5–6 experiments (5–10 mice/group); $P < 0.007$, **$P < 0.033$ *vs.* WT values.

mice had lowered amounts of retinol stores, but were not impaired in vitamin A status, as shown by normal weight and other processes supported by vitamin A. These data indicate that both the amount of vitamin A in the diet and the expression of *Rbp1* contribute independently to glucose tolerance.

The effect of blood glucose on pancreas 9cRA, and the inverse relationship between 9cRA and serum insulin, prompted assay of 9cRA in the $Rbp1^{-/-}$ mouse. 9cRA concentrations were higher in the $Rbp1^{-/-}$ mouse relative to WT and resisted change after the transition from the fasted to the fed state, whereas atRA remained normal (Figure 27.6A). 9cRA was the only directly active pancreas retinoid that differed from WT in the $Rbp1^{-/-}$ mouse.

27.7 Function of 9cRA in the Pancreas Islet β-Cell

9cRA impairs the activities of Glut2 and glucokinase (GK) (non-genomic mechanism), and decreases expression of *Pdx-1* and *HNF4α* in islets and insulinoma cells (Kane *et al* 2010. In the $Rbp1^{-/-}$ mouse, blood glucose was higher than WT in both the fasted and fed states, and serum insulin in the fed state did not increase to the same extent as in WT (Figure 27.6B). In addition, pancreas islets of $Rbp1^{-/-}$ mice had *Pdx-1*, *Glut2* and *GK* mRNA expression reduced 26–53% relative to WT (Kane *et al* 2011). Defects in *GK* and *Pdx-1* cause two of the six diseases known as maturity onset diabetes of the young (MODY) 2 and 4, respectively (Hattersley and Pearson 2006). A defect in *HNF4α* causes MODY1. Pancreatic and duodenal homeobox 1 (Pdx-1) stimulates insulin biosynthesis, and Glut2 and GK catalyze the uptake and rate-limiting steps, respectively, of glucose metabolism (Muoio and Newgard 2008). Hepatocyte nuclear factor 4α (HNF4α) regulates insulin release through controlling mitochondrial glucose metabolism and *HNF1α*, *Glut2* and insulin gene expression (Bartoov-Shifman *et al* 2002; Wang *et al* 2000).

Insulin and glucagon secretion from the β- and α-cells, respectively, of the endocrine pancreas inter-relate. Insulin secretion attenuates glucagon secretion. The resulting increase in the insulin/glucagon ratio decreases liver gluconeogenesis, which contributes to a decrease in blood glucose (Bansal and Wang 2008; Ramnanan *et al* 2010; Shah *et al* 1999). The inverse relationship

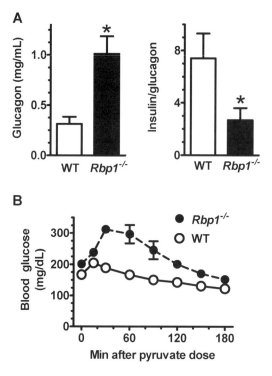

Figure 27.7   Effects of *Rbp1* on glucagon secretion and hepatic gluconeogenesis. (A) Serum glucagon and the insulin/glucagon ratio of fasted and fed mice: *$P<0.003$ *vs.* WT (8 mice per group). (B) Blood glucose after a pyruvate challenge with fasted mice: two-way ANOVA; $P < 0.0001$ (5 mice per group).

between insulin and 9cRA, the rapid and intense decrease in 9cRA prompted by glucose, and the relatively high 9cRA in the fed $Rbp1^{-/-}$ mouse suggest that 9cRA would affect glucagon (if only indirectly). In fact, the fed $Rbp1^{-/-}$ mice have 3-fold higher glucagon levels and a 3-fold lower insulin/glucagon ratio than WT (Figure 27.7A). This would contribute to higher blood glucose, partially through increased liver gluconeogenesis. A pyruvate tolerance test (PTT), which assesses conversion of pyruvate into glucose and thereby reveals rates of gluconeogenesis, showed a 50% increase [area under the curve (AUC)] in the $Rbp1^{-/-}$ mouse relative to WT (Figure 27.7B). This confirmed an increase in gluconeogenesis. The fairly rapid decrease in glucose with time after the pyruvate dose in the $Rbp1^{-/-}$ mouse, however, indicates a lack of insulin resistance.

Elevated glucagon also promotes hepatic fatty acid oxidation. The $Rbp1^{-/-}$ mouse relies disproportionately on fatty acids, as opposed to glucose, as fuel relative to WT.

27.8 Conclusions

9cRA effects many biological actions as an activator of six nuclear receptors and possibly through non-genomic actions (Cheng *et al* 2008). As the formulation alitretinoin, 9cRA ameliorates chronic dermatitis and T-cell lymphoma. 9cRA alters energy metabolism, reduces ischemic brain injury in a rat model, and immunosuppresses human dendritic cells (Shen *et al* 2009; Zapata-Gonzalez *et al* 2007). Notably, the biological actions of atRA and 9cRA overlap only partially. For example, 9cRA inhibits, and atRA induces, acini differentiation in embryo day 11 pancreatic organ cultures (Kadison *et al* 2001; Kobayashi *et al* 2002).

Data to date indicate that Crbp1 modulates pancreas 9cRA concentrations and support the hypothesis that 9cRA attenuates glucose-stimulated insulin secretion (GSIS), which decreases the insulin/glucagon ratio. The rapid, temporary and very large decrease in 9cRA caused by glucose and the inverse relationship between 9cRA and insulin suggests that 9cRA attenuates GSIS. A reasonable hypothesis that models these data would have 9cRA reducing GSIS when glucose concentrations are relatively low to prevent hypoglycemia, and decreasing the transcription of genes necessary for insulin biosynthesis as a longer-term modulation (Figure 27.8). An increase in glucose decreases 9cRA and relieves 9cRA inhibition of Glut2 and GK activities (non-genomic actions of 9cRA), allowing a greater increase in GSIS during a relatively short time. The temporary decrease in 9cRA also would relieve transcription suppression of genes that promote insulin biosynthesis, allowing a burst of insulin biosynthesis to replace the pre-formed insulin secreted from granules. These data are consistent with a recent report that 9cRA, functioning *via* RXR, suppresses GSIS (Miyazaki *et al* 2010). Overall, 9cRA seems to prevent hypoglycemia during low blood glucose by modulating insulin secretion and biosynthesis.

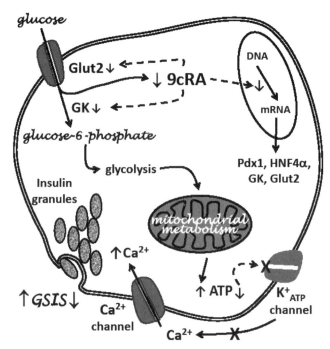

Figure 27.8   9cRA contributes to regulating GSIS. Solid arrows indicate actions of
glucose. Dashed lines and Xs indicate actions of 9cRA. 9cRA decreases
the activities of Glut2 and GK, thereby reducing glucose uptake and
metabolism into glucose-6-phosphate. Mitochondrial metabolism of
glucose-6-phosphate generates ATP, which blocks the K^+_{ATP} channel.
This prompts increased intracellular Ca^{2+} uptake, which stimulates
insulin secretion from insulin containing granules. The decrease in ATP
caused by 9cRA allows the K^+_{ATP} channel to remain open, which
decreases intracellular Ca^{2+} thereby decreasing GSIS. This short-term
action of 9cRA probably has non-genomic mechanisms. 9cRA also
regulates transcription of genes necessary for insulin biosynthesis and
secretion: *Pdx-1*, *HNF4α*, *GK*, and *Glut2*.

Summary Points

- atRA and 9cRA regulate gene transcription by activating nuclear receptors,
  but also have non-transcriptional mechanisms of action.
- 9cRA and atRA have overlapping but different biological actions.
- 9cRA has been detected only in pancreas.
- 9cRA functions as a pancreas autacoid that contributes to regulating GSIS.
- atRA and 9cRA regulate expression of genes crucial for pancreas
  development and insulin secretion.
- 9cRA may prevent hypoglycemia during low blood glucose by modulating
  insulin secretion and biosynthesis.
- Vitamin A status influences glucose disposal.

Key Facts

Key Facts of Glucose-stimulated Insulin Secretion (GSIS)

- GSIS relies on glucose metabolism, and the production of ATP, but products in addition to ATP also seem to stimulate GSIS.
- Glut2 is the main but not sole transporter of glucose entry into pancreas islet β-cells.
- The activity of GK is considered the rate-limiting step in GSIS.
- The anti-diabetic drug glipizide blocks the potassium channel, thereby stimulating insulin release.

Definitions of Words and Terms

Gluconeogenesis: The metabolic path that converts products of amino acid metabolism into glucose, which occurs primarily in the liver. Pyruvate is a major intermediate in gluconeogenesis.

Glucose-stimulated insulin secretion (GSIS): The process whereby glucose uptake and metabolism by the pancreas islet β-cell induces insulin secretion from granules that house pre-formed insulin.

List of Abbreviations

atRA	all-*trans*-retinoic acid
9cRA	9-*cis*-retinoic acid
13cRA	13-*cis*-retinoic acid
Crbp1	cellular retinol-binding protein, type 1
CVA	diet copious in vitamin A
9	13dcRA, 9,13-di-*cis*-retinoic acid
DIO	diet-induced obesity
GK	glucokinase
GSIS	glucose-stimulated insulin secretion
GTT	glucose tolerance test
HNF4α	hepatocyte nuclear factor 4α
HPLC	high-performance liquid chromatography
i.p.	intraperitoneal
IU	international units
LC	liquid chromatography
MODY	maturity onset diabetes of the young
MS	mass spectrometry
MS/MS	tandem mass spectrometry
Pdx-1	pancreatic and duodenal homeobox 1
Q	quadrupole
RA	retinoic acid

RAR   retinoic acid receptor
RXR   retinoid X receptor
VAD   vitamin A-deficient diet

References

Bansal, P. and Wang, Q., 2008. Insulin as a physiological modulator of glucagon secretion. American Journal of Physiology, Endocrinology and Metabolism. 295: E751–E761.

Bartoov-Shifman, R., Hertz, R., Wang, H., Wollheim, C. B., Bar-Tana, J. and Walker, M. D., 2002. Activation of the insulin gene promoter through a direct effect of hepatocyte nuclear factor 4α. The Journal of Biological Chemistry. 277: 25914–25919.

Blomhoff, R. and Blomhoff, H. K., 2006. Overview of retinoid metabolism and function. Journal of Neurobiology. 66: 606–630.

Calleja, C., Messaddeq, N., Chapellier, B., Yang, H., Krezel, W., Li, M., Metzger, D., Mascrez, B., Ohta, K., Kagechika, H., Endo, Y., Mark, M., Ghyselinck, N.B., and Chambon, P., 2006. Genetic and pharmacological evidence that a retinoic acid cannot be the RXR-activating ligand in mouse epidermis keratinocytes. Genes and Development. 20: 1525–1538.

Chen, N., Onisko, B. and Napoli, J. L., 2008. The nuclear transcription factor RARα associates with neuronal RNA granules and suppresses translation. The Journal of Biological Chemistry. 283: 20841–20847.

Cheng, C., Michaels, J. and Scheinfeld, N., 2008. Alitretinoin: a comprehensive review. Expert Opinions on Investigational Drugs 17: 437–443.

Germain, P., Chambon, P., Eichele, G, Evans, R. M., Lazar, M. A., Leid, M., De Lera, A. R., Lotan, R., Mangelsdorf, D. J. and Gronemeyer, H., 2006. International Union of Pharmacology. LXIII. Retinoid X receptors. Pharmacological Reviews. 58: 760–772.

Hattersley, A. T. and Pearson, E. R., 2006. Minireview: pharmacogenetics and beyond: the interaction of therapeutic response, β-cell physiology, and genetics in diabetes. Endocrinology. 147: 2657–2663.

Kadison, A., Kim, J., Maldonado, T., Crisera, C., Prasadan, K., Manna, P., Preuett, B., Hembree, M., Longaker, M. and Gittes, G., 2001. Retinoid signaling directs secondary lineage selection in pancreatic organogenesis. Journal of Pediatric Surgery. 36: 1150–1156.

Kane, M. A. and Napoli, J. L., 2010. Quantification of endogenous retinoids. Methods in Molecular Biology. 652: 1–54.

Kane, M. A., Chen, N., Sparks, S. and Napoli, J. L., 2005. Quantification of endogenous retinoic acid in limited biological samples by LC/MS/MS. Biochemical Journal. 388: 363–369.

Kane, M. A., Folias, A. E., Wang, C. and Napoli, J. L., 2008. Quantitative profiling of endogenous retinoic acid *in vivo* and *in vitro* by tandem mass spectrometry. Analytical Chemistry. 80: 1702–1708.

Kane, M. A., Folias, A. E., Pingitore, A., Perri, M., Obrochta, K. M., Krois, C. R., Cione, E., Ryu, J. Y. and Napoli, J. L., 2010. Identification of 9-*cis*-retinoic acid as a pancreas-specific autacoid that attenuates glucose-stimulated insulin secretion. Proceedings of the National Academy of Sciences, U.S.A. 107: 21884–21889.

Kane, M. A., Folias, A. E., Pingitore, A., Perri, M., Krois, C.R., Ryu, J. Y., Cione, E. and Napoli, J.L., 2011. CrbpI modulates glucose homeostasis and pancreas 9-*cis*-retinoic acid concentrations. Molecular Cell Biology. 16: 3277–3285.

Kobayashi, H., Spilde, T. L., Bhatia, A. M., Buckingham, R. B., Hembree, M. J., Prasadan, K., Preuett, B. L., Imamura, M. and Gittes, G. K., 2002. Retinoid signaling controls mouse pancreatic exocrine lineage selection through epithelial–mesenchymal interactions. Gastroenterology. 123: 1331–1340.

Mark, M., Ghyselinck, N. B. and Chambon, P., 2006. Function of retinoid nuclear receptors: lessons from genetic and pharmacological dissections of the retinoic acid signaling pathway during mouse embryogenesis. Annual Reviews of Pharmacology and Toxicology. 46: 451–480.

Miyazaki, S., Taniguchi, H., Moritoh, Y., Tashiro, F., Yamamoto, T., Yamato, E., Ikegami, H., Ozato, K. and Miyazaki, J., 2010. Nuclear hormone retinoid X receptor (RXR) negatively regulates the glucose-stimulated insulin secretion of pancreatic β-cells. Diabetes. 59: 2854–2861.

Muoio, D. M. and Newgard, C. B., 2008. Mechanisms of disease: molecular and metabolic mechanisms of insulin resistance and β-cell failure in type 2 diabetes. Nature Reviews in Molecular and Cell Biology. 9: 193–205.

Napoli, J. L., 2012. Physiological insights into all-*trans*-retinoic acid biosynthesis. Biochimica et Biophysica Acta. 31: 3277–3285.

Newcomer, M. E., Jamison, R. S. and Ong, D. E., 1998. Structure and function of retinoid-binding proteins. Subcellular Biochemistry. 30: 53–80.

Noy, N., 2007. Ligand specificity of nuclear hormone receptors: sifting through promiscuity. Biochemistry. 46: 13461–13467.

Ramnanan, C. J., Edgerton, D. S., Rivera, N., Irimia-Dominguez, J., Farmer, B., Neal, D. W., Lautz, M., Donahue, E. P., Meyer, C. M., Roach, P. J. and Cherrington, A.D., 2010. Molecular characterization of insulin-mediated suppression of hepatic glucose production *in vivo*. Diabetes. 59: 1302–1311.

Shah, P., Basu, A., Basu, R. and Rizza, R., 1999. Impact of lack of suppression of glucagon on glucose tolerance in humans. American Journal of Physiology. 277: E283–E290.

Shen, H., Luo, Y., Kuo, C. C., Deng, X., Chang, C. F., Harvey, B. K., Hoffer, B. J. and Wang, Y., 2009. 9-*cis*-Retinoic acid reduces ischemic brain injury in rodents via bone morphogenetic protein. Journal of Neuroscience Research. 87: 545–555.

Sidell, N., Feng, Y., Hao, L., Wu, J., Yu, J., Kane, M. A., Napoli, J. L. and Taylor, R. N., 2010. Retinoic acid is a cofactor for translational regulation of vascular endothelial growth factor in human endometrial stromal cells. Molecular Endocrinology. 24: 148–160.

Wang, H., Maechler, P., Antinozzi, P. A., Hagenfeldt, K. A. and Wollheim, C. B., 2000. Hepatocyte nuclear factor 4α regulates the expression of pancreatic β-cell genes implicated in glucose metabolism and nutrient-induced insulin secretion. The Journal of Biological Chemistry. 275: 35953–35959.

Wolf, G., 2006. Is 9-*cis*-retinoic acid the endogenous ligand for the retinoic acid-X receptor? Nutritional Reviews. 64: 532–538.

Yoshioka, M., Kayo, T. Ikeda, T. and Koizumi, A., 1997. A novel locus, Mody4, distal to D7Mit189 on chromosome 7 determines early-onset NIDDM in nonobese C57BL/6 (Akita) mutant mice. Diabetes. 46: 887–894.

Zapata-Gonzalez, F., Rueda, F., Petriz, J., Domingo, P., Villarroya, F., de Madariaga, A. and Domingo, J.C., 2007. 9-*cis*-Retinoic acid (9cRA), a retinoid X receptor (RXR) ligand, exerts immunosuppressive effects on dendritic cells by RXR-dependent activation: inhibition of peroxisome proliferator-activated receptor γ blocks some of the 9cRA activities, and precludes them to mature phenotype development. Journal of Immunology. 178: 6130–6139.

CHAPTER 28
Vitamin A and Cancer Risk

SIDDHARTHA KUMAR MISHRA AND MI KYUNG KIM*

Carcinogenesis Branch, Division of Cancer Epidemiology and Prevention,
National Cancer Center, Ilsandong-gu, Goyang-si,
Gyeonggi-do, 410-769, Republic of Korea
*E-mail: alrud@ncc.re.kr

28.1 Introduction

Vitamin A is a group of compounds vital for biological functions such as
vision, bone growth, reproduction, cell division, and cell differentiation. There
are two categories of vitamin A depending on whether the food source is an
animal or a plant. Vitamin A from animal sources is called 'preformed vitamin
A' which is absorbed in the form of retinol, the most active forms of vitamin A.
Vitamin A from plant sources is called 'provitamin A carotenoid', an orange-
pigmented molecule, mainly characterized as α- and β-carotene, and β-
cryptoxanthin, which can also be converted into retinol. Among these β-
carotene is most efficiently converted into retinol as compared with the others.
Retinol is a yellow, fat-soluble substance that is stored in tissues in the form of
retinyl ester, which can be easily converted into its active aldehyde form
(retinal) and acidic form (retinoic acid) [reviewed in Niles (2000)]. The World
Health Organization (WHO) suggested that a high intake of fruit and
vegetables probably reduces the risk of certain cancers and also the American
Cancer Society have issued advice to cancer survivors adopting fruit and
vegetable consumption.

Several studies with vitamin A and animal models have shown its connection
with susceptibility to carcinogenesis. Vitamin A could induce the differentiation of malignant leukemia cells, inhibit the growth of many tumor cells,

Food and Nutritional Components in Focus No. 1
Vitamin A and Carotenoids: Chemistry, Analysis, Function and Effects
Edited by Victor R Preedy
© The Royal Society of Chemistry 2012
Published by the Royal Society of Chemistry, www.rsc.org

down-regulate the expression of the progesterone receptor in human breast-cancer cells, and has been proven an effective therapy against promyelocytic leukemia [reviewed in Niles (2000)]. Carotenoids have been used in chemoprevention of a few human tumors with notable success and their effect is most likely exerted at the early phase of carcinogenesis by inhibiting proliferation, inducing apoptosis and differentiation, or a combination of these actions. Vitamin A has been effective in the treatment of patients with loco-regional recurrence in head and neck cancer (Jyothirmayi *et al* 1996). High consumption of carotenoids that are rich fruit and vegetable may reduce the risks of cancer in breast (Gaudet *et al* 2004; Mignone *et al* 2009), colorectal (Kim *et al* 2007; Wakai *et al* 2005), lung (Holick *et al* 2002; Ito *et al* 2005), and bladder (Hung *et al* 2006). In general, vitamin A has been found protective but showed no association with risks of prostate cancer (Ambrosini *et al* 2008; Virtamo *et al* 2003). In addition, higher serum retinol concentration was associated with a greater risk of prostate cancer in male smokers (Mondul *et al* 2011). Although the diverse effects of retinoid represent it as a potent chemotherapeutic agent, contrary reports put the role of vitamin A and its derivatives under question. In this Chapter, we review the published reports on the association of dietary carotenoids and other vitamin A derivatives with the risk of cancer.

28.2   Metabolism and Mechanisms of Action of Vitamin A

Basic scientific studies have highlighted key regulators of the retinoid signaling pathway including nuclear retinoid receptors (RRs) and their coregulators, which leads to ligand dependent transcriptional activation of target genes that ultimately signal retinoid growth and differentiation effects (Niles 2000). The two classes of RRs include retinoic acid receptors (RARs), and retinoid X receptors (RXRs). Of these, RARs have three subtypes with several isoforms (RARα, RARβ, and RARγ) and their corresponding RXRs (RXRα, RXRβ, and RXRγ). Retinoids bind their nuclear receptors through ligand-binding domains, which contain DNA-binding domains that recognize specific sequences present in genomic DNA. These interactions induce transcriptional activation or repression of retinoid response elements (RREs) that are present in the promoter region of retinoid target genes. This ultimately leads to alterations in the gene expression mediating biological effects. The RRs can form homodimers or heterodimers, and RARs heterodimerizes with RXRs, which then heterodimerize with multiple members of the steroid receptor superfamily and regulate nuclear receptor-dependent signaling. Retinoid biological signals are cell-type specific and individual RRs reveal receptor-specific developmental and differentiation defects. There are also two classes of coregulators, inhibitory corepressors and stimulatory coactivators, playing important roles in the basal transcriptional machinery of RRs. Cellular retinol-binding proteins (CRBPs) may serve as intracellular storage sites for retinoids

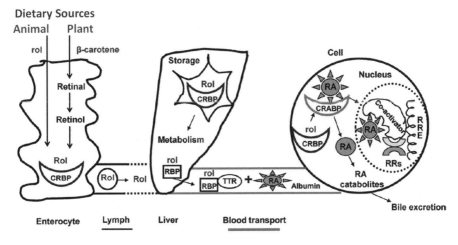

Dietary Sources

Figure 28.1 Illustration of the metabolism of vitamin A/carotenoids. Carotenoids and retinol (rol) are converted in enterocytes into retinol esters (Rol) which are transported by chylomicrons *via* the lymphatic vessels for storage, mainly in the liver (in Ito cells). Rol are converted back into rol that binds the retinol-binding protein (RBP) for peripheral tissue distribution. RBP–rol complexes are bound to transthyretin (TTR) during plasmatic transport, which enters the target cell and binds to cellular retinol-binding proteins (CRBPs) for cytoplasmic transport. In the cell, rol is irreversibly oxidized into all-*trans* retinoic acid (ATRA) which complexes with cellular retinoic acid-binding proteins (CRABPs) before entering the nucleus. There, retinoic acid (RA) serves as a ligand for retinoid receptors (RRs) dimers bound to DNA through specific retinoid response elements (RREs) located in the target genes promoters. Then, RRs dimers bind co-activators and initiate the expression of target genes by activating the transcriptional machinery (modified from Poulain *et al* 2009).

that facilitate the retinoid transport from the cytoplasm to nucleus (Poulain *et al* 2009) (Figure 28.1).

28.3 Epidemiological Studies on Vitamin A and Cancer Risk

Carotenoids are among the best-studied natural compounds for cancer prevention. There is an abundance of scientific evidence in support of the beneficial role of these phytochemicals in the prevention of several types of cancer (Figures 28.2 and 28.3).

28.3.1 Breast Cancer

Observational studies have showed lower breast-cancer risk in women with a high dietary intake of a carotenoid-rich diet. The Nurses' Health Studies

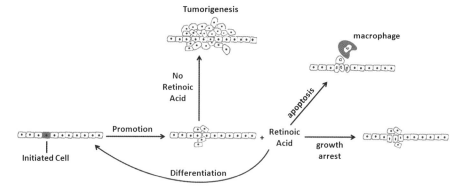

Figure 28.2   Schematic mechanism of inhibition of tumorigenesis by retinoic acid. The promotion of an initiated cell (carcinogen-induced mutation) to a small foci of abnormal cells. These abnormal cells can progress to a clinical tumor in the absence of retinoic acid, while in the presence it can arrest any further growth of the abnormal cells. It can also induce the differentiation of abnormal cells to normal counterparts, or induce the abnormal cells to undergo apoptosis. Cells that die after apoptosis are then ingested by macrophages (modified from Niles 2000).

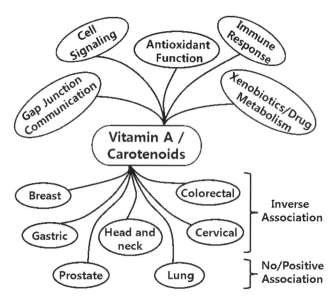

Figure 28.3   Biological functions of vitamin A and carotenoids and their associations with cancer risk. A summarized and schematic diagram for biological functions of vitamin A and carotenoids and their associations with cancer risk. The role of vitamin A in cellular system is mostly dependent on its antioxidant property against oxidative damage to DNA, cellular macromolecules and membranes. The categorization of inverse association and no/positive association is based on the reference studies on vitamin A and cancer risk cited in Tables 28.1–28.3.

showed inverse associations with dietary carotenes and breast cancer in premenopausal high alcohol-consuming women (Zhang *et al* 1999), and among smoking cohorts (Cho *et al* 2003). The intakes of fruit and vegetable were inversely associated to both pre- and postmenopausal breast-cancer risk (Gaudet *et al* 2004; Mignone *et al* 2009). Overall serum carotenoids and retinol levels were also inversely associated to the risk of breast cancer, yet some studies found no clear association between carotenoids and breast-cancer risk (Maillard *et al* 2010; Terry *et al* 2002).

28.3.2 Gastric Cancer

Dietary carotenoids may have protective roles against oxidative damage and reduce the risk of gastric cancer by virtue of their known antioxidants properties. The Swedish adults' cohort study found that high intakes of vitamin A, retinol, and carotenoids may reduce the risk of gastric cancer (Larsson *et al* 2007). Similarly, a study among Japanese people with known *Helicobacter pylori* infection showed a significant association between plasma carotenoids levels and gastric cancer risk (Persson *et al* 2008). A combination of β-carotene with selenium and vitamin E showed beneficial effects on mortality due to gastric cancer, even up to 10 years after the cessation of therapy (Qiao *et al* 2009).

28.3.3 Cervical Cancer

Vitamin A has shown promising effects in inhibiting human papilloma virus (HPV)-associated proliferation of cervical epithelial cells, which promotes the use of retinoid therapy in inhibiting the progression of early cervical lesions into cancer. Cervical intraepithelial neoplasia (CIN) patients of grade 2 and 3 treated with all-*trans* retinoic acid (ATRA) showed higher histologic regression in CIN grade 2 (Meyskens and Manetta 1995). Carotenoids were also associated with a significant decrease in the clearance time of type-specific HPV infection, particularly during the early stages of infection (Goodman *et al* 2007). We also have reported inverse association between β-carotene and the risk of cervical cancer (Kim *et al* 2010). However, Siegel *et al* (2010) found no significant association between circulating retinoic acid and early events in cervical carcinogenesis.

28.3.4 Head and Neck Cancer

Carotenoids have been applied as an experimental chemopreventative measure to prevent the progression of precancerous lesions to cancer, in head-and-neck cancer (HNC) patients. Isotretinoin (13-*cis*-retinoic acid, 13-cRA) significantly decreased the incidence of second primary tumors (SPT) in patients with HNC, resulting in regression of leukoplakia. The low dose of 13-cRA reduced the occurrence of SPTs (Minard *et al* 2006) and high dose could decrease the incidence of SPT in patients after curative therapy of the initial primary tumor (Lippman *et al* 1993). Jyothirmayi *et al* (1996) reported inverse association between retinyl plamitate level and HNC risk. A recent report also supports that 13-cRA is highly effective chemoprevention agent for the risk of SPT in HNC patients (Lee *et al* 2011).

28.3.5   Colorectal Cancer

Epidemiological investigations suggest that higher intake or biochemical status of carotenoids might be associated with the reduced risk of colorectal cancer (CRC). The Japan Collaborative Cohort (JACC) study reported that higher level of serum total carotenoids was associated with a decreased risk of CRC in men. However, in women, it was instead related to an increased risk (Wakai *et al* 2005). We observed that low serum β-carotene may be inversely related to the risk of CRC (Kim *et al* 2007). However, the Alpha-Tocopherol, Beta-Carotene Cancer Prevention Trial (ATBC) Study have previously found no effect of carotenoids on CRC in male smokers (Albanes *et al* 2000).

28.3.6   Lung Cancer

Several β-carotene-based clinical trials against lung cancer have showed unexpected and conflicted observations. The ATBC Study observed a beneficial association between several dietary carotenoids and the lung cancer risk in male smokers (Holick *et al* 2002). The JACC study observed significantly lower risk of lung cancer with higher carotenoids in men, but not in women (Ito *et al* 2005). However, high-dose β-carotene in high-risk people has been associated with increased lung cancer risk, however, the effect in the general population is unclear. The VITamins And Lifestyle (VITAL) cohort study showed that longer duration of use of individual β-carotene and retinol was associated with elevated risk of total lung cancer in smokers (Satia *et al* 2009). The β-Carotene and Retinol Efficacy Trial (CARET) study showed contrary results in smoking cohorts, and was stopped 21 months prior to completion because of the increased risk of lung-cancer incidence and mortality in the treated group (Omenn *et al* 1996).

28.3.7   Prostate Cancer

The antioxidant properties of carotenoids support the preventive role in prostate cancer, however, this is controversial. The ATBC Study on male smokers observed a significant reduction in prostate cancer and a lower mortality in those receiving α-tocopherol but not with carotenes (Virtamo *et al* 2003). A nested case-controlled study extended epidemiological evidence that serum carotenoids were not associated with the risk of prostate cancer (Gill *et al* 2009). The CARET study also showed no association with total prostate cancer risk and β-carotene and retinyl palmitate. Vitamin A supplementation had a relatively beneficial association in men with aggressive prostate cancer but these associations disappeared in the post-intervention period (Neuhouser *et al* 2009). Recently, a large study revealed that higher serum retinol was associated with elevated risk of prostate cancer (Mondul *et al* 2011) (Tables 28.1–28.3).

28.4 Genetic Polymorphisms Associated with Vitamin A and Cancer Risk

Although a comprehensive review of epidemiological evidence relating to the protective role of vitamin A against cancers suggests a putative protective role, as discussed earlier, it remains an open question as to whether consumption of carotenoid-rich fruit and vegetables may reduce the risk. One possible explanation for the inconsistent evidence across epidemiological studies may be variations in genes that are directly and indirectly involved in carcinogenesis in relation to cancer risk. Genetic polymorphisms associated with the exposure to environmental risk factors for cancer can affect the cancer susceptibility of each individual. Recently, Lee *et al* (2011) identified 13 loci in a majority subgroup of patients at a high risk of SPT/recurrence in whom 13-cRA was protective. More than 70% patients with common genotype of rs3118570 in RXRA were at 3-fold increased risk. The CDC25C:rs6596428 and JAK2:rs1887427 are two other genetic loci with major roles in prognosis and response of 13-cRA. Patients with all three common genotypes (RXRA:rs3118570, JAK2:rs1887427, and CDC25C:rs6596428) showed higher reduction in SPT/recurrence following 13-cRA chemoprevention and these showed individual prognostic and predictive abilities (Lee *et al* 2011). The GST-M1 null genotype was associated with an increased risk for SPTs, simultaneously non-null GST genotypes with a decreased risk for SPTs as compared to null GST-M1 and non-null GST-T1 (Minard *et al* 2006). Ferrucci *et al* (2009) identified that a G allele at rs6564851, near the β-carotene 15,15′-mono-oxygenase 1 (BCMO1) gene was associated with higher level of carotenes (α and β). BCMO1 catalyzes the first step in the conversion of dietary provitamin A carotenoids into vitamin A in the small intestine. The antioxidant enzyme manganese superoxide dismutase (MnSOD) has an allele with a Val at amino acid position 16 that encodes a protein with 30–40% lower activity compared with the MnSOD Ala variant, hence possibly increasing susceptibility to oxidative stress. Han *et al* (2007) found an inverse association between carotenoids intake and the MnSOD polymorphism with the risk of squamous cell carcinoma, which was limited to the Val carriers, and had no association among AA genotype. Rather, epidemiological studies suggest that the Ala allele is associated with a higher risk of prostate cancer. However, Mikhak *et al* (2008) found no association between the MnSOD genotypes and risk of prostate cancer in association with plasma carotenoid concentrations.

Table 28.1 Comparative analysis of studies with "inverse association" between dietary and blood vitamin A and cancer risk. All the data on characteristics were collected from reference studies. CI, confidence interval; HR, hazard ratio; NA, not available; OR, odds ratio; RR, relative risk. P values < 0.05 were considered significant.

Study design	Cancer site	Subjects	Vitamin A component and significance	Reference
Case-control	Breast	Cases, $n = 5707$; Controls, $n = 6389$	Dietary β-carotene, OR: 0.81, 95% CI: 0.68–0.98, P trend $= 0.009$	Mignone *et al* 2009
		Cases, $n = 1463$; Controls, $n = 1500$	Dietary carotenoids, OR: 0.66, 95% CI: 0.48–0.90, P trend $= 0.03$	Gaudet *et al* 2004
	Colorectal	Cases, n=31; Controls, n=34	Serum β-carotene, OR: NA	Kim *et al* 2007
		Cases, n=116; Controls, $n = 298$	Serum carotenoid, OR: 0.34, 95% CI: 0.11–1.0, P trend $= 0.04$	Wakai *et al* 2005
	Lung	Cases, n=211; Controls, $n = 487$	Serum carotenenoids, OR: 0.35, 95% CI: 0.17–0.74, P trend $= 0.01$	Ito *et al* 2005
	Cervical	Cases, $n = 144$; Controls, $n = 288$	Dietary β-carotene, OR: 0.48, 95% CI: 0.26–0.88, P trend $= <0.05$	Kim *et al* 2010
	Gastric	Cases, $n = 511$; Controls, $n = 511$	Dietary β-carotene, OR: 0.46, 95% CI: 0.28–0.75, P trend $= <0.01$	Persson *et al* 2008
Cohort	Breast	$n = 90\ 655$	Dietary vitamin A, OR: 0.33, 95% CI: 0.16–0.70, P trend $= 0.001$	Cho *et al* 2003
		$n = 83\ 234$	Dietary β-carotene, OR: 0.83, 95% CI: 0.66–1.04, P trend $= 0.04$	Zhang *et al* 1999
	Gastric	$n = 1199$	Supplement β-carotene, HR: 0.89, 95% CI: 0.79–1.0, P trend $= 0.04$	Qiao *et al* 2009
		$n = 82\ 002$	Dietary vitamin A, OR: 0.53, 95% CI: 0.32–0.89, P trend $= 0.02$	Larsson *et al* 2007
	Head-and-neck	$n = 106$	Supplement retinyl plamitate, OR: NA	Jyothirmayi *et al* 1996

Table 28.1 (*Continued*)

Study design	Cancer site	Subjects	Vitamin A component and significance	Reference
		$n = 70$	Supplement 13-*cis*-retinoic acid, OR: NA	Lippman *et al* 1993
	Lung	$n = 27\ 084$	Dietary carotenoids, RR: 0.84, 95% CI: 0.72–0.99, P trend = 0.005	Holick *et al* 2002
	Cervical	$n = 122$	Dietary β-carotene, HR: 0.86, 95% CI: 0.52–1.41, P trend = 0.007	Goodman *et al* 2007

Table 28.2 Comparative analysis of studies with "no association" between dietary and blood vitamin A and cancer risk. All the data on characteristics were collected from reference studies. CI, confidence interval; HR, hazard ratio; OR, odds ratio; RR, relative risk. P values < 0.05 were considered significant.

Study design	Cancer site	Subjects	Vitamin A component and significance	Reference
Case-control	Prostate	Cases, $n = 467$; Controls, $n = 936$	Serum β-Carotene, OR: 0.81, 95% CI: 0.55–1.18, P trend = 0.40	Gill *et al* 2009
		Cases, $n = 97$; Controls, $n = 1888$	Dietary β-carotene, RR: 0.96, 95% CI: 0.58–1.61, P trend = 0.86	Ambrosini *et al* 2008
Cohort	Breast	$n = 19\ 934$	Dietary carotenoids, OR: 0.74, 95%CI 0.47–1.16, $P = 0.38$	Maillard *et al* 2010
		$n = 56\ 837$	Dietary β-carotene, HR: 1.01, 95% CI: 0.07–1.33, P trend = 0.98	Terry *et al* 2002
	Lung	$n = 22\ 071$	Dietary β-carotene, RR: 1.08, 95% CI: 0.95–1.22, P trend = 0.25	Hennekens *et al* 1996
	Cervical	$n = 643$	Serum RA, OR: 1.28, 95% CI: 0.66–2.47, P trend = 0.39	Siegel *et al* 2010
	Colorectal	$n = 135$	Supplement β-carotene, OR: 1.05, 95% CI: 0.75–1.47, P trend = 0.78	Albanes *et al* 2000
Intervention	Prostate	$n = 12\ 000$	Supplement β-carotene, RR: 0.75, 95% CI: 0.51–1.09, $P = >0.05$	Neuhouser *et al* 2009

Table 28.3 Comparative analysis of selected cohort studies with "positive association" between dietary vitamin A and cancer risk. All the data on characteristics were collected from reference studies. CI, confidence interval; HR, hazard ratio; OR, odds ratio; RR, relative risk. *P* values < 0.05 were considered significant.

Cancer site	Subjects	Vitamin A component and significance	Reference
Lung	n = 77 126	Dietary vitamin A, OR: 1.53, 95% CI: 1.12–2.08, *P* trend = 0.004	Satia *et al* 2009
	n = 18 314	Dietary β-carotene, OR: 1.36, 95% CI: 1.07–1.73, *P* trend = 0.01	Omenn *et al* 1996
Prostate	n = 29 104	Dietary vitamin A, HR: 1.19, 95% CI: 1.03–1.36, *P* trend = 0.009	Mondul *et al* 2011
	n = 29 133	Supplement β-carotene, RR: 1.07, 95% CI: 1.02–1.12, *P* = >0.05	Virtamo *et al* 2003

28.5   Conclusions

Fruit and vegetables are good sources of carotenoids, having been extensively studied for their role in prevention of cancers. Results of investigations suggest that vitamin A and its metabolites act as chemopreventive agent to suppress the tumor-promotion phase of carcinogenesis. Carotenoids have shown significant chemopreventive activities on head-and-neck, breast, bladder, cervical, and gastric cancers. Of note, high doses of vitamin A may be toxic, and long-term use of high-dose supplements may increase the risk of lung cancer. The fascinating possibility that vitamin A or β-carotene intake is more beneficial in lung cancer among smokers than non-smokers needs cautious confirmation. Similarly, neither dietary nor supplemental vitamin A intake was related to prostate cancer risk. In summary, epidemiological studies suggest the importance of vitamin A in the prevention of certain cancers, inclusive of some contrary reports. Studies conducted on association of vitamin A and cancer risk revealed that dietary intake of certain fruit and vegetables rich in carotenoids were inversely associated with the risk of certain cancer. The findings indicate the necessity for well-designed future studies that take into consideration of subject selection, end-point measurements, and the level of carotenoids being tested. Also, extensive studies may provide insight into the mechanism underlying biological actions of carotenoids and their interactions with the risk of cancer.

Summary Points

- Vitamin A and its metabolites apparently act as chemopreventive agents against cancer at cellular and molecular levels.
- Carotenoids block tumor promotion by inhibiting proliferation, inducing apoptosis, inducing differentiation, or a combination of these actions.

- Carotenes (α and β) were inversely associated with breast-cancer risk.
- Serum retinol was associated with decreasing risk of colorectal cancer, supporting its protective effect.
- Overall, vitamin A is protective against HPV-associated cervical cancer and favors retinoid therapy against cervical lesions.
- Clinical trials using β-carotene against lung cancer showed unexpected and conflicted observations.
- Neither dietary nor supplemental vitamin A intake was associated to prostate cancer risk.
- Dietary carotenoids in plasma levels and their properties are affected by common genetic variation.

Key Facts

- Retinoids, vitamin A, and natural and synthetic derivatives of vitamin A are the best studied of many agents that have been evaluated for preventing cancer.
- The American Cancer Society have issued advice to cancer survivors, particularly of breast, lung, and colorectal cancers, to adopt the general cancer preventative recommendations for fruit and vegetable consumption, although there are few studies that have examined whether this improves cancer survival.
- The Cancer Council supports the Australian Dietary Guidelines that recommend eating plenty of fruit and vegetables, and the population recommendation of at least two servings of fruit and five servings of vegetables daily.
- The International Agency for Research on Cancer concluded that 5–12% of cancers could be attributed to low fruit and vegetable consumption.
- Many cancer associations worldwide, together with a number of National Dietary Guideline committees, have recommended a daily intake of 5–7 servings of fruit and vegetables to reduce cancer risk.
- The World Health Organization suggested that a high intake of fruit and vegetables probably reduced the risk of certain cancers and recommended an intake of at least 400 g of fruit and vegetables daily.

Definitions of Words and Terms

Alpha-Tocopherol, Beta-Carotene Cancer Prevention Trial (ATBC): A cancer prevention trial conducted by the U.S. National Cancer Institute (NCI) and the National Public Health Institute of Finland. The study comprised a group of male smokers given α-tocopherol, β-carotene, both, or a placebo.

β-Carotene and Retinol Efficacy Trial (CARET): A randomized, double-blind, placebo-controlled trial of the cancer prevention efficacy and safety of a daily combination of β-carotene and retinyl palmitate in people with a high risk of lung cancer.

Case-control study: A type of epidemiological, clinical study design typically used for retrospective and prospective studies. In this study, people with a disease (cases) are matched with people who do not have the disease (controls).

Confidence interval (CI): a particular kind of interval estimate of a population parameter and is used to indicate the reliability of an estimate.

Hazard ratio (HR): The effect of an explanatory variable on the hazard or risk of an event in a survival analysis. HRs are commonly used in clinical trials involving survival data, and allows hypothesis testing.

Intervention study: Studies in which individuals are assigned by an investigator based on a protocol to receive specific interventions. Subjects may receive diagnostic, therapeutic or other types of interventions.

Japan Collaborative Cohort (JACC): A collaborative cohort study for the evaluation of cancer risk, conducted by the Ministry of Education, Science, Sports and Culture of Japan.

Odds ratio (OR): A measure of effect size, describing the strength of association or non-independence between two binary data values.

Prospective study: A cohort study that follows over time a group of similar individuals (cohorts) who differ with respect to certain factors under study to determine how these factors affect rates of a certain outcome.

Relative risk (RR): The risk of an event (or of developing a disease) relative to exposure. RR is a ratio of the probability of the event occurring in the exposed group *versus* a non-exposed group.

VITamins And Lifestyle (VITAL): A cohort study of dietary supplements and cancer risk, conducted by Fred Hutchinson Cancer Research Center, USA, comprising a cohort of men and women, with a high percentage of supplement users.

List of Abbreviations

ATBC	Alpha-Tocopherol, Beta-Carotene Cancer Prevention Trial
ATRA	all-*trans* retinoic acid
BCMO1	β-carotene 15,15′-mono-oxygenase 1
CARET	β-Carotene and Retinol Efficacy Trial
CIN	cervical intraepithelial neoplasia
13-cRA	13-*cis*-retinoic acid
CRBP	cellular retinol-binding protein
CRC	colorectal cancer
HNC	head-and-neck cancer
HPV	human papilloma virus
JACC	Japan Collaborative Cohort
MnSOD	manganese superoxide dismutase
RAR	retinoic acid receptor
RR	retinoid receptor

RRE	retinoid response element
RXR	retinoid X receptor
SPT	second primary tumor

References

Albanes, D., Malila, N., Taylor, P. R., Huttunen, J. K., Virtamo, J., Edwards, B. K., Rautalahti, M., Hartman, A. M., Barrett, M. J., Pietinen, P., Hartman, T. J., Sipponen, P., Lewin, K., Teerenhovi, L., Hietanen, P., Tangrea, J. A., Virtanen, M., and Heinonen, O. P. 2000. Effects of supplemental α-tocopherol and β-carotene on colorectal cancer: results from a controlled trial (Finland). Cancer Causes and Control. 11: 197–205.

Ambrosini, G. L., de Klerk, N. H., Fritschi, L., Mackerras, D., and Musk, B. 2008. Fruit, vegetable, vitamin A intakes, and prostate cancer risk. Prostate Cancer and Prostatic Diseases. 11: 61–66.

Cho, E., Spiegelman, D., Hunter, D. J., Chen, W. Y., Zhang, S. M., Colditz, G. A., and Willett, W. C. 2003. Premenopausal intakes of vitamins A, C, and E, folate, and carotenoids, and risk of breast cancer. Cancer Epidemiology, Biomarkers & Prevention. 12: 713–720.

Ferrucci, L., Perry, J. R. B., Matteini, A., Perola, M., Tanaka, T., Silander, K., Rice, N., Melzer, D., Murray, A., Cluett, C., Fried, L. P., Albanes, D., Corsi, A. M., Cherubini, A., Guralnik, J., Bandinelli, S., Singleton, A., Virtamo, J., Walston, J., Semba, R. D., and Frayling, T. M. 2009. Common variation in the β-carotene 15,15'-monooxygenase 1 gene affects circulating levels of carotenoids: A Genome-wide Association Study. The American Journal of Human Genetics. 84: 123–133.

Gaudet, M. M., Britton, J. A., Kabat, G. C., Steck-Scott, S., Eng, S. M., Teitelbaum, S. L., Terry, M. B., Neugut, A. I., and Gammon, M. D. 2004. Fruits, vegetables, and micronutrients in relation to breast cancer modified by menopause and hormone receptor status. Cancer Epidemiology, Biomarkers & Prevention. 13: 1485–1494.

Gill, J. K., Franke, A. A., Morris, J. S., Cooney, R. V., Wilkens, L. R., Le Marchand, L., Goodman, M. T., Henderson, B. E., and Kolonel, L. N. 2009. Association of selenium, tocopherols, carotenoids, retinol, and 15-isoprostane F_{2t} in serum or urine with prostate cancer risk: the multiethnic cohort. Cancer Causes Control. 20: 1161–1171.

Goodman, M. T., Shvetsov, Y. B., McDuffie, K., Wilkens, L. R., Zhu, X., Franke, A. A., Bertram, C. C., Kessel, B., Bernice, M., Sunoo, C., Ning, L., Easa, D., Killeen, J., Kamemoto, L., and Hernandez, B. Y. 2007. Hawaii Cohort Study of serum micronutrient concentrations and clearance of incident oncogenic human papillomavirus infection of the cervix. Cancer Research. 67: 5987–5996.

Han, J., Colditz, G. A., and Hunter, D. J. 2007. Manganese superoxide dismutase polymorphism and risk of skin cancer (United States). Cancer Causes Control. 18: 79–89.

Hennekens, C. H., Buring, J. E., Manson, J. E., Stampfer, M., Rosner, B., Cook, N. R., Belanger, C., LaMotte, F., Gaziano, J. M., Ridker, P. M., Willett, W., and Peto, R. 1996. Lack of effect of long term supplementation with beta-carotene on the incidence of malignant neoplasms and cardiovascular disease. The New England Journal of Medicine. 334: 1145–1149.

Holick, C. N., Michaud D. S., Stolzenberg-Solomon, R., Mayne, S. T., Pietinen, P., Taylor, P. R., Virtamo, J., and Albanes, D. 2002. Dietary carotenoids, serum β-carotene, and retinol and risk of lung cancer in the Alpha-Tocopherol, Beta-Carotene Cohort Study. American Journal of Epidemiology. 156: 536–547.

Hung, R. J., Zhang, Z. F., Rao, J. Y., Pantuck, A., Reuter, V. E., Heber, D., and Lu, Q. Y. 2006. Protective effects of plasma carotenoids on the risk of bladder cancer. The Journal of Urology. 176: 1192–1197.

Ito, Y., Wakai, K., Suzuki, K., Ozasa, K., Watanabe, Y., Seki, N., Ando, M., Nishino, Y., Kondo, T., Ohno, Y., and Tamakoshi, A.; JACC Study Group. 2005. Lung cancer mortality and serum levels of carotenoids, retinol, tocopherols, and folic acid in men and women: a case-control study nested in the JACC Study. Journal of Epidemiology. 15: S140–149.

Jyothirmayi, R., Ramadas, K., Varghese, C., Jacob, R., Nair, M. K., and Sankaranarayanan, R. 1996. Efficacy of vitamin A in the prevention of locoregional recurrence and second primaries in head and neck cancer. Oral Oncology, European Journal of Cancer. 32: 373–376.

Kim, J., Kim, M. K., Lee, J. K., Kim, J-H., Son, S. K., Song, E. S., Lee, K. B., Lee, J. P., Lee, J. M., and Yun, Y. M. 2010. Intakes of vitamin A, C, and E, and β-carotene are associated with risk of cervical cancer: a case-control study in Korea. Nutrition and Cancer. 62: 181–189.

Kim, M. K., Ahn, S. Y., and Lee-Kim, Y. C. 2001. Relationship of serum α-tocopherol, carotenoids and retinol with the risk of breast cancer. Nutrition Research. 21: 797–809.

Kim, M. K., Choi, K. Y., Lee, W. C., and Park, J. H. Y. 2007. Low serum β-carotene is associated with the incidence of colorectal adenoma. Nutrition Research. 27: 127–132.

Larsson, S. C., Bergkvist, L., Naslund, I., Rutegard, J., and Wolk, A. 2007. Vitamin A, retinol, and carotenoids and the risk of gastric cancer: a prospective cohort study. The American Journal of Clinical Nutrition. 85: 497–503.

Lee, J. J., Wu, X., Hildebrandt, M. A. T., Yang, H., Khuri, F. R., Kim, E., Gu, J., Ye, Y., Lotan, R., Spitz, M. R., and Hong, W. K. 2011. Global assessment of genetic variation influencing response to retinoid chemo-prevention in head and neck cancer patients. Cancer Prevention Research. 4: 185–193.

Lippman, S. M., Bataakis, J. G., Toth, B. B., Weber, R. S., Lee, J. J., Martin, J. W., Hays, G. L., Goepfert, H., and Hong, W. K. 1993. Comparison of low-dose isotretinoin with β-carotene to prevent oral carcinogenesis. The New England Journal of Medicine. 328: 15–20.

Maillard, V., Kuriki, K., Lefebvre, B., Boutron-Ruault, M. C., Lenoir, G. M., Joulin, V., Clavel-Chapelon, F., and Chajès, V. 2010. Serum carotenoid, tocopherol and retinol concentrations and breast cancer risk in the E3N-EPIC study. International Journal of Cancer. 127:1188–1196.

Meyskens, F. L. and Manetta, A. 1995. Prevention of cervical intraepithelial neoplasia and cancer. The American Journal of Clinical Nutrition. 62: s1417–1419.

Mignone, L. I., Giovannucci1, E., Newcomb, P. A., Titus-Ernstoff, L., Trentham-Dietz, A., Hampton, J. M., Willett, W. C., and Egan, K. M. 2009. Dietary carotenoids and the risk of invasive breast cancer. International Journal of Cancer: 124: 2929–2937.

Mikhak, B., Hunter, D. J., Spiegelman, D., Platz, E. A., Wu, K., Erdman, J. W. Jr., and Giovannucci, E. 2008. Manganese superoxide dismutase (MnSOD) gene polymorphism, interactions with carotenoid levels and prostate cancer risk. Carcinogenesis. 29: 2335–2340.

Minard, C. G., Spitz, M. R., Wu, X., Hong, W. K., and Etzel, C. J. 2006. Evaluation of glutathione s-transferase polymorphisms and mutagen sensitivity as risk factors for the development of second primary tumors in patients previously diagnosed with early-stage head and neck cancer. Cancer. 106: 2636–2644.

Mondul, A. M., Watters, J. L., Mannisto, S., Weinstein, S. J., Snyder, K., Virtamo, J., Albanes, D. 2011. Serum Retinol and Risk of Prostate Cancer. American Journal of Epidemiology. 173: 813–821.

Neuhouser, M. L., Barnett, M. J., Kristal, A. R., Ambrosone, C. B., King, I. B., Thornquist, M., and Goodman, G. G. 2009. Dietary supplement use and prostate cancer risk in the Carotene and Retinol Efficacy Trial. Cancer Epidemiology, Biomarkers & Prevention. 18: 2202–2206.

Niles, R. M. 2000. Recent advances in the use of vitamin A (retinoids) in the prevention and treatment of cancer. Nutrition 16: 1084–1090.

Omenn, G. S., Goodman, G. E., Thornquist, M. D., Balmes, J., Cullen, M. R., Glass, A., Keogh, J. P., Meyskens, F. L. Jr., Valanis, B., and Williams, J. H. Jr. 1996. Risk factors for lung cancer and for intervention effects in CARET, the beta-carotene and retinol efficacy trial. Journal of the National Cancer Institute. 88: 1550–1559.

Persson, C., Sasazuki, S., Inoue, M., Kurahashi, N., Iwasaki, M., Miura, T., Ye, W., and Tsugane, S.; JPHC Study Group. 2008. Plasma levels of carotenoids, retinol and tocopherol and the risk of gastric cancer in Japan: a nested case–control study. Carcinogenesis. 29: 1042–1048.

Poulain, S., Evenou, F., Carré, M. C., Corbel, S., Vignaud, J. M., Martinet, N. 2009. Vitamin A/retinoids signalling in the human lung. Lung Cancer. 66:1–7.

Qiao, Y-L., Dawsey, S. M., Kamangar, F., Fan, J-H., Abnet, C. C., Sun, X. D., Johnson, L. L., Gail, M. H., Dong, Z. W., Yu, B., Mark, S. D., and

Taylor, P. R. 2009. Total and cancer mortality after supplementation with vitamins and minerals: follow-up of the Linxian General Population Nutrition Intervention Trial. Journal of the National Cancer Institute. 101: 507–518.

Satia, J. A., Littman, A., Slatore, C. G., Galanko, J. A., and White, E. 2009. Long-term Use of β-carotene, retinol, lycopene, and lutein supplements and lung cancer risk: results from the VITamins And Lifestyle (VITAL) Study. American Journal of Epidemiology. 169: 815–828.

Siegel, E. M., Salemi, J. L., Craft, N. E., Villa, L. L., Ferenczy, A. S., Franco, E. L., and Giuliano, A. R. 2010. No association between endogenous retinoic acid and human papillomavirus clearance or incident cervical lesions in Brazilian women. Cancer Prevention Research. 3: 1007–1014.

Terry, P., Jain, M., Miller, A. B., Howe, G. R., and Rohan, T. E. 2002. Dietary carotenoids and risk of breast cancer. The American Journal of Clinical Nutrition. 76: 883–888.

Virtamo, J., Pietinen, P., Huttunen, J. K., Korhonen, P., Malila, N., Virtanen, M. J., Albanes, D., Taylor, P. R., and Albert, P.; ATBC Study Group. 2003 Incidence of cancer and mortality following alpha-tocopherol and beta-carotene supplementation: a postintervention follow-up. The Journal of the American Medical Association. 290: 476–485.

Wakai, K., Suzuki, K., Ito, Y., Kojima, M., Tamakoshi, K., Watanabe, Y., Toyoshima, H., Hayakawa, N., Hashimoto, S., Tokudome, S., Suzuki, S., Kawado, M., Ozasa, K., and Tamakoshi, A.; Japan Collaborative Cohort Study Group. 2005. Serum carotenoids, retinol, and tocopherols, and colorectal cancer risk in a Japanese cohort: Effect modification by sex for carotenoids. Nutrition and Cancer. 51: 13–24.

Zhang, S., Hunter, D. J., Forman, M. R., Rosner, B. A., Speizer, F. E., Colditz, G. A., Manson, J. E., Hankinson, S. E., and Willett, W. C. 1999. Dietary carotenoids and vitamins A, C, and E and risk of breast cancer. Journal of the National Cancer Institute. 91: 547–556.

CHAPTER 29

Vitamin A and Immune Function

CHARLES B. STEPHENSEN

Immunity and Disease Prevention Research Unit, Western Human Nutrition Research Center, Agricultural Research Service, United States Department of Agriculture, 430 West Health Sciences Drive, University of California, Davis, CA 95616, USA
E-mail: Charles.Stephensen@ars.usda.gov

29.1 Introduction and Historical Perspective

The biological activities of vitamin A in promoting growth and sustaining vision were described nearly a century ago (Wolf 2001) (Table 29.1). Within a few years vitamin A was additionally termed the "anti-infective vitamin" because deficient animals were found to develop symptomatic infections of mucosal surfaces at a much greater rate than non-deficient animals (Green and Mellanby 1928). These infections were thought to result from compromised barrier defenses resulting from the patchy squamous metaplasia that was described at respiratory and intestinal epithelial surfaces. As a result of such observations, treatment studies were conducted into the 1940s to determine if vitamin A-rich preparations (*e.g.*, cod liver oil) could be used to treat respiratory and enteric infections in humans. Some successes were reported but results were mixed overall (Semba 1999). Interest in such studies waned with the advent of antibiotics. The mechanisms underlying this increase in risk were not understood at that time, primarily because the science of immunology was only in its infancy. In addition, the mechanism of action of vitamin A outside of the visual cycle was not defined until 1987 when retinoic acid was found to be the ligand for the retinoic acid receptor (RAR). Three genes encode the three RAR subtypes: α, β, and γ (Germain *et al* 2006). This discovery showed

Food and Nutritional Components in Focus No. 1
Vitamin A and Carotenoids: Chemistry, Analysis, Function and Effects
Edited by Victor R Preedy
Published by the Royal Society of Chemistry, www.rsc.org

Table 29.1   Chronology of seminal discoveries in vitamin A research.

Seminal events in vitamin A research	*Year*
Hippocrates recommends eating raw liver to treat night blindness	$\sim$400 BC
Vision and growth effects of vitamin A identified in experimental animals	1913–1920
Vitamin A called the "anti-infective vitamin"	1928
Clinical trials of vitamin A-rich preparations to treat infectious diseases	1930s
Discovery of the retinoic acid receptor (RAR)	1987
Community trials show vitamin A decreases mortality from infections	1990s
Characterization of specific effects of vitamin A on immune system	1980s–present

that vitamin A could regulate gene expression and thus affect many cellular processes known to be responsive to vitamin A, such as cellular differentiation. This mechanism appears to account for the effects of vitamin A in the immune system. Other mechanisms may also be a factor, such as direct interaction of RARs with other transcription factors to modify their activity. Retinoic acid is produced by specific cell types in the immune system and acts to regulate gene expression in an autocrine and paracrine manner, thereby affecting the immune response.

29.2   Vitamin A Deficiency and Childhood Mortality

Vitamin A deficiency is common in many developing countries (Figure 29.1). Worldwide, 190 million infants and young children have low serum retinol concentrations indicative of inadequate vitamin A stores (<0.70 μmol L^{-1}), with the majority living in South Asia and Africa.

Controlled intervention trials conducted in the 1990s demonstrated that treatment of vitamin A deficiency in populations at risk of vitamin A deficiency in Africa and Asia decreases mortality from early childhood infections by $\sim$24% relative to subjects receiving placebo (Table 29.2). In particular, deaths from diarrhea were averted, as well as episodes of diarrhea severe enough to require hospitalization or clinic attendance (Imdad *et al* 2010). Vitamin A supplementation is thus currently recommended by the World Health Organization for young children aged 6 months to 5 years of age in populations in which vitamin A deficiency is common. Providing vitamin A supplements at <6 months of age has had mixed results and is not currently recommended to decrease infant mortality. While the specific biological mechanisms underlying this protective effect of vitamin A have not been clearly defined in human studies, it is likely that providing vitamin A supplements to deficient individuals restores impaired immune function, thus improving recovery from infections and decreasing the risk of death.

Countries and areas with survey data and regression-based estimates: Pregnant women

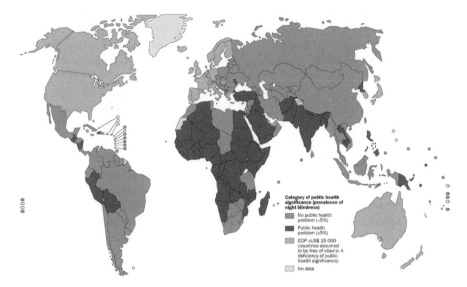

Figure 29.1 Map of vitamin A deficiency in pregnant women. Vitamin A deficiency is based on the prevalence of night blindness. The map was provided with the permission from the World Health Organization (http://www.who.int/vmnis/database/vitamina/status/en/index.html).

Table 29.2 Effects of vitamin A supplementation, relative to placebo, in infants and young children 6–59 months of age at risk of vitamin A deficiency (Imdad *et al* 2010).

Health indicators where changes were seen with moderate or high confidence	Change (%)
Total mortality	24 ↓
Diarrhea-related mortality	28 ↓
Measles-related mortality	20 ↓
Symptomatic measles infection	50 ↓
Night blindness (*i.e.*, low sensitivity of retina to light)	68 ↓
Corneal signs of vitamin A deficiency (Bitot's spots)	55 ↓
Prevalence of vitamin A deficiency defined by serum retinol concentration	29 ↓

29.3 The Immune System

The principal function of the human immune system is to protect us from death caused by pathogenic parasites and micro-organisms. Immunologists currently think of the immune system as having two components: "innate" and "adaptive" (Murphy *et al* 2011), although the two work together as an integrated whole. The innate system is evolutionarily older than the adaptive system and it is fully functional at birth. Innate immune cells (*e.g.*, neutrophils, eosinophils, macrophages, and antigen-presenting cells such as dendritic cells) use a diverse group of receptors [*e.g.*, Toll-like receptors (TLRs)] to recognize and respond to signature molecules from classes of micro-organisms (*e.g.*, flagella from some bacteria, cell-wall carbohydrates from yeast, RNA from viral genomes). These protective responses (*e.g.*, phagocytosis, production of cytokines to promote inflammation) are essentially the same for all individuals within a species and do not vary after exposure to specific pathogens. Tissue damage from trauma, such as surgery or physiological damage, such as the development of lesions in the coronary arteries due to a poor diet, can also trigger the innate immune response. The adaptive immune system is different in that the host's response adapts to a specific pathogen (*e.g.*, measles virus specifically and not RNA viruses in general) in order to develop "immunologic memory" that will respond more quickly and more efficiently the next time the same pathogen is encountered. This adaptation includes production of antigen-specific T- and B-lymphocytes, as well as antibodies. The innate immune system can help steer the development of the adaptive immune response by interacting with T- and B-lymphocytes, the principal cellular components of the adaptive immune system. Thus, individuals have different levels of adaptive immunity depending on their exposure history. Such an adaptive immune response occurs more rapidly after the second exposure to an antigen than it does after the first exposure. Thus, the first encounter with a childhood pathogen (*e.g.*, measles) can make a child quite ill, but subsequent infections will likely go unnoticed if an effective adaptive immune response has developed.

29.4 Vitamin A and Innate Immunity

29.4.1 Epithelial Surfaces

The first lines of defense against pathogens for any animal are its epithelial tissues. Vitamin A deficiency causes squamous metaplasia of the respiratory epithelium. That is, the ciliated, columnar epithelial cells that help protect the respiratory tract from infection are replaced by squamous epithelium (which is normally limited to the skin) during vitamin A deficiency. This is most pronounced when these cells are challenged by external factors, including cigarette smoke or a respiratory infection. Squamous metaplasia is not limited to the respiratory tract and is seen at mucosal epithelial surfaces throughout the body, including the gastrointestinal tract and urinary bladder. Mucus-producing goblet cells are also lost from mucosal epithelial surfaces as a result of vitamin A deficiency. Squamous metaplasia and decreased mucus production can decrease protection of epithelial surfaces from bacterial pathogens. These changes may increase adherence of some pathogens, as can occur in the respiratory tract and urinary bladder, and can increase tissue damage by viral infection in the gut. These epithelial changes may increase the risk of invasive diseases, as shown in Figure 29.2, which could be life threatening (Stephensen 2001).

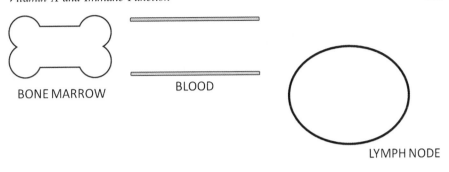

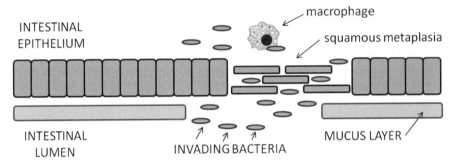

Figure 29.2 Bacterial translocation from the gut into the intestinal submucosal tissue. Bacteria from the intestinal lumen penetrate *via* an area of squamous metaplasia resulting from vitamin A deficiency. Squamous metaplasia disrupts the barrier by decreasing mucus production and disrupting the epithelial layer consisting of columnar epithelium with tight junctions. The function of resident tissue macrophages may also be impaired by vitamin A deficiency.

29.4.2 Granulocytes

Granulocytes are a group of white blood cells of the innate immune system that include neutrophils, eosinophils, basophils, and mast cells. These cells have a variety of responses to infection but primarily phagocytose and kill bacteria, or secrete anti-microbial compounds to kill bacteria outside the cell. These cells also produce soluble mediators of inflammation (*e.g.*, cytokines, chemokines, leukotrienes, prostaglandins) that regulate immune responses (Murphy *et al* 2011). Neutrophils are a part of the initial response of the innate immune system to invasive infection that will result when microbial pathogens cross a mucosal barrier.

Retinoic acid is required for normal differentiation of neutrophils, and dietary vitamin A deficiency impairs development of mature neutrophils in rodents and results in circulating neutrophils with significantly impaired ability to phagocytose and kill bacteria (Stephensen 2001). One study in vitamin A-deficient children reported improved neutrophil phagocytosis with vitamin A supplementation (Jimenez *et al* 2010). The mechanism of impaired granulopoesis in vitamin A deficiency is not well-described, although many genes involved in neutrophil development are responsive to retinoic acid. This impairment in neutrophil development implies that a reduced resistance to invasive bacterial disease may result from vitamin A deficiency. Consistent with this prediction, the clearance of bacteria from the blood is impaired in vitamin A-deficient rats (Wiedermann *et al* 1996). Increased or unchanged numbers of neutrophils have been reported from vitamin A-deficient rodents but neutrophil function was not examined in the studies describing neutrophilia (Stephensen 2001). In rats, a paradoxical increase in vitamin A levels is seen in the bone marrow during development of vitamin A deficiency (Twining *et al* 1996), perhaps suggesting a compensatory response by the deficient animal to maintain normal development of neutrophils and other cells of the immune system in bone marrow. Thus vitamin A deficiency impairs neutrophil function, although, under some circumstances, the numbers of neutrophils are not depressed in the blood (Figure 29.3).

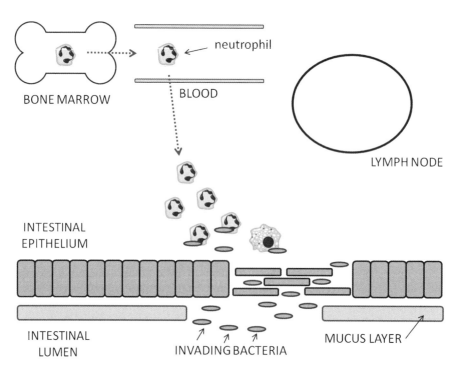

Figure 29.3   Vitamin A deficiency impairs neutrophil response to bacterial infection. Inflammation triggered by bacterial infection will stimulate production of neutrophils in the bone marrow and their migration from the bloodstream to the site of infection. The function of these granulocytes may be impaired by vitamin A deficiency.

29.4.3 Natural Killer Cells

Natural killer (NK) cells are cytotoxic lymphocytes that play an important role in the protection against viral infections before the development of an adaptive immune response (Murphy *et al* 2011). Vitamin A-deficient animals have lower numbers of NK cells with a decreased ability to kill damaged or virus-infected cells. Treatment with retinoic acid restores the NK cell population and increases cytotoxic activity (Ross and Stephensen 1996). Thus, as vitamin A deficiency impairs NK cell function, this deficit may result in a decreased ability to clear infections once they occur.

29.4.4 Monocytes/Macrophages

Monocytes develop in the bone marrow and travel through the bloodstream to tissue sites, where they differentiate into macrophages. Some macrophages are found in healthy tissues and their numbers increase during most infections, particularly infection by viruses and other intracellular pathogens. Their preferred method of killing is through phagocytosis, as is seen with neutrophils. Macrophages also secrete many cytokines that act to promote inflammation by attracting other immune cells (Murphy *et al* 2011). The number of macrophages in secondary lymphoid tissues may be increased by vitamin A deficiency in rodents. When such animals are treated with retinoic acid, the number of monocyte decreases (Miller and Kearney 1998). This apparently paradoxical observation may result from increased production of some cytokines [*e.g.*, interleukin (IL)-12 and interferon (IFN)-γ] in vitamin A deficiency that can promote macrophage-mediated inflammation, thus potentially increasing macrophage numbers in tissues. Vitamin A deficiency also impairs the ability of macrophages to ingest bacteria, which is enhanced with vitamin A supplementation (Stephensen 2001). Thus macrophage function is impaired by vitamin A deficiency, but counter-balancing effects may minimize the impact on macrophage-mediated inflammation (Figure 29.2).

29.5 Vitamin A and Adaptive Immunity

29.5.1 Antigen-presenting Cells

Antigen-presenting cells, such as dendritic cells, are found in peripheral tissues and in lymph nodes. When they encounter microbial products during an infection or immunization, they are activated to take up and process antigen, produce specific cytokines (which may vary depending on the type of microbe encountered) and other signaling molecules (*e.g.*, retinoic acid, in some cases) and migrate to the draining lymph node. Activation will increase the expression of the major histocompatibility complex (MHC) and co-stimulatory molecules on dendritic cells to enhance presentation of antigen and activation of antigen-specific T-cells. The pattern of cytokines produced by the activated dendritic cells helps to regulate T-cell differentiation [*e.g.*, IL-12 will promote T helper 1 (Th1) cell development] (Murphy *et al* 2011).

Vitamin A has diverse effects on dendritic cell development, but the most important feature of dendritic cells with regard to vitamin A biology is that they produce retinoic acid, the active metabolite of vitamin A, from retinol (Duriancik *et al* 2010; Manicassamy and Pulendran 2009). Retinoic acid produced by dendritic cells can have autocrine effects on the dendritic cells themselves (*e.g.*, to affect antigen presentation and dendritic cell migration to lymph nodes) but the most pronounced effects are on lymphocytes that interact with dendritic cells during antigen presentation to stimulate development of memory T- and B-lymphocytes during the initiation of an adaptive immune response (as shown in Figure 29.4). These effects are discussed below.

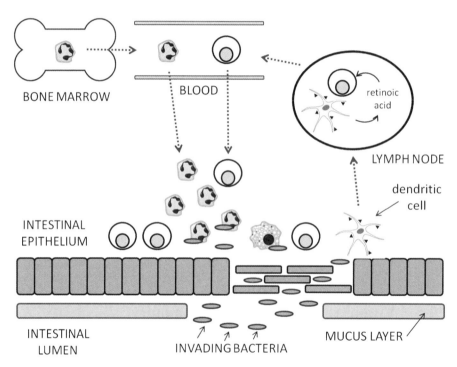

Figure 29.4   Retinoic acid is produced by dendritic cells in intestinal lymph nodes. Dendritic cells acquire antigen (small triangles) from invading pathogens and carry it to the draining lymph nodes where it is presented to T-cells to elicit an adaptive immune response. Intestinal dendritic cells also produce retinoic acid from vitamin A to stimulate development of specific subsets of T-cells, including Treg cells, and to induce expression of cell-surface molecules (including CCR9 and α4β7 integrin) that allow migration of these developing lymphocytes back to the intestine to respond to infection.

29.5.2 Thymic Function

The thymus is an important organ of the immune system where T-cell development occurs. The rate of production T-cells by the thymus peaks within the first few months of life but does persist at a lower level into adulthood (Murphy *et al* 2011). Retinoic receptors are expressed in the thymus and thymocytes development and survival are dependent on retinoic acid signaling during selection of mature T cells in the thymus. These data indicate that vitamin A is important for normal thymic development (Engedal 2011). Impairment of thymic development by vitamin A deficiency in infancy could result in long-term loss of T-cell diversity and thus potentially diminish the efficacy of the adaptive immune response to infections later in life.

29.5.3 Peripheral T-cells

Following thymic development, retinoic acid is also required for normal proliferation and survival of T cells in response to antigenic stimulation in the periphery (Engedal 2011). In addition, retinoic acid directly modulates development of T-cell differentiation following antigen exposure (Iwata 2009; Mucida *et al* 2009; Stephensen 2001). Naïve cluster of differentiation antigens 4^+ ($CD4^+$) Th cells differentiate into many phenotypes based on the need to deal with different types of pathogens (Murphy *et al* 2011). Current well-established subsets of memory/effector Th cells include the following: Th1 cells produce the effector cytokine IFN-γ that activates macrophages to kill intracellular pathogens (*e.g.*, *Mycobacterium tuberculosis* and some *Salmonella* species) and promote anti-viral responses (*e.g.*, development of cytotoxic $CD8^+$ T cells). Production of IL-27 and IL-12 by dendritic cells drives Th1 development (as well as production of IFN-γ itself by Th1 and other cell types). Th1 cell responses can be enhanced by vitamin A deficiency (Cantorna *et al* 1994), perhaps primarily due to the ability of retinoic acid to decrease IFN-γ production, although this pattern is not unvarying and may depend on the patterns of cytokines produced in response to specific conditions (*i.e.*, specific infections or types of vaccination). Th2 cells produce IL-4, IL-5 and IL-13, and promote "weep and sweep" responses in the gut and eosinophilic inflammation to expel metazoan parasites. The production of IL-4 by dendritic cells or other cell types drives Th2 development. Retinoic acid enhances Th2 development *ex vivo* in the presence of IL-4. Th17 cells produce the effector cytokines IL-17A and IL-22 which promote epithelial production of anti-bacterial peptides to kill extracellular bacteria and chemokines to attract neutrophils, which phagocytose and kill such bacteria. IL-17A also promotes neutrophil differentiation in the bone marrow. Th17 development is promoted by IL-6 from dendritic cells working together with transforming growth factor-β (TGF-β), which can be produced by many cell types, particularly in gut lymphoid tissue. Cytokines produced by Th17 cells, including IL-23, are also needed to sustain Th17 development. Inducible regulatory (iTreg) cells also develop in the periphery after encountering antigen [whereas natural Treg cells (nTreg) develop in the thymus in response to self-antigen] and act to inhibit rather than promote inflammation. This development of iTreg cells is an inherent regulatory component of adaptive immunity to control inflammation in order to prevent excessive pathology. TGF-β in the absence of inflammatory cytokines drives iTreg development, but data also show that retinoic acid acts in concert with TGF-β to enhance iTreg development. Both TGF-β and retinoic acid (Sun *et al* 2007) are produced by immune cells in the gut and mesenteric lymph nodes, which are key sites of Treg development (Mucida *et al* 2009).

29.5.4   Vitamin A and Mucosal Targeting of Immune Cells

A final key vitamin A-dependent aspect of maintenance of the mucosal immune system is the ability of lymphocytes first exposed to antigen at mucosal sites to return to mucosal sites as effector or memory cells (Murphy *et al* 2011). The teleogical explanation for this return is to allow these gut-derived lymphocytes to be present at sites where their cognate pathogens are likely to re-appear. Interestingly, retinoic acid produced by CD103$^+$ dendritic cells in the gut facilitates such return by inducing the expression of CC-chemokine receptor 9 (CCR9) and the α4β7 integrin dimer on the surface of T- and B-lymphocytes undergoing differentiation in the gut. CCR9 responds to CC-chemokine ligand 25 (CCL25), which is constitutively expressed in the intestine, and α4β7 integrin binds to mucosal addressin cell adhesion molecule-1 (MAdCAM-1), which is expressed on vascular endothelium associated with intestinal lymphoid tissue (Gorfu *et al* 2009). These two molecules combine to provide a gut-homing signature for lymphocytes. Acquisition of CCR9 and α4β7 expression *in vitro* by T-cells is blocked by inhibition of aldehydyde dehydrogenase activity in dendritic cells, and by RAR-α antagonists and disruption of the RAR-α in T-cells, and can be induced by exogenous retinoic acid (Coombes *et al* 2007; Iwata and Yokota 2011; Molenaar *et al* 2011). Dietary vitamin A deficiency decreases retinoic acid production by gut dendritic cells (Jaensson-Gyllenback *et al* 2011; Molenaar *et al* 2011) and inflammation may also have a similar effect (Laffont *et al* 2010). In addition, development of mucosally targeted CD103$^+$ dendritic cells expressing retinaldehyde dehydrogenase occurs in the bone marrow, and this development is also directed by production of retinoic acid, in this case by bone marrow cells (Feng *et al* 2010).

29.5.5   B-cells and Antibody Responses

Naïve B-cells develop into antibody-producing plasma cells and memory B-cells following appropriate exposure to antigen. Thus B cells are responsible for development of the humoral immune response (Murphy *et al* 2011). Some antibody responses develop without T-cell help (*e.g.*, bacterial polysaccharides), and these are not impaired by vitamin A deficiency but many humoral responses require such help. Vitamin A deficiency generally impairs T-cell-mediated antibody responses, particularly Th2-dependent antibody responses, such as immunoglobulin E (IgE) and IgG1 responses (Ertesvag *et al* 2009; Ross *et al* 2009). Antibody responses promoted by Th1 cells may not be affected or can be slightly increased by vitamin A deficiency in mice (Stephensen 2001).

IgA antibody is secreted across mucosal surfaces to neutralize pathogens in the respiratory, urogenital, and intestinal tracts. It is a crucial element of the adaptive immune response protecting against such mucosal pathogens (Murphy *et al* 2011). Vitamin A deficiency impairs the serum and secretory IgA response in the gut and respiratory tracts (Stephensen 2001). This phenomenon is partially explained by diminished Th2 development and mucosal targeting of lymphocytes, but vitamin A also promotes IgA responses by enhancing class-switching to IgA by plasma cells (Watanabe *et al* 2010). Thus vitamin A deficiency impairs this crucial protective mechanism at mucosal surfaces.

29.6   Conclusions

Although vitamin A was termed "the anti-infective vitamin" in 1928, it was not until 60 years later, following the discovery of the retinoic acid receptors and the development of cellular immunological techniques, that specific mechanisms of action were identified in the immune system. Currently, most of the mechanistic information is available from studies in cell culture and in experimental animal models. While it is known that vitamin A deficiency increases the risk of death from childhood infections, it is not yet known whether the mechanisms of action of retinoic acid in the immune system of mice will be confirmed in human studies. Future work should examine the role of vitamin A in human immunity, and, in particular, determine if this knowledge can be used to improve vaccine efficacy and to better target nutrition programs to minimize deaths from infectious diseases.

Summary Points

- Vitamin A deficiency was first shown to impair resistance to infections in experimental animal studies in 1928 and was termed "the anti-infective vitamin".
- Community-based intervention trials in the 1980s and 1990s demonstrated that vitamin A supplementation of children 6 months to 5 years of age who were at risk of vitamin A deficiency decreased mortality from infectious diseases by 30%.
- Development of cells of the innate immune system in the bone marrow—including neutrophils and macrophages—requires retinoic acid. Vitamin A deficiency impairs the development of these cells, making them less able to kill pathogenic microorganisms.
- Development and survival of mature T-cells in the thymus and mature natural killer (NK) cells from the bone marrow requires retinoic acid.

- Dendritic cells stimulate the development of antigen-specific memory T-cells by presenting unique antigens from micro-organisms or vaccines. During this process, dendritic cells from the intestinal immune system also produce retinoic acid that acts on T-cells to increase expression of mucosal homing receptors to allow these cells to return to the original site of antigen presentation to protect against recurrent infection.
- Retinoic acid produced by dendritic cells during development of memory T-cells also helps direct memory T-cell development toward a particular phenotype. In particular, retinoic acid plus transforming growth factor-β (TGF-β) induce development of inducible T-regulatory (iTreg) cells. Retinoic acid also enhances the development of T-helper 2 (Th2) cells in the presence of interleukin-4 (IL-4).
- B-cells develop into antibody-secreting plasma cells, sometimes with the help of T-cells. Vitamin A deficiency impairs the development of T-cell-dependent serum (IgG, IgE) and secretory (IgA) antibody responses. This is due to the effect of retinoic acid on T-cells but retinoic acid also directly affects B-cell development and survival.

Key Facts

Key Facts of Vitamin A and the Immune System

- All vertebrates have both innate and adaptive immune systems.
- The innate immune system recognizes broad classes of potentially pathogenic organisms (*e.g.*, viruses with RNA genomes) without previous exposure.
- The adaptive immune response is specific to a particular pathogen (*e.g.*, influenza virus), develops only after the initial exposure, and is thus synonymous with "immunological memory".
- All vertebrates utilize the vitamin A metabolite retinoic acid to regulate gene expression *via* the retinoic acid receptor (RAR).
- Retinoic acid is produced by dendritic cells in the immune system and acts in an autocrine and paracrine manner to regulate immune function.
- The cardinal sign of vitamin A deficiency is xerophthalmia (nutritional blindness) but impaired immunity also occurs in children with xerophthalmia, although it is not as easily identified.

Definitions of Words and Terms

Basophil: These are found in blood; they are granulocytes that enter tissue in response to parasitic infections.

B-cell: A lymphocyte normally found in blood and lymph nodes; cell-surface B-cell receptor (BCR) is a membrane-anchored Ig that recognizes foreign antigens: after antigenic stimulation B-cells develop into antibody-secreting plasma cells that are found in bone marrow and at sub-mucosal surfaces.

Dendritic cells: These cells function as "antigen-presenting cells"; delivering antigen from periphery to lymphocytes in draining lymph nodes.

Eosinophil: These are found in blood; they are granulocytes that enter tissue to mediate inflammation in response to parasitic infections and allergies, including asthma.

Macrophages: Phagocytic cells found in tissues involved in defense against micro-organisms and in "sterile" inflammation initiated by tissue damage (*e.g.*, wound or plaque in coronary artery).

Mast cell: A granulocyte found in tissues primarily at submucosal sites; responds to some antigens, including allergens, *via* IgE molecules on surface of mast cell; this activation triggers release of mediators that triggers local and systemic inflammation, including anaphylaxis.

Monocytes: These are found in blood and differentiate into macrophages on entering tissues.

Natural Killer (NK) cell: A lymphocyte found in blood and tissues; does not have antigen-specific cell-surface receptor; recognizes and kills virus-infected and other "stressed" or damaged cells *via* a change in expression of cell surface receptors.

Neutrophils: These are the principal phagocytic cells in blood; they are granulocytes which enter tissue in response to inflammation to kill invading bacteria by phagocytosis (ingestion), oxidative metabolism and secretion of anti-bacterial peptides.

Retinoic acid receptor (RAR): A ligand-activated nuclear receptor that binds to specific DNA repeats in the regulatory region of genes to regulate genes in a vitamin A-responsive manner.

T-cell: A lymphocyte normally found in blood and lymph nodes, as well as in tissue at sites of inflammation; cell-surface T-cell receptor (TCR) recognizes peptide antigens; CD8$^+$ "killer" T-cells recognize and kill virus-infected host cells; CD4$^+$ helper T-cells produce cytokines that stimulate development of CD8$^+$ T-cells, B-cells and stimulate protective responses of some myeloid cells, including macrophages.

List of Abbreviations

CCR	chemokine receptor
CD	cluster of differentiation antigens
IFN	interferon
Ig	immunoglobulin
IL	interleukin
iTreg	inducible T regulatory
NK cell	natural killer cell
RAR	retinoic acid receptor
TGF	transforming growth factor
Th	T-helper

References

Cantorna, M. T., Nashold, F. E. and Hayes, C. E., 1994. In vitamin A deficiency multiple mechanisms establish a regulatory T helper cell imbalance with excess Th1 and insufficient Th2 function. Journal of Immunology. 152: 1515–1522.

Coombes, J. L., Siddiqui, K. R., Arancibia-Carcamo, C. V., Hall, C. J., Sun, M., Belkaid, Y. and Powrie, F., 2007. A functionally specialized population of mucosal CD103+ DCs induces Foxp3+ regulatory T cells via a TGF-βand retinoic acid-dependent mechanism. Journal of Experimental Medicine. 204: 1757–1764.

Duriancik, D. M., Lackey, D. E. and Hoag, K. A., 2010. Vitamin A as a regulator of antigen presenting cells. Journal of Nutrition. 140: 1395–1399.

Engedal, N., 2011. Immune regulator vitamin A and T cell death. Vitamins and Hormones. 86: 153–178.

Ertesvag, A., Naderi, S. and Blomhoff, H. K., 2009. Regulation of B cell proliferation and differentiation by retinoic acid. Seminars in Immunology. 21: 36–41.

Feng, T., Cong, Y., Qin, H., Benveniste, E. N. and Elson, C. O., 2010. Generation of mucosal dendritic cells from bone marrow reveals a critical role of retinoic acid. Journal of Immunology. 185: 5915–5925.

Germain, P., Chambon, P., Eichele, G., Evans, R. M., Lazar, M. A., Leid, M., De Lera, A. R., Lotan, R., Mangelsdorf, D. J. and Gronemeyer, H., 2006. International Union of Pharmacology. LX. Retinoic acid receptors. Pharmacological Reviews. 58: 712–725.

Gorfu, G., Rivera-Nieves, J. and Ley, K., 2009. Role of β7 integrins in intestinal lymphocyte homing and retention. Current Molecular Medicine. 9: 836–850.

Green, H. N. and Mellanby, E., 1928. Vitamin a as an anti-infective agent. British Medical Journal. 2: 691–696.

Imdad, A., Herzer, K., Mayo-Wilson, E., Yakoob, M. Y. and Bhutta, Z. A., 2010. Vitamin A supplementation for preventing morbidity and mortality in children from 6 months to 5 years of age. Cochrane Database Systemic Reviews. 8: CD008524.

Iwata, M., 2009. The roles of retinoic acid in lymphocyte differentiation. Seminars in Immunology. 21: 1.

Iwata, M. and Yokota, A., 2011. Retinoic Acid production by intestinal dendritic cells. Vitamins and Hormones. 86: 127–152.

Jaensson-Gyllenback, E., Kotarsky, K., Zapata, F., Persson, E. K., Gundersen, T. E., Blomhoff, R. and Agace, W. W., 2011. Bile retinoids imprint intestinal CD103(+) dendritic cells with the ability to generate gut-tropic T cells. Mucosal Immunology. 4: 438–447.

Jimenez, C., Leets, I., Puche, R., Anzola, E., Montilla, R., Parra, C., Aguilera, A. and Garcia-Casal, M. N., 2010. A single dose of vitamin A improves haemoglobin concentration, retinol status and phagocytic function of neutrophils in preschool children. British Journal of Nutrition. 103: 798–802.

Laffont, S., Siddiqui, K. R. and Powrie, F., 2010. Intestinal inflammation abrogates the tolerogenic properties of MLN CD103+ dendritic cells. European Journal of Immunology. 40: 1877–1883.

Manicassamy, S. and Pulendran, B., 2009. Retinoic acid-dependent regulation of immune responses by dendritic cells and macrophages. Seminars in Immunology. 21: 22–27.

Miller, S. C. and Kearney, S. L., 1998. Effect of *in vivo* administration of all *trans*-retinoic acid on the hemopoietic cell populations of the spleen and bone marrow: profound strain differences between A/J and C57BL/6J mice. Laboratory Animal Science. 48: 74–80.

Molenaar, R., Knippenberg, M., Goverse, G., Olivier, B. J., de Vos, A. F., O'Toole, T. and Mebius, R. E., 2011. Expression of retinaldehyde dehydrogenase enzymes in mucosal dendritic cells and gut-draining lymph node stromal cells is controlled by dietary vitamin A. Journal of Immunology. 186: 1934–1942.

Mucida, D., Park, Y. and Cheroutre, H., 2009. From the diet to the nucleus: vitamin A and TGF-β join efforts at the mucosal interface of the intestine. Seminars in Immunology. 21: 14–21.

Murphy, K. P., Travers, P., Walport, M. and Janeway, C., 2011. Janeway's Immunobiology. Garland Science, New York, USA.

Ross, A. C. and Stephensen, C. B., 1996. Vitamin A and retinoids in antiviral responses. FASEB Journal. 10: 979–985.

Ross, A. C., Chen, Q. and Ma, Y., 2009. Augmentation of antibody responses by retinoic acid and costimulatory molecules. Seminars in Immunology. 21: 42–50.

Semba, R. D., 1999. Vitamin A as "anti-infective" therapy, 1920–1940. Journal of Nutrition. 129: 783–791.

Stephensen, C. B., 2001. Vitamin A, infection, and immune function. Annual Reviews in Nutrition. 21: 167–192.

Sun, C. M., Hall, J. A., Blank, R. B., Bouladoux, N., Oukka, M., Mora, J. R. and Belkaid, Y., 2007. Small intestine lamina propria dendritic cells promote *de novo* generation of Foxp3 T reg cells via retinoic acid. Journal of Experimental Medicine. 204: 1775–1785.

Twining, S. S., Schulte, D. P., Wilson, P. M., Fish, B. L. and Moulder, J. E., 1996. Retinol is sequestered in the bone marrow of vitamin A-deficient rats. Journal of Nutrition. 126: 1618–1626.

Watanabe, K., Sugai, M., Nambu, Y., Osato, M., Hayashi, T., Kawaguchi, M., Komori, T., Ito Y.and Shimizu, A., 2010. Requirement for Runx proteins in IgA class switching acting downstream of TGF-β1 and retinoic acid signaling. Journal of Immunology. 184: 2785–2792.

Wiedermann, U., Tarkowski, A., Bremell, T., Hanson, L. A., Kahu, H. and Dahlgren, U. I., 1996. Vitamin A deficiency predisposes to *Staphylococcus aureus* infection. Infection and Immunity. 64: 209–214.

Wolf, G., 2001. The discovery of the visual function of vitamin A. Journal of Nutrition. 131: 1647–1650.

CHAPTER 30

Vitamin A and Brain Function

CHRISTOPHER R. OLSON AND CLAUDIO V. MELLO*

Department of Behavioral Neuroscience, Oregon Health and Science
University, 3181 SW Sam Jackson Park Road, Portland, OR 97239, USA
*E-mail: melloc@ohsu.edu

30.1 Introduction

Among its various roles as a micronutrient, vitamin A (vit. A; retinol) is
necessary for juvenile brain development and maintenance of adult neuronal
phenotypes. Metabolism of vit. A produces all-*trans* retinoic acid (ATRA),
which is unique as a diet-derived activator of transcription factors: the retinoic
acid receptors (RARs) and retinoid X receptors (RXRs). These form dimers of
RAR–RXR or RXR–RXR to regulate the expression of a large number of
genes. In fact, the specificity of ATRA action in different tissues may largely
depend on the subsets of RARs and RXRs, as well as the respective cofactors
and coactivators that are co-expressed in any given tissue. Through its action
on gene transcription, ATRA can regulate cell proliferation, differentiation
and apoptosis in various neural systems, but its essential role in juvenile neural
development and adult neural maintenance has more recently come to light.
Importantly, the cellular processes governed by ATRA may have strong effects
on behavior and cognition, and these are reviewed here.

In humans, vit. A deficiency (VAD) manifests mainly as decreased stature
and vision problems (Underwood and Arthur, 1996); an effect on cognition
has not been explicitly demonstrated, as its consequences to cognition may be
subtle and difficult to detect across cultures where VAD is endemic. Yet, use of
the anti-acne medication Isotretinoin (a retinoic acid analogue) has been linked
with mood disorders (Hull and D'Arcy 2005), resulting in a potential

Food and Nutritional Components in Focus No. 1
Vitamin A and Carotenoids: Chemistry, Analysis, Function and Effects
Edited by Victor R Preedy
© The Royal Society of Chemistry 2012
Published by the Royal Society of Chemistry, www.rsc.org

contraindication. In light of the proposed use of vit. A and its retinoid derivatives to treat conditions that range from psychiatric disorders to certain cancers, a complete understanding of its effect in the brain is needed.

30.2 Altered Vitamin A Signaling Affects Spatial Learning in Rodents

A number of rodent studies have linked vit. A nutrition or altered retinoid signaling with cognitive behavior and its neuronal underpinnings. The first suggestion that vit. A nutrition might be relevant to spatial memory came from a study of knockout mice that lacked RARβ or RXRγ (Chiang *et al* 1998). These receptors are in high abundance in the hippocampus. Mice lacking RARβ or RXRγ have normal development and neuronal architecture but demonstrate deficits in cognitive tasks such as exploring a maze or swimming the Morris water maze. Furthermore, both RARβ and RXRγ knockout mice lack long-term depression (LTD) in hippocampal CA1 cells, and RARβ knockouts lack long-term potentiation (LTP), which possibly explains the deficits in learning and memory. A recent study that used appetitive motivational rewards showed deficits in spatial and working memory as well as in object recognition in RXRγ single and RARβ/RXRγ double mutants (Wietrzych *et al* 2005), but, in contrast to Chiang *et al* (1998), did not detect deficiencies in RARβ single mutants.

Subsequent to Chiang *et al* (1998) a number of studies have examined diet-induced VAD in mice and rats. Because of the long time required to deplete liver retinol, the onset of VAD pathology is gradual—VAD diets beginning at parturition in mice have observable phenotypic affects after 12 weeks, including clouded eyes, hunched posture, low body fat and keratinized epithelial tissue, but with considerable variation in individual onset of symptoms (Misner *et al* 2001). Yet, the earlier onset of hippocampal LTD and LTP deficits is a more reliable indicator of the effects of VAD, as they occur even in animals that do not show outward phenotypic effects of VAD. When the VAD diet is started at a later age of 3 weeks, behavioral effects can be observed after 39 weeks, where mice fail to learn the locations of food rewards in a radial arm maze (Etchamendy *et al* 2003).

The return of cognitive function following VAD has been shown in some studies, but not in others. Resumption of vit. A nutrition allows the return of hippocampal LTD and LTP after as little as 48 h (Misner *et al* 2001), however, in experiments that subject relatively young-aged mice to VAD, cognitive function is not rescued by return to regular vit. A nutrition or supplementation with ATRA (Etchamendy *et al* 2003). In a similar study in rats, VAD started at 3 weeks of age is reversible (Cocco *et al* 2002), suggesting that species differences may exist in susceptibility to VAD. Furthermore, because brain development progresses significantly throughout juvenile stages in rodents, the effects of VAD may differ between young and adult animals. For instance, when a VAD diet is started in adult rats (Bonnet *et al* 2008; Cocco *et al* 2002) or in old mice (Etchamendy *et al* 2001), age-related cognition deficits are reversible by administering ATRA. Thus, VAD may have longer-lasting effects on younger brains that are still undergoing some development compared with in older brains.

The retinoic acid isomer, 13-*cis* retinoic acid (Isotretinoin or Accutane; a treatment for acne vulgaris) administered daily at clinical doses to mice reduces the incorporation of new neurons into the hippocampus and affects spatial learning as measured by their ability to run a radial maze (Crandall *et al* 2004). The similarity of this result to studies of VAD strongly suggests that retinoid levels in the brain must be kept in balance for adequate cognitive health to be maintained. Applications of therapeutic retinoids have been proposed for age-related cognitive deficits and for dementias, particularly for Alzheimer's disease. In a mouse model of Alzheimer's disease, ATRA administration has a rescue effect on learning deficits (Ding *et al* 2008), but because cognition is also improved by ATRA administration in regular aged mice (Etchamendy *et al* 2001), it remains hard to disentangle the effects on Alzheimer's-related cognitive deficits from those on age-related deficits. However, Ding *et al* (2008) also found fewer β-amyloid deposits, the hallmark physical symptom of Alzheimer's disease, in the brains of animals treated with the ATRA rescue compared with controls. Interestingly, rats subjected to VAD for a year also show deposits of β-amyloid into their cerebral blood vessels (Corcoran *et al* 2004), establishing a possible connection between vit. A nutrition and the onset of this disease.

Some of the effects of certain environmental pollutants that affect cognition may also be mediated through vit. A signaling pathways. Dioxins are known to impair thinking ability, and dioxin-induced memory deficits that impair the ability of mice to swim the Morris water maze have been reversed with the daily administration of ATRA (Brouillette and Quirion 2008). Dioxin disrupts the delivery of vit. A to the brain *via* transthyretin (TTR), the main carrier protein that binds the retinol–retinol-binding protein complex in blood (Brouillette and Quirion 2008).

Recently, ethanol exposure was shown to elevate retinoic acid levels in the brains of mice (Kane *et al* 2010). Long-term ethanol abuse has long been associated with cognitive deficits in humans. As the ethanol detoxification enzymes are closely related to the enzymes in the vit. A signaling pathway, the up-regulation of these shared enzymatic activities by ethanol may therefore have indirect effects on retinoic signaling.

30.3 Altered Vitamin A Signaling Affects Vocal Learning in Songbirds

Juvenile songbirds learn their song by imitating an adult in a prolonged process that involves the acquisition of a long-lasting auditory memory of adult song followed by a sensory–motor practice phase, when birds are able to change their own vocalizations to match the memorized tutor song. This process of vocal acquisition has served as a model for understanding the neural basis of human speech acquisition and complex learning (Jarvis 2004). Bird song thus stands in contrast to studies of spatio-relational memory in murines where learning can occur over a relatively short period and depends primarily on hippocampus function. The ability of a bird to produce song after a prolonged learning period can thus be used as a reliable measurement of a complex learned trait in an experimental paradigm. Bird song is a strong indicator of a bird's reproductive fitness, used to attract the attention of females and establish territory, and consequently birds are highly motivated to sing. This is in contrast to studies in rodents, where the motivation to escape a maze is frequently driven by hunger or fear.

In zebra finches (*Taeniopygia gutatta*; the predominant model species) young males are sensitive to tutor song between ~20–60 days old, during which they form an auditory memory of a particular song. This acquisition period overlaps with a sensory–motor practice phase between ~40–90 days where they initially demonstrate highly variable, incomplete subsongs that contain recognizable elements of the tutor's song, and this behavior is comparable to the babbling of human toddlers prior to speaking their first words. During a later plastic song phase, the song learning circuitry presumably compares the bird's own vocalizations, through auditory feedback, to the memorized tutor song, gradually reinforcing the connections that are necessary for the production of full adult song. This period, thought to be one of heightened neuronal plasticity in the song system, ends by ~90 days (corresponding to sexual maturity) when song "crystallizes", having high complexity and similarity to the original tutor song, and low variation across renditions.

To produce song, songbirds rely on discrete pallial, striatal and thalamic nuclei that exist in two complementary circuits (Figure 30.1A). In general the vocal–motor pathway is necessary for production of crystallized song during adulthood, while the anterior forebrain pathway is necessary for the juvenile learning of song. Lesions to the anterior forebrain pathway [*e.g.* lateral magnocellular nucleus of the anterior nidopallium (LMAN)] in adult finches have little effect on already established song, but in juveniles, lesions prematurely reduce song variability (Kao and Brainard 2006). In contrast, lesions to adult HVC do not necessarily inactivate song, but birds tend to lose song structure and stereotypy and instead produce song characteristic of juvenile subsong (Aronov *et al* 2008). During the juvenile learning period lesions of HVC have no effect on subsong, which is driven primarily by LMAN (Aronov *et al* 2008).

The retinoic acid synthesizing enzyme *ALDH1A2* is strongly expressed in X-projecting neurons of HVC and LMAN (Figures 30.1A and 30.1B; Denisenko-Nehrbass *et al* 2000), thus retinoid signaling seems to have a prominent role in regulating song learning in the anterior forebrain pathway. Local application of disulfiram (a broad inhibitor of ALDHs to inhibit retinoic acid synthesis) to HVC of ~35 day old finches results in a failure of song to crystallize at maturity compared with controls with disulfiram localized in a non-song

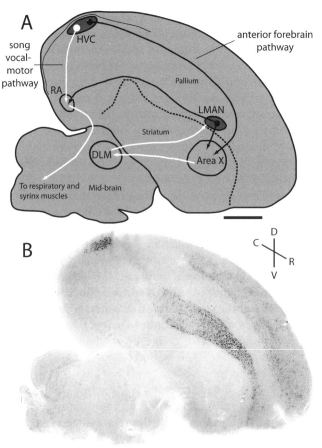

Figure 30.1   Retinoid signaling in the avian song system. (A) The HVC (a proper
name) sits at the top of the song system and sends projections to both
the song vocal–motor pathway and the anterior forebrain pathway. The
vocal–motor projection innervates the robust nucleus of the arcopallium
(RA), and from there the pathway continues on to brainstem nuclei that
control respiration and syrinx muscles. The anterior projection from
HVC innervates the striatal song nucleus, area X (X), which is part of a
loop consisting of the consecutive projections from X to the medial part
of the dorsal lateral thalamic nucleus (DLM), from DLM to the lateral
magnocellular nucleus of the anterior nidopallium (LMAN), and from
LMAN back to X. The projection neurons in LMAN have bifurcating
axons that also innervate RA in the vocal–motor pathway. Neurons in
HVC and LMAN that innervate X are significant to retinoid signaling.
Projections that express *ALDH1A2* (black) include HVC neurons that
innervate X and LMAN neurons that bifurcate to innervate both X and
RA. Other projections, including the HVC to RA projections, lack
ALDH1A2 expression (white). (B) Expression of *ALDH1A2* in the avian
brain, particularly in the pallial song nuclei HVC and LMAN.
Parasagital view with dorsal–ventral and rostral–caudal axes shown.
Scale bar = 1 mm.

region of the brain (Denisenko-Nehrbass *et al* 2000). Otherwise, plastic song continues throughout the juvenile song learning period and contains recognizable elements of adult tutor song. Thus, blocking retinoic acid production in HVC to some extent seems to extend juvenile plasticity similarly to ablation of HVC. More importantly, it may disrupt the bird's ability to use sensorimotor feedback to learn a song from a memory, but it does not seem to affect the recall or memory storage. Adult birds treated with disulfiram implants show no effects on their already crystallized songs.

In a related study, juvenile zebra finches administered large daily doses of ATRA showed a behavioral result similar to that of dilsulfiram application, which blocked ATRA synthesis in HVC, *i.e.* song remained variable into adulthood (Wood *et al* 2008). Thus, up or down changes in retinoid signaling in the song system have similar phenotypic effects. However, there are no clear effects on the establishment or recall of the tutor memory. ALDH1A2 expression in HVC is detectable from an early age prior to the beginning of the song learning period and persists into adulthood (Kim and Arnold 2005), However, we find that both the number of ALDH1A2-positive cells in HVC and the HVC volume based on ALDH1A2 expression increases gradually during the song learning period, peaks at ∼50 days of age, and then declines somewhat as birds reach sexual maturity at ∼90 days (Olson *et al* 2011). It seems quite likely that dynamic changes in retinoid levels over the song learning period, driven by regulation of ALDH1A2 expression, may act as an important modulator of behavioral plasticity observed in finches during the song-learning period.

30.4 Vitamin A Metabolism is Regulated Within the Brain

The control of retinoid signaling is achieved by balancing a number of binding proteins, catalytic enzymes, and degradation enzymes, as well as the presence and abundance of receptors and cofactors (Figure 30.2). It is the site of action where most biochemical processing occurs to control levels of retinoic acid based on the cell's needs. It is thought that these enzymes may work in concert as oxidizers or reducers, and that redundancy among many of these, particularly short-chain dehydrogenases/reductases (DHRSs), allows for the steady but controlled supply of retinoic acid to the tissues. The exception may be retinol dehydrogenase (RDH10), which has strong expression in the adult brain (Wang *et al* 2011), and is the only DHRS known to result in lethality when knocked out in mice (Sandell *et al* 2007). Binding proteins have been proposed to have regulatory function, but their actual effects on retinoid-related phenotypes is not resolved because knockout experiments with cellular retinol-binding protein (CRBP) (Ghyselinck *et al* 1999) and cellular retinoic acid-binding proteins (CRABP) I/II (Lampron *et al* 1995) result in mice that appear to have normal development and behavior. Their role in the brain may be to attenuate fluctuations in retinoids, particularly during conditions of

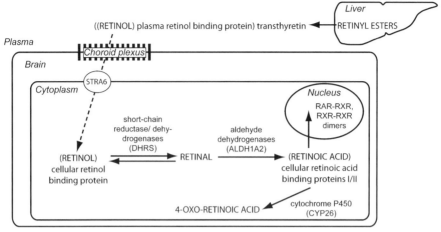

Figure 30.2   Movement and metabolism of retinoid substrates from the liver to the brain. Vit. A is acquired from the diet and converted into retinyl esters in the liver, where it is stored in large quantities. When needed, it is converted back into retinol and liberated into the circulation for delivery to tissues, including brain. In the cytoplasm, retinol is initially converted into retinal (an intermediate aldehyde) by any of several enzymes from the short-chain dehydrogenase/reductase (DHRS) family such as retinol dehydrogenases (RDH10, RDH2 or RDH14), or other DHRS enzymes (*e.g.* DHRS9) that are known to act in the vit. A pathway. This initial step is reversible, and thus its rate and direction also depend in part on the concentration gradient between retinol and retinal. For instance, some DHRS enzymes are believed to be stronger oxidizers (forward direction, *e.g.* RDH10), while others are thought to be better reducers (*e.g.* DHRS9). Dehydrogenation of retinal to produce ATRA is then achieved by one of three retinal dehydrogenases in the aldehyde dehydrogenase (ALDH) gene family, and in the brain that is usually ALDH1A2. Carrying proteins for retinoids exist at all stages of the vit. A signaling pathway, beginning with plasma retinol-binding protein (RBP) that carries retinol in the blood to the tissues, cellular retinol-binding protein (CRBP) and cellular retinoic acid-binding proteins I and II, (CRABPI/II). In addition to binding proteins, transport proteins facilitate the movement of retinoids into the nervous system, including transthyretin (TTR), which is expressed in the choroid plexus and here binds the retinol–RBP complex, thus allowing entry into the brain (Brouillette and Quirion 2008). STRA6 (stimulated by retinoic acid 6), expressed throughout the brain, is a membrane-bound protein that binds the retinol–RBP complex for transport into the cell (Kawaguchi and Sun 2010). For vit. A to have a biological effect retinoid receptors (RAR and RXRs) need to be present in the cell nucleus. Finally, cytochrome P450s (CYPs) irreversibly metabolize ATRA into 4-oxo-retinoic acid and other retinoid metabolites (Catharine-Ross and Zolfaghari 2011).

VAD where CRBP knockout mice very quickly develop symptoms of the disease (Ghyselinck *et al* 1999). The distribution and disruption of these transport proteins may determine patterns of vit. A production in the brain.

Finally, removal of ATRA by cytochrome P450s (*e.g.* CYP26) may create local sinks of ATRA with consequences to retinoid signaling (Catharine-Ross and Zolfaghari 2011).

Of the three RARs, -α, -β and –γ, and three RXRs, -α, -β and –γ, all are expressed in the brain. Together they form multiple combinations of RXR–RXR homodimers and RAR–RXR heterodimers, resulting in several retinoic acid-related transcription factors. RARs and RXRs show a broad distribution across the rodent (Zetterström *et al* 1999) and songbird (Jeong *et al* 2005) brains, and regions associated with learning and memory show high expression, for instance, in rodent hippocampus (Chiang *et al* 1998). Songbirds have high RAR levels in area X of the striatum (X) and neighboring striatal tissue, suggesting retinoid signaling in this part of the anterior pathway is important for song learning (Jeong *et al* 2005). RXRs are not retinoid-exclusive, but may form dimers with several receptors for other ligands, including peroxisome proliferator-activated receptors (PPARs), liver X receptors and vitamin D receptors, among others that are expressed in the brain. These other receptors may compete with RARs in formation of retinoid-responsive dimers—namely PPARγ–RXRγ heterodimers in the rat hippocampus, which compete with RARs for available binding sites (Moreno *et al* 2004). Single knockout mutations of RARs and RXRs physiologically and behaviorally differ from each other (Chiang *et al* 1998; Wietrzych *et al* 2005), and the promiscuity RXRs to dimerize with these other receptors may explain this difference.

ATRA signaling is commonly understood developmentally where an ATRA diffusion gradient across a developing body plan or a limb bud determines the dorsal–ventral axis of the body, or the medial–lateral axis of a limb. However, in contrast to developing embryos, mature nervous systems are structurally very complex, even at small scales, and thus the mechanism of ATRA signaling may operate differently than in early development, and these mechanisms have yet to be fully elucidated. Despite broad brain distribution of receptors, catalytic enzymes and binding proteins, ATRA levels vary among different brain regions in mice, with high levels in the hippocampus, thalamus, striatum and olfactory bulb relative to the brain as a whole (Kane *et al* 2008). In the murine hippocampus, ATRA signaling may be achieved by a paracrine metabolism of retinol to ATRA in astrocytes, which have higher activity of metabolic enzymes, RDH10, DHRS9 and ALDH1A2 (Figure 30.3). ATRA may then diffuse to adjacent hippocampal neurons (Figure 30.3B), which themselves have minimal expression of these catalytic enzymes (Wang *et al* 2011).

In the finch song system, ALDH1A2 is expressed in large projection neurons of HVC that project to X, as well as the large neurons of LMAN that dually project to robust nucleus of the arcopallium (RA) and X (Figure 2B; Denisenko-Nehrbass *et al* 2000) rather than in glial cells. In this system, paracrine signaling may indeed regulate function within HVC and LMAN, yet the delivery of ATRA to X, where ALDH1A2 expression does not occur, may

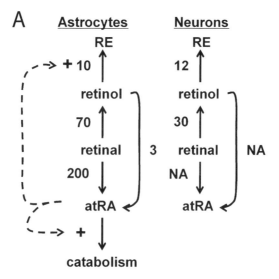

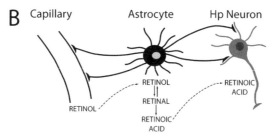

Figure 30.3   Proposed model of retinoid signaling in the murine hippocampus. (A) Hippocampal glial astrocytes (in culture) show greater metabolism compared to hippocampal neurons. Values represent rates of each step (pmol/10^6 cells/4 h) and dotted lines show ATRA feedback on vit. A metabolism and ATRA degradation. Note the rate-limiting step that occurs between retinol and retinal, which is thought to be achieved by balancing synthetic enzymes. This research was originally published in The Journal of Biological Chemistry. Wang C, Kane M.A. and Napoli J.L. Multiple retinol and retinal dehydrogenases catalyze all-*trans*-retinoic acid biosynthesis in astrocytes. Journal of Biological Chemistry 2011; 286: 6542-6553. © The American Society for Biochemistry and Molecular Biology. (B) The dual role of astrocytes in transferring retinoids from the circulation to hippocampal (Hp) neurons and in metabolizing retinol to ATRA.

also be required. Absent the projection neurons that supply X, no other source of retinoic acid would exist, yet this nucleus is relatively rich in retinoic acid receptors (Jeong *et al* 2005), as well as the catalytic enzymes RDH10 and DHRS9 that are expected to produce its substrate (Wood *et al* 2008; Olson and Mello, unpublished data). Furthermore, supplemental ATRA down-regulates DHRS9 expression in X and the striatum more so than the rest of the brain

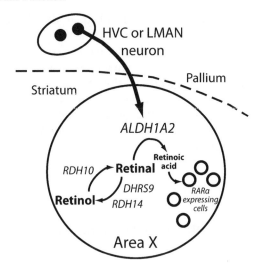

Figure 30.4 Proposed model of retinoid signaling in area X of the striatum of the avian song system. Here the retinol to retinal pathway is present, as well as an abundance of RARs, but the final step in ATRA synthesis is controlled remotely by projection neurons from the HVC or LMAN.

(Wood *et al* 2008). Hence, there exists the possibility that retinoid signaling occurs across major brain regions *via* axonal projections into the striatum (Figure 30.4). This difference in ATRA synthesis between rodents and birds illustrates two independent evolutionary examples of how retinoid signaling in the brain is a useful mechanism to finely control neuronal plasticity in terms of dendritic remodeling, synaptogenesis, neurogenesis and membrane potentials that affect LTP and depression—traits that are essential for memory formation and learning.

Summary Points

- Vit. A-related cognitive deficits in humans are difficult to uncover due to difficulty in measuring human cognition across cultures where VAD occurs, thus studies of vit. A nutrition in animal models are needed.
- Knockout mutations of retinoic acid receptors, VAD, and high doses of retinoids impair spatial and working memory of rodents.
- VAD may be reversed in some models, but not others. Younger animals may be more vulnerable to the disease than older.
- Supplemental retinoids decrease cognitive performance in normal rodents.
- Supplemental retinoids slow age-related cognitive declines in rodents, and slow advanced cognitive declines in a rodent model of Alzheimer's disease.
- Songbirds with blocked retinoid signaling fail to learn full adult male song.
- Songbirds supplemented with high doses of retinoic acid fail to learn song, similar to birds where retinoid signaling is blocked.

- The vit. A signaling pathway is controlled at the site of action, where regulation of ATRA levels is achieved by balancing a large number of binding proteins, synthesis enzymes, catalytic enzymes and receptors.
- There is redundancy in the vit. A signaling pathway, as many knockouts (loss of function) for the genes in this pathway do not affect vit. A-dependent phenotypes. Exceptions may include RDH10 and ALDH1A2.
- Deficiencies and excesses in vit. A result in cognitive impairments, suggesting that carefully controlled levels of ATRA are highly important for proper brain function.

Key Facts

Key Facts of Altered Vitamin A Signaling Affects Spatial Learning in Rodents

- The hippocampus functions to establish long-term and short-term working memory, as well as spatial memory.
- Knockout mutations of the retinoid receptors RARβ or RXRγ provided the first evidence that vit. A signaling affects cognitive tasks that require spatial and relational memory in rodents.
- Studies of vit. A deficiency (VAD) and studies of supplemental retinoids show similar declines in the ability of rodents to complete tasks that require spatial and relational memory.
- Vit. A deficiency related deficits in cognitive ability are reversible in some conditions, but not others. These include increased vulnerability at young ages, as well as possible species differences between rats and mice.
- Rodent models of Alzheimer's disease have shown improvement with therapeutic doses of ATRA, but the effect of these treatments is difficult to disentangle from a similar delay in cognitive decline that is noted in normally aged rodents that have been administered therapeutic retinoids.

Key Facts of Altered Vitamin A Signaling Affects Vocal Learning in Songbirds

- Bird song is a complex learned behavior that serves as a model to understand the neuronal underpinnings of human vocal learning.
- Songbirds establish a memory from a vocal tutor during the sensory acquisition period, then enter a sensory motor-practice period where they exhibit highly variable subsong, followed by maturation to predictable, but complex song, with high similarity to the original tutor song.
- The song system consists of two complementary brain pathways that control: (1) the learning of song during the juvenile period, and (2) the production of song in adults.

- Blocking the production of ATRA in the song system during the juvenile sensory motor-practice period results in the failure to mature highly stereotyped and highly complex song, but it does not affect a bird's ability to hear or recall a previously heard song. Blocking ATRA synthesis in adult songbirds has no effect on their song.
- Excess ATRA inhibits the maturation to adult song. Instead, songs remain highly variable and have low complexity. These results suggest that careful control of the vit. A signaling pathway is necessary to learn a complex learned trait such as song.

Key Facts of Vitamin A Metabolism is Regulated within the Brain

- Retinol is delivered to the brain, where it is converted into the bioactive form ATRA in a two-step process *via* an intermediate compound, retinal.
- The careful control of ATRA levels in the brain is achieved by several enzymes, binding proteins and degradation proteins to keep ATRA levels within a narrow range.
- Retinoic acid receptors RARα, RARβ, RARγ, RXRα, RXRβ and RXRγ form multiple combinations of RXR–RXR homodimers and RAR–RXR heterodimers, resulting in several retinoic acid-related transcription factors in the brain. RXRs may also dimerize with a number of other nuclear receptors to influence non-retinoic acid transcription pathways.
- Knockouts in mice for RARβ or RXRγ, genes that are prevalent in the hippocampus, result in cognitive deficits, but knockouts for cellular retinol-binding protein (CRBP) and cellular retinoic acid-binding proteins I and II (CRABPI/II) have no effect on cognitive tasks unless under conditions of VAD.
- In the mammalian hippocampus, astrocytes have the enzymes for the production of ATRA and likely influence neighboring neurons in a paracrine manner.
- In the avian song system, large projection neurons produce ATRA, and here ATRA may have an autocrine or paracrine effect, or it may exert its effects in remote regions innervated by these cells.

Definitions of Words and Terms

Aldehyde dehydrogenase 1A2 (ALDH1A2): The terminal enzyme of ATRA synthesis in the brain.

All-trans retinoic acid (ATRA): The vit. A derivative with the strongest biological effect in most tissues.

Animal model: A species or strain that is best suited to the study of a given problem; for instance, a disease such as Alzheimer's disease, or trait such as vocal learning.

Avian song system: The network of song nuclei in the songbird brain that control the learning and production of song.

Cellular retinoic acid binding protein (CRABP): A binding protein that binds ATRA in the cellular cytoplasm.

Cellular retinol binding protein (CRBP): A binding protein that binds vit. A in the cellular cytoplasm.

Complex learned behavior: A behavior with complex motor output that requires a period of practice to learn.

Hippocampus: The mammalian brain center responsible for spatial memory, and short- and long-term memory storage.

Retinol: The alcohol form of vitamin A.

Retinol-binding protein (RBP): A binding protein for the transport of vit. A in the blood.

Spatial memory: The recorded memory about the spatial arrangement of one's environment.

Therapeutic retinoids: Vit. A and its derivatives that are used to treat disease.

Vitamin A deficiency (VAD): A disease state whose symptoms are caused by dietary deficits in vitamin A.

List of Abbreviations

ATRA	all-*trans* retinoic acid
ALDH1A2	aldehyde dehyrogenase 1A2
CRABP	cellular retinoic acid-binding protein
CRBP	cellular retinol-binding protein
DHRS	short-chain dehydrogenase/reductase
LMAN	lateral magnocellular nucleus of the anterior nidopallium
LTD	long-term depression
LTP	long-term potentiation
PPAR	peroxisome proliferator-activated receptor
RA	robust nucleus of the arcopallium
RAR	retinoic acid receptors
RBP	retinol-binding protein
RDH	retinol dehydrogenase
RXR	retinoid X receptors
TTR	transthyretin
VAD	vitamin A deficiency
vit. A	vitamin A
X	area X of the striatum

Acknowledgements

Financial support during preparation of this Chapter includes NIH grants F32NS062609 to C.R.O. and R24GM092824 to C.V.M.

References

Aronov, D., Andalman, A. S. and Fee, M. S., 2008. A specialized forebrain circuit for vocal babbling in the juvenile songbird. Science. 320: 630–634.

Bonnet, E., Touyarot, K., Alfos, S., Pallet, V., Higueret, P. and Abrous, D. N., 2008. Retinoic acid restores adult hippocampal neurogenesis and reverses spatial memory deficit in vitamin A deprived rats. PLoS One. 3: e3487.

Brouillette, J. and Quirion, R., 2008. The common environmental pollutant dioxin-induced memory deficits by altering estrogen pathways and a major route of retinol transport involving transthyretin. NeuroToxicology. 29: 318–327.

Catharine-Ross, A. and Zolfaghari, R., 2011. Cytochrome P450s in the regulation of cellular retinoic acid metabolism. Annual Review of Nutrition. 31: 65–87.

Chiang, M.-Y., Misner, D., Kempermann, G., Schikorski, T., GiguFre, V., Sucov, H. M., Gage, F. H., Stevens, C. F. and Evans, R. M., 1998. An essential role for retinoid receptors RARβ and RXRγ in long-term potentiation and depression. Neuron. 21: 1353–1361.

Cocco, S., Diaz, G., Stancampiano, R., Diana, A., Carta, M., Curreli, R., Sarais, L. and Fadda, F., 2002. Vitamin A deficiency produces spatial learning and memory impairment in rats. Neuroscience. 115: 475–482.

Corcoran, J. P. T., So, P.-L. and Maden, M., 2004. Disruption of the retinoid signaling pathway causes a deposition of amyloid β in the adult rat brain. European Journal of Neuroscience. 20: 896–902.

Crandall, J., Sakai, Y., Zhang, J., Koul, O., Mineur, Y., Crusio, W. E. and McCaffery, P., 2004. 13-cis-retinoic acid suppresses hippocampal cell division and hippocampal-dependent learning in mice. Proceedings of the National Academy of Science USA. 101: 5111–5116.

Denisenko-Nehrbass, N. I., Jarvis, E., Scharff, C., Nottebohm, F. and Mello, C. V., 2000. Site-specific retinoic acid production in the brain of adult songbirds. Neuron. 27: 359–370.

Ding, Y., Qiao, A., Wang, Z., Goodwin, J.S., Lee, E.-S., Block, M. L., Allsbrook, M., McDonald, M. P. and Fan, G.-H., 2008. Retinoic acid attenuates β-amyloid deposition and rescues memory deficits in an Alzheimer's disease transgenic mouse model. Journal of Neuroscience. 28: 11622–11634.

Etchamendy, N., Enderlin, V., Marighetto, A., Vouimba, R.-M., Pallet, V., Jaffard, R. and Higueret, P., 2001. Alleviation of a selective age-related relational memory deficit in mice by pharmacologically induced normalization of brain retinoid signaling. Journal of Neuroscience. 21: 6423–6429.

Etchamendy, N., Enderlin, V., Marighetto, A., Pallet, V., Higueret, P. and Jaffard, R., 2003. Vitamin A deficiency and relational memory deficit in adult mice: relationships with changes in brain retinoid signaling. Behavioral Brain Research. 145: 37–49.

Ghyselinck, N. B., Båvik, C, Sapin, V., Mark, M., Bonnier, D., Hindelang, C., Dierich, A., Nilsson, C. B., Håkansson, H., Sauvant, P., Azaïs-Braesco, V.,

Frasson, M., Picaud, S. and Chambon, P., 1999. Cellular retinol-binding protein I is essential for vitamin A homeostasis. EMBO Journal. 18: 4903–4914.

Hull, P.R. and D'Arcy, C., 2005. Acne, depression and suicide. Dermatologic Clinics. 23: 665–674.

Jarvis, E. D., 2004. Learned birdsong and the neurobiology of human language. Annals of the New York Academy of Sciences. 1016: 749–777.

Jeong, J. K., Velho, T. A. F. and Mello, C. V., 2005. Cloning and expression analysis of retinoic acid receptors in the zebra finch brain. The Journal of Comparative Neurology. 489: 23–41.

Kane, M. A., Folias, A. E., Wang, C. and Napoli, J.L., 2008. Quantitative profiling of endogenous retinoic acid *in vivo* and *in vitro* by tandem mass spectrometry. Analytical Chemistry. 80: 1702–1708.

Kane, M. A., Folias, A. E., Wang, C. and Napoli, J.L., 2010. Ethanol elevates physiological all-trans-retinoic acid levels in select loci through altering retinoid metabolism in multiple loci: a potential mechanism of ethanol toxicity. The FASEB Journal. 24: 823–832.

Kao, M. H. and Brainard, M. S., 2006. Lesions of an avian basal ganglia circuit prevent context-dependent changes to song variability. Journal of Neurophysiology. 96: 1441–1455.

Kawaguchi, R. and Sun, H., 2010. Techniques to study specific cell-surface receptor mediated cellular vitamin A uptake. In: Sun, H. and Travis, G. H. (ed.) Retinoids, Vol. 652. Humana Press, pp. 341–361.

Kim, Y.-H. and Arnold, A. P., 2005. Distribution and onset of retinaldehyde dehydrogenase (zRalDH) expression in zebra finch brain: lack of sex difference in HVC and RA at early posthatch ages. Journal of Neurobiology. 65: 260–268.

Lampron, C., Rochette-Egly, C., Gorry, P., Dolle, P., Mark, M., Lufkin, T., LeMeur, M. and Chambon, P., 1995. Mice deficient in cellular retinoic acid binding protein II (CRABPII) or in both CRABPI and CRABPII are essentially normal. Development. 121: 539–548.

Misner, D. L., Jacobs, S., Shimizu, Y., de Urquiza, A.M., Solomin, L., Perlman, T., De Luca, L. M., Stevens, C. F. and Evans, R. M., 2001. Vitamin A deprivation results in reversible loss of hippocampal long-term synaptic plasticity. Proceedings of the National Academy of Science USA. 98: 11714–11719.

Moreno, S., Farioli-Vecchioli, S. and Cerù M. P., 2004. Immunolocalization of peroxisome proliferator-activated receptors and retinoid X receptors in the adult rat CNS. Neuroscience. 123: 131–145.

Olson, C. R., Vianney, P., Jeong, J., Prahl, D. and Mello, C. V., 2011. Organization and development of zebra finch HVC and paraHVC based on expression of zRalDH, an enzyme associated with retinoic acid production. The Journal of Comparative Neurology. 519: 148–161.

Sandell, L. L., Sanderson, B. W., Moiseyev, G., Johnson, T., Mushegian, A., Young, K., Rey, J.-P., Ma, J.-X., Staehling-Hampton, K. and Trainor,

P. A., 2007. RDH10 is essential for synthesis of embryonic retinoic acid and is required for limb, craniofacial, and organ development. Genes and Development. 21: 1113–1124.

Underwood, B. A. and Arthur, P., 1996. The contribution of vitamin A to public health. The FASEB Journal. 10: 1040–1048.

Wang, C., Kane, M. A. and Napoli, J. L., 2011. Multiple retinol and retinal dehydrogenases catalyze all-*trans*-retinoic acid biosynthesis in astrocytes. Journal of Biological Chemistry. 286: 6542–6553.

Wietrzych, M., Meziane, H., Sutter, A., Ghyselinck, N., Chapman, P. F., Chambon, P. and Krężel, W., 2005. Working memory deficits in retinoid X receptor γ-deficient mice. Learning and Memory. 12: 318–326.

Wood, W. E., Olson, C. R., Lovell, P. V. and Mello, C. V., 2008. Dietary retinoic acid affects song maturation and gene expression in the song system of the zebra finch. Developmental Neurobiology. 68: 1213–1224.

Zetterström, R. H., Lindqvist, E., De Urquiza, A. M., Tomac, A., Eriksson, U., Perlmann, T. and Olson, L. 1999. Role of retinoids in the CNS: differential expression of retinoid binding proteins and receptors and evidence for presence of retinoic acid. European Journal of Neuroscience. 11: 407–416.

CHAPTER 31

The Importance of Vitamin A during Pregnancy and Childhood: Impact on Lung Function

HANS K. BIESALSKI* AND DONATUS NOHR

Department of Biological Chemistry and Nutrition, University Hohenheim, Garbenstr. 30, D-70593 Stuttgart, Germany
*E-mail: biesal@uni-hohenheim.de

31.1 Introduction

During pregnancy and lactation, the demand for micronutrients increases. If an adequate supply is not ensured, this might have a negative impact on the mother and the developing child. Doyle *et al* (2001) clearly showed that young, pregnant and breastfeeding women have to be seen as a group at risk in terms of the intake of micronutrients of animal origin with focus on folic acid and iron. The authors concluded that, especially for women with multiple births, the plasma values of micronutrients decrease rapidly and therefore fall within the critical area. A low supply of micronutrients, in particular vitamin A, is a risk factor regarding fetal development.

On average, vitamin A intake should be one-third higher during pregnancy and during the breastfeeding period. Due to the importance of vitamin A for lung development and maturation, sufficient intake should be especially ensured during the second and third pregnancy trimester. The best source of preformed vitamin A is animal liver. The German Federal Institute for Consumer Protection and Veterinary Medicine (BgVV 1995) and other authorities, however, advise pregnant women not to consume animal liver.

Food and Nutritional Components in Focus No. 1
Vitamin A and Carotenoids: Chemistry, Analysis, Function and Effects
Edited by Victor R Preedy
Published by the Royal Society of Chemistry, www.rsc.org

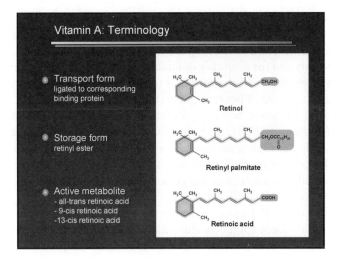

Figure 31.1 To understand the different roles of vitamin A in the human body, it is necessary to discriminate between the different types of the vitamin. Based on IUPAC definitions, the term retinoids defines all vitamin A derivatives. In contrast biologists understand retinoids as RA and its derivatives (active compound not present in human diet but available as a drug) and vitamin A is taken as the term for retinol and retinyl ester (preformed Vitamin A present in human diet).

Thus, despite existing sources, the intake of vitamin A is low during pregnancy which might have consequences.

In developing countries low vitamin A supply due to missing sources is a wide-spread problem. As the child is dependent on the mother in terms of its vitamin A supply during the newborn period, great importance is placed on the mother's vitamin A intake during pregnancy for the later supply of the child, especially because the liver store of the child only lasts for a couple of days and can be quickly emptied due to sudden strains or absorption dysfunction. Insufficient intake by the mother during pregnancy also influences the postpartum supply of the child through breast milk. In different studies, it was shown that the risk of bronchopulmonary dysplasia (BPD) in preterm infants showing insufficient vitamin A supply is significantly increased. However, in order to explore a potential supply bottleneck, vitamin A and β-carotene intake should be investigated and risk groups identified to avoid creating a problem, which should not exist in industrialized countries.

31.2 The Vicious Cycle

The vicious cycle of vitamin A deficiency (VAD) is of great importance in developing countries with poor sources for preformed vitamin A (liver, liver products, fat fish). Low intake by the mother contributes to the development of VAD. Development of VAD is exaggerated in cases of frequent pregnancies in particular short birth rates. As a consequence the newborn has poor vitamin A stores and develops rapid VAD, which cannot be counteracted *via* vitamin A supply through breast milk, because there is no vitamin A from the mother. VAD of the newborn and developing child is a leading cause for mortality below the age of 5 years and for mortality of females during or just after delivery. If the child survives, especially in the case of a female, the VAD is transferred to the next generation in case of pregnancy and closes the vicious cycle.

In developed countries, however, an inadequate dietary intake might occur, despite sufficient availability of vitamin A sources due to imbalanced diet and misleading political announcements. For example, if it is claimed that liver contains a lot of contaminations or is too high in vitamin A [not retinoic acid (RA)], this results in a nearly complete avoidance of the major dietary source for preformed vitamin A. (Figure 31.1)

As a consequence the more or less pronounced VAD is transferred from mother to child. Depending on the vitamin A status of the mother, fetal development might be impaired, in particular lung development is affected. If the newborn is not sufficiently supplied with vitamin A *via* breast milk, the risk for respiratory tract infections and complications during viral infections (*e.g.* measles) increases. To overcome problems of inadequate vitamin A supply, it is of great importance to interrupt this vicious cycle.

31.2.1   Dietary Vitamin A Supply

Vitamin A is obtained from the diet either as pre-formed vitamin A in the form of retinol or retinyl esters, or as provitamin A carotenoids. The highest content of preformed vitamin A can be found found in liver and liver oils of marine animals.

Yellow and green leafy vegetables provide significant amounts of provitamin A carotenoids but zero preformed vitamin A. However, high doses (minimum of 6 mg day^{-1}) see Table 31.1 of provitamin A are needed to substitute preformed retinol (Grune *et al* 2010).

Vitamin A plays a key role in ocular retinoid metabolism and visual function, as well as in cellular differentiation related to embryonic develop-

Table 31.1   Dietary sources of preformed vitamin A. The recommendation of 1 mg of vitamin A day^{-1} can be achieved with the following sources.

Source	Amount
Liver	10 g
Liver paté	100 g
Butter	150 g
Tuna	200 g
Cheese (fat)	300 g
Fish	2 000 g
Full fat milk	3 000 ml

ment, in particular lung maturation and immunity. Fetal and neonatal vitamin A status depends on the maternal vitamin A status. The fetal/neonatal synthesis of retinol-binding protein (RBP) is not sufficient to ensure continuous supply from liver stores. Thereby, maternal vitamin A supply is of essential importance for adequate fetal supply, growth and development. An inadequate supply of the fetus during pregnancy is associated with malformations, preterm birth, low birth weight and low neonatal liver stores. Low vitamin A status of the newborn appears to contribute to the risk for BPD (chronic lung disease). Low neonatal liver stores and a low supply during lactation also appear to increase the risk for infectious diseases (Strobel *et al* 2007).

The German Nutrition Society (DGE) recommends a 40% increase in vitamin A intake for pregnant women and a 90% increase for breastfeeding women. However, pregnant women or those considering becoming pregnant are generally advised to avoid the intake of vitamin A- rich liver and liver foods, based upon unsupported scientific findings. As a result, the provitamin A carotenoid β-carotene remains their essential source of vitamin A. Basic sources of provitamin A are orange and dark green vegetables, followed by fortified beverages which represent between 20% and 40% of the daily supply. The average intake of β-carotene in Germany is approx. 1.5–2 mg day^{-1}. Assuming a vitamin A conversion rate for β-carotene for juices of 4:1, and fruit and vegetables between 12:1 and 26:1; the total vitamin A contribution from β-carotene intake represents 10 to 15% of the recommended daily input (DRI).

The American Academy of Pediatrics (AAP) cites vitamin A as one of the most critical vitamins during pregnancy and the breastfeeding period, especially in terms of lung function and maturation (Heinig, 1998). If the vitamin A supply of the mother is inadequate, her supply to the fetus will also be inadequate, as will later be her milk. These inadequacies cannot be compensated by postnatal supplementation. A clinical study in pregnant women with short birth intervals or multiple births showed that almost one-third of the women showed plasma retinol levels below 1.4 μmol L^{-1} which can be taken as borderline deficiency (Schulz *et al* 2006). Despite the fact that vitamin A and β-carotene rich food is generally available, risk groups for low vitamin A supply exist in the Western world.

Although 25% of the population have an inadequate retinol supply according to a report of the German Federation Office for Risk Assessment (BfR), advice against consumption of liver, the major dietary source of preformed vitamin A, during pregnancy is widely provided. Alternative strategies to augment maternal vitamin A status are either selective supplementation with retinol or supplementation/fortification with isolated β-carotene. However, the conversion of β-carotene to retinol may be limited and varies with genetic polymorphisms frequently found in the population (Leung *et al* 2009). The polymorphism of the β-carotene cleavage enzyme β,β-carotene 15,15'-oxygenase-1 (BCO) results in a low vitamin A formation and is present according to the calculation of Lietz and co-workers in 20–40% of the white population (Grune *et al* 2010). This raises the question whether preformed vitamin A intake during pregnancy might be a risk for inadequate vitamin A supply. Excessive dietary intake has been associated with tertogenicity in humans in 20 cases (Biesalski, 1989). There is no doubt that synthetic vitamin A (RA and derivatives) which are not present in food but prescribed for the treatment of acne bears a significant risk for malformations. However, sufficient data regarding a teratogenic effect of preformed vitamin A (retinol or retinyl ester) are missing. (Azais-Bresco and Pascal 2000). In particular, there are no data showing any relationship between consumption of liver and malformations. Nevertheless, a daily intake of more than 10 000 international units (IU) (3 mg of preformed Vitamin A) should be avoided to be on the safe side. During the first trimester of pregnancy, vitamin A intake from liver or supplements is not recommended. However, especially during the third trimester, a sufficient vitamin A intake is strongly recommended. This ensures adequate liver stores of the newborn and adequate vitamin A stores in the lung to ensure sufficient maturation (Biesalski 2004).

31.3 Vitamin A During Pregnancy and Lactation

Doyle *et al* (2001) examined the diet of women who gave birth to children with low birth weight. The group with an adequate diet, compared to the group with a diet not providing adequate amounts of nutrients, had 20% lower vitamin A intake. Most probably, the actual vitamin A intake was even lower because likely a conversion factor of 1:6 was used, which overestimates the contribution of β-carotene to vitamin A supply, as further discussed below. In addition, the mean energy intake of the inadequately nourished group was only 1633 kcal day^{-1}, *i.e.*, below the 1800 kcal day^{-1}, which are considered borderline for a diet providing adequate levels of micronutrients.

We conducted a clinical pilot study in pregnant women with multiple births, Gemini or short birth intervals to evaluate the vitamin A and β-carotene supply during pregnancy and after delivery in this vulnerable population group (Schulz *et al* 2006). Twenty-nine volunteers aged between 21 and 36 years were evaluated within 48 h after delivery. A food frequency protocol considering 3 months retrospectively was obtained from all participants. In order to establish overall supply, retinol and β-carotene levels were determined in maternal plasma, cord blood and colostrum *via* high-performance liquid chromatography (HPLC) analysis. Regardless of the high-to-moderate socio-economic background, 27.6% of participants showed plasma retinol levels below 1.4 μmol L^{-1} which can be taken as borderline or marginal deficiency. In addition, 46.4% showed retinol intake <66% of DRI, and 50.0% did not consume liver at all. Despite a high total carotenoid intake of 6.9 ± 3.6 mg day^{-1}, 20.7% of mothers showed plasma levels <0.5 μmol L^{-1} β-carotene. In contrast 38% showed levels >0.5–1 μmol L^{-1} and 41% >1 μmol L^{-1}. These high plasma β-carotene levels might bet the result of a BCO polymorphism (reducing retinol formation due to impaired ß-carotene cleavage). Retinol and β-carotene levels were highly significantly correlated between maternal plasma *versus* cord blood and colostrum. In addition, significantly lower levels were found in cord blood ($31.2 \pm 13.0\%$ retinol; $4.1 \pm 1.4\%$ β-carotene) compared with maternal plasma (Figure 31.2).

The mean cord blood level of retinol (the level supplying the fetus with vitamin A) was 0.7 μmol L^{-1}. According to Godel *et al* (1996), a normal range of vitamin A in cord blood should be between 0.7–2.3 μmol L^{-1}. There is no clear consensus on the cutoff concentration for "vitamin A deficiency" in cord blood. Levels below 0.35 and 0.7 μmol L^{-1} are considered to resemble deficiency (Godel *et al* 1996). Taking 0.35 μmol L^{-1} as the cutoff level, 31.4% of newborns in the study of Schulz *et al* (2007) showed levels below. In particular, gemini showed the lowest levels. We conclude that, despite the fact that vitamin A- and β-carotene-rich food is generally available, risk groups for low vitamin A supply exist in Germany, which is representative of the Western world. (Figure 31.3)

This situation is based upon two main causes: young women and those considering pregnancy have been repeatedly advised to avoid the consumption of liver due to the uncertain vitamin A content. From the consumption of 100 g of liver, only approx. 40% of its vitamin A content is absorbed and it is absorbed more slowly than from capsules with very little (physiological) formation of RA, a formation which is tightly controlled (Blomhoff *et al*

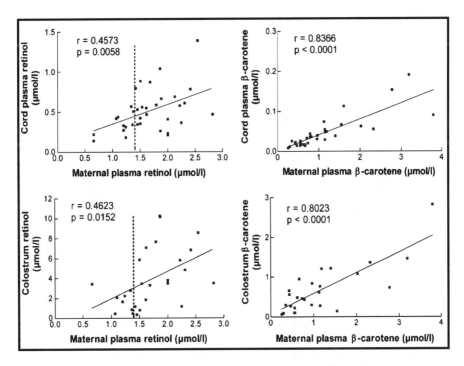

Figure 31.2 Levels of retinol and β-carotene in colostral milk, mother's plasma and umbilical cord blood. Schulz *et al.*, 2007. European Journal Nutrition. 46: 12–20, with permission.

2006). Thus the likelihood that critical vitamin A levels, in particular, the increased formation of the teratogenic compound RA, occur from a dietary intake of liver is minimal or even not present.

Doyle *et al* (2001) and our study (Schulz *et al* 2007) clearly show that young, pregnant and breastfeeding women have to be considered as a group at risk for low intake of micronutrients provided largely by animal-based foods. These groups should be advised to consume foods fortified with β-carotene or, even better, with vitamin A plus β-carotene, in order to avoid nutrition deficits which should not exist in industrialized countries.

31.3.1   Concerns Against the Recommendation to Supply Vitamin A or Vitamin A-rich Food During Pregnancy

The potential teratogenic metabolite involving vitamin A, RA, does not exist in food and will not be formed under normal circumstances beyond the physiological limit, as the metabolism of vitamin A to RA is strictly controlled at several levels. Even in cases of a continuous high intake (*e.g.* more than one portion of liver per week), the plasma level of retinol and consequently delivery to target cells will not increase.

The homoeostasis is a result of the controlled hepatic synthesis of RBP. If the supply with preformed vitamin A is low, however, RBP synthesis increases. This allows that all absorbed vitamin A will be immediately delivered to the target cells. If, however, intake is high, the RBP synthesis remains constant ensuring a constant release of vitamin A from liver stores to the target cells. This homoeostatic control can be easily determined using the relative dose

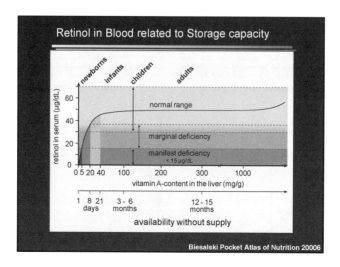

Figure 31.3   The plasma level is neither correlated to the amount of vitamin A stored in the liver, nor to the amount of ingestion *via* the diet. Only in cases when the liver stores began to become depleted the plasma level drops down.

response test. In cases of a beginning VAD, a small dose of vitamin A (3 000 IU) will result in an increase after 5 h. In contrast, in cases of a normal status the plasma levels will not increase. This control excludes an unphysiological delivery to the target cells were vitamin A can only be taken up if bound to RBP. In cases of a postprandial increase of circulating retinyl esters, there might be a delivery to target cells (Biesalski and Nohr 2004) before the chylomicrons are taken up by the liver and stored there. However, retinyl esters entering the cells are either stored there or metabolized to retinol, a step which again is strictly controlled by the intracellular level of retinol and CRBP (cytoplasmic retinol-binding protein) (Blomhoff *et al* 2006). The retinyl ester stores within different tissues undergo a strictly controlled metabolism, which seems to be linked to the cellular need. As a consequence, an increased formation of RA from retinyl esters cannot occur. Therefore the warning against the consumption of normal portions of liver (*e.g.* 100–200 g once per week) is scientifically questionable and might cause the already low amount of liver consumption to decrease further, especially among young women. The amount of liver consumed in Germany is approximately 500 g capita^{-1} year^{-1} and for young women liver consumption has practically stopped completely.

31.3.2 Relevance of the Mother's Vitamin A/β-Carotene Intake for the Fetal Vitamin A Status

Several epidemiological studies show that an insufficient intake of vitamin A poses a risk for fetal development and also during the newborn period. As the child is dependent on the mother in terms of its vitamin A supply during the newborn period, an expectant mother's vitamin A intake during pregnancy is of critical importance for the later supply to the child, as the liver stores of the baby only last for a couple of days and will be depleted quickly upon sudden strains or malabsorption states.

When assessing the liver stores in relation to the birth weight of the child and the intakes of the mother, Shah *et al* (1987) observed a significant difference between intake groups: insufficient vitamin A intake resulted in low liver stores, low birth weights and a higher risk of complications.

Fetal liver stores increase with increasing gestational age, but strongly depend on the vitamin A status of the mother. Supplementation in the second trimester of pregnancy with physiological doses of vitamin A can lead to an improvement in fetal vitamin A liver stores. Such supplementation also increases the retinol levels in the breast milk, which also positively affects the baby during the postnatal period.

There is a direct correlation between the levels of vitamin A in the mother's plasma and the concentration of retinol in the umbilical cord. Research is urgently required to validate the importance of sufficient vitamin A intakes during pregnancy and also during the breastfeeding period for ensuring a sufficient supply to the baby. In case of a poor vitamin A status of the mother, a major aspect of inadequate vitamin A supply of the fetus is a negative impact on lung maturation prior to delivery.

31.4   The Influence of Vitamin A on the Maturation and Differentiation of the Lung

Vitamin A and its major active metabolite RA have profound influence on alveolar development, maintenance and function of the lung. Type II alveolar cells synthesize and secrete surfactant. RA is able to control—dependent on its concentration—the expression of surfactant protein A (SP-A) in human fetal lung explants. Insulin, transforming growth factor β (TGF-β) and high concentrations of glucocorticoids can also down-regulate the expression of SP-A mRNA, but lower concentrations of glucocorticoids stimulate the expression of these genes. In contrast, the expression of SP-A mRNA is increased in human fetal lung explants both by hyperoxia (rats) and by dexametazone (Metzler and Snyder 1993).

Prostaglandins of the type prostaglandin E2 (PGE2) are able to increase the surfactant synthesis. Under the influence of epidermal growth factor (EGF), the formation of prostaglandins rises, especially of PGE2. On the other hand, the expression of the EGF receptor is increased by RA. EGF increases the proliferation of the lung tissues, and this leads to an amplified formation of surfactant phospholipids. RA and EGF both lead to an increase (40 and 80%, respectively) of the secretion of PGE2 in fetal lung cells of the rat *in vitro*. The combination of RA and EGF, however, leads to a more than 6-fold increase in the secretion of PGE2. Consequently, RA can interfere in the lung development by its modulating effect on the expression of EGF and the subsequent PGE2-induced surfactant formation. Sufficient and continuous availability of vitamin A (either on the blood pathway or by local storage sides) is pivotal, especially for a timely regulation of the lung development and the related formation of the active metabolite RA. During embryonic development, RA regulates cell proliferation and differentiation and regular morphogenesis (Table 31.2). In the postnatal period, RA is important for lung growth, alveolarization, and elastin formation (Maden and Hint, 2004; Bland et al, 2003).

Table 31.2   Some RA-regulated Target Genes for Lung Morphogenesis.

Genes of the HOX complex
Genes with retinoic acid-responsive elements (RAREs)
Laminin β1
Transglutaminases I and II
Elastin
Osteocalcin
Midkine
Surfactant protein B (SP-B)
Fibroblast growth factor 10 (Fgf10)

31.4.1 Vitamin A Kinetics During Fetal Lung Development

In fibroblast-like cells close to the alveolar cells and in type II cells, as well as in the respiratory epithelium, local extrahepatic stores of retinyl esters are present. The importance of these retinyl esters as acute reserve during the development of the lung becomes apparent during the late phase of gestation and the beginning of lung maturation. During this period, a rapid emptying of the retinyl ester storage in the lung of rat embryos occurs (Geevarghese and Chytil 1994). This depletion is the result of an increased RA demand in the process of the lung development, because RA is instantly needed for the process of cellular differentiation and metabolic work. Indeed, RA is important for the formation of alveoli and may rescue failed alveolar formation (Massaro and Massaro 2006; 2010). The effect of an adequate amount of RA on alveolar formation, which starts prior to birth and lasts up to the age of 8 years or even longer, has been recently documented. Checkley *et al* (2010) reported that children of mothers from a region with more or less pronounced VAD supplemented during pregnancy and for 6 months after pregnancy with 7 mg of retinol equivalents (RE) as a single oral supplement once a week had a sigfnificant better lung function at the ages of 9 to 11 years than those of mothers receiving either placebo or 42 mg of β-carotene.

The fact that the β-carotene group had no benefit regarding lung function may be due to either a poor absorption, lower cleavage rate or a polymorphism of BCO as recently discussed (Grune *et al* 2010; Leung *et al* 2009). Administration of preformed vitamin A will contribute to a more sufficient supply of the lung. The effect of vitamin A on later lung function might be a consequence of adequate alveolar formation during fetal lung development and during early childhood. However, if there was no further supplementation 6 months after delivery to the vitamin A-deficient area, how might this improvement of the lung function 10 years later be explained? One explanation might be a sufficient repletion of vitamin A storing cells in the lung of the offspring's, which may serve as storage sites for a longer time period. Retinyl ester stores have been described in lipid-laden fibroblasts and in the bronchiolar epithelium (Biesalski 1990; Shenai and Chytil 1990). These lipid interstitial cells deliver RA, which induces alveolus formation (Dirami *et al* 2004). In the alveolus, the lipid-loaden fibroblast is a major contributor to the formation of the extracellular matrix (Isakson *et al* 2004). Following hydrolysis of retinyl esters to form retinol, retinol is oxidized *via* alcohol dehydrogenase (ADH) followed by an irreversible oxidation to RA. All the steps are tightly controlled *via* intracellular binding proteins. Retinol bound to holo-CRBP is protected from degradation and delivered to lecithin:retinol acyltransferase (LRAT) for esterification (Gottesman *et al* 2001). CRBPs form a substrate-controlled network, which at least controls the delivery of RA to the nuclear-related metabolic enzymes (LRAT, ADH) *via* a feedback mechanism (Blomhoff *et al* 1990; Theodosiou *et al* 2010). This feedback network might explain why a combination of RA and retinyl palmitate given orally on postnatal days 5–7 significantly increases lung retinyl esters in neonatal rats compared to RA and vitamin A alone (Ross and Ambalavanan 2007). RA increases esterification of retinol and blocks hydrolysis of retinyl palmitate to avoid RA overload of the cells. Liver vitamin A stores, as well as plasma levels of retinol and RBP, are relatively low at birth (Ambalavanan *et al* 2005; Mupanemunda *et al* 1994; Shah and Rajalekshmi 1984; Shenai *et al* 1981). Consequently, sufficient pre-natal pulmonary retinyl ester stores and their metabolization to RA in the lung are the critical component, at least regulating the fetal lung maturation including alveolarization and at least postnatal function.

Prenatal lung development is also influenced by glucocorticoids. Steroid hormones have a similar effect on lung development as does vitamin A, and the two factors complement each other; however, this is not surprising as the receptors for steroids and retinoids belong to the same multi-receptor complex. The mode of action of glucocorticoids exists not only on the level of gene expression, but seems also to have an impact during a much earlier phase of vitamin release. The application of dexametazone leads to an increase of the maternal and fetal RBP, leading to an improvement of the vitamin A supply by channeling out *via* the normal hepatic pathway. Such an increase of the vitamin A concentration in the systemic circulation clearly diminishes the morbidity and mortality attributable to BPD (Shenai *et al* 1981; 1990) in the case of babies born prematurely.

Dexametazone and glucocorticoids not only lead to an improvement of the total vitamin A supply through a change of the release from the liver; they also influence the metabolization of the vitamin A esters stored in the lung (Geevarghese and Chytil 1994). Following administration of dexametazone, but also without steroid application, a significant reduction of retinyl esters in the maturing lung can be detected, as well as a moderate increase of retinol, the hydrolyzation product of retinyl esters. This observation may explain the therapeutic success with steroids and also their failures, in the cases of poor retinyl ester stores, during the therapy of lung distress syndrome of premature infants. A further component involved in the hydrolysis and formation of retinyl esters is the concentration of CRBP (Boerman and Napoli, 1991). A high apo-CRBP increases the activity of the retinyl ester hydrolase (REH), which subsequently results in an increase of retinol and as a consequence an increase of holo-CRBP. Liganded CRBP is responsible for the delivery of retinol to LRAT for esterification (Ross and Zolfaghari, 2004). Indeed, lipid-loaden pulmonary interstitial fibroblasts derived from perinatal rat lungs show a high CRBP concentration, which declines following formation of retinol and at least RA during the early postnatal period (McGowan *et al* 1995).

As far as an insufficient supply is concerned, inappropriate retinyl ester stores caused by a shortage of supply of the fetal lung during the late pregnancy explain why the regulatory effect of glucocorticoids and apo-CRBP for the vitamin A metabolism of the lung cells cannot take place.

31.4.2   Consequences of Marginal Deficiency

A marginal deficiency can be defined as a poor vitamin A status with missing clinical and also missing biochemical signs of the deficiency (see Figure 31.3).

Due to the strong retinol homoeostasis in human plasma, the poor status is overlooked. Clinical signs, such as night blindness or Bitot's spots of the cornea, appear later as a sign for the deficiency. However, prior to theses clinical signs, the incidence and severity of respiratory infectious diseases increases. Masuyama *et al* (1995) demonstrated that a marginal VAD may have an important impact on late lung development.

They also documented an additional aspect: retinyl-ester increased rapidly to a peak on day 17 of gestation and decreased to a minimum on day 21 of gestation. These data show that there might be a very small window during which the retinyl esters are stored in the lung shortly before they are needed. If, in the case of early delivery, the stores are not adequately filled, this might have serious consequences for lung maturation. Retinoid acid receptor α (RARα) and RARβ mRNA were detected in all samples obtained from perinatal and adult rat lung, and only a trace of RARγ mRNA was detected in the fetuses on days 15, 17 and 19 of gestation as well as in the adults. After a maternal retinol deficiency of 28 days, fetal body and lung weights were significantly lower than those of controls; the concentrations of retinyl palmitate and phosphatidylcholine (PC) in the lung after a maternal retinol deficiency of 14, 21 or 28 days were significantly lower than those of controls. Expression of RARβ mRNA in the group with 28 days of retinol deficiency was lower than in controls, that of RARα mRNA was increased and that of RARγ mRNA was not influenced by retinol deficiency. In the developing mouse embryo, RARβ expression is spatially and temporally restricted in various tissues, suggesting a role for RARβ in morphogenesis (Hind *et al* 2002). RARβ is both a strong target for RA and highly activated by treatment with exogenous RA (Kurie *et al* 2003). The rate of choline incorporation into PC in fetal lung explants was significantly higher in the group treated with RA than in controls. RA enhanced the effect of EGF on choline incorporation and prevented that of dexametazone. Taken together, a marginal deficiency results in altered expression of nuclear receptors of vitamin A with the consequence of impaired lung maturation.

31.5   The Influence of Vitamin A Supply for the Postnatal Development of the Lung in Preterm Infants

Vitamin A is essentially required for healthy development of the fetus and the newborn. A number of intervention studies indicate that the number of neural tube defects in children born to women taking multivitamin supplements compared to women taking no supplements was significantly lower. Further, insufficient micronutrient intakes by pregnant woman influence the postpartum supply of the child. Several studies have shown an increased risk of BPD in preterm infants with insufficient vitamin A status (Atkinson 2001; Chytil 1992; Verma *et al* 1996).

An adequate vitamin A intake during pregnancy is of great importance for the formation of retinyl ester stores in the developing lung. These stores are the basis for RA formation during lung maturation and postnatal function with long-term benefits as described above (Checkley *et al* 2010). However, in the case of early delivery or very low birth weight, an insufficient vitamin A supply during pregnancy might have serious consequences.

A disease observed recurrently in connection with vitamin A supply is BPD. The pathogenesis of BPD certainly depends on a multitude of factors. Some of the observed morphological changes are strongly reminiscent of VAD in the case of humans and animals. Particularly of note is the focal loss of ciliated cells with keratinizing metaplasia and necrosis of the bronchial mucosa, as well as the increase in mucous-secreting cells (Stofft *et al* 1992).

Focal keratinizing metaplasia, as may occur after a VAD, especially strengthens the assumption of an impairment of the differentiation on the level of gene expression. Since vitamin A regulates the expression of different cytokeratins and therefore influences terminal differentiation, it seems obvious to suppose the existence of common mechanisms. Consequently premature neonates, are dependent on a sufficient supply of vitamin A to ensure adequate lung maturation. The earlier a child is born before its due date, the lower are its serum retinol levels.

31.5.1 Retinol Serum Levels in Neonatals

It was shown repeatedly that serum retinol level and RBP level depend on birth weight and are significantly lower in prematures with low birth weight compared to similar aged neonates with higher birth weight (Mupanemunda *et al* 1994). In addition, mothers from low income groups had lower levels of serum vitamin A and a higher incidence of prematurity (Radhika *et al* 2002). In the liver of prematures, significantly lower retinol levels can be found in comparison to neonates (Shenai *et al* 1985). Plasma values lower than 0.70 μmol L^{-1} are not rare in this case, and they should be taken as an indicator of a relative vitamin A deficit.

Very low plasma vitamin A levels can be found recurrently in premature infants compared to term neonates (Coutsoudis *et al* 1995; Shah *et al* 1984;). This can, among other things, be attributed to the relative immaturity of the liver for the synthesis of RBPs. The neonate is almost exclusively dependent on the mother for its supply: this includes the lung retinyl esters which are either directly absorbed by the cells (from chylomicrons) or else by esterification of retinol after uptake into the cells. These lung retinyl ester stores can only be sufficiently filled if the mother guarantees an appropriate vitamin A supply during late pregnancy.

Reduced plasma levels during the first developmental months have a considerable influence on the total development of infants, as well as on their susceptibility to infections. In the case of reduced retinol plasma levels, repeated infections are more often described and they are counted among the main complications of a poor vitamin A supply in developing countries. In addition, the serum vitamin A level during infectious diseases, particularly of the respiratory tract, continues to drop (Agarwal *et al* 1996; Filteau *et al* 1993; Neuzil *et al* 1994). On the one hand, this can be explained by an increased metabolic demand and, on the other hand, by a renal elimination of retinol and of RBP during the process of acute infections (Pinnock *et al* 1989). If the retinyl ester stores of the lung are low at delivery, these storage sites can hardly be replenished, and as a consequence lung function may be impaired.

31.5.2   Relevance of Breastfeeding for the Vitamin A/β-Carotene Supply of the Newborn

Because the fetal liver only stores a small amount of vitamin A during pregnancy, almost all babies are born with marginal VAD (Bates *et al* 1983; Chappell *et al* 1985; Underwood 1994). This is usually corrected quickly *via* the vitamin A supply from the mother's milk and the extremely high vitamin A concentrations of the colostrum (up to 7 μmol L^{-1}) (Underwood 1994). However during lactation, maternal intake of vitamin A and β-carotene strongly affects the amount of these micronutrients secreted into breast milk (Allen 2005). The average American newborn has a mean liver vitamin A content of 5 μmol (assuming that the liver is approx. 4% of the body's weight). By comparison, the breastfed infant obtains approximately 310 μmol of vitamin A from mother's milk in the first 6 months. Thus, during the first 6 months, vitamin A intake from breastfeeding is usually 60 times higher than can be attained during the 9 months of pregnancy (Rasmussen 1998).

Vitamin A stores in the fetal liver accumulate during the last trimester of pregnancy but stores are related to maternal plasma concentrations (Newman 1993). In cases of zinc and VAD during pregnancy, daily supplementation with β-carotene (4.5 mg) and zinc (30 mg) improved vitamin A status in expectant mothers and then in their newborns (Dijkhuizen *et al* 2004). Breast milk concentrations of retinol and β-carotene were higher at 6 months of lactation after supplementation with β-carotene. The authors calculated the median daily retinol intake from breast milk would be 216 RE in the supplemented and 148 RE in the control group. The UK estimated average requirement and lower reference nutrient intake for retinol in this age group are 250 RE and 150 RE respectively (Azais-Braesco and Pascal 2000).

The World Health Organization (WHO) recommends a daily minimum intake of 600 μg of retinol for the baby in order to meet basic requirements (FAO/WHO 1988). For the development of liver stores, however, 1.200–1.300 μg of retinol day^{-1} during the first year and 1.4 μmol day^{-1} during the second and third year are necessary (Newman 1994). With an average milk intake of the baby of approx. 750 ml day^{-1}, the mother's milk needs to contain 1.6 μmol of retinol L^{-1} in order to provide the recommended intake requirements of the breastfed baby. From available data for Germany, the mean retinol concentration of mother's milk is approx. 2.8 μmol L^{-1}, but was less than

1.6 μmol L^{-1} in more than 20% of the women, *i.e.* concentrations which are considered critically low for the babies supply. Insufficient vitamin A intake of the baby may have serious consequences, especially regarding susceptibility to infections, the development and function of breathing organs, and the integrity of mucous membranes.

In a recent randomized controlled trial (Basu *et al* 2003), 150 women were supplemented with a single dose of retinol (209 μmol L^{-1}) soon after delivery and were advised to breastfed for 6 months. The pre-supplementation mean level of serum retinol was 0.98 μmol and breast milk retinol was 3.85 μmol. Serum and breast milk retinol increased immediately after supplementation compared to the control group, and breast milk retinol remained significantly higher for 4 months compared to the control group. Furthermore, in the supplemented group a decreased incidence and duration of various diseases was observed.

31.6   Supplementation of Newborns with Vitamin A

Different approaches exist to improve the neonatal vitamin A status. However, there are conflicting data, which at least might overlook the fact that the critical component might be the adequate dosage. High doses in newborns may be not sufficient to supply cells and tissues with adequate vitamin A for a longer time period. This is due to two different physiological aspects of the newborn:

(1)   Inadequate protein synthesis of the liver, including RBP;
(2)   Inadequate (immature) fat digestion.

(1). In case of a high dose (50 000 IU) orally, the majority of the vitamin appears as retinyl ester in chylomicrons immediately following absorption. These chylomicrons are transported to the liver. During the transport, a part of the retinyl esters can be taken up by different target tissues, as shown in a recent published report on siblings without RBP (Biesalski *et al* 1999). Within the tissues the retinyl esters can be hydrolysed and form retinol. As shown by Ross *et al* (2006), administration of retinyl esters alone did not result in a sufficient increase in lung tissues of rats. Following uptake of the chylomicron remnants into the liver, vitamin A is stored there and released under a strict homoeostatic control as retinol. In case of an immature protein synthesis, the release might be very low and inadequate. This is clearly documented in premature infants.

The existing results of two randomized double-blind controlled studies (Barreto *et al* 1994; Pearson *et al* 1992) of prematures show that supplementation with vitamin A in the study by Barreto *et al* (1994) led to a considerable reduction (55%) of the risk of being affected by chronic lung disease of prematurity. In a third study (ICGPD 1993), 12 prematures received vitamin A intravenously for a period of 28 days (400 IU day^{-1}), and during later development vitamin A was also administered orally (1500 IU day^{-1}). In the process of the supplementation, a significant change of the initially reduced plasma and RBP values occurred. The latter is an indication of an actual VAD of prematures, because an increase of retinol RBP can only be seen if a VAD really exists (principle of the relative dose response test). All trials delivered vitamin A in doses <50 000 IU to the child. The data clearly documented that a late supply in high doses might not work due to the above mentioned reasons (immaturity of RBP synthesis in the liver, distribution problems, and finally low accumulation of retinyl esters in the lung).

In case of lower doses (5 000 IU) the short-term increase of retinyl esters in the blood is not sufficient to supply much retinyl esters to target tissues. Indeed, low levels of vitamin A as usually delivered within breast milk of either supplemented or vitamin A sufficient mothers may be more effective. Maternal supplementation post-partum improves breast milk retinol and vitamin A status of the newborn (Bahl *et al* 2002; Bezerra *et al* 2009). Colostral retinol increased significantly compared to an unsupplemented control group, whereas the retinol level of mature milk (30 days after delivery) increased only slightly but significant. The impact on child vitamin A status over a time period of 6 month shows that small doses might be better to improve the status than large doses. That might be explained by the special conditions of fat absorption (including fat soluble vitamins) in newborns and infants in the first 6 month of life.

2. Fat digestion is different in newborn infants compared to adults. The activity of different lipases vary and may contribute to differences in absorption, depending on the concentration and at least way of administration *via* supplement or breast milk. The abrupt transition from carbohydrate to fat as the main energy source that occurs at birth is not matched by commensurate endogenous fat-digesting capacity in the newborn. The activity of the gastric lipase and the lipase present in human milk may compensate the low activity of the pancreatic lipase and low bile salts in newborn (Lindquist and Hernell 2010). In addition, the lingual lipase plays a dominant role in the hydrolysis of milk fat globules containing vitamin A (Hamosh *et al* 1985). This hydrolysis compensates the low luminal concentration of bile salts of newborns. Consequently, fat digestion from breast milk-derived fat is superior than fat from supplements.

The effect of a lower dose on improving vitamin A status was shown in a recent study (Ambalavanan *et al* 2003) with three different regimen in low-birth weight infants (5 000 IU vitamin A 3 times per week for 4 weeks, 10 000 IU 3 times a week and 15 000 IU once per week). The two higher doses did neither improve vitamin A status [retinol or RBP and relative dose response (RDR)] nor outcome.

In conclusion, a continuous low dose approach, with improved bioavailability might be more successful for the long term than high dosages.

Summary Points

- This chapter focuses on vitamin A and fetal lung maturation.
- Impaired lung maturation favours respiratory tract infections.

- Vitamin A is either derived from preformed vitamin A from animal sources or provitamin A from plant-derived food.
- An inadequate supply during late pregnancy results in low fetal lung vitamin A stores.
- Fetal lung vitamin A stores are needed to supply the lung during maturation with retinoic acid.
- Retinoic acid interacts with retinoic acid receptors and regulates expression of surfactant proteins and alveolar septation.
- Vitamin A-regulated lung maturation influences later lung function. Low fetal lung vitamin A stores enhance risk for e.g. BPD.

Key Facts

Key Facts of Vitamin A Action on Differentiation

- Vitamin A (retinol) is metabolized to form all-*trans* retinoic acid.
- Intracellular formation of RA is strictly controlled. Even in cases of high intake, the formation of RA will not exceed physiological levels.
- RA is delivered to the nucleus were it interacts as an activating ligand with nuclear RA receptors (RARs).
- These receptors belong to the superfamily of steroid-/thyroid-hormone receptors.
- At least 24 different types of RAR exist.
- In addition to RARs, retinoid X receptors (RXRs) are present in human cells.
- 9-*cis* RA is the ligand for RXR.
- Within a RA responsive element (RARE), the two nuclear receptors come together. Usually these receptors combine as heterodimers (RAR--RXR) and, in rare cases, as homodimers (RAR–RAR).
- RAR or RXR also heterodimerize with the nuclear vitamin D receptor (VDR).
- *Via* their nuclear receptors, cellular differentiation is regulated, including cellular function and formation of growth and transcription factors.
- Vitamin A and vitamin D control, *via* their nuclear receptors, a couple of important functions of the immune system.

Definition of Words and Terms

Cellular differentiation: The development of specialized cells from stem cells. In case of vitamin A, the main cells undergoing differentiation control *via* vitamin A are epithelial cells, *e.g.* those building the mucous membranes of the respiratory epithelium.

Lung maturation: Prior to birth, the fetal lung needs to be prepared for the first contact with inhaled air. Maturation includes formation of surfactant which is necessary to prevent collapse of the small lung alveoli.

Retinyl esters: These are esterified (mostly with palmitic acid) vitamin A (retinol). As retinyl esters, vitamin A can be stored and is chemically inert. The cleavage of retinyl esters to form retinol is the first step to activate vitamin A. Retinol itself has few functions, *e.g.* on secretory processes in the choroid plexus (liquor formation). Finally, retinoic acid is formed as the active metabolite of vitamin A.

Retinol homoeostasis: The delivery of retinol, bound to RBP from the liver to the blood stream is strictly controlled. Only in cases of deficiency, retinol plasma levels decrease. As a consequence, retinol plasma levels do not document low liver stores or inadequate supply.

Vitamin A deficiency (VAD): The first signs of VAD are night blindness, due to a low concentration of the vitamin A aldehyde, retinal, in the retina. Because retinal is needed for dark adaptation, night blindness occurs. The next clinical signs of VAD are Bitot's spots—squamous cells on the cornea and finally ulceration of the cornea with subsequent blindness. Prior to clinical signs of VAD, there is an increased susceptibility to respiratory tract infections due to impaired cellular differentiation of the respiratory mucosa.

List of Abbrevations

ADH	alcohol dehydrogenase
BCO	β,β-carotene 15,15'-oxygenase-1
BPD	bronchopulmonary dysplasia
CRBP	cytoplasmic retinol-binding protein
EGF	epidermal growth factor
IU	international units
LRAT	lecithin:retinol acyl transferase
PC	phosphatidylcholine
PGE2	prostaglandin E2
RA	retinoic acid
RAR	retinoic acid receptor
RBP	retinol-binding protein
RDA	recommended dietary allowance
RE	retinol equivalents
SP-A	surfactant protein A
VAD	vitamin A deficiency

References

Agarwal, D. K., Singh, S. V., Gupta, V. and Agarwal K. N., 1996. Vitamin A status in early childhood diarrhoea, respiratory infection and in maternal and cord blood. Journal of Tropical Pediatrics. 42: 12–14.

Allen, L. H., 2005. Multiple micronutrients in pregnancy and lactation: an overview. American Journal of Clinical Nutrition. 81: 1206–1212.

Ambalavanan, N., Wu, T. J., Tyson, J. E., Kennedy, K. A., Roane, C. and Carlo, W. A., 2003. A comparison of three vitamin A dosing regimens in extremely-low-birth-weight infants. Journal of Pediatrics. 142: 656–661.

Ambalavanan, N., Tyson, J. E., Kennedy, K. A., Hansen, N. I., Vohr, B. R., Wright, L.L. and Carlo, W. A., 2005. National Institute of Child Health and Human Development Neonatal Research Network. Vitamin A supplementation for extremely low birth weight infants: outcome at 18 to 22 months. Pediatrics. 115: 249–254.

Atkinson, S. A., 2001. Special nutritional needs of infants for prevention of and recovery from bronchopulmonary dysplasia. Journal of Nutrition. 131: 9425–9465.

Azais-Braesco, V. and Pascal, G., 2000. Vitamin A in pregnancy: requirements and safety limits. American Journal of Clinical Nutrition. 71: 1325–1333.

Bahl, R., Bhandari, N., Wahed, M. A., Kumar, G. T. and Bhan, M. K., 2002. WHO/CHD Immunization-Linked Vitamin A Group. Vitamin A supplementation of women postpartum and of their infants at immunization alters breast milk retinol and infant vitamin A status. Journal of Nutrition. 132: 3243–3248.

Barreto, M. I., Farenzena, G. G., Fiaccone, R. L., Santos, I. M. P., Assis, A. M. O., Araujo, M. P. N. and Santos, P. A. B., 1994. Effect of vitamin A supplementation on diarrhoea and acute lower-respiratory-tract infections in young children in Brazil. Lancet. 344: 228–231.

Basu, S., Sengupta, B., Roy Paladhi, P. K., 2003. Single megadose vitamin A supplementation of Indian mothers and morbidity in breastfed young infants. Postgraduate Medical Journal. 79: 397–402.

Bates, C. J., 1983. Vitamin A in pregnancy and lactation. Proceedings of the Nutrition Society. 42: 65–79.

Bezerra, D. S., Araújo, K. F., Azevêdo, G. M. and Dimenstein, R., 2009. Maternal supplementation with retinyl palmitate during immediate postpartum period: potential consumption by infants. Revista de Saúde Pública. 43: 572–579.

Bland, R. D., Albertine, K. H., Pierce, R. A., Starcher, B. C. and Carlton, D. P., 2003. Impaired alveolar development and abnormal lung elastin in preterm lambs with chronic lung injury: potential benefits of retinol treatment. Biology of the Neonate. 84: 101–102.

Blomhoff, R., Green, R. H., Berg, T. and Norum, K. R., 1990. Transport and storage of vitamin A. Science. 250: 399–404.

Boerman, M. H. and Napoli, J. L., 1991. Cholate-independent retinyl ester hydrolysis. Stimulation by apo-cellular retinol-binding protein. Journal of Biological Chemistry. 266: 22273–22278.

Biesalski, H. K., 1989. Comparative assessment of the toxicology of vitamin A and retinoids in man. Toxicology. 57: 117–161.

Biesalski, H. K., 1990. Separation of retinyl esters and their geometric isomers by isocratic adsorption high-performance liquid chromatography. Methods in Enzymology. 189: 181–189.

Biesalski, H. K. and Nohr, D., 2004. New aspects in vitamin A metabolism: the role of retinyl esters as systemic and local sources for retinol in mucous epithelia. Journal of Nutrition. 134(Suppl.): 3453–3457.

Biesalski, H. K., Frank, J., Beck, S. C., Heinrich, F., Illek, B., Reifen, R., Gollnick, H., Seeliger, M. W., Wissinger, B. and Zrenner, E., 1999. Biochemical but not clinical vitamin A deficiency results from mutations in the gene for retinol binding protein. American Journal of Clinical Nutrition. 69: 931–936. (Erratum in: American Journal of Clinical Nutrition. 2000, 71: 1010).

Bundesinstitut für Risikobewertung, 1995. Schwangere sollten weiterhin auf Leber verzichten. http://www.bfr.bund.de/de/presseinformation/1995/20/schwangere_sollten_weiterhin_auf_den_verzehr_von_leber_verzichten-775.html

Checkley, W., West, Jr, K. P., Wise, R. A., Baldwin, M. R., Wu, L., LeClerq, S. C., Christian, P., Katz, J., Tielsch, J. M., Khatry, S. and Sommer, A., 2010. Maternal vitamin A supplementation and lung function in offspring. New England Journal of Medicine. 362: 1784–1794.

Chytil, F., 1996. Retinoids in lung development. FASEB Journal. 10: 986–992.

Coutsoudis, A., Adhikari, M. and Coovadia, H.M., 1995. Serum vitamin A (retinol) concentrations and association with respiratory disease in premature infants. Journal of Tropical Pediatrics. 41: 230–233.

Filteau, S. M., Morris, S. S., Abbott, R. A., Tomkins, B. R., Kirkwood, B. R., Arthur, P., Ross, D. A., Gyapong, J. O. and Raynes, J. G., 1993. Influence of morbidity on serum retinol of children in a community-based study in northern Ghana. American Journal of Clinical Nutrition. 58: 192–197.

Dijkhuizen, M. A., Wieringa, F. T., West, C. E. and Muhilal, 2004. Zinc plus β-carotene supplementation of pregnant women is superior to β-carotene supplementation alone in improving vitamin A status in both mothers and infants. American Journal of Clinical Nutrition. 80: 1299–1307.

Dirami, G., Massaro, G. D., Clerch, L. B., Ryan, U. S., Reczek, P. R. and Massaro, D., 2004. Lung retinol storing cells synthesize and secrete retinoic acid, an inducer of alveolus formation. American Journal of Physiology. Lung Cellular and Molecular Physiology. 286: L249–L256.

Doyle, W., Srivastava, A., Crawford, M. A., Bhatti, R., Brooke, Z. and Costeloe, K. L., 2001. Inter-pregnancy folate and iron status of women in an inner-city population. British Journal of Nutrition. 86: 81–87.

Geevarghese, S. K. and Chytil, F., 1994. Depletion of retinyl esters in the lungs coincides with lung prenatal morphological maturation. Biochemical and Biophysical Research Communications. 200: 529–535.

Godel, J. C., Basu, T. K., Pabst, H. F., Hodges, R. S., Hodges, P. E. and Ng, M. L., 1996. Perinatal vitamin A (retinol) status of northern Canadian mothers and their infants. Biology of the Neonate. 69: 133–139.

Gottesman, M. E., Quadro, L. and Blaner, W. S., 2001. Studies of vitamin A metabolism in mouse model systems. Bioessays. 23: 409–419.

Grune, T., Lietz, G., Palou, A., Ross, A. C., Stahl, W., Tang, G., Thurnham, D., Yin, S. A. and Biesalski, H. K., 2010. β-Carotene is an important vitamin A source for humans. Jounral of Nutrition. 140: 2268–2285.

Hamosh, M., 1979. The role of lingual lipase in neonatal fat digestion. Ciba Foundation Symposium. 70: 69–98.

Heinig, M. J., 1998. The American Academy of Pediatrics recommendations on breastfeeding and the use of human milk. Journal of Human Lactation. 14: 2–3.

Hind, M., Corcoran, J. and Maden, M., 2002. Temporal/spatial expression of retinoid binding proteins and RAR isoforms in the postnatal lung. American Journal of Physiology. Lung Cellular and Molecular Physiology. 282: L468–L476.

Isakson, B. E., Lubman, R. L., Seedorf, G. J. and Boitano, S., 2001. Modulation of pulmonary alveolar type II cell phenotype and communication by extracellular matrix and KGF. American Journal of Cell Physiology. 281: C1291–C1299.

Italian Collaborative Group on Preterm Delivery (ICGPD), 1993. Supplementation and plasma levels of vitamin A premature newborns at risk for chronic lung disease. Developmental Pharmacology and Therapeutics. 20: 144–151.

Joint FAO/WHO Expert Consultation on Human Vitamin and Mineral Requirements, 1998. Vitamin and mineral requirements in human nutrition: report of a joint FAO/WHO expert consultation. 21–30 September 1998, Bangkok, Thailand.

Kurie, J. M., Lotan, R., Lee, J. J., Lee, J. S., Morice, R. C., Liu, D. D., Xu, X. C., Khuri, F. R., Ro, J. Y., Hittelman, W. N., Walsh, G. L., Roth, J. A., Minna, J. D. and Hong, W. K. , 2003. Treatment of former smokers with 9-*cis*-retinoic acid reverses loss of retinoic acid receptor-β expression in the bronchial epithelium: results from a randomized placebo-controlled trial. Journal of the National Cancer Institute. 95: 206–214.

Lindquist, S. and Hernell, O., 2010. Lipid digestion and absorption in early life: an update. Current Opinion in Clinical Nutrition and Metabolic Care. 13: 314–320.

Leung, W. C., Hessel, S., Méplan, C., Flint, J., Oberhauser, V., Tourniaire, F., Hesketh, J. E., von Lintig, J. and Lietz, G., 2009. Two common single nucleotide polymorphisms in the gene encoding β-carotene 15,15'-monoxygenase alter β-carotene metabolism in female volunteers. FASEB Journal. 23: 1041–1045.

Maden, M. and Hind, M., 2004. Retinoic acid in alveolar development, maintenance and regeneration. Philosophical Transactions of the Royal Society of London. Series B, Biological Sciences. 359: 799–808.

Massaro, D. and Massaro, G. D., 2006. Toward therapeutic pulmonary alveolar regeneration in humans. Proceedings of the American Thoracic Society. 3: 709–712.

Massaro, D. and Massaro, G. D., 2010. Lung development, lung function and retinoids. New England Journal of Medicine. 362: 1829–1831.

Masuyama, H., Hiramatsu, Y. and Kudo, T., 1995. Effect of retinoids on fetal lung development in the rat. Biology of the Neonate. 67: 264–273.

McGowan, S. E., Harvey, C. S. and Jackson, S. K., 1995. Retinoids, retinoic acid receptors, and cytoplasmic retinoid binding proteins in perinatal rat lung fibroblasts. American Journal of Physiology. 269: L463–L467.

Metzler, M. D. and Snyder, J. M., 1993. Retinoic acid differentially regulates expression of surfactant regulated proteins in human fetal lung. Endocrinology 133: 1990–1998.

Mupanemunda, R. H., Lee, D. S. C., Fraher, L. J., Koura, I. R., Chance, G. W., 1994. Postnatal changes in serum retinol status in very low birth weight infants. Early Human Development. 38: 45–54.

Neuzil, K. M., Gruber, W. C., Chytil, F., Stahlman, M. T., Engelhardt, B. and Graham, B. S., 1994. Serum vitamin A levels in respiratory syncytial virus infection. Journal of Pediatrics. 124: 433–436.

Newman, V., 1993. Vitamin A and breast feeding: a comparison of data from developing and developed countries. Wellstart Int., San Diego, CA, USA.

Newton, V., 1994. Vitamin A and breast feeding: a comparison of data from developing and developed countries. Food Nutrition Bulletin. 15: 161–176.

Pearson, E., Bose, C., Snidow, T., Ransom, L., Young, T., Bose, G. and Stiles, A., 1992. Trial of vitamin A supplementation in very low birth weight infants at risk for bronchopulmonary dysplasia. Journal of Pediatrics. 121: 420–427.

Quadro, L., Hamberger, L., Gottesmann, M. E., Wang, F., Colantuoni, V., Blaner, W. S. and Mendelsohn, C. L., 2005. Pathways of vitamin A delivery to the embryo: Insights from a new tunable model of embryonic vitamin A deficiency. Endocrinology. 146: 4479–4490.

Radhika, M. S., Bhaskaram, P., Balakrishna, N., Ramalakshmi, B. A., Devi, S. and Kumar B. S., 2002. Effects of vitamin A deficiency during pregnancy on maternal and child health. British Journal of Obstetrics and Gynaecology. 109: 689–693.

Rasmussen K. M., 1998. Vitamin A needs during pregnancy and lactation for the health of the mother and her fetus or infant. World Health Organization. Safe vitamin A dosage during pregnancy and lactation. Recommendations and report of a consultation.

Ross, A. C. and Zolfaghari, R., 2004. Regulation of hepatic retinol metabolism: perspectives from studies on vitamin A status. Journal of Nutrition. 134: 269–275.

Ross, A. C. and Ambalavanan, N., 2007. Retinoic acid combined with vitamin A synergizes to increase retinyl ester storage in the lungs of newborn and dexamethasone-treated neonatal rats. Neonatology. 92: 26–32.

Schulz, C., Engel, U., Kreienberg, R. and Biesalski, H. K., 2007. Vitamin A and β-carotene supply of women with gemini or short birth intervals: a pilot study. European Journal Nutrition. 46: 12–20.

Shah, R. S. and Rajalekshmi, R., 1984.Vitamin A status of the newborn in relation to gestational age, body weight, and maternal nutritional status. American Journal of Clinical Nutrition. 40: 794–800.

Shah, R. S., Rajalakshmi, R., Bhatt, R. V., Hazra, M. N., Patel, B. C., Swamy, N. B. and Patel, T. V., 1987. Liver stores of vitamin A in human fetuses in relation to gestational age, fetal size and maternal nutritional status. British Journal of Nutrition. 58: 181–189.

Shenai, J. P. and Chytil, F., 1990. Vitamin A stores in the lungs during perinatal development in the rat. Biology of the Neonate. 57: 126–132.

Shenai, J. P., Chytil, F. and Stahlman, M. T., 1985. Liver vitamin A reserves of very low birth weight neonates. Pediatric Research. 19: 892–893.

Shenai, J. P., Kennedy, K. A., Chytil, F. and Stahlman, M. T., 1987. Clinical trial of vitamin A supplementation in infants susceptible to bronchopulmonary dysplasia. Journal of Pediatrics. 111: 269–277.

Shenai, J. P., Rush, M. G., Stahlman, M. T. and Chytil, F., 1990. Plasma retinol-binding protein response to vitamin A administration in infants susceptible to bronchopulmonary dysplasia. Journal of Pediatrics. 116: 607–614.

Stofft, E., Biesalski, H. K., Zschaebitz, A. and Weiser, H., 1992. Morphological changes in the tracheal epithelium of guinea pigs in conditions of "marginal" vitamin A deficiency. International Journal of Nutrition Research. 62: 134–142.

Strobel, M., Heinrich, F. and Biesalski, H.K., 2000. Improved method for rapid determination of vitamin A in small samples of breast milk by high-performance liquid chromatography. Journal of Chromatography A. 898: 179–183.

Theodosiou, M., Laudet, V. and Schubert, M., 2010. From carrot to clinic: an overview of the retinoic acid signaling pathway. Cellular and Molecular Life Sciences. 67: 1423–1445.

Underwood, B. A., 1994. Maternal vitamin A status and its importance in infancy and early childhood. American Journal of Clinical Nutrition. 59: 517–522.

Verma, R. P., McCulloch, K. M., Worrell, L. and Vidyasagar, D., 1996. Vitamin A deficiency and severe bronchopulmonary dysplasia in very low birthweight infants. American Journal of Perinatology. 13: 389–393.

Subject Index

Illustrations and figures are in **bold**. Tables are in *italics*.
There are multiple images of chemical structure throughout the book. Where there is repetition, only the first diagram is indexed.